Palgrave Studies in Environmental Policy and Regulation

Series editor
Justin Taberham
London, UK

The global environment sector is growing rapidly, as is the scale of the issues that face the environment itself. The global population is estimated to exceed 9 billion by 2050. New patterns of consumption threaten natural resources, food and energy security and cause pollution and climate change.

Policy makers and investors are responding to this in terms of supporting green technology as well as developing diverse regulatory and policy measures which move society in a more 'sustainable' direction. More recently, there have been moves to integrate environmental policy into general policy areas rather than having separate environmental policy. This approach is called Environmental Policy Integration (EPI).

The series will focus primarily on summarising present and emerging policy and regulation in an integrated way with a focus on interdisciplinary approaches, where it will fill a current gap in the literature.

More information about this series at
http://www.palgrave.com/series/15053

Brian Joseph McFarland

Conservation of Tropical Rainforests

A Review of Financial and Strategic Solutions

palgrave
macmillan

Brian Joseph McFarland
Windham, NH, USA

Palgrave Studies in Environmental Policy and Regulation
ISBN 978-3-319-63235-3 ISBN 978-3-319-63236-0 (eBook)
https://doi.org/10.1007/978-3-319-63236-0

Library of Congress Control Number: 2017955811

Cover illustration: Jon G. Fuller/VWPics/Alamy Stock Photo.

Printed on acid-free paper

This Palgrave Macmillan imprint is published by Springer Nature
The registered company is Springer International Publishing AG
The registered company address is: Gewerbestrasse 11, 6330 Cham, Switzerland

"At first I thought I was fighting to save rubber trees, then I thought I was fighting to save the Amazon rainforest. Now I realize I am fighting for humanity."

—Chico Mendes

I would like to dedicate this book to my family and friends—particularly my son Attila, my wife Brigitta, my dad Joseph, my mom Tamara, my brother Trevor, my brother-in-law Dani, and my mother-in-law Eva—for all their love and support.

I would like to share a special dedication with my son Attila: when you are older, I hope that we, as humanity, have come up with the solutions to save tropical rainforests and mitigate global climate change and allow you to see in person all the wonderful animals that I used to read to you about. I wish the same for my future niece Adrian and my goddaughter Olivia.

I would also like to thank my former teachers Dan Bisaccio, Christopher Books, and Dr. Joseph Domask and the School for Field Studies for teaching me about the natural world, teaching me how to be an analytical thinker, and opening up my mind to a world of travels and adventures.

I would like to thank my hardworking colleagues at Carbonfund.org, particularly Jarett Emert, Linda Kelly, and Suzie Kaufman, and a particular thank you to Carbonfund.org's President and Founder Eric Carlson, along with Carbonfund.org's Chairman Paul Rowland, for all the opportunities that have been provided to me over the last 10 years at Carbonfund.org.

I would also like to say thank you to Gabriel Thoumi for all his hard work editing this book and thoughtful conversations since the day I first met him. I truly appreciate the time and effort of Jan McAlpine for writing the foreword, along with the kind endorsements by Dan Bisaccio, Scott Settelmyer (CFA, CPA), and Steven V. Soranno (CFA, CAIA).

Thank you Justin Taberham for making this book possible by connecting me with Palgrave Macmillan. Likewise, thank you to the entire team at Palgrave Macmillan for their dedication and support, particularly Chloe Fitzsimmons, Jack Redden, and Rachael Ballard. In addition, I would like to thank the team at SPi Global for their exceptional help during the copyediting and production stages.

Last, but certainly not least, I would like to dedicate this book to the world's conservationists and especially all of those who sadly were assassinated fighting to conserve forests and secure human rights:

- *Wilson Pinheiro (1980)*
- *Joy Adamson (1980)*
- *Dian Fossey (1985)*
- *Vicente Canas (1987)*
- *Francisco Alves ("Chico") Mendes Filho (1988)*
- *George Adamson (1989)*
- *Theodore Albert "Ted" Parker III (1993)*
- *Alwyn Howard Gentry (1993)*
- *Eduardo Aspiazu (1993)*
- *Carlos Catalan (1997)*
- *Medardo Varela (1998)*
- *Erwin Haroldo Ochoa López (2000)*
- *Julio Armando Vásquez Ramírez (2000)*
- *Digna Ochoa (2001)*
- *Dorothy Stang (2005)*
- *José ("Zé Claudio") Ribeiro da Silva (2011)*
- *Maria do Espírito Santo (2011)*
- *Edwin Chota Valero (2014)*
- *Leoncio Quintisima Melendez (2014)*
- *Francisco Pinedo Ramirez (2014)*

- *Jorge Ríos Pérez (2014)*
- *Berta Cáceres Flores (2016)*
- *Nelson Noé Garcia (2016)*
- *Luiz Alberto Araújo (2016)*
- *Bill Kayong (2016)*
- *Sikhosiphi Rhadebe (2016)*

Sadly, there are hundreds of others who have lost their lives fighting to conserve forests and secure human rights and I mean no disrespect if their names are not included above.

Series Foreword: Environmental Policy and Regulation

Environmental Policy is a sector focused on the commitment to developing laws and regulations regarding the environment, including air, water, land, waste, natural resources, ecosystems and biodiversity. Environmental policy can be focused on a single 'resource' or many, and can be organisational, local, regional, national or international.

The seminal report 'Our Common Future', also known as the Brundtland Report, from the United Nations World Commission on Environment and Development (WCED) was published in 1987 and noted:

> 'The world must quickly design strategies that will allow nations to move from their present, often destructive, processes of growth and development onto sustainable development paths. This will require policy changes in all countries, with respect both to their own development and to their impacts on other nations' development possibilities.'

The global environmental policy sector is growing rapidly, as is the scale of the issues that face the environment itself. The global population is estimated by the UN to approach 10 billion by 2050. New patterns of consumption threaten natural resources, food and energy security, and cause pollution and climate change.

Policy makers are responding to this challenge in terms of supporting green technology and innovation, as well as developing diverse regulatory and policy measures which move society in a more 'sustainable' direction. More recently, there have been moves to integrate environmental policy into general policy areas rather than having separate environmental policy. This approach is called Environmental Policy Integration (EPI).

This series will focus primarily on summarising present and emerging policy and regulation in an integrated way with a focus on interdisciplinary approaches, where it will fill a current gap in the literature.

London, UK Justin Taberham

Foreword by Jan McAlpine

Brian McFarland has written a precise and, at the same time, sweeping overview of the state of play for financing tropical forest conservation. For anyone wanting to understand the landscape within which forests and finance reside, this book is a must read. It provides an historical overview of financing tropical forest conservation and its players from the eighteenth century forward. It also outlines a summary of the reason why tropical forests are essential and identifies what threatens them. It goes so far as describing the ecology of tropical forests, something often missed in a discussion of forest conservation finance. While by no means exhaustive, it provides a succinct view, through the lens of conservation, to understand what went before, what is happening now, and what could take place in the future for financing for tropical forest conservation.

What is unique in this book is the approach McFarland takes in describing topics such as the history of forest management, the development of understanding the ecology of forests, and the financing of forests, inter alia. He starts with a very broad overview of each topic—for example, on financing, he provides a broad description of finance globally, irrespective of forests, covering such topics as currencies, capital markets, equities and bonds, and short-term and long-term debt, inter alia. From there, the book narrows to a description of categories of conservation finance and the frameworks within which these reside.

It is notable that McFarland emphatically points to a large number of specific causes of deforestation and forest degradation, as he says, "…for every investment in the conservation of tropical rainforests, there needs to be a root cause analysis to make sure funds are specifically targeting the underlying drivers and agents of tropical rainforest deforestation and degradation." This is considered by foresters to be a critical element in achieving sustainability of forests and has most often been ignored in the past. Changes are under way along the lines McFarland points to, for example, in the IUCN-led Global Partnership on Forest Landscape Restoration.

Unlike some other literature on the subject, which ignores this aspect, McFarland clearly believes that we must learn from the past to craft the future. To this end, he provides a wide range of case studies which describe how financing occurred in each, how effective it was, and what are the lessons learned.

McFarland points out that tropical forest conservation has been financed not only by domestic but also international sources of financing. These came over the past several hundred years from only a few sources, notably Official Development Assistance (ODA) and domestic profits from logging or deforestation (albeit little financial capture at the local level). This is a pivotal point, because for the last decade we have seen the traditional sources of financing drying up and the need to craft, facilitate, and catalyze the new and innovative financing that this book points to as possible next steps.

One must note that many of the examples given are from the United States; however, the case studies' focus is on tropical forests. As the United States has a solid history of sustainable management of forests with innovative conservation finance mechanisms, it provides an excellent basis for what has worked in the Western Hemisphere and could work elsewhere.

Commodities which have the most severe impact on deforestation and forest degradation have also been given coverage here—also a point often ignored. These include cattle (including beef), soy, palm oil, pulp and paper, and timber. In the past, logging for timber production has been the sole focus, but these chapters illustrate the more compelling role of other commodities in deforestation and forest degradation.

There is no way for this foreword to do justice to this outstanding book. It is a rich exposé of the forests and the landscape within which they reside, as well as their needs—political, policy, finance, and notably, their importance for people, not the least of which are Indigenous Peoples.

I am simply hoping that this foreword titillates enough to get you started and I will not preempt the fascinating conclusions and recommendations provided by Brian McFarland—you must read them for yourself.

—Jan L. McAlpine*

* Jan McAlpine is the former Director of the United Nations Forum on Forests. Ms. McAlpine is highly respected, internationally recognized forest policy expert, who has served from her governmental and intergovernmental leadership positions. She is perhaps best known for her work on illegal logging, having conceived of the Forest Law Enforcement and Governance (FLEG or FLEG-T) process, first in Bali, Indonesia, where the first historic declaration was signed by both producing and consuming countries of illegally harvested timber. Ms. McAlpine is also recognized for her role in designing, developing, and negotiating the Congo Basin Forest Partnership (CBFP), launched in Johannesburg, South Africa, among other noteworthy accomplishments, among them finance and forests.

Preface

The inspiration for this book comes from a lifetime of observing first-hand some of the world's most spectacular landscapes and wildlife, and then hoping my son will be able to see the same landscapes and wildlife when he is older. There are few things in life that I appreciate more than seeing wildlife in their natural landscapes. Over the years, I have snorkeled with whale sharks (*Rhincodon typus*) the size of a school bus, swam near Amazon river dolphins (*Inia geoffrensis)*, come upon fresh jaguar (*Panthera onca*) tracks, watched colorful scarlet macaws (*Ara macao*) fly overhead, and stood face to face with a wild black rhinoceros (*Diceros bicornis*).

While growing up in the small town of Amherst, New Hampshire, my life was complimented by a close family, wildlife, and a well-financed public school system. I can recall memories of my dad feeding hummingbirds out of his hand, my mom tending our vegetable garden, watching a moose (*Alces alces*) drink from our pond outside my bedroom, and fishing for largemouth bass (*Micropterus salmoides*) with my brother. Up until senior year of high school, I recall being most intrigued by science classes and conservation biology.

In parallel, toxic contamination sites in the neighboring town of Milford, New Hampshire, which may have resulted in localized illnesses such as brain cancer, stirred a desire in me to take up environmental studies and

Fig. 1 Brigitta and Brian snorkeling with whale shark (Credit: Brian Joseph McFarland)

work for a cause.[1] I considered becoming a lawyer or a medical doctor, but then I attended an experiential environmental learning program. It was during the summer after my junior year at Souhegan High School that I traveled to the Yucatan Peninsula in Mexico with our teacher Dan Bisaccio, now the Director of Science Education at Brown University, for real-world fieldwork in conjunction with the Smithsonian Institution's Monitoring & Assessment of Biodiversity (SIMAB) Program.[2] Although it took years working in a restaurant to earn the money to fund the trip, I became forever impassioned with conservation biology and our interconnected globe. For the first time, I saw poverty, I walked in a tropical rainforest, I snorkeled over a coral reef, and I watched Yucatan spider monkeys (*Ateles geoffroyi yucatanensis*) in the canopy. I started to understand the connections between poverty, slash-and-burn agriculture, commercial agriculture—particularly the global cattle trade—and tropical deforestation, tropical degradation, catastrophic climate change, and its impact on coral reefs.

Since then, I decided to travel back to Mexico and shortly thereafter to Costa Rica, Tanzania, and Kenya. During these additional education programs, I gained further insights into the interconnectivity between natural resource management, ecotourism, and sustainable economic development. Next, I chose to study under Dr. Joseph Domask at American University. During this time, I interned at the US Environmental Protection Agency's Office of Enforcement and Compliance Assurance. In addition, I attended networking events at the Brazilian Embassy, the Inter-American Development Bank, the World Wildlife Fund, Greenpeace, and the World Bank and spent a month in Brazil studying tropical ecology and income inequality. I later enrolled in a Global Environmental Policy Master's program at American University's School of International Service and in a Master's in Business Administration program at American University's Kogod School of Business with a specialization in finance in emerging markets.

I started as an intern at Carbonfund.org Foundation when I began my graduate studies, and I still work today at Carbonfund.org Foundation and its wholly owned subsidiary CarbonCo. These two climate change organizations are supporting renewable energy and energy efficiency projects, along with the development and implementation of innovative forest conservation projects in the State of Acre, Brazil.

This book reflects my background. With travels to more than 22 countries, nearly 10 years of field experience, more than 60 semi-structured, in-depth interviews, and dozens of background readings, the book focuses on some of the financial instruments that can fund the conservation of the world's last remaining tropical rainforests. The world's tropical rainforests, from the Amazon Rainforest to the Congo Basin to the rainforests of Southeast Asia, are under grave threat of deforestation and degradation. These deforestation and degradation threats include, but are not limited to: cattle ranching, commercial and slash-and-burn subsistence agriculture, unsustainable logging, forest fires, and fuelwood collection, along with megaprojects for infrastructure development such as dams, mining, oil and gas exploration, and road construction.

Tropical deforestation and forest degradation are serious global environmental issues. Conserving tropical rainforests helps mitigate global climate change by stopping deforestation and forest degradation that annually release more than three billion tonnes of carbon dioxide equivalent emissions. Conservation can improve the livelihoods of billions of locally forest-dependent people and Indigenous Peoples. Protecting tropical rainforests also maintains the global hydrological cycle. Finally, conservation can provide refuge for the world's highest levels of terrestrial biodiversity. However, conserving these tropical rainforests is contingent upon raising both sufficient and long-term international financing that is structured to support scalable, landscape-level initiatives.

This book follows a chronological order. It briefly examines the historical development of conservation finance for tropical rainforests. Conservation finance originally developed from governments in the form of domestic budgeting from taxation for the creation and maintenance of state or national parks alongside philanthropy, followed by utilizing tax deductions created for enrolling lands in conservation easements, and then by international aid to assist foreign countries with their own state or national parks. Conservation finance then evolved to include impact investing, payments for ecosystem services, debt-for-nature swaps, ecotourism, and bioprospecting. Today, conservation finance includes these approaches while it also is expanding into certified sustainable products, corporate commitments to zero deforestation supply chains, and the issuance of green bonds.

The book, in order to help an interdisciplinary audience, will begin with a context of tropical rainforest deforestation and degradation, followed by a review of tropical rainforest ecology, global environmental policy, and international finance.

Each chapter will then be dedicated to a unique conservation financing instrument. Each of these chapters will follow a similar format, starting with a brief historical overview of the instrument, followed by an explanation of how the instrument works, then by providing background information on the size of the instrument, and concluding with a presentation of case studies, a financial analysis, a policy analysis, and a future outlook for the instrument.

All 24 case studies, which span 15 countries and 4 continents, are based off the following format:

1. The Problem

    a. Identify the Problem.
    b. Explain Why the Problem Is Important.

c. How Was the Problem Identified?
d. Was the Process for Identifying the Problem Effective?

2. Steps Taken to Address the Problem
3. Results
4. Challenges and How They Were Met
5. Beyond Results
6. Lessons Learned[3]

The financial analysis will examine the return versus risks of the financial instruments. The risk categories will include:

- Business Risk;
- Strategic Risk;
- Reputation Risk;
- Liquidity Risk;
- Operational Risk;
- Legal and Regulatory Risk;
- Credit Risk; and
- Market Risk.

The policy analysis will utilize the following format:

1. Defining the Problem.
2. Establishing Goals.
3. Selecting a Policy.
4. Implementing a Policy.
5. Evaluating the Policy.[4]

I want to highlight, upfront, the following:

First, this book focuses on the conservation of tropical rainforests. While all biomes are important—from temperate rainforests to coral reefs to the Arctic tundra—tropical rainforests are the focus of this book given its unique issues.

Second, protected areas or conservation projects are often financed via complex mechanisms such as a combination of trust fund financing, domestic budgetary allocations, and revenue from ecotourism. This said, if a case study is categorized under domestic budgetary allocations, that is not to say that the protected area received exclusive financing from domestic budgetary allocations.

Third, US-based and European Union-based conservation is different than tropical rainforest conservation. For example, Indigenous Peoples and forest-dependent communities are often present within tropical rainforest conservation areas and their buffer zones. In addition, there tends to be greater income inequality and less overall wealth in the host countries of tropical rainforests. For example, the GINI coefficient—which is a leading economic indicator of wealth inequality among households—was 25.90 (in 2012) for Norway, 26.94 (in 2012) for Iceland, and 30.13 (in 2011) for Germany, while Brazil was 52.87 (in 2013), Indonesia was 35.57 (in 2010), and the Democratic Republic of the Congo was 42.10 (in 2012).[5] Competing for government budgets are other pressing domestic issues such as poverty, energy access, education, and healthcare. Furthermore, corruption tends to be higher in such areas[6] and the ease of doing business is often much more difficult.[7]

Fourth, the requirements for effective tropical rainforest conservation include the rule of law, transparency, social justice, and long-term financing requirements.

Fifth, there are sources of financing, there are conservation outcomes, and there are financing instruments that bridge the gap between the sources of financing and these conservation outcomes. These three groups can often be confused. To clarify, sources of financing can come from individuals, foundations (corporate and family), businesses (small and medium enterprises and publicly traded corporations), and governments (domestic spending and through bilateral or multilateral channels). Conservation outcomes include: state or national parks, international peace parks, private reserves, conservation concessions, conservation easements, and sustainable forestry and agriculture operations. The financing instruments that bridge the gap between the sources of financing and the conservation outcomes—which will be the focus of this book—include: debt-for-nature swaps, payments for ecosystem services, premiums for green procurements, taxation, and tax deductions. This book explores all three groups, with an emphasis on the historical evolution of financing instruments.

Sixth, unfortunately, developing long-term financing to conserve tropical rainforests may not protect such forests unless climate change is also mitigated.

Seventh, for every investment in the conservation of tropical rainforests, there needs to be a root cause analysis to make sure funds are specifically targeting the underlying drivers and agents of tropical rainforest deforestation and degradation.

Eighth, while tropical rainforests are beautiful, they are incredibly tough environments to work and live in. From poisonous snakes to mosquito-borne illnesses to the hot, humid weather, tropical rainforest work is—to say the least—challenging and yet rewarding and absolutely necessary because of the need to mitigate global climate change, to halt the catastrophic collapse of biodiversity, and to improve the livelihoods of forest-dependent peoples.

Ninth, while I truly believe that tropical rainforests have intrinsic value—due to their tremendous biodiversity and the resources they provide for billions of livelihoods—if there is no extrinsic value placed on these forests, then decision-makers, whether it be political or business, will be more apt to undertake activities and policies that will lead to the conversion of these forests to other land uses.

April 2017 Brian Joseph McFarland

Comprehensive Timeline

This is a fairly comprehensive timeline outlining when conservation organizations and associations were established, along with when famous conservationists were born. The author apologizes if not everyone and every organization is listed as this is intended to be more of an illustration. However, there is a more detailed timeline in each of the chapters.

Conservation Organizations Around the World

Nineteenth Century

- 1826: The Zoological Society of London (ZSL) is founded by Sir Thomas Stamford Raffles.[8]
- 1846: Smithsonian Institution is founded.[9]
- 1865: Massachusetts State Fish & Wildlife Agency—the first such agency in the United States—originates.[10]
- 1873: Massachusetts Fish & Game Association (MFGA), oldest incorporated conservation organization in the United States, is founded.[11]
- 1875: Led by John Aston Warder, the American Forestry Association (now known as American Forests) is founded.[12]
- 1888: The National Geographic Society is founded.[13]

- 1889: Royal Society for the Protection of Birds (RSPB) is founded.[14]
- 1891: The Trustees of Public Reservations (now the Trustees of Reservations, or TTOR), the world's first regional land trust, is established.[15]
- 1892: The Sierra Club is founded by John Muir.[16]
- 1895: Wildlife Conservation Society (WCS, originally known as the New York Zoological Society) is founded by Henry Fairfield Osborn and Andrew Haswell Green.[17]
- 1896: Massachusetts Audubon Society, the world's oldest existing Audubon Society, is created in Boston by Harriet Hemenway and Mina Hall.[18]

Twentieth Century

- 1903: The Society for the Preservation of the Wild Fauna of the Empire (now known as Fauna and Flora International or FFI) is founded.[19]
- 1905: The National Audubon Society is founded.[20]
- 1919: National Parks Conservation Association is established.[21]
- 1922: The International Council for Bird Preservation (ICBP) is founded; ICBP would rebrand as BirdLife International in 1993.[22]
- 1926: Charles Stewart Mott Foundation is founded.[23]
- 1935: The Wilderness Society is founded by Aldo Leopold, Bob Marshall, Robert Sterling Yard, and others.[24]
- 1936: The General Wildlife Federation (now known as the National Wildlife Federation or NWF) is founded by Jay Norwood Darling.[25]
- 1936: Ford Foundation is established.[26]
- 1937: Ducks Unlimited is incorporated.[27]
- 1942: Inter-American Institute for Agricultural Science (IICA) is founded. The IICA would later spin out the Centro Agronomico Tropical de Investigacion y Enseñanza (CATIE; The Tropical Agricultural Research and Higher Education Center) in 1973.[28]
- 1945: Cooperative for Assistance and Relief Everywhere (CARE) is founded.[29]
- 1946: "Barro Colorado Island became part of the Smithsonian Institution in 1946. In 1966, the organization changed its name to the Smithsonian Tropical Research Institute (STRI)."[30]

- 1946: The Ecologists Union, which would later change its name to The Nature Conservancy (TNC) in 1950, is formed.[31]
- 1947: Defenders of Wildlife (formerly known as Defenders of Fur Bearers) is founded.[32]
- 1948: International Union for the Conservation of Nature (IUCN) is founded.[33]
- 1951: The Nature Conservancy (TNC) is incorporated.[34]
- 1952: Resources for the Future (RFF) is founded.[35]
- 1961: World Wildlife Fund (WWF) is founded by Julian Huxley, Edward Max Nicholson, Peter Scott, and Guy Mountfort.[36] WWF is known as World Wildlife Fund in the United States and Canada, but is known elsewhere as the World Wide Fund for Nature.
- 1961: The African Wildlife Foundation (AWF, formerly known as the African Wildlife Leadership Foundation) is founded.[37]
- 1963: Weeden Foundation is established.[38]
- 1963: Organization for Tropical Studies (OTS) is founded.[39]
- 1964: David and Lucile Packard Foundation is established.[40]
- 1965: National Recreation and Park Association (NRPA) is formed.[41]
- 1967: Environmental Defense Fund (EDF) is founded by Art Cooley, Charlie Wurster, and Dennis Puleston.[42]
- 1970: Natural Resources Defense Council (NRDC) is founded.[43]
- 1970: MacArthur Foundation is founded.[44]
- 1971: Greenpeace is founded.[45]
- 1971: The Consultative Group on International Agricultural Research (CGIAR) is formed.[46]
- 1971: Pact International is founded.[47]
- 1971: Orangutan Research and Conservation Project (ORCP) "was the initial name of the program started in 1971 by Dr. Biruté Mary Galdikas and her former husband, Rod Brindamour." ORCP would later lead to the establishment of the Orangutan Foundation International (OFI) in 1986.[48]
- 1972: Trust for Public Land (TPL) is founded.[49]
- 1973: The Centro Agronomico Tropical de Investigacion y Enseñanza (CATIE; The Tropical Agricultural Research and Higher Education Center) is spun out of the Inter-American Institute for Agricultural Science (IICA).[50]

- 1974: The Worldwatch Institute is founded by Lester Brown.[51]
- 1974: Open Space Institute is founded.[52]
- 1976: TRAFFIC is established.[53]
- 1977: Jane Goodall Institute for Wildlife Research, Education and Conservation (JGI) is founded by Jane Goodall and Princess Genevievedi San Faustino.[54]
- 1977: The Green Belt Movement (GBM) is founded by Professor Wangari Maathai.[55]
- 1978: The World Agroforestry Centre (i.e., formerly known as The International Council for Research in Agroforestry or ICRAF) is founded.[56]
- 1980: The School for Field Studies (SFS) is founded.[57]
- 1981: Pronatura is founded.[58]
- 1982: World Resources Institute (WRI) is founded by James Gustave Speth.[59]
- 1982: Land Trust Alliance is founded.[60]
- 1983: The Fundación Defensores de la Naturaleza (FDN; Nature Defenders Foundation) is founded.[61]
- 1984: National Fish and Wildlife Foundation is founded.[62]
- 1984: SCS is founded. In 1988, SCS would become Scientific Certification Systems, Inc. (SCS) and now is SCS Global Services.[63]
- 1985: Rainforest Action Network (RAN) is founded by Randy Hayes and Mike Roselle.[64]
- 1985: Woods Hole Research Center (WHRC) is founded by George Woodwell.[65]
- 1985: Winrock International is created by Rockefeller Family by merging Winrock International Research and Training Center, the International Agricultural Development Service, and the Agricultural Development Council.[66]
- 1985: The Conservation Fund is founded.[67]
- 1986: Amigos de Sian Ka'an (Friends of Sian Ka'an) is founded.[68]
- 1986: Environmental Services, Inc. (ESI) is founded.[69]
- 1987: Conservation International (CI) is founded by Spencer Beebe and Peter A. Seligmann.[70]
- 1987: Rainforest Alliance is cofounded by Daniel Katz.[71]
- 1988: SOS Amazônia is founded.[72]

- 1988: Rainforest Conservation Fund is founded.[73]
- 1988: Rainforest Trust is founded.[74]
- 1988: Wildlife Habitat Council (WHC) is founded.[75]
- 1988: SaveNature.org is founded.[76]
- 1989: Stockholm Environmental Institute (SEI) is founded.[77]
- 1989: CERES (Coalition for Environmentally Responsible Economies) is founded by Joan Bavaria.[78]
- 1989: The Rainforest Foundation founded by "Sting and his wife Trudie Styler in 1989 focuses primarily on the protection of Indigenous Peoples rights."[79] In 1996, Rainforest Foundation Norway would become an independent foundation.[80]
- 1990: Imazon is founded.[81]
- 1990: International Institute for Sustainable Development (IISD) is founded.[82]
- 1990: The Turner Foundation is established.[83]
- 1991: Project Piaba is started.[84]
- 1992: EcoLogic Development Fund is founded by Russell Byers, Jr. and Shaun Paul.[85]
- 1993: Center for International Forestry Research (CIFOR) is founded.[86]
- 1994: Leuser International Foundation (LIF) is established.[87]
- 1994: Yayasan Keanekaragaman Hayati Indonesia (KEHATI; The Indonesian Biodiversity Foundation) is established.[88]
- 1995: The World Industry Council for the Environment (WICE) and Business Council for Sustainable Development (BCSD) merge to form the World Business Council for Sustainable Development (WBCSD). In 1990, Stephan Schmidheiny created the first Business Council for Sustainable Development (BCSD).[89]
- 1995: Instituto de Pesquisa Ambiental da Amazônia (IPAM; Amazon Environmental Research Institute in English) is founded.[90]
- 1995: Instituto de Manejo e Certificação Florestal e Agrícola (IMAFLORA) is founded.[91]
- 1995: Wildlife Alliance is founded.[92]
- 1995: The Mexican Civil Council for Sustainable Forestry is formalized.[93]
- 1996: Nature and Culture International is founded.[94]

- 1996: Fundo Brasileiro para a Biodiversidade (FUNBIO, The Brazilian Biodiversity Fund) started operations.[95]
- 1996: Amazon Watch is founded.[96]
- 1996: Doris Duke Charitable Foundation is established.[97]
- 1997: Wildlife Works is founded by Mike Korchinsky; Wildlife Works Carbon LLC would later be founded in 2008.[98]
- 1997: Future Forests is founded. Future Forests would later rebrand as The CarbonNeutral Company and then rebrand again in 2015 as Natural Capital Partners.[99]
- 1997: Emerald Planet is founded by Michael Robison.[100]
- 1997: Bonobo & Congo Biodiversity Initiative (BCBI) is created by the Zoological Society of Milwaukee (ZSM).[101]
- 1997: Sumatran Orangutan Society (SOS) is established.[102]
- 1998: Forest Trends is founded.[103]
- 1998: Conservation Strategy Fund (CSF) is founded by John Reid.[104]
- 1998: Global Nature Fund is founded.[i][i] Global Nature Fund. "About Us." Accessed August 24, 2017.http://www.globalnature.org/34517/Home/About-us/resindex.aspx•1999: Root Capital is founded.[105]
- 1999: Amazon Conservation Association (ACA) is founded.[106]
- 1999: CO_2OL-USA is founded by Keegan Eisenstadt (CO_2OL-USA would later assimilate into ClearSky Climate Solutions in 2008).[107]
- 1999: The International Small Group & Tree Planting Program (TIST) is founded.[108]
- 1999: The Programme for the Endorsement of Forest Certification (PEFC) is founded.[109]
- 1999: Mongabay is founded by Rhett A. Butler.[110]

Twenty-First Century

- 2000: Carbosur is founded.[111]
- 2000: Trucost is established.[112]
- 2000: Gordon and Betty Moore Foundation is founded.[113]
- 2000: Sumatran Rainforest Institute (SRI) is established.[114]
- 2000: Kemitraan partnership is established.[115]
- 2001: Yayasan Orangutan Sumatera Lestari—Orangutan Information Centre (YOSL—OIC) is founded by Panut Hadisiswoyo with assistance from the Sumatran Orangutan Society (SOS).[116]

- 2001: Global Canopy Programme is founded by Andrew Mitchell.[117]
- 2001: The Center for Global Development (CGD) is founded by Edward Scott Jr., C. Fred Bergsten, and Nancy Birdsall.[118]
- 2001: BlueOrchard is founded.[119]
- 2003: Carbonfund.org Foundation is established by Eric and Lesley Carlson.
- 2004: Bosques Amazonicos (BAM) is established.[120]
- 2004: Climate Focus is founded.[121]
- 2004: Instituto de Conservação e Desenvolvimento Sustentável do Amazonas (IDESAM, Institute for Conservation and Sustainable Development of Amazonas in English) is established.[122]
- 2005: New Forests is founded.[123]
- 2005: ERA Carbon Offsets is founded. In 2012, Offsetters Climate Solutions (OCS) was formed through the consolidation of ERA Ecosystem Restoration Associates Inc., Carbon Credit Corp. (CCC), and Offsetters Clean Technology.[124] In 2015, "NatureBank includes the combination of Era Ecosystem Services and Forest Finest Consulting Inc. under the NatureBank banner, while the voluntary carbon brands of Offsetters and CO_2OL will continue to operate as subsidiaries under NatureBank."[125]
- 2005: Paso Pacifico is founded.[126]
- 2006: Ecosystem Services LLC is founded.[127]
- 2006: Panthera is founded by Daphne Recanati Kaplan and Dr. Thomas S. Kaplan.[128]
- 2006: South Pole Carbon is created.[129]
- 2006: TerraCarbon is formed by Scott Settelmyer and Dr. Bernhard Schlamadinger.[130]
- 2006: Terra Global Capital is founded by Leslie Durschinger.[131]
- 2006: EnviroMarket Ltd. is founded.[132]
- 2007: Anthrotect is founded.[133]
- 2007: PT. Rimba Makmur Utama (PT. RMU) is founded.[134]
- 2007: Forest Carbon Group is founded.[135]
- 2007: Fundação Amazonas Sustentável (FAS, The Sustainable Amazonas Foundation) is founded.[136]
- 2007: EKO Asset Management Partners, LLC is founded. EKO would later merge with Wolfensohn Fund Management, L.P. in 2014 to form Encourage Capital.[137]

- 2007: International Conservation Fund of Canada (ICFC) is founded.[138]
- 2008: Climate Advisers is founded.[139]
- 2008: C-Quest Capital (CQC) is founded.[140]
- 2008: Carbon Decisions International is founded.[141]
- 2008: InfiniteEARTH is founded.[142]
- 2008: ClimateWorks Foundation is established.[143]
- 2009: BioCarbon Group is founded (initial capital raised in 2011).[144]
- 2009: The Carbon War Room is founded.[145]
- 2009: Forest Carbon Offsets LLC is established.[146]
- 2009: Climate Bonds Initiative is established.[147]
- 2010: Permian Global Advisors LLP is incorporated.[148]
- 2010: Clarmondial is established.[149]
- 2011: Althelia Ecosphere is launched.[150]
- 2011: Bolsa de Valores Ambientais Rio de Janeiro (BVRio, Rio de Janeiro Environmental Exchange) and the BVRio Institute are established.[151]
- 2013: Earth Innovation Institute (EII) is founded (created out of IPAM's International Program).[152]
- 2015: Institut Penelitian Inovasi Bumi (INOBU, Earth Innovation's Research Institute) was registered as an Indonesian, nonprofit research institute in 2015. INOBU grew out of Earth Innovation Institute's (EII) Indonesia program.[153]

• 2017: Everland is founded by Wildlife Works.Citation should be: Everland. "About Us." Accessed August 23, 2017. http:// www.everlandmarketing.com/ Associations/Networks

- 1951: Association Technique Internationale des Bois Tropicaux (ATIBT, International Tropical Timber Technical Association) is created.[154]
- 1965: Associação Nacional dos Exportadores de Cereais (ANEC, National Association of Cereal Exporters) has begun.[155]
- 1975: National Wildlife Refuge Association is founded.[156]
- 1981: Associação Brasileira das Indústrias de Óleos Vegetais (ABIOVE, Brazilian Association of Vegetable Oil Industries) is founded.[157]
- 1984: La Coordinadora de las Organizaciones Indígenas de la Cuenca Amazónica (COICA, Coordinator of Indigenous Organizations of the Amazon River Basin) is founded.[158]

- 1985: Society for Conservation Biology (SCB) is founded.[159]
- 1995: Alliance of Religions and Conservation (ARC) is founded by Prince Philip.[160]
- 1995: International Land Coalition (ILC) is founded.[161]
- 1996: Tebtebba (Indigenous Peoples' International Centre for Policy Research and Education) is established.[162]
- 1997: Environmental Markets Association (EMA) is established.[163]
- 1998: National Mitigation Banking Association (NMBA) is established.[164]
- 1999: The International Emissions Trading Association (IETA) is founded.[165]
- 1999: Latin American and Caribbean Network of Environmental Funds (RedLAC) is established.[166]
- 2000: Carbon Disclosure Project (now known as CDP) is founded.[167]
- 2002: Conservation Finance Alliance is established.[168]
- 2002: Wildlife Conservation Network (WCN) is established.[169]
- 2003: Global Footprint Network is established.[170]
- 2004: Roundtable on Sustainable Palm Oil (RSPO) is established.[171]
- 2005: Leather Working Group (LWG) is formed.[172]
- 2006: Round Table on Responsible Soy (RTRS) is established.[173]
- 2006: Conservation Finance Alliance grows out of "a pilot workshop envisioned at Lincoln Institute of Land Policy in 2006 and held at Yale School of Forestry & Environmental Studies in 2007."[174]
- 2007: Grupo de Trabalho da Pecuária Sustentável (GTPS, Brazilian Roundtable for Sustainable Livestock) is created.[175]
- 2008: International Carbon Reduction & Offset Alliance (ICROA) is founded.[176]
- 2008: The Finance Alliance for Sustainable Trade (FAST) is launched.[177]
- 2008: The Carbon Markets and Investors Association (CMIA) is formed "by the merger of The Carbon Markets Association (CMA) and International Carbon Investors and Services (INCIS)."[178] The Carbon Markets and Investors Association rebranded in October 2011 as the Climate Markets and Investment Association.[179]
- 2009: Global Impact Investing Network (GIIN) is launched.[180]
- 2009: Consumer Goods Forum is formed.[181]
- 2010: The Global Roundtable for Sustainable Beef is formed.[182]

- 2010: Climate and Land Use Alliance is established.[183]
- 2011: REDD Offset Working (ROW) Group is established.[184]
- 2011: The Practitioners' Network for Large Landscape Conservation is founded. The Network is now known as the Network for Landscape Conservation.[185]
- 2012: Tropical Forest Alliance 2020 is launched.[186]
- 2012: Code REDD is established.[187]
- 2014: First organizational meeting of the International Land Conservation Network is held.[188]
- 2016: The Conservation Finance Practitioner Roundtable is launched.[189]
- 2016: Coalition for Private Investment in Conservation (CPIC) is launched.[190]

Famous Conservationists Throughout History

Eighteenth Century

- February 13/14, 1766, to December 29, 1834: Thomas Robert Malthus.[191]
- September 14, 1769, to May 6, 1859: Alexander von Humboldt.[192]
- August 29, 1773, to May 11, 1858: Aimé Jacques Alexandre Bonpland.[193]

Nineteenth Century

- February 12, 1809, to April 19, 1882: Charles Robert Darwin.[194]
- July 12, 1817, to May 6, 1862: Henry David Thoreau.[195]
- January 8, 1823, to November 7, 1913: Alfred Russel Wallace.[196]
- February 8, 1825, to February 16, 1892: Henry Walter Bates.[197]
- April 3, 1837, to March 29, 1921: John Burroughs.[198]
- April 21, 1838, to December 24, 1914: John Muir.[199]
- September 20, 1849, to April 11, 1938: George Bird Grinnell.[200]
- December 1, 1854, to March 6, 1937: William T. Hornaday.[201]

- May 12, 1856, to September 9, 1901: Andreas Franz Wilhelm (A.F.W.) Schimper.[202]
- August 8, 1857, to November 6, 1935: Henry Fairfield Osborn.[203]
- October 27, 1858, to January 6, 1919: US President Theodore Roosevelt.[204]
- February 1, 1861, to May 17, 1945: Robert Sterling Yard.[205]
- May 5, 1865, to January 19, 1958: Cândido Rondon.[206]
- August 11, 1865, to October 4, 1946: Gifford Pinchot.[207]
- July 4, 1867, to January 22, 1930: Stephen Mather.[208]
- November 3, 1877, to November 20, 1962: Rosalie Edge.[209]
- March 6, 1879, to December 11, 1975: Emile Benton MacKaye.[210]
- February 6, 1884, to June 6, 1977: Ernest Oberholtzer.[211]
- January 11, 1887, to April 21, 1948: Aldo Leopold.[212]
- March 1, 1889, to October 21, 1963: Olaus Murie.[213]
- December 27, 1896, to March 18, 1956: Louis Bromfield.[214]
- April 4, 1899, to January 13, 1982: Sigurd Ferdinand Olson.[215]

Twentieth Century

- January 2, 1901, to November 11, 1939: Bob Marshall.[216]
- February 20, 1902, to April 22, 1984: Ansel Adams.[217]
- March 7, 1902, to November 15, 1964: Bernard Frank.[218]
- July 15, 1902, to March 8, 1968: Harvey Broome.[219]
- August 18, 1902, to October 19, 2003: Margaret (Mardy) Murie.[220]
- August 7, 1903, to October 1, 1972: Louis Seymour Bazett Leakey.[221]
- July 5, 1904, to February 3, 2005: Ernst Walter Mayr.[222]
- March 28, 1905, to June 14, 1986: Marlin Perkins.[223]
- February 3, 1906, to August 20, 1989: George Alexander Graham Adamson.[224]
- February 25, 1906, to May 5, 1964: Howard Zahniser.[225]
- May 27, 1907, to April 14, 1964: Rachel Carson.[226]
- December 9, 1908, to October 4, 1995: Paul Westmacott (P.W.) Richards.[227]

- September 14, 1909, to August 29, 1989: Sir Peter Markham Scott.[228]
- January 20, 1910, to January 3, 1980: Joy Adamson.[229]
- June 11, 1910, to June 25, 1997: Jacques-Yves Cousteau.[230]
- September 17, 1913, to August 10, 2002: Eugene Pleasants Odum.[231]
- January 12, 1915, to April 10, 2001: Dr. Richard Evans Schultes.[i][i] Harvard University Herbaria. "Richard Evan Schultes Papers." Accessed August 23, 2017. http://botlib.huh.harvard.edu/libraries/Nash/schultes.htm•April 21, 1915, to September 14, 2003: Garrett James Hardin.[232]
- June 4, 1916, to July 3, 2005: Gaylord Anton Nelson.[233]
- January 19, 1917, to May 26, 2015: Leo A. Drey.[234]
- April 17, 1918, to November 16, 1981: William Holden.[235]
- January 13, 1919, to December 1, 2001: Celia Hunter.[236]
- May 27, 1919, to November 5, 2002: Raymond F. Dasmann.[237]
- January 6, 1920, to April 19, 2004: John Maynard Smith.[238]
- May 12, 1921, to May 6, 2014: Farley McGill Mowat.[239]
- September 1, 1924, to September 11, 2002: Howard T. Odum.[240]
- January 7, 1925, to January 30, 1995: Gerald Malcolm Durrell.[241]
- May 12, 1926, to September 8, 2010: George Christopher Williams.[242]
- January 29, 1927, to March 14, 1989: Edward Abbey.[243]
- March 15, 1927, to November 30, 2014: Ian Cedric Player.[244]
- May 8, 1926, to Present: Sir David Attenborough.[245]
- June 10, 1929, to Present: Dr. Edward Osborne (E.O.) Wilson.[246]
- April 7, 1930, to November 1, 1972: Robert H. MacArthur.[247]
- January 16, 1932, to December 26, 1985: Dian Fossey.[248]
- May 29, 1932, to Present: Paul Ralph Ehrlich.[249]
- July 10, 1932, to Present: Dr. Gordon Orians.[250]
- March 28, 1934, to Present: Lester R. Brown.[251]
- April 3, 1934, to Present: Dr. Jane Goodall.[252]
- August 24, 1934, to Present: Norman Myers.[253]
- October 21, 1934, to Present: Arturo Gómez-Pompa.[254]
- June 13, 1936, to Present: Peter H. Raven.[255]
- August 1, 1936, to March 7, 2000: William D. Hamilton.[256]
- September 10, 1937, to Present: Jared Diamond.[257]

- March 5, 1938, to November 22, 2011: Lynn Margulis.[258]
- July 21, 1938, to Present: Herman Daly.[259]
- January 18, 1939, to Present: Dr. Daniel Hunt Janzen.[260]
- April 20, 1939, to Present: Gro Harlem Brundtland.[261]
- 1939 (circa) to September 7, 2006: Clem Coetzee.[262]
- April 1, 1940, to September 25, 2011: Wangari Muta Maathai.[263]
- June 28, 1940, to Present: Muhammad Yunus.[264]
- December 1, 1940, to Present: Terry L. Erwin.[265]
- 1940 (circa) to July 24, 2006: Abi Kusno Nachran.[266]
- March 29, 1941, to Present: Dr. James Hansen.[267]
- July 7, 1941, to Present: Bill Oddie.[268]
- August 22, 1941, to Present: Dr. Thomas Lovejoy.[269]
- September 10, 1941, to May 20, 2002: Stephen Jay Gould.[270]
- March 4, 1942, to Present: James Gustave Speth.[271]
- February 19, 1943, to Present: Robert Ludlow ("Bob") Trivers.[272]
- December 15, 1944, to December 22, 1988: Francisco Alves Mendes, Jr. (Chico Mendes).[273]
- December 19, 1944, to Present: Richard Leakey.[274]
- January 6, 1945, to August 3, 1993: Alwyn Howard Gentry.[275]
- May 10, 1946, to Present: Dr. Biruté Mary Galdikas.[276]
- January 2, 1947, to Present: Jack Bushnell Hanna.[277]
- March 31, 1948, to Present: Albert ("Al") Arnold Gore, Jr.[278]
- October 29, 1948, to Present: Dr. Franciscus Bernardus Maria "Frans" de Waal.[279]
- November 8, 1949, to Present: Russell Mittermeier.[280]
- December 2, 1950, to Present: Captain Paul Watson.[281]
- June 27, 1951, to Present: Dr. Adrian Forsyth.[282]
- February 14, 1953, to May 10, 2016: Ilkka A. Hanski.[283]
- April 1, 1953, to August 3, 1993: Theodore Albert "Ted" Parker III.[284]
- August 19, 1954, to Present: Daniel J. Bisaccio.[285]
- December 8, 1960, to Present: Bill McKibben.[286]
- February 22, 1962, to September 4, 2006: Steve Irwin.[287]
- July 20, 1964, to Present: Terri Irwin.[288]
- July 11, 1967, to Present: Jeff Corwin.[289]

Notes

1. Environmental Protection Agency. "Waste Site Cleanup & Reuse in New England: Fletcher's Paint Works and Storage: Milford, NH." Last modified November 23, 2016. http://yosemite.epa.gov/r1/npl_pad.nsf/f52fa5c31fa8f5c885256adc0050b631/E9C7C97347501D608525691F0063F6C8?OpenDocument
2. United Nations Educational, Scientific and Cultural Organization. "Man and the Biosphere Programme." Accessed November 23, 2016. http://www.unesco.org/new/en/natural-sciences/environment/ecological-sciences/man-and-biosphere-programme/
3. Pathfinder International. "Preparing a Case Study: A Guide for Designing and Conducting a Case Study for Evaluation Input." Accessed August 1, 2016. http://www.pathfinder.org/publications-tools/pdfs/Preparing-a-Case-Study-A-Guide-for-Designing-and-Conducting-a-Case-Study-for-Evaluation-Input.pdf
4. American University. "Tips for Writing a Policy Analysis." Accessed August 1, 2016.
5. Index Mundi. "GINI Index (World Bank Estimate)—Country Ranking." Accessed February 9, 2017. http://www.indexmundi.com/facts/indicators/SI.POV.GINI/rankings
6. Transparency International. "Corruption Perceptions Index 2016." Accessed February 8, 2017. http://www.transparency.org/news/feature/corruption_perceptions_index_2016
7. World Bank Group. "Doing Business: Economy Rankings." Accessed January 12, 2017. http://www.doingbusiness.org/rankings
8. Zoological Society of London. "Remember Sir Stamford Raffles, Founder and First President of ZSL." Last modified July 1, 2015. https://www.zsl.org/blogs/artefact-of-the-month/remembering-sir-stamford-raffles-founder-and-first-president-of-zsl
9. Smithsonian Libraries. "From Smithson to Smithsonian: The Birth of an Institution." Accessed November 23, 2016. http://www.sil.si.edu/Exhibitions/Smithson-to-Smithsonian/intro.html
10. Foster, Dave. "Meeting the Conservation Challenge in New England." In *Conservation Capital in the Americas*, edited by James N. Levitt. Massachusetts: Lincoln Institute of Land Policy, 2010. 20–21.
11. Ibid.
12. American Forests. "History." Accessed November 23, 2016. http://www.americanforests.org/discover-american-forests/our-story/

13. National Geographic Society. "This Day in History: January 13, 1888: National Geographic Society Founded." Accessed November 23, 2016. http://nationalgeographic.org/thisday/jan13/national-geographic-society-founded/
14. Royal Society for the Protection of Birds, "History of the RSPB." Accessed November 23, 2016. https://www.rspb.org.uk/whatwedo/history/
15. Foster, Dave. "Meeting the Conservation Challenge in New England." In *Conservation Capital in the Americas*, edited by James N. Levitt. Massachusetts: Lincoln Institute of Land Policy, 2010. 20–21.
16. Sierra Club. "History: Sierra Club Timeline." Accessed November 23, 2016. http://vault.sierraclub.org/history/timeline.aspx
17. Wildlife Conservation Society. "Zoos & Aquarium." Accessed November 23, 2016. https://www.wcs.org/parks
18. National Audubon Society. "History of Audubon and Science-based Bird Conservation." Accessed November 23, 2016. http://www.audubon.org/content/history-audubon-and-waterbird-conservation
19. Fauna & Flora International. "110 Years of Fauna & Flora International." Accessed November 23, 2016. http://www.fauna-flora.org/timeline/
20. National Audubon Society. "History of Audubon and Science-based Bird Conservation." Accessed November 23, 2016. http://www.audubon.org/content/history-audubon-and-waterbird-conservation
21. National Parks Conservation Association. "Our Story." Accessed November 30, 2016. https://www.npca.org/about/our-story
22. BirdLife International. "Our History." Accessed November 23, 2016. http://www.birdlife.org/worldwide/partnership/our-history
23. Charles Stewart Mott Foundation. "History and Founder." Accessed February 13, 2017. https://www.mott.org/about/history/
24. The Wilderness Society. "Wilderness Timeline." Accessed November 23, 2016. http://wilderness.org/article/wilderness-timeline
25. National Wildlife Federation. "Our History." Last modified January 19, 2011. http://www.nwf.org/News-and-Magazines/National-Wildlife/Animals/Archives/2011/NWF-75-years-timeline.aspx
26. Ford Foundation. "Our Origins." Accessed February 13, 2017. https://www.fordfoundation.org/about-us/our-origins/
27. Ducks Unlimited. "The Birth of DU." Accessed November 23, 2016. http://www.ducks.org/about-du/the-birth-of-du/page5
28. Tropics Foundation. "CATIE." Accessed November 23, 2016. http://www.tropicsfoundation.org/whoweare/history.php

29. CARE. "CARE Package." Accessed February 9, 2017. http://www.care.org/care-package
30. Smithsonian Tropical Research Institute. "History." Accessed November 23, 2016. https://www.stri.si.edu/english/about_stri/history.php
31. The Nature Conservancy. "Our History." Accessed November 23, 2016. http://www.nature.org/about-us/vision-mission/history/
32. Defenders of Wildlife. "Frequently Asked Questions." Accessed November 23, 2016. http://www.defenders.org/frequently-asked-questions
33. International Union for the Conservation of Nature. "About." Accessed November 23, 2016. https://www.iucn.org/about
34. The Nature Conservancy. "Our History." Accessed November 23, 2016. http://www.nature.org/about-us/vision-mission/history/
35. Resources for the Future. "About: History." Accessed April 5, 2017. http://www.rff.org/about
36. World Wildlife Fund. "WWF from 1961 to 2006." Accessed November 23, 2016. http://wwf.panda.org/who_we_are/history/wwf_conservation_1961_2006/
37. African Wildlife Foundation. "Our History." Accessed November 23, 2016. http://www.awf.org/about/history
38. Weeden Foundation. "Weeden Foundation Mission Statement." Accessed February 13, 2017. http://www.weedenfdn.org/Weeden-Foundation-Mission-Statement.htm
39. Organization for Tropical Studies. "About Us." Accessed April 5, 2017. http://www.ots.ac.cr/index.php?option=com_content&task=view&id=55&Itemid=258
40. David and Lucile Packard Foundation. "Our History." Accessed February 13, 2017. https://www.packard.org/about-the-foundation/history/
41. National Recreation and Park Association. "About NRPA." Accessed November 23, 2016. http://www.nrpa.org/About-National-Recreation-and-Park-Association/
42. Environmental Defense Fund. "Our Story: How EDF Got Started." Accessed November 23, 2016. https://www.edf.org/about/our-history
43. Natural Resources Defense Council. "About Us." Accessed November 23, 2016. https://www.nrdc.org/about
44. MacArthur Foundation. "Our History." Accessed February 13, 2017. https://www.macfound.org/about/our-history/
45. Greenpeace. "About Us: History." Accessed November 23, 2016. http://www.greenpeace.org/canada/en/About-us/History/

46. The Consultative Group on International Agricultural Research. "The CGIAR at 40 and Beyond." Accessed November 23, 2016. http://www.cgiar.org/www-archive/www.cgiar.org/pdf/cgiar%4040_final_LOWRES.pdf
47. Pact International. "About Us." Accessed November 23, 2016. http://www.pactworld.org/our-promise
48. Orangutan Foundation International. "History of OFI." Accessed March 21, 2017. https://orangutan.org/about/history-of-ofi-2/
49. Trust for Public Land. "Our History." Accessed November 23, 2016. https://www.tpl.org/about/history
50. Tropics Foundation. "CATIE." Accessed November 23, 2016. http://www.tropicsfoundation.org/whoweare/history.php
51. Worldwatch Institute. "History." Accessed November 23, 2016. http://www.worldwatch.org/mission
52. Open Space Institute. "About the Open Space Institute." Accessed November 23, 2016. http://www.osiny.org/site/PageServer?pagename=AboutOSI_Mission
53. TRAFFIC. "What We Do." Accessed April 2, 2017. http://www.traffic.org/overview/
54. Jane Goodall Institute. "History of the Jane Goodall Institute." Accessed November 23, 2016. http://janegoodallug.org/about-us/history/
55. The Green Belt Movement. "Our History." Accessed February 9, 2017. http://www.greenbeltmovement.org/who-we-are/our-history
56. World Agroforestry Centre. "History of ICRAF and the World Agroforestry Centre." Accessed November 23, 2016. http://www.worldagroforestry.org/about/history
57. The School for Field Studies. "Our History." Accessed November 23, 2016. http://www.fieldstudies.org/about/history
58. San Diego Natural History Museum & PRONATURA. "PRONATURA." Accessed November 23, 2016. http://sdnhm.org/oceanoasis/behindthescenes/pronatura.html
59. World Resources Institute. "The WRI Story: 30 Years of Big Ideas." Accessed November 23, 2016. http://www.wri.org/wri-story-30-years-big-ideas
60. Land Trust Alliance. "About Us." Accessed November 23, 2016. http://www.landtrustalliance.org/about-us
61. Nature Defenders Foundation. "About Us." Accessed March 24, 2017. https://defensores.org.gt/en#conocenos

62. National Fish and Wildlife Foundation. "About National Fish and Wildlife Foundation." Accessed February 15, 2017. http://www.nfwf.org/whoweare/Pages/home.aspx#.WKT9G39WfeQ
63. SCS. "Company." Accessed February 20, 2017. https://www.scsglobalservices.com/company
64. Rainforest Action Network. "RAN'S Greatest Hits: 1985–2010." Accessed November 23, 2016. http://www.ran.org/ran_s_greatest_hits_1985_2010
65. Woods Hole Research Center. "About Us." Accessed November 23, 2016. http://whrc.org/about-whrc/
66. Winrock International. "Winrock History." Accessed November 23, 2016. http://www.winrock.org/winrock-history/
67. The Conservation Fund. "History." Accessed November 30, 2016. http://www.conservationfund.org/about-us/history
68. Amigos de Sian Ka'an. "¿What We Do and How We Work?" Accessed February 12, 2017. http://amigosdesiankaan.org/en/who-we-are/what-we-do-and-how-we-work
69. Environmental Services, Inc. "Our Company History and Values." Accessed February 20, 2017. http://www.esinc.cc/about-us/company-profile
70. Conservation International. "25 and Counting: Conservation International Marks Anniversary Milestone." Last modified January 30, 2012. http://www.conservation.org/NewsRoom/pressreleases/Pages/25_and_Counting_Conservation_International_Marks_Anniversary_Milestone.aspx
71. Rainforest Alliance. "Our History." Accessed November 23, 2016. http://www.rainforest-alliance.org/about/history
72. SOS Amazônia. "Nossa história." Accessed November 23, 2016. http://sosamazonia.org.br/conteudo/quem-somos/nossa-historia/
73. Rainforest Conservation Fund. "About RCF." Accessed November 30, 2016. http://www.rainforestconservation.org/25-2-2/
74. Charity Navigator. "Rainforest Trust: Mission." Accessed February 20, 2017. https://www.charitynavigator.org/index.cfm?bay=search.summary&orgid=11733
75. Wildlife Habitat Council. "Welcome to the Wildlife Habitat Council." Accessed March 27, 2017. http://wildlifehc.conservationregistry.org/
76. SaveNature.org. "History and Milestones." Accessed April 11, 2017. http://www.savenature.org/content/about/History

77. Stockholm Environment Institute. "About SEI Asia." Accessed November 23, 2016. https://www.sei-international.org/asia/about-sei-asia
78. Ceres. "Exxon Valdez Oil Spill Still Leaves a Painful Legacy." Last modified March 2014. http://www.ceres.org/about-us/our-history/exxon-valdez-oil-spill-still-leaves-a-painful-legacy
79. Martin, *On the Edge*, 180.
80. Rainforest Foundation Norway. "About Rainforest Foundation Norway: History." Accessed March 30, 2017.http://www.regnskog.no/en/about-rainforest-foundation-norway
81. Imazon. "About Us: Foundation." Accessed November 23, 2016. http://imazon.org.br/about-us/foundation/?lang=en
82. International Institute for Sustainable Development. "Sustainable Development Timeline—2012—IISD." Accessed November 23, 2016. https://www.iisd.org/pdf/2002/sd_timeline2002.pdf
83. Turner Foundation. "Our Story." Accessed February 13, 2017. http://www.turnerfoundation.org/our-story/
84. Project Piaba. "Background about the Fishery and History." Accessed April 10, 2017. http://projectpiaba.org/who-we-are/history/
85. EcoLogic Development Fund. "History of EcoLogic." Accessed November 23, 2016. http://www.ecologic.org/about-us/history-of-ecologic/
86. Center for International Forestry Research. "History and Mission of CIFOR." Accessed November 23, 2016. http://www.cifor.org/publications/Html/AR-98/His-Mission.html
87. Leuser International Foundation. "History." Accessed March 26, 2017. http://www.leuserfoundation.org/index.php/aboutlif/history
88. Yayasan Keanekaragaman Hayati Indonesia. "History of KEHATI Foundation." Accessed March 26, 2017. http://www.kehati.or.id/about-us-2/
89. World Business Council for Sustainable Development. "History." Accessed November 23, 2016. http://www.wbcsd.org/Overview/About-us
90. Amazon Environmental Research Institute. "Nossa História." Accessed November 23, 2016. http://ipam.org.br/sobre-o-ipam/
91. Brazil de Souza, Bruno of IMAFLORA. Interviewed by Brian McFarland. October 2016.
92. Wildlife Alliance. "Who We Are: History." Accessed March 8, 2017. http://www.wildlifealliance.org/who-we-are/

93. Consejo Civil Mexicano para la Silvicultura Sostenible. "¿QUIÉNES SOMOS?" Accessed April 10, 2017. http://www.ccmss.org.mx/quienes-somos/
94. Nature and Culture International. "Celebrating 20 Years of Conservation." Accessed November 23, 2016. https://natureandculture.org/20th-anniversary/
95. FUNBIO. "About Us." Accessed November 28, 2016. http://www.funbio.org.br/en/o-funbio/quem-somos/
96. Amazon Watch. "About: Our Mission." Accessed March 26, 2017. http://amazonwatch.org/about
97. Doris Duke Charitable Foundation. "Mission & History." Accessed February 13, 2017. http://www.ddcf.org/about-us/mission-and-history/
98. Wildlife Works. "About the Founder." Accessed November 23, 2016. http://www.wildlifeworks.com/company/about-founder.php
99. Natural Capital Partners. "Our Story." Accessed November 23, 2016. http://www.naturalcapitalpartners.com/company
100. Emerald Planet. "Welcome to Emerald Planet." Accessed November 23, 2016. http://www.theemeraldplanet.com/index.html
101. Bonobo & Congo Biodiversity Initiative. "History with Bonobos." Accessed February 20, 2017. http://www.bonoboconservation.com/about/background.php
102. The Orangutan Project. "Sumatran Orangutan Society/Orangutan Information Centre." Accessed March 26, 2017. https://www.orangutan.org.au/projects/our-projects/sumatran-orangutan-societyorangutan-information-centre/
103. Forest Trends. "Financial Information." Accessed November 23, 2016. http://www.forest-trends.org/page.php?id=189&name=Financial%20Information
104. Conservation Strategy Fund. "About." Accessed November 30, 2016. http://conservation-strategy.org/en/page/about-conservation-strategy-fund
105. Root Capital. "About Us." Accessed November 23, 2016. https://www.rootcapital.org/about-us/our-team
106. Amazon Conservation Association. "Overview." Accessed November 23, 2016. http://www.amazonconservation.org/about/index.html
107. Eisenstadt, Keegan of ClearSky Climate Solutions. Interviewed by Brian McFarland. October 2016.

108. TIST. "Home." Accessed February 9, 2017. https://www.tist.org/i2/index.php
109. PEFC. "Our History." Accessed March 15, 2017. https://www.pefc.org/about-pefc/who-we-are/history
110. Mongabay. "About Mongabay." Accessed March 27, 2017. https://www.mongabay.com/about/
111. Carbosur. "La Empresa." Accessed November 23, 2016. http://www.carbosur.com.uy/about/
112. Trucost. "Company History." Accessed November 23, 2016. http://www.trucost.com/about-trucost/company-history/
113. Foundation Directory Online. "Gordon and Betty Moore Foundation." Accessed February 15, 2017. https://fconline.foundationcenter.org/grantmaker-profile?key=MOOR151
114. Sumatran Rainforest Institute. "Home." Accessed March 26, 2017. http://sumatranrainforest.org/
115. Kemitraan. "Our History." Accessed April 10, 2017. http://www.kemitraan.or.id/our-history
116. Hadisiswoyo, Panut of YOSL—OIC. Interviewed by Brian McFarland. November 2016.
117. Tropical Landscapes Summit 2015. "Andrew Mitchell." Accessed November 23, 2016. http://www.summit2015.org/speakers/item/387-andrew-mitchell
118. The Center for Global Development. "About CGD." Accessed November 23, 2016. http://www.cgdev.org/page/about-cgd
119. BlueOrchard. "About Us." Accessed November 30, 2016. http://www.blueorchard.com/about-us/blue-orchard/
120. Bosques Amazonicos. "Home." Accessed November 23, 2016. http://www.bosques-amazonicos.com/
121. Climate Focus. "About Us." Accessed November 23, 2016. http://www.climatefocus.com/about-us
122. Institute for the Conservation and Sustainable Development of Amazonas. "About IDESAM." Accessed November 23, 2016. http://www.idesam.org.br/sobre-o-idesam/?lang=en
123. New Forests. "Our Roots." Accessed November 23, 2016. https://www.newforests.com.au/#roots
124. Era Ecosystem Services. "Era Over the Years." Accessed November 23, 2016. http://www.eraecosystems.com/about-us/corporate-timeline/

125. NatureBank. "Introducing NatureBank." Accessed November 23, 2016. https://www.naturebank.com/about-us/current-news/introducing-naturebank/
126. Paso Pacifico. "Mission Statement." Accessed November 23, 2016. http://www.pasopacifico.org/our-mission.html
127. Castro de la Mata V., Gonzalo. Email Message to author. March 27, 2017.
128. Panthera. "Thomas S. Kaplan, PhD." Accessed November 23, 2016. https://www.panthera.org/people/thomas-s-kaplan-phd
129. South Pole Carbon. "Our Story." Accessed November 23, 2016. http://www.thesouthpolegroup.com/why-spg/history
130. TerraCarbon. "About Us." Accessed November 23, 2016. http://terra-carbon.com/about/
131. Terra Global Capital. "Team and Advisors." Accessed November 23, 2016. http://www.terraglobalcapital.com/team-and-advisors
132. LinkedIn. "EnviroMarket." Accessed April 11, 2017. https://www.linkedin.com/company/enviromarket
133. Anthrotect. "About Us: History." Accessed March 8, 2017. https://www.anthrotect.com/about.html
134. PT. RMU. "Katingan Peatland Restoration and Conservation Project." Accessed November 28, 2016. http://www.v-c-s.org/wp-content/uploads/2016/10/CCB_PROJ_DESC_1477_11MAY16.pdf. 29.
135. Chatellier, Jeffrey of Forest Carbon. Interviewed by Brian McFarland. October 2016.
136. The Sustainable Amazonas Foundation. "FAS." Accessed November 23, 2016. http://fas-amazonas.org/a-fas/?lang=en
137. Encourage Capital. "Our History." Accessed January 17, 2017. http://encouragecapital.com/our-firm/our-history/
138. International Conservation Fund of Canada. "About ICFC." Accessed April 11, 2017. http://icfcanada.org/about-us/about-icfc
139. Thoumi, Gabriel. Interviewed by Brian McFarland. February 2017.
140. C-Quest Capital. "History." Accessed November 23, 2016. http://www.cquestcapital.com/about/history/
141. Carbon Decisions International. "About." Accessed March 26, 2017. http://www.carbondecisions.com/en-us/about/
142. Lemons, Todd. Interviewed by Brian McFarland. March 2017.
143. ClimateWorks Foundation. "History." Accessed February 13, 2017. http://www.climateworks.org/about-us/our-history/

144. Ebeling, Johannes of BioCarbon Group. Interviewed by Brian McFarland. October 2016.
145. The Carbon War Room. "Frequently Asked Questions: When did the Carbon War Room start?" Accessed February 13, 2017. http://carbon-warroom.com/what-we-do/faq
146. Forest Carbon Offsets LLC. "Home." Accessed April 10, 2017. http://www.forestcarbonoffsets.net/
147. LinkedIn. "Climate Bonds Initiative: About Us." Accessed April 11, 2017. https://www.linkedin.com/company-beta/1032198/
148. Companies House. "Permian Global Advisors LLP: Company number OC355305." Accessed March 26, 2017. https://beta.companieshouse.gov.uk/company/OC355305
149. Clarmondial. "About Us." Accessed April 10, 2017. https://www.clar-mondial.com/about-us/
150. Ecosystem Marketplace. "Althelia Raises $80 Million For REDD and Ecosystem Services." Accessed December 13, 2016. http://www.ecosys-temmarketplace.com/articles/althelia-raises-80-million-for-br-redd-and-ecosystem-services/
151. Moura Costa, Pedro of BVRio. Interviewed by Brian McFarland. March 2017.
152. Earth Innovation Institute. "Our History." Accessed November 23, 2016. http://earthinnovation.org/about/our-history/
153. Institut Penelitian Inovasi Bumi. "Our History." Accessed November 23, 2016. http://www.inobu.org/about/our-history.html
154. ATIBT. "About Us." Accessed November 28, 2016. https://www.atibt.org/en/about-us/
155. ANEC. "History." Accessed November 28, 2016. http://www.anec.com.br/en/2015-09-28-19-08-08/history
156. National Wildlife Refuge Association. "About the Refuge System and FWS." Accessed December 6, 2016. http://refugeassociation.org/about/about-the-refuge-system/
157. ABIOVE. "About ABIOVE." Accessed November 28, 2016. http://www.abiove.org.br/site/index.php?page=history&area=OS0xLTU
158. Coordinator of Indigenous Organizations of the Amazon River Basin. "Historia de la COICA." Accessed November 23, 2016. http://www.coica.org.ec/index.php/sobre-coica/historia
159. Society for Conservation Biology. "What is SCB." Accessed November 23, 2016. https://conbio.org/about-scb/who-we-are

160. Alliance of Religions and Conservation. "About ARC." Accessed November 23, 2016. http://www.arcworld.org/about.asp?pageID=2
161. International Land Coalition. "Who we are." Accessed March 21, 2017. http://www.landcoalition.org/en/who-we-are
162. Tebtebba. "Who We Are." Accessed November 23, 2016. http://tebtebba.org/index.php/content/who-we-are
163. Environmental Markets Association. "About EMA." Accessed November 23, 2016. http://www.emahq.org/about-us/about-ema
164. National Mitigation Banking Association. "About." Accessed November 23, 2016. http://mitigationbanking.org/index.php/about/
165. International Emissions Trading Association. "Our History." Accessed November 23, 2016. http://www.ieta.org/Our-History
166. Latin American and Caribbean Network of Environmental Funds. "Home." Accessed November 30, 2016. http://redlac.org/en/
167. CDP. "Staff: Paul Dickinson." Accessed November 23, 2016. https://www.cdp.net/en/info/staff
168. Conservation Finance Alliance. "About the CFA." Accessed November 23, 2016. http://www.conservationfinance.org/history.php
169. Wildlife Conservation Network. "Meet Our Team." Accessed February 9, 2017. https://wildnet.org/about/our-team
170. Global Footprint Network. "At a Glance." Accessed November 23, 2016. http://www.footprintnetwork.org/pt/index.php/GFN/page/at_a_glance/
171. Roundtable on Sustainable Palm Oil. "About Us." Accessed November 23, 2016. http://www.rspo.org/about
172. Leather Working Group. "Background of the LWG." Accessed November 23, 2016. http://www.leatherworkinggroup.com/about/background.htm
173. Round Table on Responsible Soy. "History." Accessed November 23, 2016. http://www.responsiblesoy.org/about-rtrs/history/?lang=en
174. Conservation Finance Network. "About Conservation Finance Network." Accessed November 23, 2016. http://www.conservationfinancenetwork.org/about-cfn
175. Brazilian Roundtable on Sustainable Livestock. "About the GTPS." Accessed November 23, 2016. http://www.pecuariasustentavel.org.br/en/about/
176. International Carbon Reduction & Offset Alliance. "About ICROA." Accessed November 23, 2016. http://www.icroa.wildapricot.org/About-ICROA

177. Finance Alliance for Sustainable Development. "History of FAST." Accessed November 23, 2016. http://www.fastinternational.org/fast/history/
178. MarketsWiki. "Carbon Markets and Investors Association." Accessed November 23, 2016. http://www.marketswiki.com/wiki/Carbon_Markets_and_Investors_Association
179. MarketsWiki. "Climate Markets & Investment Association." Accessed November 23, 2016. http://www.marketswiki.com/wiki/Climate_Markets_%26_Investment_Association
180. Global Impact Investing Network. "The GIIN History." Accessed December 6, 2016. https://thegiin.org/giin/history
181. Consumer Goods Forum. "Our History." Accessed November 23, 2016. http://www.theconsumergoodsforum.com/about-the-forum/our-history
182. World Wildlife Fund. "Beef: Our Approach." Accessed November 23, 2016. http://wwf.panda.org/what_we_do/how_we_work/our_global_goals/markets/mti_solutions/certification/agriculture/beef/
183. The Bridgespan Group. "The Climate and Land Use Alliance." Accessed February 15, 2017. https://www.bridgespan.org/bridgespan/Images/articles/lessons-in-funder-collaboration/case-study-climate-and-land-use.pdf
184. REDD Offset Working Group. "About ROW." Accessed November 23, 2016. http://stateredd.org/about-row/
185. Bateson, Emily M. of the Network for Landscape Conservation. Email message to author. March 24, 2017.
186. Tropical Forest Alliance 2020. "Our History." Accessed November 23, 2016. https://www.tfa2020.org/about-tfa/objectives/
187. Code REDD. "IRS Form 990: 2014." Accessed December 2, 2016. http://www.guidestar.org/FinDocuments/2014/460/888/2014-460888828-0c0d4182-9.pdf
188. Lincoln Institute of Land Policy. "International Land Conservation Network: Why We Exist." Accessed April 2, 2017. http://landconservationnetwork.org/sites/default/files/Why%20We%20Exist_0.pdf
189. United States Department of Agriculture. "Conservation Finance Practitioner Roundtable." Accessed November 23, 2016. http://www.nrcs.usda.gov/wps/portal/nrcs/detail/national/technical/emkts/?cid=nrcseprd660806
190. IUCN. "New Coalition Launches to Scale Private Conservation Investment at IUCN World Conservation Congress." Accessed December 12, 2016. https://www.iucn.org/news/new-coalition-launches-scale-private-conservation-investment-iucn-world-conservation-congress

191. MacRae, Donald Gunn. "Thomas Robert Malthus." *Encyclopedia Britannica*. Accessed November 23, 2016. https://www.britannica.com/biography/Thomas-Malthus
192. Kellner, Charlotte. "Alexander von Humboldt." *Encyclopedia Britannica*. Accessed November 23, 2016. https://www.britannica.com/biography/Alexander-von-Humboldt
193. Australian National Herbarium. "Bonpland, Aime Jacques Alexandre (1773–1858)." Accessed November 23, 2016. https://www.anbg.gov.au/biography/bonpland-aime-jacques.html
194. Desmond, Adrian J. "Charles Darwin." *Encyclopedia Britannica*. Accessed November 23, 2016. https://www.britannica.com/biography/Charles-Darwin
195. The Editors of Encyclopedia Britannica. "Henry David Thoreau." *Encyclopedia Britannica*. Accessed November 23, 2016. https://www.britannica.com/biography/Henry-David-Thoreau
196. The Alfred Russel Wallace Website. "Biography of Wallace." Accessed November 23, 2016. http://wallacefund.info/content/biography-wallace
197. Editors of Encyclopedia Britannica. "H.W. Bates." *Encyclopedia Britannica*. Accessed March 8, 2017. https://www.britannica.com/biography/H-W-Bates
198. The Editors of Encyclopedia Britannica. "John Burroughs." *Encyclopedia Britannica*. Accessed November 23, 2016. https://www.britannica.com/biography/John-Burroughs
199. The Editors of Encyclopedia Britannica. "John Muir." *Encyclopedia Britannica*. Accessed November 23, 2016. https://www.britannica.com/biography/John-Muir
200. US Geological Survey. "George Bird Grinnell." Accessed November 23, 2016. http://www.pwrc.usgs.gov/resshow/perry/bios/grinnellgeorge.htm
201. US Scouting Service Project. "Biography of William T. Hornaday." Accessed November 23, 2016. http://usscouts.org/history/hornadaybiography.asp
202. Editors of Encyclopedia Britannica. "Andreas Franz Wilhelm Schimper." *Encyclopedia Britannica*. Accessed November 23, 2016. https://www.britannica.com/biography/Andreas-Franz-Wilhelm-Schimper
203. The Editors of Encyclopedia Britannica. "Henry Fairfield Osborn." *Encyclopedia Britannica*. Accessed November 23, 2016. https://www.britannica.com/biography/Henry-Fairfield-Osborn

204. Cooper, John Milton. "Theodore Roosevelt." *Encyclopedia Britannica*. Accessed November 23, 2016. https://www.britannica.com/biography/Theodore-Roosevelt
205. Wilderness.net. "Robert Sterling Yard: Bringing People to Wilderness." Accessed November 23, 2016. http://www.wilderness.net/NWPS/Yard
206. The Editors of Encyclopedia Britannica. "Cândido Rondon." *Encyclopedia Britannica*. Accessed November 23, 2016. https://www.britannica.com/biography/Candido-Rondon
207. Editors of Encyclopedia Britannica. "Gifford Pinchot." *Encyclopedia Britannica*. Accessed November 23, 2016. https://www.britannica.com/biography/Gifford-Pinchot
208. Undiscovered Yosemite. "Yosemite National Park History: The Story Of Stephen Mather, The Father Of The National Park Service." Accessed November 28, 2016. http://www.undiscovered-yosemite.com/yosemite-national-park-history.html
209. Biography.com. "Rosalie Edge." Accessed November 23, 2016. http://www.biography.com/people/rosalie-edge-072715
210. Wilderness.net. "Benton MacKaye: A Wilderness Visionary." Accessed November 23, 2016. http://www.wilderness.net/NWPS/MacKaye
211. Ernest Oberholtzer Foundation. "Oberholtzer—Timeline: Overview." Accessed November 28, 2016. http://eober.org/Oberholtzer/Timeline.shtml
212. Aldo Leopold Foundation. "The Leopold Legacy." Accessed November 23, 2016. http://www.aldoleopold.org/AldoLeopold/leopold_bio.shtml
213. Wilderness.net. "Olaus and Mardy Murie: Alaska's Passionate Protectors." Accessed November 23, 2016. http://www.wilderness.net/NWPS/Murie
214. Editors of Encyclopedia Britannica. "Louis Bromfield." Encyclopedia Britannica. Accessed November 23, 2016. https://www.britannica.com/biography/Louis-Bromfield
215. Lake Superior Magazine. "Sigurd F. Olson: A Brief Biography." Last modified June 25, 2013. http://www.lakesuperior.com/travel/wisconsin/352sigurd-olson-bio/
216. Wilderness.net. "Bob Marshall." Accessed November 23, 2016. http://www.wilderness.net/nwps/Marshall
217. Szarkowski, John. "Ansel Adams." Encyclopedia Britannica. Accessed November 23, 2016. https://www.britannica.com/biography/Ansel-Adams-American-photographer

218. Today in Science History. "Bernard Frank." Accessed March 26, 2017.https://todayinsci.com/F/Frank_Bernard/FrankBernard-OurNeedForWater.htm
219. Wilderness.net. "Harvey Broome." Accessed November 23, 2016. http://www.wilderness.net/NWPS/Broome
220. Wilderness.net. "Olaus and Mardy Murie: Alaska's Passionate Protectors." Accessed November 23, 2016. http://www.wilderness.net/NWPS/Murie
221. The Editors of Encyclopedia Britannica. "Louis S.B. Leakey." Encyclopedia Britannica. Accessed November 23, 2016. https://www.britannica.com/biography/Louis-S-B-Leakey
222. The Editors of Encyclopedia Britannica. "Ernst Mayr." *Encyclopedia Britannica*. Accessed November 28, 2016. https://www.britannica.com/biography/Ernst-Mayr
223. Saint Louis Zoo. "Marlin Perkins." Accessed November 23, 2016. https://www.stlzoo.org/about/history/marlinperkins/
224. Father of Lions. "George Adamson." Accessed November 23, 2016. http://www.fatheroflions.org/GeorgeAdamson_Information.html
225. Wilderness.net. "Howard Zahniser: Author of the Wilderness Act." Accessed November 23, 2016. http://www.wilderness.net/nwps/zahniser
226. The Editors of Encyclopedia Britannica. "Rachel Carson." *Encyclopedia Britannica*. Accessed November 23, 2016. https://www.britannica.com/biography/Rachel-Carson
227. The Independent. "Obituary: Professor P.W. Richards." Last modified October 10, 1995. http://www.independent.co.uk/news/people/obituary-professor-p-w-richards-1577031.html
228. The Editors of Encyclopedia Britannica. "Sir Peter Markham Scott." *Encyclopedia Britannica*. Accessed November 23, 2016. https://www.britannica.com/biography/Peter-Markham-Scott
229. The Editors of Encyclopedia Britannica. "Joy Adamson." *Encyclopedia Britannica*. Accessed November 23, 2016. https://www.britannica.com/biography/Joy-Adamson
230. The Editors of Encyclopedia Britannica. "Jacques Cousteau." *Encyclopedia Britannica*. Accessed November 23, 2016. https://www.britannica.com/biography/Jacques-Cousteau
231. The Editors of Encyclopedia Britannica. "Eugene Pleasants Odum." *Encyclopedia Britannica*. Accessed November 28, 2016. https://www.britannica.com/biography/Eugene-Pleasants-Odum

232. The University of California Senate. "In Memoriam: Garrett James Hardin." Accessed November 23, 2016. http://senate.universityofcalifornia.edu/_files/inmemoriam/html/garretthardin.htm
233. The Editors of Encyclopedia Britannica. "Gaylord Anton Nelson." *Encyclopedia Britannica*. Accessed November 23, 2016. https://www.britannica.com/biography/Gaylord-Nelson
234. Conservation Federation Missouri. "A Biography of Leo A. Drey." Accessed November 23, 2016. http://www.confedmo.org/a-biography-of-leo-a-drey/
235. The Editors of Encyclopedia Britannica. "William Holden." *Encyclopedia Britannica*. Accessed November 23, 2016. https://www.britannica.com/biography/William-Holden
236. Wilderness.net. "Celia Hunter: An Alaskan Preservation Pioneer." Accessed November 23, 2016. http://www.wilderness.net/NWPS/Hunter
237. Jarrell, Randall. *Raymond F. Dasmann: A Life in Conservation Biology*. California: Regional History Project, University Library, University of California, Santa Cruz, 2000. http://escholarship.org/uc/item/4j9397s9#page-10
238. The Editors of Encyclopedia Britannica. "John Maynard Smith." *Encyclopedia Britannica*. Accessed November 28, 2016. https://www.britannica.com/biography/John-Maynard-Smith
239. Rafferty, John P. "Farley McGill Mowat." *Encyclopedia Britannica*. Accessed November 23, 2016. https://www.britannica.com/biography/Farley-McGill-Mowat
240. The Editors of Encyclopedia Britannica. "Howard T. Odum." *Encyclopedia Britannica*. Accessed November 28, 2016. https://www.britannica.com/biography/Howard-Thomas-Odum
241. The Editors of Encyclopedia Britannica. "Gerald Malcolm Durrell." *Encyclopedia Britannica*. Accessed November 23, 2016. https://www.britannica.com/biography/Gerald-Malcolm-Durrell
242. The Editors of Encyclopedia Britannica. "George Christopher Williams." *Encyclopedia Britannica*. Accessed November 28, 2016. https://www.britannica.com/biography/George-C-Williams
243. The Editors of Encyclopedia Britannica "Edward Abbey." *Encyclopedia Britannica*. Accessed November 28, 2016. https://www.britannica.com/biography/Edward-Abbey
244. Martin, Douglas. "Ian Player Is Dead at 87; Helped to Save Rhinos." *NY Times*. Accessed November 28, 2016. http://www.nytimes.com/2014/12/04/world/africa/ian-player-conservationist-who-helped-save-white-rhinos-dies-at-87.html?_r=0

245. The Editors of Encyclopedia Britannica. "Sir David Attenborough." *Encyclopedia Britannica*. Accessed November 23, 2016. https://www.britannica.com/biography/David-Attenborough
246. Ruse, Michael. "Edward O. Wilson." *Encyclopedia Britannica*. Accessed November 23, 2016. https://www.britannica.com/biography/Edward-O-Wilson
247. Encyclopedia.com. "MacArthur, Robert Helmer." Accessed November 23, 2016. http://www.encyclopedia.com/topic/Robert_MacArthur.aspx
248. The Dian Fossey Gorilla Fund International. "Dian Fossey—Biography." Accessed November 23, 2016. https://gorillafund.org/dian_fossey_bio
249. The Editors of Encyclopedia Britannica. "Paul R. Ehrlich." *Encyclopedia Britannica*. Accessed November 23, 2016. https://www.britannica.com/biography/Paul-R-Ehrlich
250. Alchetron. "Gordon Orians." Accessed November 23, 2016. http://alchetron.com/Gordon-Orians-303456-W
251. Famous Scientists. "Lester R. Brown." Accessed November 28, 2016. http://www.famousscientists.org/lester-r-brown/
252. The Editors of Encyclopedia Britannica. "Jane Goodall." *Encyclopedia Britannica*. Accessed November 23, 2016. https://www.britannica.com/biography/Jane-Goodall
253. University of Oxford. "Professor Norman Myers." Accessed November 23, 2016. https://www.biodiversity.ox.ac.uk/people/professor-norman-myers/
254. MyHeritage. "Arturo Gómez Pompa." Accessed November 23, 2016. https://www.myheritage.com/research/record-10182-3067520/arturo-gomez-pompa-in-biographical-summaries-of-notable-people
255. Discover Life. "Peter Raven." Accessed November 23, 2016. http://www.discoverlife.org/who/CV/Raven,_Peter.html
256. Segerstrale, Ullica C. "William Donald Hamilton." *Encyclopedia Britannica*. Accessed March 8, 2017. https://www.britannica.com/biography/William-Donald-Hamilton
257. Jared Diamond. "About Me." Accessed November 23, 2016. http://www.jareddiamond.org/Jared_Diamond/About_Me.html
258. Tao, Amy. "Lynn Margulis." *Encyclopedia Britannica*. Accessed November 23, 2016. https://www.britannica.com/biography/Lynn-Margulis

259. Prabook. "Herman Edward Daly." Accessed November 23, 2016. http://prabook.com/web/person-view.html?profileId=199202
260. Kyoto Prize. "Daniel Hunt Janzen." Accessed November 23, 2016. http://www.kyotoprize.org/en/laureates/daniel_hunt_janzen/
261. The Editors of Encyclopedia Britannica. "Gro Harlem Brundtland." *Encyclopedia Britannica*. Accessed November 23, 2016. https://www.britannica.com/biography/Gro-Harlem-Brundtland
262. ResponseABILITY Alliance. "Clem Coetzee." Accessed November 28, 2016. http://www.responseabilityalliance.com/html/clem_coetzee.html
263. The Editors of Encyclopedia Britannica. "Wangari Maathai." *Encyclopedia Britannica*. Accessed November 23, 2016. https://www.britannica.com/biography/Wangari-Maathai
264. The Editors of Encyclopedia Britannica. "Muhammad Yunus." *Encyclopedia Britannica*. Accessed November 23, 2016. https://www.britannica.com/biography/Muhammad-Yunus
265. Smithsonian National Museum of Natural History. "SI Entomology." Accessed November 23, 2016. http://entomology.si.edu/staffpages/erwint.htm
266. MyHeritage. "Abi Kusno Nachran." Accessed November 28, 2016. https://www.myheritage.com/research/record-10182-629575/abi-kusno-nachran-in-biographical-summaries-of-notable-people
267. The Nation Institute. "The Ridenhour Courage Prize: James Hansen." Accessed March 20, 2017. http://www.ridenhour.org/prizes_courage_2013.html
268. British Bird Lovers. "Profile: Bill Oddie." Accessed November 23, 2016. http://www.britishbirdlovers.co.uk/articles/profile-bill-oddie
269. Tyler Prize. "2001 Tyler Laureates." Accessed November 23, 2016. http://tylerprize.usc.edu/laureates/tyler2001.html
270. Editors of Encyclopedia Britannica. "Stephen Jay Gould." *Encyclopedia Britannica*. Accessed March 8, 2017. https://www.britannica.com/biography/Stephen-Jay-Gould
271. The American Presidency Project. "Council on Environmental Quality Nomination of James Gustave Speth to Be a Member." Accessed April 13, 2017. http://www.presidency.ucsb.edu/ws/?pid=7126
272. Alchetron. "Robert Trivers." Accessed November 28, 2016. http://alchetron.com/Robert-Trivers-865596-W
273. Ramalho, Tania. "Chico Mendes." *Encyclopedia Britannica*. Accessed November 23, 2016. https://www.britannica.com/biography/Chico-Mendes

274. The Editors of Encyclopedia Britannica. "Richard Leakey." *Encyclopedia Britannica.* Accessed November 23, 2016. https://www.britannica.com/biography/Richard-Leakey
275. MyHeritage. "Alwyn Howard Gentry." Accessed March 26, 2017. https://www.myheritage.com/research/record-10182-1747682/alwyn-howard-gentry-in-biographical-summaries-of-notable-people
276. Orangutan Foundation International. "Dr. Biruté Mary Galdikas." Accessed November 23, 2016. https://orangutan.org/about/dr-birute-mary-galdikas/
277. Schreiber, Barbara A. "Jack Hanna." *Encyclopedia Britannica.* Accessed November 23, 2016. https://www.britannica.com/biography/Jack-Hanna
278. The Editors of Encyclopedia Britannica. "Al Gore." Accessed March 20, 2017. https://www.britannica.com/biography/Al-Gore
279. Veritas. "Frans de Waal." Accessed November 23, 2016. http://nl.veritas.org/speakers/frans-de-waal/
280. OMICS International. "Russell Mittermeier." Accessed December 6, 2016. http://research.omicsgroup.org/index.php/Russell_Mittermeier
281. Sea Shepard Conservation Society. "Captain Paul Watson's Biography." Accessed November 23, 2016. http://www.seashepherd.org/who-we-are/captain-watsons-biography.html
282. Scholastic. "Adrian Forsyth." Accessed April 10, 2017. https://www.scholastic.com/teachers/authors/adrian-forsyth/
283. National Academy of Sciences. "Ilkka A. Hanski." Accessed November 28, 2016. http://www.nasonline.org/member-directory/deceased-members/20022451.html
284. Robbins, Mark B., Gary R. Graves and J.V. Remsen, Jr. "In Memoriam: Theodore A. Parker III (1953–1993)." *Ornithological Monographs.* Volume 48 (1997). Accessed March 26, 2017. https://repository.si.edu/bitstream/handle/10088/2020/Robbins_Remsen_Graves--Parker_memoriam--Ornithological_M.pdf?sequence=1. 1–5.
285. Bisaccio, Daniel J. of Brown University. Interviewed by Brian McFarland. July 2016.
286. Texas Archival Resources Online. "Bill McKibben." Accessed March 8, 2017. http://www.lib.utexas.edu/taro/tturb/00154/trb-00154.html
287. Duignan, Brian. "Jeff Irwin." *Encyclopedia Britannica.* Accessed November 23, 2016. https://www.britannica.com/biography/Steve-Irwin

288. Biography.com. "Terri Irwin." Accessed November 23, 2016. http://www.biography.com/people/terri-irwin-244192
289. Biography.com. "Jeff Corwin." Accessed November 23, 2016. http://www.biography.com/people/jeff-corwin-244191
290. To review the official US Government *Finance and Carbon Markets Lexicon* used by the US Agency for International Development's Forest Carbon, Markets and Communities program, see here for English and here for Spanish.

Acknowledgments

The remarks that follow should not be attributed to any of the following interviewees, unless otherwise noted. Nevertheless, I am very grateful for their comments, the thought-provoking conversations we had, and their dedication to the field:

- Adam Schoenberg
- Agustin Silvani
- Alicia Entem
- Amos Bien
- Ane Broch Graver
- Brett Byers
- Cecilia (Sid) Embree
- Chris Stone
- Daniel Bisaccio
- Datuk Sam Mannan
- Dave Viale
- Deb Joyce
- Dr. Allard Blom
- Dr. Craig Kauffman
- Dr. Daniel H. Janzen
- Dr. David Romo Vallejo
- Dr. Gay Edward Reinartz

- Dr. Jeff Waldon
- Dr. John Tobin-de la Puente
- Dr. Matthew C. Larsen
- Dr. Pedro Moura Costa
- Dr. Pete Lowry
- Dr. Rosa Lemos de Sá
- Dr. Théophile Zognou
- Edit Kiss
- Eric Jaeschke
- Fabian Huwyler
- Gabriel Eickhoff
- Gabriel Thoumi
- Gabriela Burian
- Gonzalo Merediz Alonso
- Greg Fishbein
- James Eaton
- Jan Willem den Besten
- Jarett Emert
- Javier Marquez
- Jeff Chatellier
- Jillian Gladstone
- Johannes Ebeling
- Jorge Cantuarias
- Kathleen Friday
- Keegan Eisenstadt
- Kenneth Johnson
- Manoel Serrão Borges de Sampaio
- Maria Kolos
- María Pía García-Godos Román
- Michael Conner
- Mike Korchinsky
- Oscar Nuñez
- Panut Hadisiswoyo
- Paul Herbertson
- Per Fredrik Ilsaas Pharo
- Peter Graham

- Rezal Kusumaatmadja
- Robert Weary
- Romas Garbaliauskas
- Scott Dowd
- Scott Settelmyer
- Shawn McMahon
- Stephen Donofrio
- Steven Iobst
- Tanja Havemann
- Ted Cheeseman
- Todd Lemons

I would like to thank Gabriel Thoumi once again who provided numerous rounds of comments and thoughtful insights to help strengthen this book.

In addition, I would like to acknowledge the forward-thinking companies, civil servants, local communities, and dedicated professionals working around-the-clock to address tropical rainforest deforestation and degradation. Thank you for all that you do.

Contents

About the Author

Brian McFarland is currently the Portfolio Director at Carbonfund.org Foundation and the Project Origination Director at CarbonCo, the wholly owned subsidiary of Carbonfund.org.

At Carbonfund.org, he identifies climate change mitigation projects in the energy efficiency, renewable energy, and forestry sectors, conducts due diligence on such projects, and then structures the financial support and manages the project portfolio. This multimillion dollar project portfolio includes approximately 75 tree planting and carbon reduction projects across 30+ US states and 15+ countries.

At CarbonCo, Brian designed, financed, and now advises on the implementation of four Reducing Emissions from Deforestation and Degradation (REDD+) projects conserving ~300,000 hectares of tropical rainforests, while reducing 16 million tonnes of carbon dioxide emissions and benefitting up to 600 local community members. This includes spearheading the first-ever, dual Verified Carbon Standard and Climate, Community and Biodiversity Standard validated and verified REDD+ project in the State of Acre, Brazil.

Brian is a certified Project Management Professional (PMP) by the Project Management Institute (PMI), certified Greenhouse Gas Inventory Quantifier (GHG-IQ) from CSA Standards, and a Certified Sustainability Professional from the International Society of Sustainability Professionals (ISSP).

Brian McFarland earned a dual graduate degree in Business Administration and Global Environmental Policy from American University. Brian's graduate thesis was entitled, *Origins, Development and Potential of the International REDD+ Market.* Brian has also published 22 articles and a previous book entitled, *REDD+ and Business Sustainability: A Guide to Reversing Deforestation for Forward Thinking Companies.*

While finishing his Psychology and International Development undergraduate degree from Clark University, Brian conducted authentic environmental fieldwork in Mexico, Costa Rica, Kenya, and Brazil. Such fieldwork included addressing human-wildlife conflicts, working on sustainable community development projects and biodiversity monitoring. During graduate school, Brian also volunteered for projects with the Smithsonian Institution, the United Nations Global Compact, and the US Department of State.

The opinions expressed in this book are those of the author and do not necessarily represent the views of Carbonfund.org or CarbonCo.

Acronyms

AB	Assembly Bill
ABIOVE	Associação Brasileira das Indústrias de Óleos Vegetais (Brazilian Association of Vegetable Oil Industries)
ACA	Amazon Conservation Association
ACES	A Community of Ecosystem Services
ACG	Area de Conservación Guanacaste (of Costa Rica)
ACR	American Carbon Registry
ACT	Amazon Cooperation Treaty
ADB	Asian Development Bank
AfDB	African Development Bank
AFOLU	Agriculture, Forestry and Other Land Uses
AIM	Alternative Investment Market
AMTAVE	La Asociación Mexicana de Turismo de Aventura y Ecoturismo (Mexican Association of Adventure Touriusm and Ecotourism)
ANEC	Associação Nacional dos Exportadores de Cereais (National Association of Cereal Exporters)
ANSI	American National Standards Institute
APEC	Asia-Pacific Economic Cooperation
APP	Asia Pulp and Paper Group
ARB	Air Resources Board (of California)
ARC	Alliance of Religions and Conservation
ARPA	Amazon Region Protected Areas

ASCAF	Agricultural Supply Chain Adaptation Facility
ATIBT	Association Technique Internationale des Bois Tropicaux (International Tropical Timber Technical Association)
AUM	Assets Under Management
AWF	African Wildlife Foundation
BAM	Bosques Amazonicos
BBOP	Business and Biodiversity Offsets Program
BCBI	Bonobo & Congo Biodiversity Initiative
BCI	Barro Colorado Island
BCN	Biodiversity Conservation Network (of USAID)
BCNM	Barro Colorado Nature Monument
BCSD	Business Council for Sustainable Development
BIOFIN	Biodiversity Finance Initiative (of UNDP)
BLM	Bureau of Land Management
BNDES	Banco Nacional de Desenvolvimento Econômico e Social (of Brazil; National Bank of Economic and Social Development)
BNS	Basic Necessities Survey
BOLFOR	Bolivia Sustainable Forest Management Project
BTF	Bhutan Trust Fund for Environmental Conservation
BVRio	Bolsa de Valores Ambientais Rio de Janeiro (Rio de Janeiro Environmental Exchange)
C2D	Debt Development Contract
CA	California
CAFI	Central African Forest Initiative
CAFs	Certificados de Abonos Forestales (Forest Bond Certificates)
CAR	Cadastro Ambiental Rural (of Brazil; Rural Environmental Registry)
CAR	Central African Republic
CAR	Climate Action Reserve
CARE	Cooperative for Assistance and Relief Everywhere
CARPE	Central Africa Regional Program for the Environment
CATIE	Centro Agronomico Tropical de Investigacion y Enseñanza (The Tropical Agricultural Research and Higher Education Center)
CBD	Convention on Biological Diversity (of United Nations)
CCAR	California Climate Action Registry

CCBA/CCBS	Climate, Community and Biodiversity Alliance/ Climate, Community and Biodiversity Standard
CCC	Carbon Credit Corp.
CCIR	Certificado de Cadastro de Imóvel Rural (of Brazil; Certificate of Registration of Rural Property)
CCX	Chicago Climate Exchange
CDM	Clean Development Mechanism
CDP	Carbon Disclosure Project (now known as CDP)
CEF	Closed-End Fund or Closed-Ended Fund
CERES	Coalition for Environmentally Responsible Economies
CGD	Center for Global Development
CGIAR	Consultative Group on International Agricultural Research
CI	Conservation International
CICERO	Center for International Climate and Environmental Research
CIFOR	Center for International Forestry Research
CINRAM	Center for Integrated Natural Resources and Agricultural Management (of the University of Minnesota)
CITES	Convention on International Trade in Endangered Species of Wild Fauna and Flora
CLDP	Conservation and Limited Development Project
CLDs	Conservation Leadership Dialogues
CMA	Carbon Market Association
CMIA	Carbon Markets and Investors Association (now known as the Climate Markets and Investment Association)
CO_2e	Carbon Dioxide Equivalent (Emissions)
COICA	La Coordinadora de las Organizaciones Indígenas de la Cuenca Amazónica (Coordinator of Indigenous Organizations of the Amazon River Basin)
COMIFAC	Commission des Forêts d'Afrique Centrale (Central African Forest Commission)
CONABIO	Comisión Nacional para el Conocimiento y Uso de la Biodiversidad (Biodiversity National Council of Mexico)

CONANP	Comisión Nacional de Áreas Naturales Protegidas (National Commission of Protected Natural Areas)
CONAP	Consejo Nacional de Areas Protegidas (of Guatemala; National Council for Protected Areas)
COP	Conference of the Parties
CORSIA	Carbon Offsetting and Reduction Scheme for International Aviation
CPIC	Coalition for Private Investment in Conservation
CQC	C-Quest Capital
CRAF	Contrato de Desenvolvimento e Venda de Cotas de Reserva Ambiental para Entrega Futura (of Brazil; Contracts of Development and Sale of Legal Reserve Credits)
CRAs	Cotas de Reserva Ambiental (of Brazil; Environmental Reserve Credits)
CREP	Conservation Reserve Enhancement Program
CREST	Center for Responsible Travel
CRP	Conservation Reserve Program
CSAs	Certificados de Servicios Ambientales (Environmental Services Certificates)
CSD	Commission on Sustainable Development (of the United Nations)
CSF	Conservation Strategy Fund
CSP	Conservation Security Program
CSR	Corporate Social Responsibility
CTFS-ForestGEO	Center for Tropical Forest Science and Forest Global Earth Observatories (of the Smithsonian Tropical Research Institute)
DBH	Diameter at Breast Height
DFI	Development Finance Institution
DIB	Development Impact Bond
DOC	Department of Conservation (of New Zealand)
DOD	Disbursed and Outstanding Debt
DOPA	Digital Observatory for Protected Areas
DRC	Democratic Republic of the Congo

DSIRE	Database of State Incentives for Renewables & Efficiency
EAI	Enterprise for the Americas Initiative
EBRD	European Bank for Reconstruction and Development
EDF	Environmental Defense Fund
EIB	Environmental Impact Bond
EIB	European Investment Bank
EII	Earth Innovation Institute
EMA	Environmental Markets Association
EPRI	Electric Power Research Institute
EQIP	Environmental Quality Incentives Program
EROS	Earth Resources Observation and Science
ESG	Environmental, Social and Governance
ESI	Environmental Services, Inc.
ETF	Exchange Traded Fund
ETFRN	European Tropical Forest Research Network
EU ETS	European Union Emissions Trading System
EU	European Union
FAN	Fundación Amigos de la Naturaleza (of Bolivia; Friends of Nature Foundation)
FAO	Food and Agriculture Organization of the United Nations
FAS	Fundação Amazonas Sustentável (The Sustainable Amazonas Foundation)
FASB	Financial Accounting Standards Board
FAST	Finance Alliance for Sustainable Trade
FCMC	Forest Carbon, Markets and Communities (of USAID)
FCPF	Forest Carbon Partnership Facility (of World Bank)
FDN	Fundación Defensores de la Naturaleza (Nature Defenders Foundation)
FECAR	Federación de Comunidades Agropecuarias de Rurrenabaque (The Federation of Agricultural Communities of Rurrenabaque)
FFI	Fauna and Flora International
FLEGT	Forest Law Enforcement, Governance and Trade (of the European Union) Regulation
FOLU	Forestry and Other Land Use

FONAFIFO	Fondo Nacional de Financiamiento Forestal (of Costa Rica; The National Forestry Financing Fund)
FONAG	Fondo para la Protección del Agua (of Ecuador; The Water Protection Fund)
FPIC	Free, Prior, and Informed Consent
FRM	Financial Risk Manager
FSC	Forest Stewardship Council
FSM	Federated States of Micronesia
FUNBIO	Fundo Brasileiro para a Biodiversidade (The Brazilian Biodiversity Fund)
GARP	Global Association of Risk Professionals
GATT	General Agreement on Tariffs and Trade
GBIF	Global Biodiversity Information Facility
GBM	Green Belt Movement
GCF	Governors' Climate & Forests Task Force
GCF	Green Climate Fund
GCP	Global Canopy Programme
GDFCF	Guanacaste Dry Forest Conservation Fund
GDP	Gross Domestic Product
GEF	Global Environmental Facility
GEMS	Global Environmental Monitoring Systems
GHG	Greenhouse Gas
GHG-IQ	Greenhouse Gas Inventory Quantifier
GIIN	Global Impact Investing Network
GMO	Genetically Modified Organism
GRASP	Great Ape Survival Partnership
GRID	Global Information Resource Database (of UNEP)
GSTC	Global Sustainable Tourism Council
Gt	Gigaton
$GtCO_2eq$	Gigaton of Carbon Dioxide Equivalent
GTPS	Grupo de Trabalho da Pecuária Sustentável (of Brazil; Roundtable for Sustainable Livestock)
HDI	Human Development Index
IAATO	International Association of Antarctica Tour Operators
IBAMA	Instituto Brasileiro do Meio Ambiente e dos Recursos Naturais Renováveis (Brazilian Institute of the Environment and Renewable Natural Resources)

IBAT	Integrated Biodiversity Assessment Tool
iBOL	International Barcode of Life Project
ICAO	International Civil Aviation Organization
ICBP	International Council for Bird Preservation
ICCN	Institut Congolais pour la Conservation de la Nature (Congolese Institute for the Conservation of Nature)
ICE	Intercontinental Exchange
ICFC	International Conservation Fund of Canada
ICRAF	International Council for Research in Agroforestry (Now known as World Agroforestry Centre)
ICROA	International Carbon Reduction & Offset Alliance
IDB	Inter-American Development Bank
IDESAM	Instituto de Conservação e Desenvolvimento Sustentável do Amazonas (Institute for Conservation and Sustainable Development of Amazonas)
IETA	International Emissions Trading Association
IFAD	International Fund for Agricultural Development
IFC	International Finance Corporation
IFF	Intergovernmental Forum on Forests
IFI	International Financial Institution
IFRS	International Financial Reporting Standards
IICA	Inter-American Institute for Agricultural Science
IISD	International Institute for Sustainable Development
ILC	International Land Coalition
ILO	International Labor Organization
IMAFLORA	Instituto de Manejo e Certificação Florestal e Agrícola
INBio	Instituto Nacional de Biodiversidad (of Costa Rica; National Institute of Biodiversity)
INCIS	International Carbon Investors and Services
INCRA	Instituto Nacional de Colonização e Reforma Agrária (of Brazil; National Institute for Colonization and Agrarian Reform)
INDC	Intended Nationally Determined Contributions
INOBU	Institut Penelitian Inovasi Bumi (Earth Innovation's Research Institute)
INPE	Instituto Nacional de Pesquisas Espaciais (Brazilian National Institute of Space Research)

IP	Intellectual Property
IPAM	Instituto de Pesquisa Ambiental da Amazônia (Amazon Environmental Research Institute)
IPF	Intergovernmental Panel on Forests
IRR	Internal Rate of Return
IRS	Internal Revenue Service (of the United States)
ISI	Import Substitution Industrialization
ISO	International Organization for Standardization
ISSP	International Society of Sustainability Professionals
ITMOs	Internationally Transferred Mitigation Outcomes
ITR	Imposto Territorial Rural (of Brazil; Rural Land Tax)
ITT	Ishpingo-Tambococha-Tiputini
ITTO	International Tropical Timber Organization
IUCN	International Union for the Conservation of Nature
JGI	Jane Goodall Institute for Wildlife Research, Education and Conservation
KAZA	Kavango—Zambezi Transfrontier Conservation Area
KBAs	Key Biodiversity Areas
KCSO	Kosrae Conservation & Safety Organization
KFS	Kenya Forest Service
KIRMA	Kosrae Island Resource Management Authority
KWS	Kenya Wildlife Service
LEED	Leadership in Energy & Environmental Design
LGX	Luxembourg Green Exchange
LIBOR	London Interbank Offered Rate
LIP	Landowner Incentive Program
LLC	Limited Liability Company
LLP	Limited Liability Partnership
LOI	Letter of Intent
LSE	London Stock Exchange
LWCF	Land and Water Conservation Fund
LWG	Leather Working Group
MAB	Man and the Biosphere
MBF	Madagascar Biodiversity Fund
MBG	Missouri Botanical Garden
MCHT	Maine Coast Heritage Trust

MCL	Monteverde Conservation League
MCT	Micronesia Conservation Trust
MDG	Millennium Development Goal
MFGA	Massachusetts Fish & Game Association
MOU	Memorandum of Understanding
MPA	Marine Protected Area
MRV	Monitoring (or Measuring), Reporting, and Verification
MT	Megaton
Mt	Metric Tonne or Metric Ton
$mtCO_2e$	Metric Tons of Carbon Dioxide Equivalent Emissions
NAFTA	North American Free Trade Agreement
NASA	National Aeronautics and Space Administration
NAWCA	North American Wetland Conservation Act
NBSAPs	National Biodiversity Strategies and Action Plans
NDC	Nationally Determined Contributions
NGO	Nongovernmental Organization
NICFI	Norway's International Climate and Forest Initiative
NMBA	National Mitigation Banking Association
NMTC	New Market Tax Credit
NORAD	Norwegian Agency for Development Cooperation
NOX	Nitrogen Oxide
NPV	Net Present Value
NRCS	Natural Resources Conservation Service
NRDC	Natural Resources Defense Council
NRPA	National Recreation and Park Association
NTFPs	Non-Timber Forest Products
NWF	National Wildlife Federation
NY	New York
OCF	Observatório do Código Florestal (of Brazil; Forest Code Observatory)
OCS	Offsetters Climate Solutions
ODA	Official Development Assistance
OECD	Organization for Economic Co-operation and Development
OFI	Orangutan Foundation International
ONF	Office National des Forêts (of France; National Forests Office)

ONFI	Office National des Forêts International (The International Branch of the French National Forestry Office)
OPIC	Overseas Private Investment Corporation
ORCP	Orangutan Research and Conservation Project
OTS	Organization for Tropical Studies
PBF	Performance-Based Financing
PDs/PDDs	Project Documents/Project Design Documents
PEFC	The Programme for the Endorsement of Forest Certification
PES	Payments for Ecosystem Services
PFT	Pacific Forest Trust
PMI	Project Management Institute
PMP	Project Management Professional
PPCDAm	Plano de Ação para Prevenção e Controle do Desmatamento na Amazônia Legal (of Brazil; The Action Plan for Prevention and Control of Deforestation in the Legal Amazon)
PPP	Public-Private Partnership
PRA	Participatory Rural Assessment or Appraisal
PRI	Program-Related Investment
PRODES	Projeto de Monitoramento do Desflorestamento na Amazônia Legal (Legal Amazon Deforestation Monitoring Project)
QIT	Quebec Iron & Titanium
QMM	QIT Madagascar Minerals
RAN	Rainforest Action Network
RBF	Results-Based Financing
RCF	Rainforest Conservation Fund
REDD/REDD+	Reducing Emissions from Deforestation and forest Degradation. The "+" designates REDD and the additional activities of forest conservation, sustainable forest management, and the enhancement of carbon stocks.
RedLAC	Latin American and Caribbean Network of Environmental Funds
REIT	Real Estate Investment Trust

RFF	Resources for the Future
RFP	Request for Proposal
ROW	REDD Offset Working Group
RRTTLLU	Risk, Return, Time (Horizon), Taxes, Liquidity, Legal and Unique
RSPB	Royal Society for the Protection of Birds
RSPO	Roundtable on Sustainable Palm Oil
RTRS	Round Table on Responsible Soy
SCB	Society for Conservation Biology
SCS	Scientific Certification Systems, Inc. (now doing business as SCS Global Services)
SDGs	Sustainable Development Goals (of the United Nations)
SEI	Stockholm Environmental Institute
SES	REDD+ Social and Environmental Safeguards (SES) Standard
SFD	Sabah Forestry Department
SFI	Sustainable Forestry Initiative
SFS	School for Field Studies
SIB	Social Impact Bond
SIMAB	Smithsonian Institution's Monitoring & Assessment of Biodiversity Program
SINAC	Sistema Nacional de Áreas de Conservación (Costa Rica's National System of Conservation Areas)
SISA	Sistema de Incentivos para Serviços Ambientais (of Acre, Brazil; The State System of Incentives for Environmental Services)
SLOSS	Single Large or Several Small
SMART	Spatial Monitoring and Reporting Tool
SOS	Sumatran Orangutan Society
SOX	Sulfur Oxide
SPV	Special Purpose Vehicle
SRI	Sumatran Rainforest Institute
STRI	Smithsonian Tropical Research Institute
TAF	Technical Assistance Facility
TBPA	Transboundary Protected Area
TEEB	The Economics of Ecosystems and Biodiversity
TFCA	Transfrontier Conservation Area

TFCA	Tropical Forest Conservation Act (of the United States)
TFT	The Forest Trust
TIES	The International Ecotourism Society
TIMO	Timber Investment Management Organization
TIST	The International Small Group & Tree Planting Program
TNC	The Nature Conservancy
TNS	Tri-National de la Sangha
TPA	Totally Protected Areas
TPL	Trust for Public Land
TRASE	Transparency for Sustainable Economies
TREES	Tropical Ecosystems Environment Observation by Satellite
TTOR	Trustees of Reservations
UCs	Compensação em Unidades de Conservação (of Brazil; Consolidation of Conservation Areas)
(U)HNWI	(Ultra)-High-Net-Worth Individual
UK	United Kingdom
UMD	University of Maryland
UN	United Nations
UNC	University of North Carolina
UNCED	United Nations Conference on Environment and Development
UNDP	United Nations Development Programme
UNDRIP	United Nations Declaration on the Rights of Indigenous Peoples
UNEP	United Nations Environment Programme
UNESCO	United Nations Educational, Scientific and Cultural Organization
UNFCCC	United Nations Framework Convention on Climate Change
UNFF	United Nations Forum on Forests
UNWTO	United Nations World Tourism Organization
US/USA	United States/United States of America
USAID	United States Agency for International Development
USD	United States Dollar
USDA FS	United States Department of Agriculture Forest Service
USDA	United States Department of Agriculture
USGBC	US Green Building Council

VCS/VCSA	Verified Carbon Standard/Verified Carbon Standard Association
VERs	Verified Emission Reductions
VPA	Voluntary Partnership Agreements
VVB	Validation/Verification Body
WAVES	Wealth Accounting and the Valuation of Ecosystem Services (of the World Bank)
WBCSD	World Business Council for Sustainable Development
WCED	World Commission on Environment and Development
WCN	Wildlife Conservation Network
WCS	Wildlife Conservation Society
WHC	Wildlife Habitat Council
WHIP	Wildlife Habitat Incentives Program
WHRC	Woods Hole Research Center
WICE	World Industry Council for the Environment
WOCAT	World Overview of Conservation Approaches and Technologies
WRI	World Resources Institute
WRP	Wetlands Reserve Program
WTO	World Trade Organization
WWF	World Wildlife Fund (also known as World Wide Fund for Nature)
Y.E.L.A.	Yela Environment Landowners Authority
YOSL—OIC	Yayasan Orangutan Sumatera Lestari—Orangutan Information Centre
ZAR	South African Rand
ZSL	Zoological Society of London
ZSM	Zoological Society of Milwaukee To review the official US Government Finance and Carbon Markets Lexicon used by the US Agency for International Development's Forest Carbon, Markets and Communities program, see here (http://www.fcmcglobal.org/documents/FinanceandCarbonMarketsLexiconFinal40clean.pdf) for English and here (http://www.fcmcglobal.org/documents/Carbonolexico-Spanish%20V1-0.pdf) for Spanish. PLEASE MAKE SURE THIS STARTS ON SEPARATE LINE AS IT IS NOT PART OF THE ACRONYM ZSM

List of Figures

List of Tables

1

Executive Summary

Rainforests are the most biologically diverse terrestrial ecosystem on Earth and provide a wide range of critical ecosystem services from regulating water cycles to recycling nutrients, to the storage of carbon. According to research by Dr. Claude Martin:

> The area covered by tropical rainforests at that time {Humboldt and Bonpland's expedition in 1799–1804} was still close to the 1.6 billion hectares believed to be the original maximum distribution areas {…}. Today less than half this area remains as undisturbed forest {less than 800 million hectares}—nobody knows exactly how much—and about another quarter survives as fragmented and degraded forest.[1]

This said, significant financial resources from both the private and public sectors need to be mobilized to reverse this trend of tropical rainforest deforestation, while at the same time to restore degraded areas. As of January 2016, it is estimated that approximately "$52 billion per year flows to conservation projects {i.e., not just rainforest conservation}, the bulk of it is public and philanthropic funds."[2] In addition, approximately USD$8.2 billion was invested by the private sector from 2004 to 2015 in the following "three groups of conservation-oriented investments:

B.J. McFarland, *Conservation of Tropical Rainforests*, Palgrave Studies in Environmental Policy and Regulation, https://doi.org/10.1007/978-3-319-63236-0_1

- Sustainable food and fiber production (e.g., forestry, agriculture, and fisheries);
- Habitat conservation (e.g., mitigation banking and forest carbon trading); and
- Water quality and quantity protection (e.g., watershed protection and water rights trading)."[3]

However, going forward, "{…} to finance the preservation of the world's precious ecosystems {including tropical rainforests} will require $200 billion to $300 billion in additional capital, and private investment capital may be the main source of additional capital. Attracting such a level of private capital will require attractive risk-adjusted rates of return, in addition to clear and measurable conservation impacts."[4] As stated by the former Commission on Climate and Tropical Forests, approximately USD$30 billion a year "will be required to reduce emissions from tropical deforestation by half in 2020."[5]

The history of conservation finance arguably dates back to 1634, when the Boston Commons was established and likewise, it is considered "the oldest public open space in the United States. Indeed, it is very likely the oldest public open space in the English-speaking world."[6] The Bogd Khan Uul National Park in Mongolia was established "by the Mongolian government in 1778, {however} it was originally chartered by Ming Dynasty officials in the 1500s as an area to be kept off limits to extractive uses, protected for its beauty and sacred nature."[7]

The first documented protected area, according to the World Database on Protected Areas, appears to be a nature park in Russia known as the Donskoj rybny zapovednik (Donskoe zapretnoe rybnoe prostranstvo) which was established in 1819.[8]

The first state park appears to be Yosemite National Park in California, United States, which was first designated a state park in 1864,[9] while the first national park—aside from Bogd Khan Uul National Park—appears to be Yellowstone National Park in the current US states of Wyoming, Montana, and Idaho in 1872.[10]

In 1891, the Trustees of Public Reservations (now the Trustees of Reservations, or TTOR) became the world's first regional land trust,[11] which established the world's first conservation easement in Boston, Massachusetts.[12]

It is difficult to discern when the first government international budgetary expenditures for conservation were allocated. However, it is likely this mechanism was first established in the late nineteenth century or early twentieth century, when the world's first international conservation organizations were established. This includes the Society for the Preservation of the Wild Fauna of the Empire (now known as Fauna and Flora International or FFI), which was founded in 1903,[13] and the International Council for Bird Preservation (ICBP; the ICBP would rebrand as BirdLife International in 1993) was established in 1922.[14] Similarly, Waterton Glacier International Peace Park was established in 1932 between the United States of America and Canada and marked the world's first international peace park.[15]

Modern-day impact investing is comparable to socially responsible investing, which took "{…} off in the 1970s as part of grassroots campaigns to boycott firms in South Africa {as a result of Apartheid} as well as the nuclear weapons and tobacco industries."[16]

In the late 1970s, the concept of payments for ecosystem services (PES, referred to earlier as payments for environmental services) begins to appear.[17]

Low-impact tourism likely began earlier, but in 1983, the term "ecotourism" is first coined by Mexican architect Héctor Ceballos-Lascuráin.[18]

Debt-for-nature swaps came along shortly thereafter with the first such debt-for-nature swaps conducted in 1987 in both Bolivia and Ecuador.[19]

Then in 1991, the world's first major bioprospecting deal was struck between Merck & Co. and Costa Rica's Instituto Nacional de Biodiversidad (INBio).[20]

The sustainable, green procurement of locally produced commodities—particularly by Indigenous Peoples and forest-dependent communities—has existed for thousands of years. However, the origin of commercial, green procurement with independent certification standards likely started with the creation of the Forest Stewardship Council (FSC)[21] and the Sustainable Forestry Initiative (SFI) in 1994.[22]

The first widely recognized green bonds were issued in 2007 and 2008 by the World Bank[23] and European Investment Bank.[24]

With this historical overview in mind, it appears as though the two most promising conservation financing instruments are the payments for ecosystem services' Reducing Emissions from Deforestation and Degradation (REDD+) mechanism, along with no-deforestation supply chain commitments and green procurement. Thus, the no-deforestation supply chain commitments and green procurement directly engage companies that, whether directly or indirectly, are responsible for much of the deforestation in tropical rainforests.

REDD+ is promising due to its pay-for-performance model, its incorporation into international agreements (most notably the Paris Agreement), and its compatibility with other potential financing mechanisms such as the International Finance Corporation's (IFC) Green Bond, the Althelia Fund, and the United Nations' Green Climate Fund.

Debt-for-nature swaps, bioprospecting, and ecotourism—while important—are relatively small mechanisms. Public finance, via domestic and international budgetary allocations, will remain very important, but the future of conservation finance will need a large expansion of private sector financing. This said, green bonds are another, very promising financing instrument, but are not currently being designed and structured at scale for the forestry sector.

Notes

1. Martin. *On the Edge*. XXI.
2. Huwyler, Fabian, Jürg Käppeli and John Tobin. "Conservation Finance: From Niche to Mainstream: The Building of an Institutional Asset Class." Last modified January 2016. https://www.credit-suisse.com/media/assets/corporate/docs/about-us/responsibility/banking/conservation-finance-en.pdf. 3.
3. Hamrick, Kelley. "State of Private Investment in Conservation 2016." *Ecosystem Marketplace*. December 2016. http://www.forest-trends.org/documents/files/doc_5474.pdf#. vii.

4. Huwyler, Fabian, Jürg Käppeli and John Tobin. "Conservation Finance: From Niche to Mainstream: The Building of an Institutional Asset Class." Last modified January 2016. https://www.credit-suisse.com/media/assets/corporate/docs/about-us/responsibility/banking/conservation-finance-en.pdf. 3.
5. The Commission on Climate and Tropical Forests. "Protecting the Climate Forests: Why Reducing Tropical Deforestation is in America's Vital National Interest." October 2009. Accessed April 13, 2017. http://www.climateforestscommission.org/documents/cctf-report.pdf. 44.
6. Levitt. *From Walden to Wall Street*. 4.
7. The Nature Conservancy. "The World's Oldest National Park: Ghosts of Monks and Red Deer." Last modified November 10, 2009. http://blog.nature.org/conservancy/2009/11/10/worlds-oldest-national-park-mongolia-nature-charles-bedford-bogdkhan/
8. IUCN and UNEP-WCMC. "The World Database on Protected Areas (WDPA) [Online]." December 13, 2016. Cambridge, UK: UNEP-WCMC, 2016. www.protectedplanet.net
9. California Department of Parks and Recreation. "Yosemite 'State Park'." Accessed November 23, 2016. http://www.150.parks.ca.gov/?page_id=27539
10. National Park Service. "Yellowstone National Park: Frequently Asked Questions: History." Accessed November 23, 2016. https://www.nps.gov/yell/learn/historyculture/history-faqs.htm
11. Foster, Dave. "Meeting the Conservation Challenge in New England." In *Conservation Capital in the Americas*, edited by James N. Levitt. Massachusetts: Lincoln Institute of Land Policy, 2010. 20–21.
12. The Nature Conservancy. "Easements 101." Accessed December 13, 2016. https://www.nature.org/about-us/private-lands-conservation/conservation-easements/conservation-easements-101.pdf
13. Fauna & Flora International. "110 Years of Fauna & Flora International." Accessed November 23, 2016. http://www.fauna-flora.org/timeline/
14. BirdLife International. "Our History." Accessed November 23, 2016. http://www.birdlife.org/worldwide/partnership/our-history
15. UNESCO. "Waterton Glacier International Peace Park." Accessed December 13, 2016. http://whc.unesco.org/en/list/354
16. Daily, Gretchen C. and Katherine Ellison. *The New Economy of Nature*. 116.

17. Gómez-Baggethun, Erik, Rudolf de Groot, Pedro L. Lomas and Carlos Montes. "The History of Ecosystem Services in Economic Theory and Practice: From Early Notions to Markets and Payment Schemes." *Ecological Economics*. 69(6). April 2010. 1209–1218. doi: 10.1016/j.ecolecon.2009.11.007; https://www.researchgate.net/publication/46490541_The_History_of_Ecosystem_Services_in_Economic_Theory_and_Practice_From_Early_Notions_to_Markets_and_Payment_Schemes
18. EcoClub. "International Ecotourism Monthly." Year 7. Issue 85. Accessed December 14, 2016. http://ecoclub.com/news/085.pdf. 2.
19. World Wildlife Fund. *Guide to Conservation Finance*. 41.
20. The North American Congress on Latin America. "Bioprospecting and Biopiracy in the Americas." Accessed December 13, 2016. http://nacla.org/news/bioprospecting-and-biopiracy-americas
21. Forest Stewardship Council. "Our History." Accessed November 23, 2016. https://ic.fsc.org/en/about-fsc/our-history
22. Sustainable Forestry Initiative. "Background." Accessed November 23, 2016. http://www.sfiprogram.org/about-us/basics-of-sfi/
23. The World Bank. "World Bank Green Bonds: Green Bond Issuances to Date." Accessed December 13, 2016. http://treasury.worldbank.org/cmd/htm/GreenBondIssuancesToDate.html
24. Climate Bonds Initiative. "History." Accessed December 13, 2016. https://www.climatebonds.net/market/history

2

The Context of Tropical Rainforest Deforestation and Degradation

> "The American consumer, seeking a good-quality hamburger at 'reasonable,' i.e. non-inflationary, price, is not aware of the spinoff consequences of his actions far away from his homeland."
>
> —Dr. Norman Myers[1]

Where Are the World's Tropical Rainforests?

Tropical rainforests are globally threatened and often rare ecosystems. According to The Nature Conservancy, only 2% of the world's total surface area is home to rainforests, while rainforests are home to 50% of the world's plants and animals. However, "every second, a slice of rainforest the size of a football field is mowed down. That's 86,400 football fields of rainforest per day, or over 31 million football fields of rainforest each year."[2] Thus the area destroyed each year is equal to an area greater than the size of the United Kingdom.

According to research by Dr. Claude Martin,

> The area covered by tropical rainforests at that time {Humboldt and Bonpland's expedition in 1799–1804} was still close to the 1.6 billion hectares believed to

B.J. McFarland, *Conservation of Tropical Rainforests*, Palgrave Studies in Environmental Policy and Regulation, https://doi.org/10.1007/978-3-319-63236-0_2

> be the original maximum distribution areas {…}. Today less than half this area remains as undisturbed forest {less than 800 million hectares}—nobody knows exactly how much—and about another quarter survives as fragmented and degraded forest.[3]

The world's tropical rainforests are located along the equator and between the Southern Hemisphere's Tropic of Capricorn (23°26′13.9″ South) and the Northern Hemisphere's Tropic of Cancer (23°26′13.9″ North). Further, tropical rainforests can also be classified based on the Köppen classification of climates[4] and the Food and Agricultural Organization (FAO) of the United Nations' designation of tropical rainforest ecological zones (Fig. 2.1).[5]

Tropical rainforests are biologically diverse, mostly evergreen forests that are located near the equator. There are also temperate rainforests which are located in countries far from the equator such as Canada, Russia, and in the Pacific Northwest of the United States. Rainforests annually receive rainfall in excess of 200 centimeters (80 inches).[6] It is important to note there are other types of forest ecosystems (i.e., mangrove forests), and not just rainforests, located around the equator.

The 3 main rainforest regions are Afrotropical, Neotropical, and Indo-Pacific. It is estimated that a total of 65–70 countries within these 3 regions have biomes classified as humid tropical forests or tropical moist forests.[7,8]

Neotropical

The countries throughout the Neotropical region, which spans the Caribbean Sea, Central and South America, include (Fig. 2.2):

- Argentina
- Belize
- Bolivia
- Brazil
- Colombia
- Costa Rica

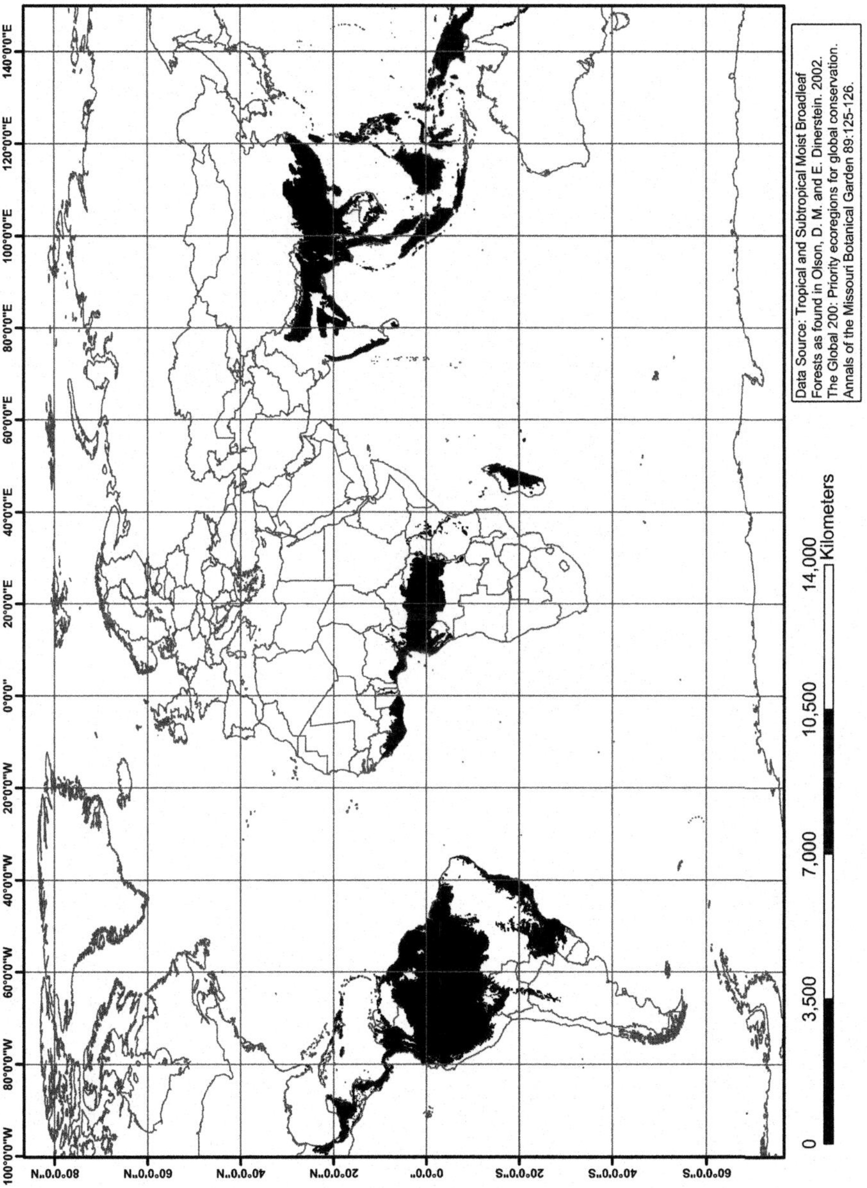

Fig. 2.1 Tropical rainforests of the world (Credit: James M. Eaton)

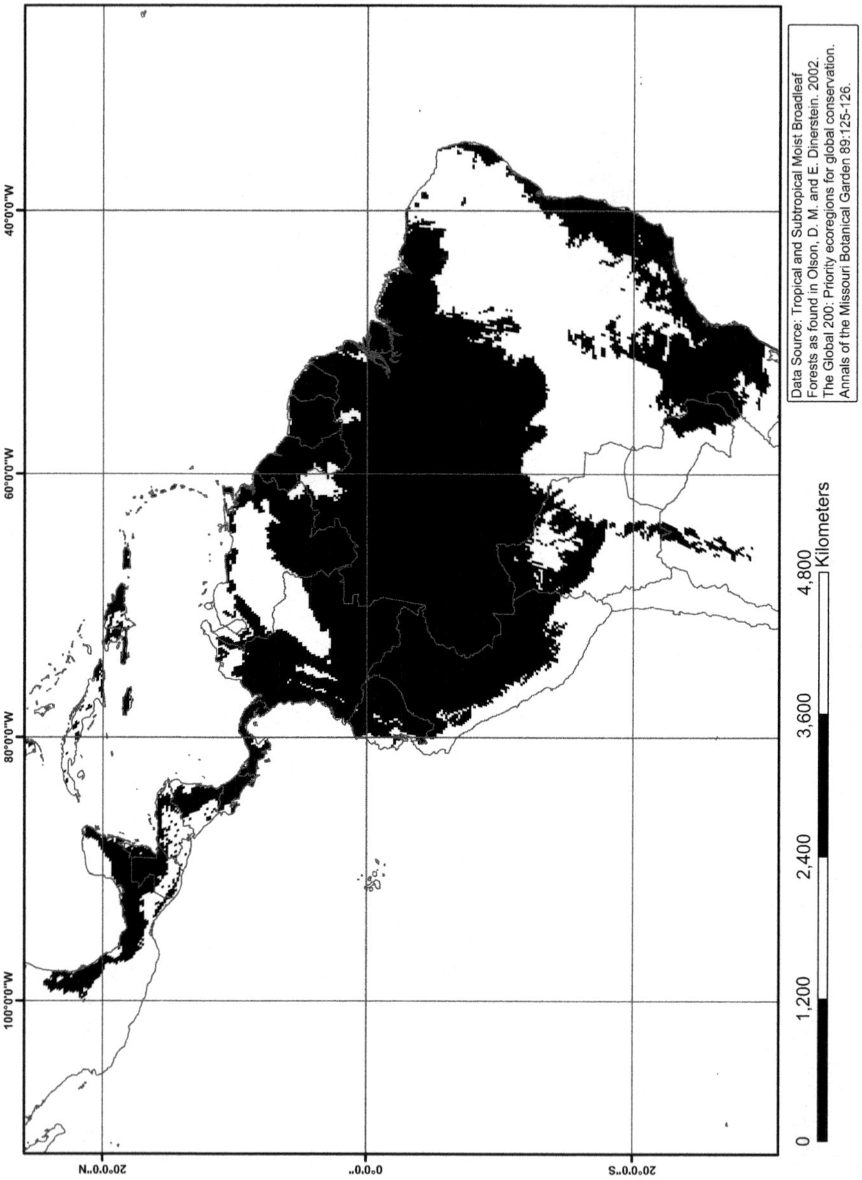

Fig. 2.2 Neotropical rainforests (Credit: James M. Eaton)

- Cuba
- Dominican Republic
- Ecuador
- El Salvador
- Guatemala
- Guiana (French overseas department)
- Guyana
- Haiti
- Honduras
- Jamaica
- Mexico
- Nicaragua
- Panama
- Paraguay
- Peru
- Suriname
- Trinidad and Tobago
- Venezuela[9,10]

Afrotropical

The countries throughout the Afrotropical region, which span throughout Sub-Saharan Africa from the Atlantic Ocean to the Indian Ocean, include (Fig. 2.3):

- Angola and Cabinda
- Benin
- Burundi
- Cameroon
- Central African Republic
- Côte d'Ivoire (Ivory Coast)
- Democratic Republic of Congo
- Djibouti
- Equatorial Guinea
- Gabon

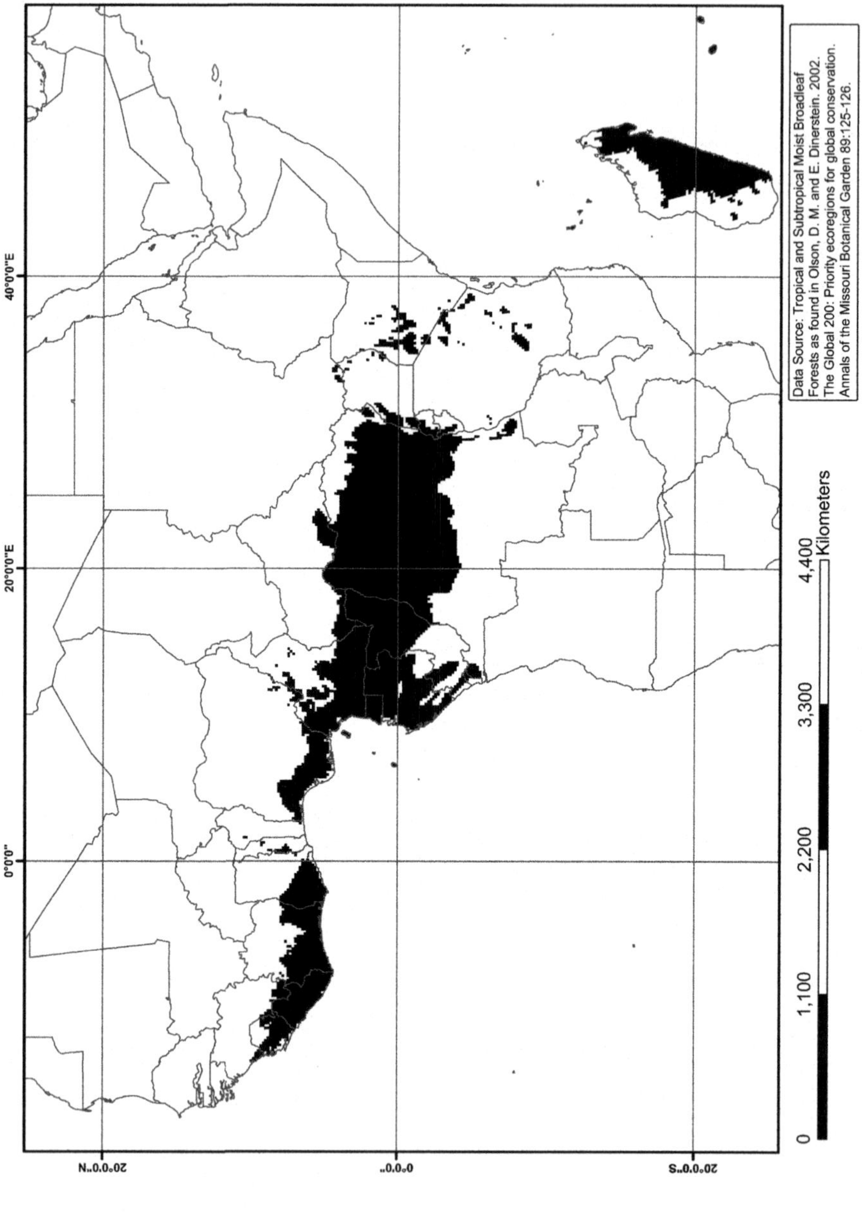

Fig. 2.3 Afrotropical rainforests (Credit: James M. Eaton)

- Gambia
- Ghana
- Guinea
- Guinea-Bissau
- Kenya
- Liberia
- Madagascar
- Malawi
- Mauritius
- Mozambique
- Nigeria
- Republic of the Congo
- Réunion
- Rwanda
- São Tomé and Príncipe
- Senegal
- Seychelles
- Sierra Leone
- Somalia
- Tanzania
- Togo
- Uganda
- Zambia
- Zimbabwe[11,12]

Indo-Pacific

The countries throughout the Indo-Pacific region, which spans across mainland Southeast Asia and throughout the various archipelagos, include the following countries. Please note there are thousands of islands in the Indo-Pacific region and the author apologizes to those islands that are not listed (Fig. 2.4):

- Australia
- Bangladesh
- Bhutan

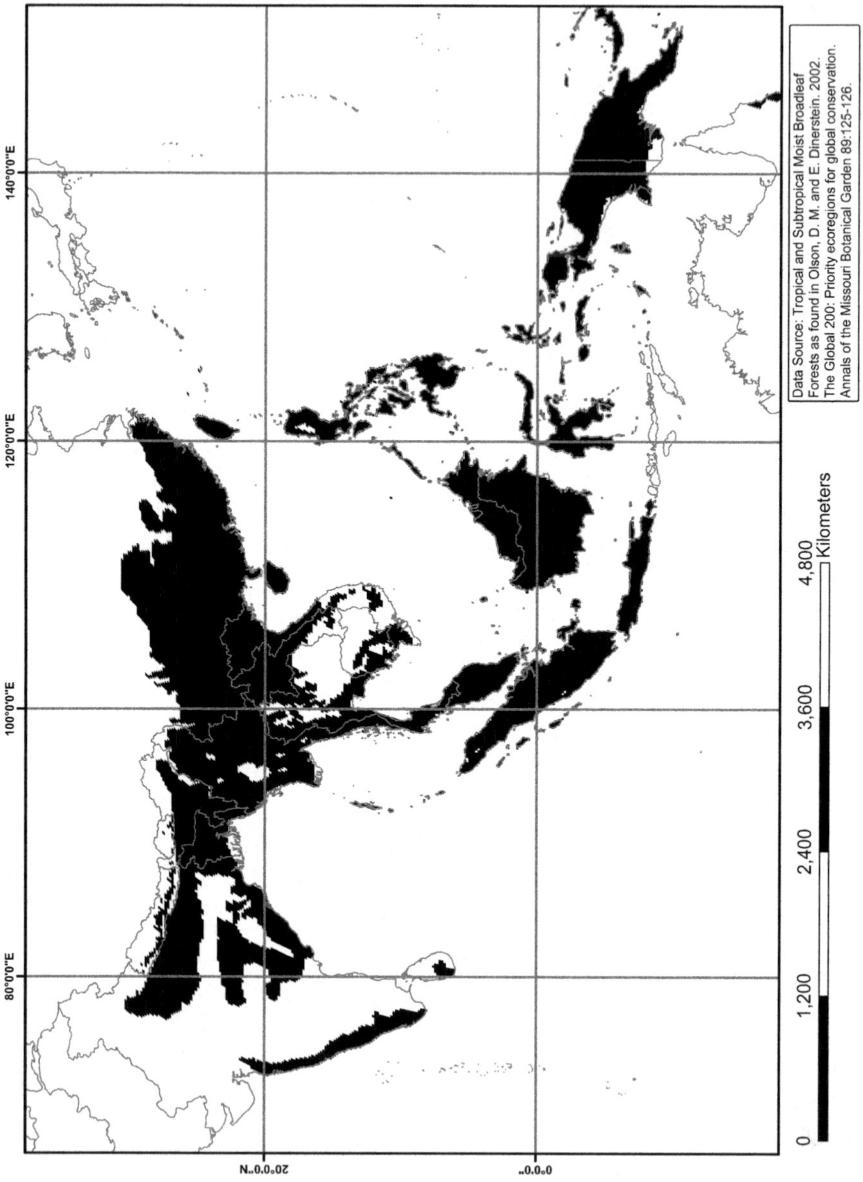

Fig. 2.4 Indo-Pacific rainforests (Credit: James M. Eaton)

- Brunei
- Cambodia
- China
- Cook Islands
- East Timor (Timor-Leste)
- Fiji
- Guam and Hawaii, United States
- India
- Indonesia
- Laos
- Malaysia
- Marquesas Islands
- Micronesia
- Myanmar
- Nepal
- New Caledonia
- Papua New Guinea
- Philippines
- Samoan Islands
- Singapore
- Solomon Islands
- Sri Lanka
- Taiwan
- Thailand
- Tonga Islands
- Tuamotu Archipelago
- Tubuai Islands
- Vietnam[13,14]

The largest tropical rainforest in the world is the Amazon Rainforest, which is primarily located in Brazil. The second largest tropical rainforest is the Congo Basin, which is primary located in the Democratic Republic of the Congo (DRC). The next largest tropical rainforest is located in Australia and throughout Southeast Asia, particularly on the island of Borneo. For comparative purposes, the Amazon Rainforest, which is larger than the Congo and Indonesian Rainforests combined, is approximately the size of the contiguous United States.[15]

Table 2.1 Countries with largest areas of forest in 2015 and the extent of deforestation

	Country	Area of forest in 2015 (1000 ha)	Deforestation 2000–2010 (1000 ha/yr)	Deforestation 2000–2010 (%)	Deforestation 2010–2015 (1000 ha/yr)	Deforestation 2010–2015 (%)
1	Russian Federation	814,931	586.7	0.1	−41.0	0.0
2	Brazil	493,538	−2,281.6	−0.4	−984	−0.2
3	Canada	347,069	−50.0	0.0	−46.6	0.0
4	United States	310,095	518.4	0.2	275.0	0.1
5	China	208,321	2,361.0	1.3	1,542.2	0.8
6	DRC	152,578	−311.4	−0.2	−311.4	−0.2
7	Australia	124,751	−563.0	−0.4	308.0	0.2
8	Indonesia	91,010	−497.7	−0.5	−684.4	−0.7
9	Peru	73,973	−133.6	−0.2	−167.6	−0.2
10	India	70,682	440.0	0.7	178.4	0.3
11	Mexico	66,040	−135.8	−0.2	−91.6	−0.1
12	Colombia	58,502	−316.3	−0.5	−26.7	0.0
13	Angola	57,856	−124.8	−0.2	−124.8	−0.2
14	Bolivia	54,764	−388.2	−0.7	−289.0	−0.5
15	Zambia	48,635	−166.6	−0.3	−166.6	−0.3
16	Venezuela	46,683	−164.6	−0.3	−164.4	−0.3
17	Tanzania	46,060	−400.0	−0.8	−372.0	−0.8
18	Mozambique	37,940	−221.6	−0.6	−206.4	−0.5
19	Papua New Guinea	33,559	−2.7	0.0	−2.8	0.0
20	Myanmar	29,041	−309.5	−0.9	−546.4	−1.8

Source: Food and Agriculture Organization of the United Nations

The following table outlines the countries with the largest forest areas in hectares (ha) as of 2015, along with the extent of deforestation in hectares and as a percent for the time periods 2000–2010 and 2010–2015.[16] Please note these historical figures are to be taken as estimates because accurately measuring forest cover can be challenging, in part, due to different classifications or definitions of forest (Table 2.1).

While forests are largely privately owned in countries such as the United States, Japan, and the United Kingdom, "today, central governments own almost 80 percent of the remaining intact forests in developing countries."[17] For instance:

- United States: 179 million ha privately owned versus 130 million ha publicly owned;
- Japan: 14.8 million ha privately owned versus 10.2 million ha publicly owned;
- United Kingdom: 2.2 million ha privately owned versus 868,000 ha publicly owned;
- Brazil: 102 million ha privately owned versus 308 million ha publicly owned;
- Indonesia: 12 million ha privately owned versus 82 million ha publicly owned;
- Democratic Republic of the Congo: 0 ha reported as privately owned versus 154 million ha publicly owned.[18]

Forest ownership and management by central governments have sometimes resulted in "mismanagement, heavy political and economic pressures to allow cutting and immigration, and widespread corruption, cronyism, and illegal logging."[19] As defined by the Organisation for Economic Co-operation and Development (OECD), central governments—also known as federal governments—are the governing body that has "the authority to impose taxes on all residents and non-resident units engaged in economic activities within the country."[20] Subnational jurisdictions, in contrast with the central or federal government, have varying degrees of power depending on the host country. Subnational jurisdictions include state or provincial governments, municipal or county governments, and city governments.

Another important land steward is Indigenous Peoples "and other local communities {who} formally own or have legal rights to manage more than 270 million hectares of forest in Latin America—almost 40 percent of the total forest area. Indigenous Peoples have also claimed significant areas of forest that they do not yet have formal rights to."[21] Historically, the forests managed by Indigenous Peoples often have lower rates of deforestation.

The following table presents for the top five largest tropical forest countries (Table 2.2):[22]

- The main economic threats of deforestation and degradation;[23]
- Their main exports;[24]
- The private landownership and public landownership in 2010;[25]

Table 2.2 Country profiles on top five countries with largest tropical forests

	Country	Main economic threats of deforestation and degradation	Main exports	Private landownership in 2010	Public landownership in 2010	Land designated for or owned by Indigenous Peoples and communities	Land under protection in 2014 (%)
1	Brazil	Cattle ranching, infrastructure projects, and commercial agriculture (e.g., soybeans)	Transport equipment, iron ore, soybeans, footwear, coffee, and automobiles	102.49 million hectares	308.09 million hectares	191.82 million hectares	28.4
2	DRC	Subsistence agriculture, logging, and fuelwood collection	Diamonds, copper, gold, cobalt, wood products, crude oil, and coffee	0 hectares	154.14 million hectares	0 hectares	12.1
3	Indonesia	Commercial agriculture (e.g., palm oil) and timber extraction	Mineral fuels, animal or vegetable fats (includes palm oil), electrical machinery, rubber, machinery and mechanical appliance parts	12.34 million hectares	82.10 million hectares	350,000 hectares	14.7

(continued)

Table 2.2 (continued)

	Country	Main economic threats of deforestation and degradation	Main exports	Private landownership in 2010	Public landownership in 2010	Land designated for or owned by Indigenous Peoples and communities	Land under protection in 2014 (%)
4	Peru	Cattle ranching, infrastructure projects, commercial and subsistence agriculture, and mining (particularly illegal gold mining)	Copper, gold, lead, zinc, tin, iron ore, molybdenum, silver; crude petroleum and petroleum products, natural gas; coffee, asparagus and other vegetables, fruit, apparel and textiles, fishmeal, fish, chemicals, fabricated metal products and machinery, alloys	13.13 million hectares	61.68 million hectares	44.56 million hectares	31.4
5	India	Commercial and subsistence agriculture, mining, and logging	Petroleum products, precious stones, vehicles, machinery, iron and steel, chemicals, pharmaceutical products, cereals, and apparel	9.77 million hectares	60.02 million hectares	130,000 hectares	5.3

Source: FAO, the US Central Intelligence Agency, and the World Bank

- The land designated for, or owned by, Indigenous Peoples and communities;[26] and
- The land under protection in 2014.[27]

It is also important to note that technology—including remote sensing and satellite imagery, advancements in data storage, transmission and computation, satellite phones, and drones for conservation—has greatly improved the world's ability to quantify, assess, and monitor the world's tropical rainforests. This includes, but is certainly not limited to, the Global Forest Watch platform,[28] the US Geological Survey's Earth Resources Observation and Science (EROS) archive of free Landsat images,[29] and PRODES (Projeto de Monitoramento do Desflorestamento na Amazônia Legal or Legal Amazon Deforestation Monitoring Project in English), which is "the world's most extensive national forest monitoring program."[30]

Where Did Tropical Rainforest Deforestation and Degradation Historically Occur?

For an excellent resource on the historical causes of deforestation and degradation across the tropics, please see Dr. Claude Martin's book entitled, *On the Edge: The State and Fate of the World's Tropical Rainforests*. The following excerpt is from Dr. Martin's book:

> Rice cultivation started to take a heavy toll when immigrant peasant farmers cleared large areas in the lowland forests of Myanmar, Thailand, Cambodia, Laos, and Vietnam. These countries became an important rice exporting region—the rice bowl of Asia. Other Southeast Asian countries followed, deforesting large areas of Sumatra and Java. Over ten million hectares of forests was cleared before 1920, although records of how this happened seem to be inconclusive. The Philippines, which had been under more than 300 years of Spanish colonial rule, also lost forest in the central places of Luzon, mainly to plantations of commercial crops such as tobacco and sugar cane. The colonial influence on the expansion of commercial crops was also noticeable in Assam, India where the British East India Company expanded its tea estates at the cost of tropical forests. In West Africa, mainly in Ghana and Nigeria, rainforests were cleared in the last quarter of the nineteenth century

for some of the first commercial cocoa and oil palm plantations. The rainforests of Latin America, on the other hand, had hardly been affected by commercial agriculture before 1920, with the exception of some coffee plantations, for example in the areas northwest of São Paulo. Most of the deforestation in South and Central America was caused by shifting agriculture, which usually affected relatively small areas compared to cultivation in South and Southeast Asia. A notable exception to this pattern of moderate deforestation was the Brazilian Atlantic coastal forests, which had already been decimated in the initial phases of colonization in the Portuguese-controlled areas. {...}

At the global level, the level of deforestation remained modest until about 1920, as agricultural expansion happened mainly in temperate countries until that time. The world population had grown from 1 billion in 1800 to 1.9 billion by 1920. {...} Through the 1920s and 1930s, tropical deforestation accelerated massively to almost eight million hectares per year. {...} In the aftermath of World War II, from 1950 to 1980, tropical deforestation increased even further. An estimated eleven million hectares of tropical rainforest disappeared every year in the 1970s—at least, this was the widely cited figure from a first assessment published by the Food and Agriculture Organization of the United Nations (FAO) in 1976.[31]

For another comprehensive resource on the historical overview of deforestation, see *Deforesting the Earth: From Prehistory to Global Crisis, An Abridgment*, by Michael Williams.[32]

Where Is Tropical Rainforest Deforestation and Degradation Occurring Today?

Between 2000 and 2012, "the world has been losing about 0.43% of its remaining tropical rainforests per year: about 1.2 times the size of Switzerland (each year). If the decline in the global deforestation rate were to continue in a linear fashion, it would reach the level of current reforestation (about 0.13%) a few years after 2030 (Fig. 2.5)."[33]

A root cause analysis of why this tropical rainforest deforestation and degradation is still occurring today reveals several compounding and interconnected factors. As Nigel Sizer, the former director of The Nature Conservancy's (TNC) Forest Conservation program in Jakarta, Indonesia,

Fig. 2.5 Deforestation in Kalimantan, Indonesia (Credit: Brian Joseph McFarland)

explained, "There is not just one factor that is leading to this wholesale slaughter of the Indonesian forests; it's a combination of weak governance, poor law enforcement, ill-advised government policies, and lack of alternative income opportunities for the rural poor."[34]

Similarly, after World War II, tropical rainforest deforestation and degradation accelerated due to a variety of forces including, but not limited to:

- The creation of the Bretton Woods Institutions (i.e., the World Bank and the International Monetary Fund) and their early financing of megaprojects;
- The establishment of many new, independent countries and their focus on import substitution industrialization (ISI) coupled with export-oriented economies of their raw commodities;
- The oil embargoes of the 1970s and the ensuing third world debt crisis of the 1980s;
- Rural-to-urban migration, rising incomes, and increasing consumption levels of beef and other consumer goods (e.g., chocolate, coffee, and paper);

- Advancement of technology and machinery, including the advancement of earthmoving equipment for mining, road construction, and timber extraction.

For an excellent analysis of the role of megaprojects, import substitution industrialization (ISI), the oil embargoes of the 1970s and the ensuing third world debt crisis of the 1980s, and foreign debt's contribution to commodity export-oriented economies, see *Trouble in Paradise: Globalization and Environmental Crises in Latin America* by J. Timmons Roberts and Nikki Demetria Thanos. Similarly and as explained by Dr. Thomas Lovejoy:

> The forest has shrunk in the face of voracious appetites for tropical resources (hardwoods, fossil fuels, and minerals), advancing agricultural frontiers (no matter how ill-advised), and mostly ill-conceived settlement initiatives.
>
> {…} much of Brazil's Atlantic forests are gone, along with 20 percent of the Amazon and most of the lowland forests of Malaysia and Indonesia. The Congo Basin has suffered multiple inroads (for mining, oil and gas, timber, roads, and dams), and much of the West African forest has been obliterated. The picture, happily, is not entirely negative. More than half of the Amazon is now under some form of protection. Protected areas (however vulnerable) have been created in the Congo Basin, Southeast Asia, and Indonesia. National capacity in science and conservation has grown impressively.[35]

Yet, tropical rainforest deforestation and degradation persists today and will surely continue in the future as the FAO:

> {…} projects a 60 percent increase in the world demand for agricultural products between 2015 and 2050, a consequence of human population increase, urbanization, and dietary changes, particularly the rapid increase of meat and palm oil consumption in emerging economies. The world may literally eat up the tropical rainforests, these being the only large areas still available for agricultural expansion. Unfortunately, it is more profitable to produce a few tons of beef or animal feed per hectare than to maintain that hectare as intact tropical rainforest.[36]

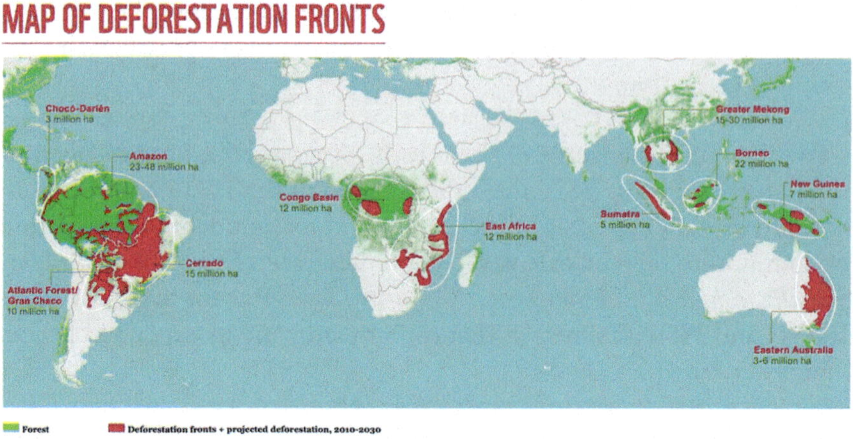

Fig. 2.6 Map of Deforestation Fronts (Credit: © WWF's Living Forest Report 2015)[39]

Much of this tropical rainforest deforestation and degradation is concentrated in select regions, such as Brazil's Arc of Deforestation and within Indonesia's Riau Province in Sumatra, and results from a few select agricultural soft commodities including cattle, soy, and palm oil. For example, "from 1996 to 2005 the Brazilian states in the arc of deforestation—Mato Grosso, Rondônia, and Pará—accounted for 85 percent of deforestation in the Amazônia Legal region, with an average of 1.66 million hectares of forest loss per year."[37] Similarly, "it revealed that Sumatra and Kalimantan together lost 5.39 million hectares or 1.15 percent annually of forest cover between 2000 and 2008—more than the area of Costa Rica. Almost half this loss happened in just two provinces: Riau in Sumatra and Central Kalimantan."[38] The following map depicts these deforestation fronts. The conservation financing instruments that will be outlined in this book should target these specific regions while also ensuring leakage does not displace such tropical rainforest deforestation and degradation to other places (Fig. 2.6).

The following maps depict the extent of forest cover loss from 2000 to 2014 in the Neotropical, Afrotropical, and Indo-Pacific rainforests (Figs. 2.7, 2.8, 2.9):

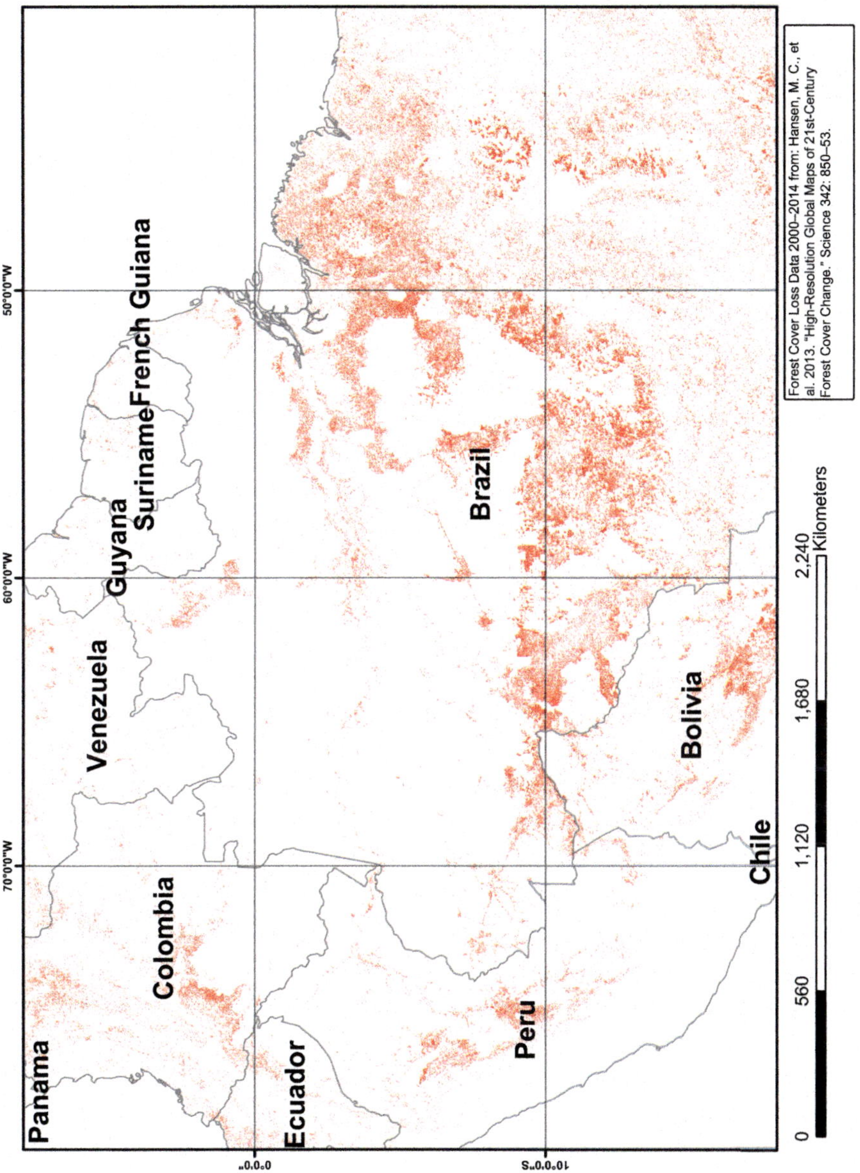

Fig. 2.7 Forest cover loss of Neotropical rainforests (Credit: James M. Eaton)

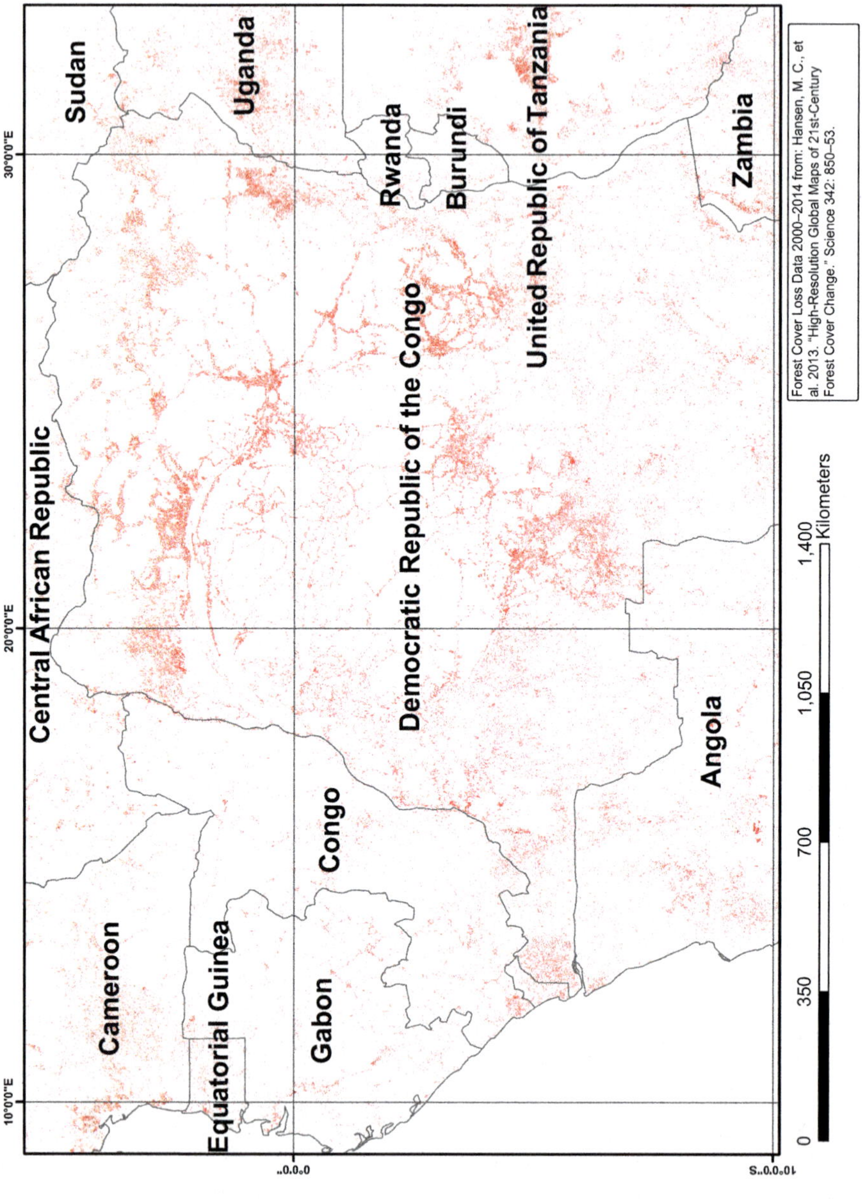

Fig. 2.8 Forest cover loss of Afrotropical rainforests (Credit: James M. Eaton)

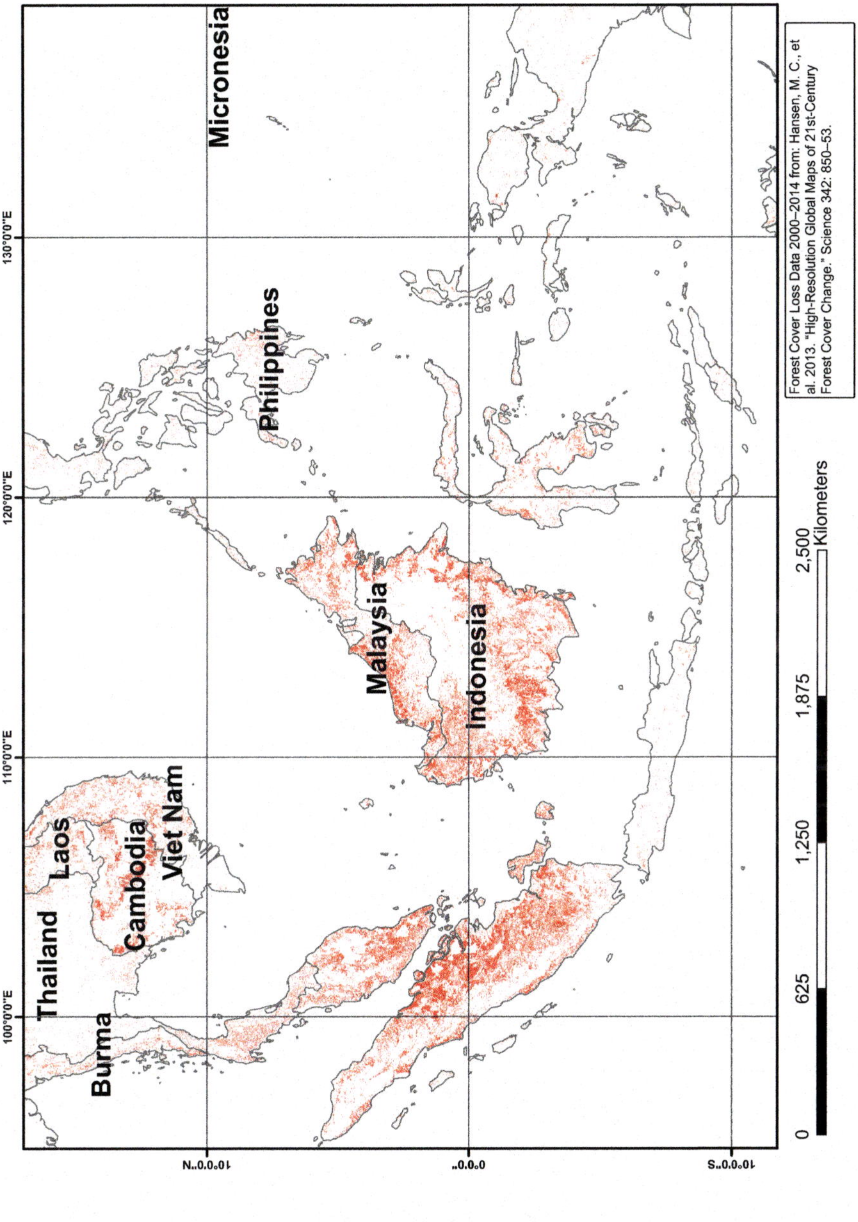

Fig. 2.9 Forest cover loss of Indo-Pacific rainforests (Credit: James M. Eaton)

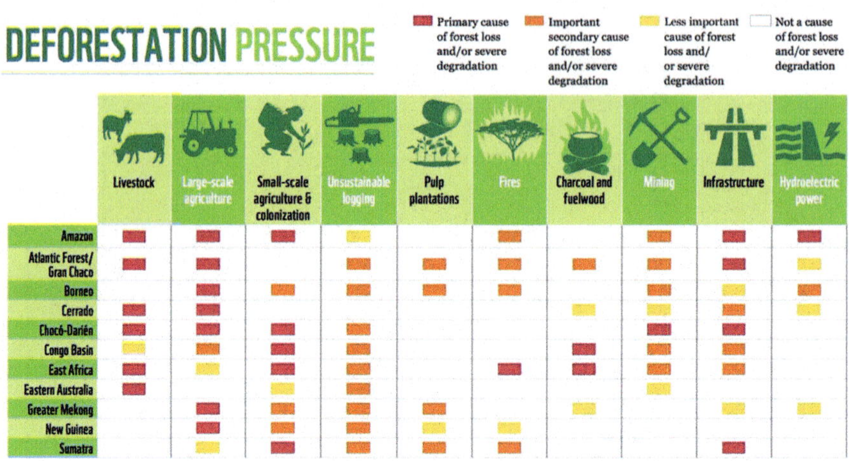

Fig. 2.10 Global deforestation pressures (Credit: © WWF's Living Forest Report 2015)[40]

Three main commodities primarily linked to tropical rainforest deforestation and degradation are cattle, palm oil, and soy (Fig. 2.10).

Cattle

Especially throughout Central and South America, along with Southeast Asia, "{…} studies confirm that commercial agriculture, including cattle ranching, is now the most important proximate driver of deforestation and has overtaken subsistence agriculture everywhere except in Africa."[41] Commercial agriculture's role in tropical rainforest deforestation and degradation—as will be explained below—is mainly due to soybeans and palm oil (Fig. 2.11).

The world in 2014 had an estimated 1.5 billion head of cattle,[42] which collectively with other livestock (e.g., buffalo, chickens, and pigs), are responsible for approximately 14.5% of global greenhouse gas (GHGs) emissions, which is greater than the GHG emissions from the entire transportation sector.[43]

Cattle ranching throughout Central and South America tends to result in the export of beef to Hong Kong, Russia, and the United States, or internally consumed in countries such as Argentina, Brazil, and Uruguay.

Fig. 2.11 Cattle ranching and deforestation in the Brazilian Amazon (Credit: Brian Joseph McFarland)

Brazil is the second largest producer of beef in the world, and this cattle ranching is largely within the Arc of Deforestation including the states of Mato Grosso and Pará. The top 10 largest consumers of beef in 2015[44] were given in Table 2.3.

Soybeans

Similar to cattle, the largest region facing tropical rainforest deforestation and degradation as a result of soybean cultivation is the Brazilian Amazon and particularly the state of Mato Grosso. The expansion of soy into South America, particularly Brazil and Argentina, "{...} started in the early 1990s with the rapid increase of meat consumption and the demand for protein-rich feed for chickens, pigs, and cattle. Fish meal, an important cattle feed, was no longer available after the collapse of the Peruvian anchovy fisheries."[45] In conjunction, "soy cultivation raised land prices and displaced cattle ranching deeper into the Amazon."[46]

Table 2.3 Top global beef consumers

	Country	Beef consumption (kilograms per capita)		Country	Total annual beef consumption (thousands of tonnes)
1	Uruguay	46.4	1	United States	11,376.2
2	Argentina	40.4	2	EU-28	7,765.3
3	Paraguay	25.6	3	China	7,504.2
4	United States	24.7	4	Brazil	7,172.6
5	Brazil	24.2	5	Argentina	2,506.6
6	Australia	22.8	6	Russia	2,484.7
7	Israel	20.2	7	Pakistan	1,706.8
8	Canada	17.4	8	Mexico	1,587.8
9	Kazakhstan	16.9	9	Egypt	1,319.8
10	Chile	15.0	10	Japan	1,210.4

Source: Organisation for Economic Co-operation and Development

The following table depicts the largest domestic consumers of soybeans in 2016[47] and the largest producers of soybeans in 2015 (Table 2.4)[48]:

Palm Oil

The world's largest producers of palm oil are Indonesia and Malaysia, which is often in direct conflict with the conservation of Southeast Asia's tropical rainforests and peatlands. Since palm oil is in a wide variety of consumer products, the oil is increasingly consumed around the world and particularly in India. As Dr. Claude Martin explains:

> Today, palm oil is everywhere; besides being used as cooking oil, it is used in thousands of products in the food industry and also in soap, detergent, lubricants, cosmetics, and as biofuel. It makes up the largest volume of vegetable oil produced, at 34 percent, and palm kernel oil adds another 4 percent to the global palm oil bounty. It is the cheapest edible oil and by far the most productive: plantations yield three to eight times as much oil by area than any other oil plant. Currently more than 70 percent of the world's palm oil is used for food. The main producers, Indonesia and Malaysia, in addition to their exports, meet a large part of their domestic demand for vegetable oils with palm oil, but there are plans underway to

Table 2.4 Top soybean consumers and producers

	Country	Annual domestic consumption (thousands of tonnes)		Country	Annual production (thousands of tonnes)
1	China	16,200	1	United States	106,958.9
2	United States	9,275	2	Brazil	96,806.5
3	Brazil	6,320	3	Argentina	60,000.0
4	India	5,400	4	China	11,000.0
5	Argentina	2,815	5	Paraguay	9,200.0
6	EU-27	2,030	6	India	7,100.0
7	Mexico	1,110	7	Canada	6,235.0
8	Bangladesh	825	8	Ukraine	3,693.0
9	Iran	800	9	Uruguay	3,200.0
10	Egypt	760	10	Russia	2,800.3

Source: IndexMundi and Organisation for Economic Co-operation and Development

Table 2.5 Top palm oil consumers and producers

	Country	Annual domestic consumption (thousands of tonnes)		Country	Annual production (thousands of tonnes)
1	India	10,200	1	Indonesia	35,000
2	Indonesia	9,100	2	Malaysia	20,000
3	EU-27	6,520	3	Thailand	2,300
4	China	5,050	4	Colombia	1,280
5	Pakistan	3,245	5	Nigeria	970
6	Malaysia	3,170	6	Ecuador	560
7	Thailand	2,185	7	Honduras	545
8	Egypt	1,600	8	Papua New Guinea	522
9	Nigeria	1,530	9	Ghana	520
10	Bangladesh	1,500	10	Guatemala	515

Source: IndexMundi and Organisation for Economic Co-operation and Development

convert millions of hectares of tropical rainforest into additional palm oil plantations, and a major driver for this is the demand for biofuels.[49]

Rounding out the top 10 palm oil producing countries[50] and consuming countries[51] in 2016 were as follows (Table 2.5):

Fig. 2.12 Deforestation for palm oil plantations in Central Kalimantan, Indonesia (Credit: Brian Joseph McFarland)

While some palm oil plantations are established on degraded lands, like the following picture illustrates, much of the palm oil plantations are at the expense of tropical rainforests (Fig. 2.12):

Large-scale commercial cultivation of palm oil is now beginning to expand into parts of Africa—such as Cameroon and Gabon—that will threaten the Congo Rainforest. For example, in December 2016:

> Société Financière des Caoutchoucs SA ('Socfin'), based in Luxembourg, raised €80 million in senior unsecured 4 percent 5-year debt to finance land-use expansion for its vertically integrated palm oil and rubber production in Africa and Asia.
>
> In 2015, SocFin had 186,767 hectares under cultivation; 120,780 ha of oil palm trees and 65,987 ha of rubber trees. The company has operations in Liberia, Côte d'Ivoire, Nigeria, Cameroon, Congo DRC, Sierra Leone, Indonesia, Cambodia, Ghana, and Sao Tome.
>
> These €80 million in funds are not ring-fenced and are assumed to be used to finance SocFin's current expansion in Sierra Leone and elsewhere. ING facilitated this first public debt offering by SocFin. It is unknown if

ING's participation required SocFin to include climate and social risk safeguards into the use of funds. ING is a member of the Roundtable of Sustainable Palm Oil.[52]

Timber, Pulp, and Paper

Pulp and paper are consumed around the world as toilet paper, paper towels, computer paper, packaging, and notepads. While such raw materials can be sourced from plantations or temperate forests, pulp and paper can also come from tropical rainforests.

There are concerns about planted, plantation forests due to the establishment of monocultures (i.e., especially if planted at the expense of biologically diverse, natural forests), water quantity and quality issues, and the use of petro-based chemicals (e.g., fertilizers, insecticides, and rodenticides). However, these plantation forests may reduce the pressure on natural forests, especially for paper, pulp, and timber. The following table outlines the countries with the largest area of planted forests as of 2015 (Table 2.6)[53]:

The world's largest exporters of wood, not necessarily tropical hardwoods, based on 2015 US dollar values, were as follows (Table 2.7)[54]:

Although not ranked among the top 10 countries, timber, particularly hardwoods, is a large export commodity for many, unlisted tropical rain-

Table 2.6 Countries with largest area of planted forests

	Country (world ranking)	Planted forest in 2015 (1000 ha)		Country (world ranking)	Planted forest in 2015 (1000 ha)
1	China	78,982	11	Sudan	6,121
2	United States	26,364	12	Germany	5,295
3	Russia	19,841	13	Indonesia	4,946
4	Canada	15,784	14	Ukraine	4,860
5	Sweden	13,737	15	Thailand	3,986
6	India	12,031	16	Vietnam	3,663
7	Japan	10,270	17	Turkey	3,386
8	Poland	8,957	18	Sweden	2,909
9	Brazil	7,736	19	Czech Republic	2,643
10	Finland	6,775	20	Australia	2,017

Source: Food and Agriculture Organization of the United Nations

Table 2.7 Top wood producers

	Country (world ranking)	Export value in 2015 (billions of USD$)		Country (ranking among tropical rainforest countries)	Export value in 2015 (billions of USD$)
1	United States	USD$42.3	1	China	USD$31.3
2	China	USD$31.3	2	Mexico	USD$8.8
3	Germany	USD$27.1	3	India	USD$5.1
4	United Kingdom	USD$18.2	4	Australia	USD$4.9
5	France	USD$16.3	5	Vietnam	USD$3.4
6	Japan	USD$14.7	6	Indonesia	USD$2.9
7	Italy	USD$12.2	7	Thailand	USD$2.8
8	Netherlands	USD$12.11	8	Malaysia	USD$2.5
9	Canada	USD$12.07	9	Singapore	USD$2.4
10	Belgium	USD$9.0	10	Philippines	USD$2.1

Source: The World Bank

forest countries. Due to the low density and high diversity of tropical rainforests, only about 1 to 3 commercial hardwood trees are found per hectare throughout most tropical rainforests.[55] The extraction of such trees often subsidizes the resulting cattle ranch or palm oil plantation. In addition, while there are few commercial hardwoods per hectare, there are other trees that might be used for staging or other short-term uses (e.g., biofuel) which in turn subsidize access to the hardwoods and the future clearing for a cattle ranch or palm oil plantation (Fig. 2.13).

One issue with logging is that a significant portion of the globally traded tropical hardwoods are illegal. For instance:

> A detailed study of illegal logging in five tropical countries (Indonesia, Malaysia, Ghana, Cameroon, and Brazil) published by Chatham House in 2010 found that illegal logging has been declining since about 2004. But at the time this study was published, illegal harvesting still represented 35–72 percent of all logging in the Brazilian Amazon, 22–35 percent in Cameroon, 59–65 percent in Ghana, 40–61 percent in Indonesia, and 14–25 percent in Malaysia. Illegal logging affects not only tropical countries, but industrialized countries as well. Worldwide, more than one hundred million cubic meters of timber are still being cut illegally each year, leading to the degradation or destruction of five million hectares of forest. The economic damage of illegal logging is equally important; it is

Fig. 2.13 Commercial logging in Brazil (Credit: Brian Joseph McFarland)

estimated at USD$ 10–15 billion worth of stolen timber per year, plus an additional USD$ 5 billion in lost royalties and taxes.[56]

Below is a picture of a temporary camp established to illegally log hardwoods (Fig. 2.14):

Timber also plays a significant role in forest degradation and the ensuing GHG emissions. As reported by researchers from Winrock International:

We estimated annual emissions of 2.1 billion tons of carbon dioxide {from forest degradation}, of which 53% were derived from timber harvest, 30% from woodfuel harvest and 17% from forest fire. These percentages differed by region: timber harvest was as high as 69% in South and Central America and just 31% in Africa; woodfuel harvest was 35% in Asia, and just 10% in South and Central America; and fire ranged from 33% in Africa to only 5% in Asia. Of the total emissions from deforestation and forest degradation, forest degradation accounted for 25%. In 28 of the 74 countries, emissions from forest degradation exceeded those from deforestation.[57]

Fig. 2.14 Illegal logging camp in Central Kalimantan, Indonesia (Credit: Brian Joseph McFarland)

Forest Fires

Forest fires—depending on the fire's severity—are also another factor that can result in forest degradation or full-scale deforestation. While some of these forest fires are naturally started, many of the fires—as the picture below depicts—are started by people to clear land for agriculture (Fig. 2.15).

One of the worst cases of forest fires recently occurred in Indonesia. While there are some disputes over the exact figures,[58] it is largely accepted that from September to October 2015, the fires in Indonesia resulted in more daily greenhouse gas emissions than the entire European Union,[59] impacted millions of people, and cost the region an estimated USD$16 billion (Fig. 2.16).[60]

More specifically, per Chain Reaction Research:

> From July 1st to October 20th, 2015, 97,000 fires were detected in Indonesia, and over two million hectares of forest caught fire, one-third of which were High Carbon Stock peat forests.[61]

Fig. 2.15 Fire used for land clearing in Brazil (Credit: Brian Joseph McFarland)

Fig. 2.16 Haze in South Kalimantan, Indonesia (September 2015) (Credit: Brian Joseph McFarland)

In 2016, Harvard and Columbia Universities published estimates suggesting that the Southeast Asian 2015 forest fires and haze may have resulted in 100,000 fatalities in Indonesia, Malaysia, and Singapore. The study only focused on "fine particulate matter," referred to as PM2.5. PM2.5 can penetrate the lungs and other organs, causing respiratory disease, birth defects, premature delivery of babies, premature death, asthmas, and lung cancer.

Often, but not always, palm oil plantations and smallholders clear existing forests using fire. These fires create haze, a toxic chemical soup that can include hundreds of airborne dangerous chemicals.

Other estimates suggest that during "fire season" in El Nino years, 11 percent of the population in SE {Southeast} Asia have been repeatedly exposed to pollution levels significantly above World Health Organization safety limits.

Competing figures compiled by SE Asian governments were significantly lower. The differences are due to many factors. Specifically, the number of deaths is based on complex estimates of the increases of potential PM2.5 related illnesses versus baseline.[62]

Unfortunately, forest fires are not just a phenomenon of Southeast Asia. In 2015, there were more than 2.3 million fires detected in the Brazilian Amazon.[63] Similarly, between January and March 2016, an approximate total of 15,000 hectares (37,050 acres) were burned in the Republic of the Congo, representing one of the largest fires ever detected in the rainforests of Central Africa.[64]

Population and Consumption

Another factor contributing to tropical rainforest deforestation and degradation is population. As explained by Dr. Claude Martin:

> {...} population density is certainly not appropriate as a single indicator, but it is far from irrelevant. In the next thirty-five years population growth is projected to peak everywhere except in Africa, where the population will double from 1.17 billion in 2015 to 2.39 billion in 2052 and is expected to rise to 4.2 billion in 2100. Western, central, and eastern African countries show some of the highest growth rates at 2–3 percent.[65]

Table 2.8 Most populated countries and fastest population growth rates

	Country	Population total in 2015		Country	Population growth in 2015 (%)
1	China	1.37 billion	1	Oman	5.8
2	India	1.31 billion	2	Lebanon	4.2
3	United States	321.4 million	3	Niger	4.0
4	Indonesia	257.6 million	4	Kuwait	3.6
5	Brazil	207.8 million	5	South Sudan	3.5
6	Pakistan	188.9 million	Tied-6	Burundi	3.3
7	Nigeria	182.2 million	Tied-6	Chad	3.3
8	Bangladesh	161 million	Tied-6	Uganda	3.3
9	Russia	144.1 million	Tied-9	Angola	3.2
10	Mexico	127.0 million	Tied-9	Iraq	3.2
–	–	–	Tied-9	Gambia	3.2

Source: The World Bank

The top 10 most populated countries in 2015,[66] along with the countries with the fastest population growth in 2015,[67] were as follows (Table 2.8):

It is important to understand that forests provide livelihoods to billions of people. If these people are no longer supported by forests—whether due to deforestation, forest degradation, or a depletion of local resources (e.g., food and water)—many of these people will be forced to move to nearby cities. Yet, this rural-to-urban migration will place further pressure on cities—many of which already lack adequate housing, employment opportunities, and water availability. This said, many of the world's megacities are already located in developing countries with tropical rainforests such as Mexico City (Mexico) and São Paulo (Brazil), and the following table outlines the largest cities around the world (Table 2.9)[68]:

While not technically in the top 20, two additional large cities located in tropical rainforest countries that should be noted are Jakarta, Indonesia, and Lima, Peru.

In addition to the population increase among rural communities living adjacent to tropical rainforests, there is also concern about the population increase among urban consumers of commodities that directly result in tropical rainforest deforestation and degradation.

Table 2.9 World's largest cities

	City	Country	Population (2016) (million)		City	Country	Estimated population (2030) (million)
1	Tokyo	Japan	38.14	1	Tokyo	Japan	37.19
2	Delhi	India	26.45	2	Delhi	India	36.06
3	Shanghai	China	24.48	3	Shanghai	China	30.75
4	Mumbai (Bombay)	India	21.36	4	Mumbai (Bombay)	India	27.80
5	São Paulo	Brazil	21.30	5	Beijing	China	27.71
6	Beijing	China	21.24	6	Dhaka	Bangladesh	27.37
7	Mexico City	Mexico	21.16	7	Karachi	Pakistan	24.84
8	Osaka	Japan	20.34	8	Cairo	Egypt	24.50
9	Cairo	Egypt	19.13	9	Lagos	Nigeria	24.24
10	New York-Newark	United States	18.60	10	Mexico City	Mexico	23.87
11	Dhaka	Bangladesh	18.24	11	São Paulo	Brazil	23.44
12	Karachi	Pakistan	17.12	12	Kinshasa	Democratic Republic of the Congo	20.00
13	Buenos Aires	Argentina	15.33	13	Osaka	Japan	19.98
14	Calcutta	India	14.98	14	New York-Newark	United States	19.89
15	Istanbul	Turkey	14.37	15	Calcutta	India	19.09
16	Chongqing	China	13.74	16	Guangzhou, Guangdong	China	17.57
17	Lagos	Nigeria	13.66	17	Chongqing	China	17.38
18	Manila	Philippines	13.13	18	Buenos Aires	Argentina	16.96
19	Guangzhou, Guangdong	China	13.07	19	Manila	Philippines	16.76
20	Rio de Janeiro	Brazil	12.98	20	Istanbul	Turkey	16.69

Source: United Nations

Infrastructure

Infrastructure, such as roads, can have both a direct and indirect impact on tropical rainforests. For instance, it was reported in August 2017 that Bolivian "President Evo Morales enacted a new law opening the way for the 190-mile (300km) road through the Isiboro Sécure Indigenous Territory and National Park, known as Tipnis, its Spanish acronym. The road will divide the park in two and strip it of the protections won in 2011 {…}."[i][i] Collyns, Dan. "Bolivia approves highway through Amazon biodiversity hotspot." *The Guardian*. August 15, 2017. https://www.theguardian.com/environment/2017/aug/15/bolivia-approves-highway-in-amazon-biodiversity-hotspot-as-big-as-jamaica. A frequently seen fish-bone pattern results where a main road is constructed and then side roads are built shortly thereafter. The main road, for example, may have been constructed to access a remote timber concession. Side roads are then often established by legal or illegal settlements and this increased access results in tropical rainforest deforestation and degradation. Interestingly,

> The pattern of deforestation hot spots reveals that deforestation is largely determined by proximity to urban settlements and roads. The opening up of remote areas by timber concessions that can only be accessed on long-distance timber roads, on the other hand, does not lead to higher deforestation. Beyond a distance of nine to twelve hours of travel time from a city (not very far with the means of transportation available), deforestation in the Congo Basin drops to almost zero, showing that the theory of timber extraction always leads to deforestation by slash-and-burn farming does not hold true for remote forest areas in the Congo Basin.[69]

Such roads are not only built for timber access but also to connect different regions within a country or to connect countries (i.e., the Trans-Amazonia Highway or the Pan-American), along with to access mining and energy infrastructure (Fig. 2.17).

Large-scale, unsustainable infrastructure projects not only contribute to tropical rainforest deforestation and degradation, but they sometimes also infringe on the benefits of local communities and Indigenous Peoples.[70]

Fig. 2.17 Coal extraction in Central Kalimantan, Indonesia (Credit: Brian Joseph McFarland)

Why Are Tropical Rainforests Valuable?

Tropical rainforests have inherent value as the most biologically diverse terrestrial ecosystem on Earth. Their beauty and complexity are unmatched, as is the human need to know that such wild places still exist today. Complementing such inherent value are the invaluable resources—such as water, oxygen, and medicine—that are generated by these rainforests for local communities to people living a half a world away (Fig. 2.18).

In the seminal article, entitled "The value of the world's ecosystem services and natural capital," the global extrinsic value of the biosphere was estimated to be in the range of USD$16–54 trillion per year, with an average value of USD$33 trillion per year. This average value is approximately twice the size of the US economy in 2015. Specifically looking at tropical forests and particularly due to their role in climate regulation, erosion control, nutrient cycling, waste treatment, raw material, and recreation, tropical forests were valued at USD$2,007 per hectare per year for a grand estimated, annual total of USD$3.813 trillion in ecosystem services.[71]

Fig. 2.18 Kapok tree (*Ceiba pentandra*) in Acre, Brazil (Credit: Brian Joseph McFarland)

According to The Economics of Ecosystems and Biodiversity (TEEB), ecosystem services can be categorized as provisioning services, regulating services, habitat or supporting services, and cultural services.[72] TEEB was initiated as a global study in 2007 by the G8+5 countries (Canada, France, Germany, Italy, Japan, Russia, the United Kingdom, and the United States, plus Brazil, China, India, Mexico, and South Africa) to analyze "the global economic benefit of biological diversity, the costs of the loss of biodiversity and the failure to take protective measures versus the costs of effective conservation."[73] TEEB has now evolved to focus on country-level studies and on work around natural capital accounting for businesses.[74] TEEB for Business Coalition would later become the Natural Capital Coalition in 2014.[75]

Provisioning Services

Provisioning services include the raw materials, freshwater, and medicinal resources.

Raw materials, such as timber, provide direct benefits to local communities for the construction of homes and canoes while also contributing billions of dollars in global trade. In addition to timber, there are a wide variety of non-timber forest products (NTFPs) such as fruits, vegetables, nuts, spices, and game meat. For example:

> A number of research projects have attempted to compare the economics of sustainable extraction versus more destructive land uses. The most publicized study, a 1989 *Nature* piece by respected botanists Peters, Gentry, and Mendelson, cataloged the marketable fruits, cocoa, oils, rubber, and medicines from tiny rainforest patches in the Peruvian Amazon. The group reported that profits from harvest of these products are two to three times greater over time than logging or clearing the forest for pasture. Seventy-two species on the site yielded products with actual market values in the neighboring city of Iquitos, Peru. Factoring in heavy transport costs and the cost of saving 25 percent of fruit harvests for regeneration, Peters et al. found that one plot could sustainably net U.S.$322 per hectare per year. Studies of Açaí palm fruit harvesting near Belém have also shown sustainable profitability without cutting forests.[76]

The Amazon Rainforest is the world's largest tropical rainforest, and the Amazon River Basin is one of the largest sources of freshwater on Earth. It is estimated the Amazon River Basin contains "12–20 percent of the world's freshwater. The Amazon biome drives atmospheric circulation in the tropics by absorbing energy and recycling about half the rainfall. This biosphere-atmosphere 'pump' is an important control mechanism for rainfall patterns in other parts of the continent, particularly its central and southern zones."[77]

Rainforests also help regulate the water cycle as trees can act as a sponge to moderate the release of water. In some circumstances, as the following picture illustrates, deforestation and degradation can lead to the loss of top soil, erosion, and local flooding (Fig. 2.19).

There is an abundance of medicinal plants in tropical rainforests, including quite possibly numerous yet-to-be discovered species. In the tropical rainforests of Acre, Brazil—as throughout much of the world—local communities utilize the forests as a pharmacy, and these are some of the many species that are locally used:

- Copaíba: This plant is an anti-inflammatory.
- Quina-Quina: This plant has many uses including reducing fever and helping heal malaria.

Fig. 2.19 Flooding of Juruá River in Cruzeiro do Sul, Acre, Brazil (Credit: Brian Joseph McFarland)

- Jatobá: This plant helps anemia by increasing red blood cells.
- Andiroba: This plant can be used as a mosquito repellent and for gastrointestinal issues.
- Casca do cedro: This plant helps to alleviate toothaches and fever.

Globally, "at least forty-seven major drugs had come from rain-forest plant species, with an accumulated net worth of $147 billion."[78]

Regulating Services

Regulating services, as described by TEEB, include the regulation of local climate and air quality, carbon sequestration and storage, moderation of extreme events, erosion prevention and maintenance of soil fertility, pollination, and biological control.

For instance, "all told, the harvest of about one in three food crops worldwide—from Alfalfa to watermelons—is possible thanks to the work of pollinators."[79] Such pollinators include a wide variety of tropical species including bats, bees, birds, and insects. In addition to pollination is the critical work of seed dispersal done by birds, monkeys, tapirs, and elephants. In fact, the Brazilian biologist Dr. Patricia Medici refers to the tapir as the "Gardener of the Forest" for their seed dispersal role.[80]

According to the Infectious Disease Emergence and Economics of Altered Landscapes (IDEEAL) Project of the EcoHealth Alliance,

> Land-use change is a significant driver of emerging infectious disease. Over 60 percent of emerging infectious diseases over the past six decades, including SARS, H1N1/Swine Flu, and HIV-AIDS have originated in animals, with nearly half of these linked to changes in land-use.[81]

The outbreak in 2013–2015 of the Ebola virus in West Africa is at least partially attributable to deforestation and forest fragmentation.[82]

Furthermore, forests are both a source of GHG emissions and a sink for GHG emissions, in the sense that forests store a significant amount of carbon. It is estimated for "the decade 2001–2010, total GHG FOLU {Forestry and Other Land Use} emissions were 3.2 $GtCO_2eq/yr$ {3.2 Gigatons of carbon dioxide equivalent emissions per year} including deforestation (3.8 $GtCO_2eq/yr$), forest degradation and forest management (−1.8 $GtCO_2eq/yr$), biomass fires including peatland fires (0.3 $GtCO_2eq/yr$), and drained peatlands (0.9 $GtCO_2eq/yr$)."[83] This source of GHG emissions—much of which is from the conversion of tropical rainforests to commercial agriculture and cattle ranches—is second only to the power sector and has literally more emissions than the entire world's transportation sector and all of its cars, buses, planes, and ships.

One of the world's largest carbon sinks was just discovered in the Congo Basin. British researchers "found 145,000 square kilometres of peatland—an area larger than England—in the forests of the central Congo basin. The reservoir of compressed plant material holds at least 30 billion metric tons of carbon. And this pristine and undisturbed sink of peat is the equivalent of about two decades of fossil fuel combustion in the United States."[84]

Habitat or Supporting Services

Habitat or supporting services include habitat for species, along with the maintenance of genetic diversity. Tropical rainforests host the greatest abundance of terrestrial biodiversity on Earth. Yet,

> Nature everywhere is under siege. Each year the world loses some 30 million acres of tropical forest, an area slightly larger than Pennsylvania. At this rate, the last rain-forest tree will bow out—dead on arrival at a sawmill or in a puff of smoke—around the middle of the twenty-first century. Biodiversity—short for biological diversity, the amazing variety of life on Earth—is being reduced to the lowest levels in human history. *Homo sapiens* has already wiped out one-quarter of all bird species, and an estimated 11 percent more are on the path to extinction, along with 24 percent of mammal and 11 percent of plant species. One-quarter of the world's coral reefs have been destroyed, with many others undergoing serious decline.[85]

Similarly, "throughout the world, with every passing hour an estimated two to three species are committed to extinction, mostly as victims of habitat destruction and the spread of non-native species, which kill or compete with natives."[86] Sadly, we are currently in the sixth great extinction period known as the Holocene extinction period. The previous five periods of mass extinction took place during the following eras: Ordovician-Silurian, the Late Devonian, Permian, Triassic-Jurassic, and Cretaceous-Tertiary.[87] The last mass extinction prior to the arrival of humans was known as the Cretaceous-Tertiary, which took place approximately 65 million years ago and resulted in the end of the dinosaurs.

These species have inherent value, but are also valued by local communities as a source of food and are valued by people around the world as demonstrated by the money spent on tourism, movies (e.g., Lorax), books (*The Jungle Book*), and school visits to zoos and aquariums. Yet, as proclaimed by Dr. E.O. Wilson of Harvard University, "this mass extinction is the folly our descendants will be least likely to forgive."[88]

As described by WWF, "ecoregions that represented the most distinctive examples of biodiversity for a given major habitat type were identified within each biogeographic realm. They were chosen based on the following parameters:

- Species richness;
- Endemism;
- Higher taxonomic uniqueness (e.g., unique genera or families, relict species or communities, primitive lineages);

- Extraordinary ecological or evolutionary phenomena (e.g., extraordinary adaptive radiations, intact large vertebrate assemblages, presence of migrations of large vertebrates); and
- Global rarity of the major habitat type."[89]

Such WWF ecoregions include the world's tropical rainforests, and similarly, Conservation International has designated 36 biodiversity hotspots (Fig. 2.20).

Many of these biodiversity hotspots include tropical rainforests such as the Caribbean Islands, Indo-Burma, Mesoamerica, Tumbes-Chocó-Magdalena, and Sundaland.

Cultural Services

Cultural services include recreation, mental, and physical health; tourism; aesthetic appreciation and inspiration for culture, art, and design; spiritual experience; and sense of place.

As Dr. Martin asks, "Are these not the places where our closest relatives still survive?"[91] Likewise, chimpanzees, orangutans, and gorillas live among these tropical rainforests and are under increasing threats due to tropical rainforest deforestation and degradation.

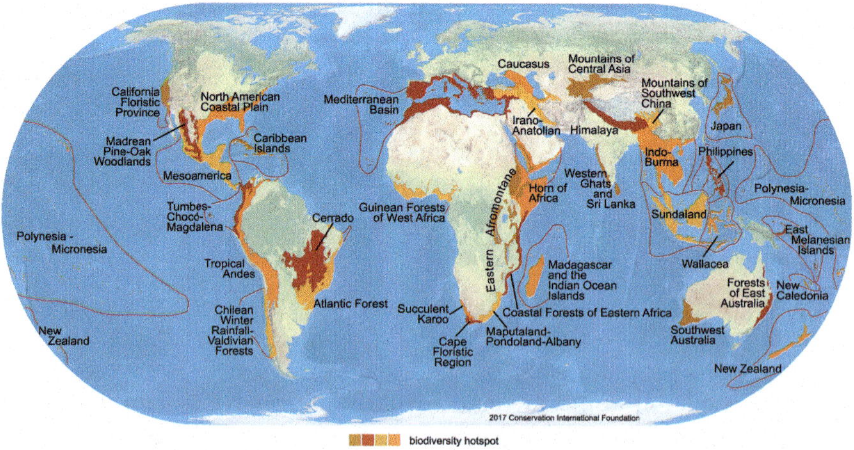

Fig. 2.20 Biodiversity hotspots (Open Source Map—Credit: Conservation International)[90]

Furthermore, if the tropical rainforest are lost, we will lose a great number of Indigenous Peoples and forest-dependent peoples. At least, "sixty-seven uncontacted tribes have been identified in Amazonia indigenous reservations,"[92] and "linguists have recognized 850 indigenous languages, which makes the country {Papua New Guinea} the most linguistically diverse country in the world by far."[93]

Other Resources on the Context of Deforestation and Degradation

Amazon Deforestation, Once Tamed, Comes Roaring Back

- https://www.nytimes.com/2017/02/24/business/energy-environment/deforestation-brazil-bolivia-south-america.html

Call for Parks Red List

- https://martynmurray.wordpress.com/2016/06/30/call-for-a-parks-red-list/

Chain Reaction Research

- https://chainreactionresearch.com/

Code REDD Videos

- http://www.coderedd.org/see-all-videos/

Conservation Biology Institute's Data Basin

- https://databasin.org/

Forest Geo-Wiki

- https://geo-wiki.org/branches/forest/

Global Footprint Network

- http://data.footprintnetwork.org/

Global Forest Watch

- http://www.globalforestwatch.org/

GRID-Arendal

- http://www.grida.no/about/

The Little Book of Big Deforestation Drivers

- Executive Summary: http://d6.globalcanopy.org/materials/little-book-big-deforestation-drivers-executive-summary
- Full Report: http://d6.globalcanopy.org/LittleBookofDrivers

Tools for Landscapes for People, Food and Nature

- http://peoplefoodandnature.org/learning-network/find-tools/

United Nations Environment Programme's Global Information Resource Database (GIRD)

- http://www.grid.unep.ch/index.php?lang=en

United States Department of Agriculture's Economic Research Service

- https://www.ers.usda.gov/

United States Department of Agriculture's Foreign Agricultural Service

- https://www.fas.usda.gov/

WWF's Database for Jurisdictional Risk Assessment

- https://www.worldwildlife.org/publications/database-for-jurisdictional-risk-assessment

Notes

1. Myers. *The Sinking Ark*. 144.
2. The Nature Conservancy. "Rainforests: Exotic, Diverse and Highly Threatened." Accessed November 23, 2016. http://www.nature.org/ourinitiatives/urgentissues/rainforests/rainforests-facts.xml
3. Martin. *On the Edge*. XXI.
4. Peel, M.C., B.L. Finlayson and T.A. McMahon. "Updated World Map of the Köppen-Geiger Climate Classification." *Hydrol. Earth Syst. Sci.* 11. 2007. 1633–1644.
5. Food and Agriculture Organization of the United Nations. "Ecological Zones: Brazil." Accessed November 23, 2016. http://www.fao.org/forestry/country/19971/en/bra/
6. National Geographic. "Encyclopedia: Rain Forest." Accessed February 19, 2017. http://www.nationalgeographic.org/encyclopedia/rain-forest/
7. Sommer, Adrian. "Attempt at an Assessment of the World's Tropical Moist Forests." *Food and Agriculture Organization of the United Nations*. Accessed November 23, 2016. http://www.fao.org/docrep/k0050e/k0050e03.htm#attempt%20at%20an%20assessment%20of%20the%20world%27s%20tropical%20moist%20forests
8. Martin. *On the Edge*. 14.

9. Sommer, Adrian. "Attempt at an Assessment of the World's Tropical Moist Forests." *Food and Agriculture Organization of the United Nations*. Accessed November 23, 2016. http://www.fao.org/docrep/k0050e/k0050e03.htm#attempt%20at%20an%20assessment%20of%20the%20world%27s%20tropical%20moist%20forestsNote. "In general, only those countries entirely located within the geographical tropical zone (exceptions are Australia, Argentina, and Paraguay) have been included. Some other countries in the subtropical zone (part of China, Bhutan, and Nepal), partially covered by a kind of moist tropical forest, have not been taken into account. On the other land, India and Burma have been included as a whole, with their subtropical part."
10. Butler, Rhett. "Where Rainforests are Located: Biogeographical Tropical Forest Realms." Last modified July 22, 2007. Accessed April 9, 2017. http://rainforests.mongabay.com/0102.htm
11. Sommer, Adrian. "Attempt at an Assessment of the World's Tropical Moist Forests." *Food and Agriculture Organization of the United Nations*. Accessed November 23, 2016. http://www.fao.org/docrep/k0050e/k0050e03.htm#attempt%20at%20an%20assessment%20of%20the%20world%27s%20tropical%20moist%20forestsNote. "In general, only those countries entirely located within the geographical tropical zone (exceptions are Australia, Argentina, and Paraguay) have been included. Some other countries in the subtropical zone (part of China, Bhutan, and Nepal), partially covered by a kind of moist tropical forest, have not been taken into account. On the other land, India and Burma have been included as a whole, with their subtropical part."
12. Butler, Rhett. "Where Rainforests are Located: Biogeographical Tropical Forest Realms." Last modified July 22, 2007. Accessed April 9, 2017. http://rainforests.mongabay.com/0102.htm
13. Sommer, Adrian. "Attempt at an Assessment of the World's Tropical Moist Forests." *Food and Agriculture Organization of the United Nations*. Accessed November 23, 2016. http://www.fao.org/docrep/k0050e/k0050e03.htm#attempt%20at%20an%20assessment%20of%20the%20world%27s%20tropical%20moist%20forestsNote. "In general, only those countries entirely located within the geographical tropical zone (exceptions are Australia, Argentina, and Paraguay) have been included. Some other countries in the subtropical zone (part of China, Bhutan, and Nepal), partially covered by a kind of moist tropical forest, have not been taken into account. On the other land, India and Burma have been included as a whole. With their subtropical part."

14. Butler, Rhett. "Where Rainforests are Located: Biogeographical Tropical Forest Realms." Last modified July 22, 2007. Accessed April 9, 2017. http://rainforests.mongabay.com/0102.htm
15. Mongabay. "10 Facts about the Amazon Rainforest." Last modified January 23, 2016. http://rainforests.mongabay.com/amazon/amazon-rainforest-facts.html
16. Food and Agriculture Organization of the United Nations. "Global Forest Resources Assessment 2015: Desk Reference." Accessed November 23, 2016. http://www.fao.org/3/a-i4808e.pdf. 9–14.
17. Speth, James Gustave and Peter M. Haas. *Global Environmental Governance*. 30.
18. Food and Agriculture Organization of the United Nations. "Global Forest Resources Assessment 2015: Desk Reference." Accessed November 23, 2016. http://www.fao.org/3/a-i4808e.pdf. 235–240.
19. Speth, James Gustave and Peter M. Haas. *Global Environmental Governance*. 30.
20. OECD. "Glossary of Terms: Central Government." Last modified April 8, 2013. https://stats.oecd.org/glossary/detail.asp?ID=306
21. Kaimowitz, David. "Indigenous Peoples and Deforestation in Latin America." In *On the Edge*, written by Claude Martin. Berkeley: Greystone Books, 2015. 168.
22. Butler, Rhett. "Largest Area of Tropical Forest, by Country." Last modified January 11, 2016. Accessed April 10, 2017. http://rainforests.mongabay.com/deforestation_forest.html
23. Thoumi, Gabriel. "Economic Drivers Of Deforestation: Sectors Exposed To Sustainability And Financial Risks." *Nasdaq*, August 3, 2016. Accessed February 20, 2017. http://www.nasdaq.com/article/economic-drivers-of-deforestation-sectors-exposed-to-sustainability-and-financial-risks-cm660031
24. US Central Intelligence Agency. "The World Factbook." Accessed December 1, 2016. https://www.cia.gov/library/publications/the-world-factbook/geos/br.html; https://www.cia.gov/library/publications/the-world-factbook/geos/cg.html; https://www.cia.gov/library/publications/the-world-factbook/geos/id.html; https://www.cia.gov/library/publications/the-world-factbook/geos/pe.html; https://www.cia.gov/library/publications/the-world-factbook/geos/in.html
25. Food and Agriculture Organization of the United Nations. "Global Forest Resources Assessment 2015: Desk Reference." Accessed November 23, 2016. http://www.fao.org/3/a-i4808e.pdf. 235–240.

26. Rights and Resources Initiative. *Who Owns the World's Land? A Global Baseline of Formally Recognized Indigenous and Community Land Rights*. Washington, DC: RRI. 2015. Accessed March 29, 2017. http://www.rightsandresources.org/wp-content/uploads/GlobalBaseline_web.pdf. 6–8.
27. World Bank. "Terrestrial Protected Areas (% of Total Land Area)." Accessed February 20, 2017. http://data.worldbank.org/indicator/ER.LND.PTLD.ZS?name_desc=false
28. Global Forest Watch. "About." Accessed December 6, 2016. http://www.globalforestwatch.org/about
29. US Geological Society. "Earth Resources Observation and Science (EROS) Center: Find Data." Accessed December 6, 2016. http://eros.usgs.gov/find-data
30. Martin. *On the Edge*. 33.
31. Martin. *On the Edge*. 3–4.
32. Williams, Michael. *Deforesting the Earth: From Prehistoric to Global Crisis, An Abridgement*. Illinois: The University of Chicago Press, 2006.
33. Martin, *On the Edge*. 247.
34. Ginn. *Investing in Nature*. 108.
35. Lovejoy, Thomas. E. "Foreword." In *On the Edge*, written by Claude Martin. Berkeley: Greystone Books, 2015. IX–X.
36. Martin. *On the Edge*. xviii
37. Martin. *On the Edge*. 100.
38. Martin. *On the Edge*. 48.
39. World Wildlife Fund. "WWF Living Forests Report: Chapter 5." Accessed November 23, 2016. http://awsassets.panda.org/downloads/lfr_chapter_5_executive_summary_final.pdf
40. World Wildlife Fund. "WWF Living Forests Report: Chapter 5." Accessed November 23, 2016. http://awsassets.panda.org/downloads/lfr_chapter_5_executive_summary_final.pdf
41. Martin. *On the Edge*. 66.
42. FAO. "Live Animals: World + (Total): Cattle: 2014." Accessed April 13, 2017. http://www.fao.org/faostat/en/#data/QA
43. Gerber, P.J., H. Steinfeld, B. Henderson, A. Mottet, C. Opio, J. Dijkman, A. Falcucci and G. Tempio. "Tackling Climate Change through Livestock—A Global Assessment of Emissions and Mitigation Opportunities." *Food and Agriculture Organization of the United Nations (FAO)*, Rome. 2013. Accessed April 13, 2017. http://www.fao.org/3/i3437e.pdf. 15.
44. OECD. "Meat Consumption." Accessed February 20, 2017. https://data.oecd.org/agroutput/meat-consumption.htm
45. Martin. *On the Edge*. 97.

46. Martin. *On the Edge*. 98.
47. IndexMundi. "Soybean Oil Domestic Consumption by Country in 1000 MT." Accessed February 20, 2017. http://www.indexmundi.com/agriculture/?commodity=soybean-oil&graph=domestic-consumption
48. OECD. "Crop Production: Soybean, Thousand Tonnes, 2015." Accessed February 20, 2017. https://data.oecd.org/agroutput/crop-production.htm#indicator-chart
49. Martin. *On the Edge*. 91.
50. IndexMundi. "Palm Oil Production by Country in 1000 MT." Accessed February 20, 2017. http://www.indexmundi.com/agriculture/?commodity=palm-oil
51. IndexMundi. "Palm Oil Domestic Consumption by Country in 1000 MT." Accessed February 20, 2017. http://www.indexmundi.com/agriculture/?commodity=palm-oil&graph=domestic-consumption
52. Thoumi, Gabriel. "The Chain: Indonesia Strengthens Peatlands Moratorium; ING Places €80 Million in Debt for SocFin's Palm Oil and Rubber Expansion." *Chain Reaction Research*. Last modified December 7, 2016. https://chainreactionresearch.com/2016/12/07/the-chain-indonesia-strengthens-peatlands-moratorium-ing-places-e80-million-in-debt-for-socfins-palm-oil-and-rubber-expansion/
53. Food and Agriculture Organization of the United Nations. "Global Forest Resources Assessment 2015: Desk Reference." Accessed March 24, 2017. http://www.fao.org/3/a-i4808e.pdf. 52–57.
54. World Bank's World Integrated Trade Solution. "Wood." Accessed February 20, 2017. http://wits.worldbank.org/CountryProfile/en/Country/WLD/Year/2015/TradeFlow/Export/Partner/by-country/Product/44-49_Wood
55. Martin. *On the Edge*. 72.
56. Martin. *On the Edge*. 78.
57. Pearson, Timothy R. H., Sandra Brown, Lara Murray and Gabriel Sidman. "Greenhouse Gas Emissions from Tropical Forest Degradation: An Underestimated Source." *Carbon Balance and Management*. 12(3). 2017. 1. doi:10.1186/s13021-017-0072-2. http://rdcu.be/pfzy
58. Eco-Business. "Indonesia Study Disputes UN Data on Peat Fire Emissions." Last modified July 25, 2016. http://www.eco-business.com/news/indonesia-study-disputes-un-data-on-peat-fire-emissions/?utm_medium=email&utm_campaign=July%2027%20newsletter&utm_content=July%2027%20newsletter+Version+B+CID_3967427e26718a3032e32d81092a4c21&utm_source=Campaign%20Monitor&utm_

term=Indonesia%20study%20disputes%20UN%20data%20on%20peat%20fire%20emissions
59. Huijnen, V., M. J. Wooster, J. W. Kaiser, D. L. A. Gaveau, J. Flemming, M. Parrington, A. Inness, D. Murdiyarso, B. Main and M. van Weele. "Fire Carbon Emissions Over Maritime Southeast Asia in 2015 Largest Since 1997." Scientific Reports 6, Article number: 26886 (2016). doi:10.1038/srep26886.
60. World Bank. "The Cost of Fire: An Economic Analysis of Indonesia's 2015 Fire Crisis." Accessed February 2016. http://documents.worldbank.org/curated/en/776101467990969768/pdf/103668-BRI-Cost-of-Fires-Knowledge-Note-PUBLIC-ADD-NEW-SERIES-Indonesia-Sustainable-Landscapes-Knowledge-Note.pdf. 1.
61. Wakker, Eric, Jan Willem van Gelder, Joel Finkelstein and Ben Cushing. "New Analysis: Fires Burn Southeast Asian Assets." Last modified December 1, 2015. *Chain Reaction Research*. https://chainreactionresearch.com/2015/12/01/indonesia-fires-analysis/
62. Thoumi, Gabriel. "The Chain: BlackRock Calls Out Deforestation as Climate Risk; SE Asia 2015 Haze Death Toll Could Top 100,000; Sime Darby Launches Responsible Agriculture Charter." *Chain Reaction Research*. Last modified September 22, 2016. https://chainreactionresearch.com/2016/09/22/the-chain-blackrock-calls-out-deforestation-as-climate-risk-se-asia-2015-haze-death-toll-could-top-100000-sime-darby-launches-responsible-agriculture-charter/
63. Rodrigues de Mello, Natália Girão. "The Alarming Number of Fires in the Brazilian Amazon." *Mongabay*. Last modified September 8, 2016. https://news.mongabay.com/2016/09/the-alarming-number-of-fires-in-the-brazilian-amazon/
64. Erickson-Davis, Morgan. "Massive Wildfire Rips through Congo Rainforest—Is Logging to Blame?" *Mongabay*. Last modified March 23, 2016. https://news.mongabay.com/2016/03/massive-wildfire-rips-through-congo-rainforest-is-logging-to-blame/
65. Martin. *On the Edge*. 249.
66. World Bank. "Population, Total." Accessed February 20, 2017. http://data.worldbank.org/indicator/SP.POP.TOTL?order=wbapi_data_value_2011+wbapi_data_value+wbapi_data_value-last&sort=desc&year_high_desc=true
67. World Bank. "Population Growth (Annual %)." Accessed February 20, 2017. http://data.worldbank.org/indicator/SP.POP.GROW?order=wbapi_data_value_2011+wbapi_data_value+wbapi_data_value-last&sort=desc&year_high_desc=true

68. United Nations Data Booklet. "The World's Cities in 2016." Accessed March 15, 2017. http://www.un.org/en/development/desa/population/publications/pdf/urbanization/the_worlds_cities_in_2016_data_booklet.pdf. 4.
69. Martin. *On the Edge*. 70.
70. Sullivan, Zoe. "Brazil's Dispossessed: Belo Monte Dam Ruinous for Indigenous Cultures." *Mongabay*. December 6, 2016. https://news.mongabay.com/2016/12/brazils-dispossessed-belo-monte-dam-ruinous-for-indigenous-cultures/
71. Costanza, Robert, Ralph d'Arge, Rudolf de Groot, Stephen Farber, Monica Grasso, Bruce Hannon, Karin Limburg, Shahid Naeem, Robert V. O'Neill, Jose Paruelo, Robert G. Raskin, Paul Sutton and Marjan van den Belt. "The Value of the World's Ecosystem Services and Natural Capital." *Nature*. 387. 15 May 1997. http://www.esd.ornl.gov/benefits_conference/nature_paper.pdf. 253–260.
72. The Economics of Ecosystems and Biodiversity. "Ecosystem Services." AccessedNovember23,2016.http://www.teebweb.org/resources/ecosystem-services/
73. TEEB. "The Initiative." Accessed February 20, 2017. http://www.teebweb.org/about/the-initiative/
74. TEEB. "Areas of Work." Accessed February 20, 2017. http://www.teebweb.org/areas-of-work/
75. Natural Capital Coalition. "History, Vision & Mission." Accessed December 13, 2016. http://naturalcapitalcoalition.org/who/history-vision-mission/
76. Roberts, J. Timmons and Nikki Demetria Thanos. *Trouble in Paradise*. 139–140.
77. Maretti, Cláudio. "The Amazon: There is Hope!—If We All Do the Right Thing." In *On the Edge*, written by Claude Martin. Berkeley: Greystone Books, 2015. 204–205.
78. Daily, Gretchen C. and Katherine Ellison. *The New Economy of Nature*. 172.
79. Daily, Gretchen C. and Katherine Ellison. *The New Economy of Nature*. 6.
80. Blue Channel 24. "The Tapir or Danta, the Unknown Gardener from South American Forests." Last modified December 22, 2011. https://bluechannel24web.wordpress.com/2011/12/22/the-tapir-or-danta-the-unknown-gardener-from-south-american-forests/
81. EcoHealth Alliance. "IDEEAL." Accessed March 23, 2017. http://www.ecohealthalliance.org/program/ideeal
82. Rulli, Maria Cristina, Monia Santini, David T. S. Hayman and Paolo D'Odorico. "The Nexus between Forest Fragmentation in Africa and

Ebola Virus Disease Outbreaks." *Scientific Reports*. 7. Article Number: 41613 (2017). doi:10.1038/srep41613. http://www.nature.com/articles/srep41613

83. Smith, P., M. Bustamante, H. Ahammad, H. Clark, H. Dong, E.A. Elsiddig, H. Haberl, R. Harper, J. House, M. Jafari, O. Masera, C. Mbow, N.H. Ravindranath, C.W. Rice, C. Robledo Abad, A. Romanovskaya, F. Sperling and F. Tubiello. "Agriculture, Forestry and Other Land Use (AFOLU)." In *Climate Change 2014: Mitigation of Climate Change. Contribution of Working Group III to the Fifth Assessment Report of the Intergovernmental Panel on Climate Change*, edited by Edenhofer, O., R. Pichs-Madruga, Y. Sokona, E. Farahani, S. Kadner, K. Seyboth, A. Adler, I. Baum, S. Brunner, P. Eickemeier, B. Kriemann, J. Savolainen, S. Schlömer, C. von Stechow, T. Zwickel and J.C. Minx. Cambridge, United Kingdom and New York, NY, USA: Cambridge University Press, 2014. 829. https://www.ipcc.ch/pdf/assessment-report/ar5/wg3/ipcc_wg3_ar5_chapter11.pdf
84. Radford, Time. "Vast Carbon Sink Found in Congo Basin." *Climate News Network*. Last modified January 15, 2017. http://climatenewsnetwork.net/20810-2/
85. Daily, Gretchen C. and Katherine Ellison. *The New Economy of Nature*. 7–8.
86. Daily, Gretchen C. and Katherine Ellison. *The New Economy of Nature*. 143.
87. BBC Nature. "Big Five Mass Extinction Events." Accessed March 29, 2017. http://www.bbc.co.uk/nature/extinction_events
88. Daily, Gretchen C. and Katherine Ellison. *The New Economy of Nature*. 157.
89. World Wildlife Fund. "Role of the Global Ecoregions and How They are Selected." Accessed November 23, 2016. http://wwf.panda.org/about_our_earth/ecoregions/about/role/
90. Naturevolution. "Who We Are." Accessed November 23, 2016. http://www.naturevolution.org/en/qui-sommes-nous/
91. Martin. *On the Edge*. XVII.
92. Martin. *On The Edge*. XX.
93. Martin. *On The Edge*. 87.

3

Tropical Rainforest Ecology

"There is a rule that all naturalists know and talk about who work in rainforests. The plant or animal species that catches your eye at this moment you may not see again that day, or week, or even year. It may never reveal itself to you again, no matter how long and hard you search."[1]
—Dr. E.O. Wilson

Introduction

The term biological diversity was first "coined by the conservation biologist Raymond F. Dasmann in 1968."[2] Biological diversity (also known as biodiversity) is one of the central components of tropical ecology. There are 3 types of diversity: genetic diversity, species diversity, and ecosystem diversity.

Ecosystem Diversity

Ecosystem diversity, specifically with respect to tropical rainforests, refers to the variety of ecosystems and micro-ecosystems that exist. A tropical rainforest may consist of lowland tropical rainforests, along with

B.J. McFarland, *Conservation of Tropical Rainforests*, Palgrave Studies in Environmental Policy and Regulation, https://doi.org/10.1007/978-3-319-63236-0_3

mountainous cloud forests and freshwater ecosystems. In addition, a single tree in the tropical rainforest may provide a variety of micro-ecosystems from the canopy of an emergent tree species, to its understory, and further down to below the subsoil.

Ecosystems are complex systems, and as a result of deforestation and degradation, these complex ecosystems become isolated islands surrounded by changed landscapes. This leads to habitat fragmentation. This, in turn, reduces both the ability of animals to migrate and of tree species to adapt to climate change through ecological shifts. This means species are unable to move to a different latitude, longitude, or altitude to find a more suitable habitat.

Species Diversity

It is estimated that there are approximately 8.7 ± 1.3 million species that call Earth their home.[3] This species diversity, referring to all the different plants and animals, is unsurpassed—except for coral reefs—in the tropical rainforests (Fig. 3.1).

For example, the Amazon Rainforest has the world's greatest species diversity as it "{…} contains the world's richest diversity of primates, birds, butterflies, and many other groups of plants and animals. Estimations of freshwater fish species continue to grow beyond the current number of about 9,000. To date, at least 40,000 plant species have been identified."[4] Similarly, "in the 1980s Terry L. Erwin from the Smithsonian Institution collected about 1,200 different beetle species from a single Central American tree species (*Luehea seemannii*) {…}."[5] Dr. E.O. Wilson "collected forty-three species of ant from a single tree in the Tambopata Reserve in Peru—almost as many as the entire indigenous ant fauna of the United Kingdom!"[6]

Often, tropical rainforests will exhibit low species density with high species diversity. Likewise, there are lots of different species but few members of the same species will be found in close proximity to one another. Such diversity helps to contribute to the stability of the ecosystem. This said, there are a wide variety of roles or niches, and the diversity helps to fulfill the niche partitioning that takes place.

Fig. 3.1 A flowering bromeliad in Mexico (Credit: Brian Joseph McFarland)

Furthermore, diversity can help contribute to food security. Take, for instance, that there are more than 4,000 species of potatoes,[7] the US Department of Agriculture's Plant Introduction Station in Iowa has nearly 20,000 samples or "accessions" of corn,[8] and there are more than 90,000 species of rice (both wild and cultivated varieties).[9]

Species diversity also relates to the terms endemic species, threatened and endangered species, keystone or flagship species, indicator species, and charismatic species.

Endemic species are species found in a single location, which could be a specific tree or within a distinct region like the Amazon Rainforest, and found nowhere else in the world. For example, jaguars (*Panthera onca*) are only found throughout the Americas, while orangutans (*Pongo pygmaeus* and *Pongo abelii*) and Sumatran rhinoceroses (*Dicerorhinus sumatrensis*) are only found in a few rare patches of tropical rainforests in Southeast Asia.

Threatened, endangered, and critically endangered species are species in population decline. The International Union for the Conservation of Nature (IUCN) has developed a comprehensive Red List for species and a Red List for habitats. The IUCN was established in 1948. It is a "membership Union uniquely composed of both government and civil society organizations."[10] This includes a wide range of central governments (e.g., Denmark, France, Germany, Spain, and the United States), multilateral agencies (e.g., the FAO, UNEP, UNDP, and the World Bank), nonprofit organizations (e.g., BirdLife International, Conservation International, TNC, and WWF), and foundations (e.g., Ford Foundation, Gordon and Betty Moore Foundation, and the John D. and Catherine T. MacArthur Foundation).[11]

Unfortunately, countless species within tropical rainforests have already gone extinct or face extinction. The IUCN Red List of 2013 "included 489 'critically endangered' amphibian species. Almost a third—32.4 percent—of amphibians are now threatened with extinction, many of them in tropical forest areas, particularly in Latin America. As many as 159 amphibian species may already be extinct."[12]

Keystone or flagship species are species that are considered especially important for a given ecosystem. While it can be argued that all species are important for a given ecosystem, some species such as seed dispersers and pollinators have a multiplying effect on other species. For instance, lowland tapirs are especially valuable dispersers of large seeds throughout the Amazon Rainforest, while apex predators—such as jaguars (*Panthera onca*) and mountain lions (*Puma concolor*)—help stabilize populations of species at lower trophic levels.

Indicator species are species that can indicate pressure on a given ecosystem. For example, "with their permeable skin, amphibians are supremely sensitive to disturbances in the air and water. They are one of the first animals to disappear when habitat is threatened."[13] In addition, the presence or absence of certain benthic macroinvertebrates in freshwater ecosystems can signal the presence of pollutants such as particulate matter coming from smokestacks.

Charismatic species are symbolic species that represent an ecosystem and can help raise awareness and financing for the preservation of the entire ecosystem and thus conserve a multitude of lesser-known species. Such charismatic species include African elephants (*Loxodonta africana*),

Fig. 3.2 Scarlet macaw (Credit: Brian Joseph McFarland)

Bengal tigers (*Panthera tigris tigris*), blue whales (*Balaenoptera musculus*), common bottlenose dolphins (*Tursiops truncatus*), common chimpanzees (*Pan troglodytes*), eastern gorillas (*Gorilla beringei*), jaguars (*Panthera onca*), polar bears (*Ursus maritimus*), and scarlet macaws (*Ara macao*). Similarly, the symbol of the World Wildlife Fund (WWF) is the giant panda (*Ailuropoda melanoleuca*) (Fig. 3.2).

There are complex, symbiotic interactions among all of the species—whether they are more common species, indicator species, or endangered species—throughout tropical rainforests. These symbiotic interactions include neutralism, competition, mutualism, protocooperation, commensalism, amensalism, parasitism, and predation.[14]

Lastly, it is important to understand the migration patterns of these animals. For instance, "fifty-one percent of all North American migrant bird species—including many kinds of protected waterfowl—fly to wintering areas in the Neotropics. Panama's birds, then, are also *our* birds."[15]

Genetic Diversity

Genetic diversity refers to the number of individuals within a given species. It is vitally important to have a sufficient number of breeding pairs to maintain a healthy, sustainable population. According to Dr. E.O. Wilson, the greatest contributors to the loss of biodiversity—including genetic diversity—are the loss of habitat, invasive species, pollution, human overpopulation, and overharvesting.[16] While invasive species are a major issue around the world, there are also introduced, alien, and/or exotic species which are foreign species that are not necessarily as detrimental as invasive species.

Due to habitat loss via tropical rainforest deforestation and degradation, there can be "empty forests." As explained by Kent H. Redford, "We should not let a forest full of trees fool us into believing all is well."[17] In addition, "larger clearings would change the soil and microclimatic conditions to such an extent that the seedlings and dormant seeds of closed forest trees would not be able to survive."[18] This reduces genetic diversity and the reason when lost, tropical rainforests may never again be able to be reestablished.

Connectivity to Landscape

While it is important to have national parks and conservation projects, it is important to establish large-scale, landscape-level conservation initiatives. As reported by Dr. Claude Martin,

> The thirty-two year investigation concluded that nature reserves in Amazonia should ideally be in the range of thousands to tens of thousands of square kilometers to provide sufficient security for the survival of the many rare species and species with a patchy distribution. Reserves should also cover a variety of different river basins as well as climatic zones and soil types.[19]

One such model is to establish core protected areas with surrounding areas of low-impact activities such as sustainably managed forests, extractive and/or Indigenous Peoples' reserves, and ecotourism models. These low-impact regions or "working landscapes" could also include silviculture and agroforestry activities that obtain independently verified and certified chain of custody such as via the Forest Stewardship Council (FSC) or the Roundtable on Sustainable Palm Oil (RSPO). As explained by William Ginn,

> Ideally these would be surrounded by a buffer of sustainably managed forests protected by easements and voluntary certification efforts. There are no absolute guidelines in setting these goals: bigger buffer areas of protected working forest may mean that a particular reserve can be smaller and still be effective and conversely, a protected area that is surrounded by development needs to be much bigger to achieve its ecological goals. It is also clear that whatever the minimum, bigger is better when it comes to protected areas.[20]

Similarly,

> Several contemporary management models are trying to address the shortcomings of the traditional park system. The United Nations Educational, Scientific and Cultural Organization (UNESCO) Man and the Biosphere's (MAB) World Network of Biosphere Reserves is by far the most internationally successful. Biosphere reserves are a management model of concentric zones, consisting of a human-free protected "core" area surrounded by a "buffer zone" for research, recreation, ecotourism, and environmental education and an outer "transition" area for human habitation, private enterprise, and so on.[21]

One way to frame the design of protected areas is SLOSS, which stands for single large or several small. It is best to have a single large protected area with interconnected biological or wildlife corridors and to have all of them surrounded by low-impact activities (Fig. 3.3).[22]

Biological or wildlife corridors are connections established between protected areas. One such corridor established throughout Central America and Mexico is the Mesoamerican Biological Corridor.[23]

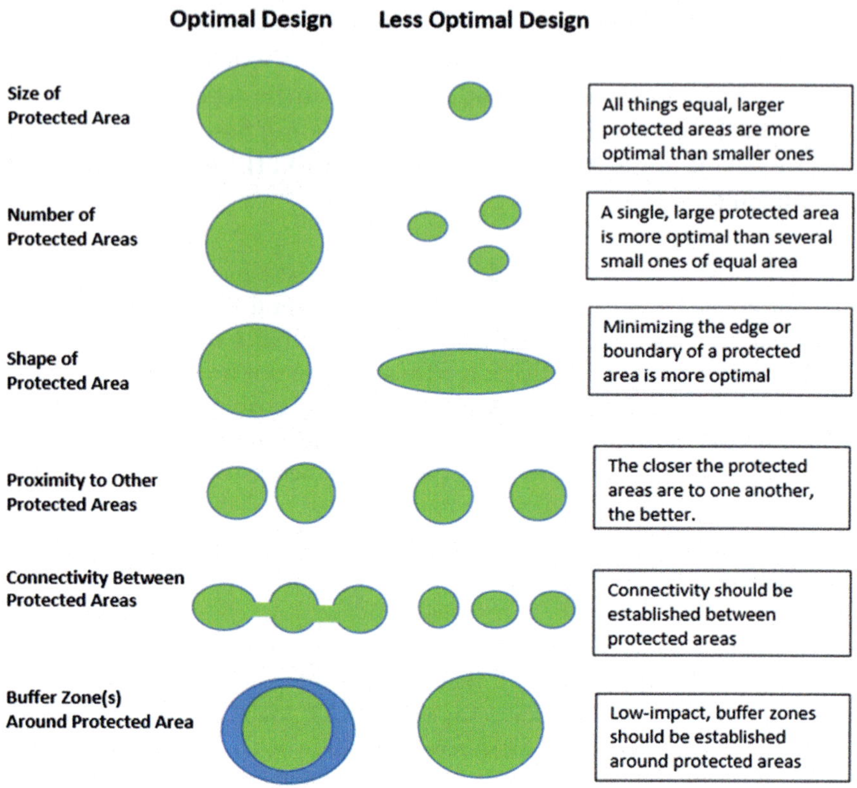

Fig. 3.3 Optimal design of protected areas (Credit: Brian Joseph McFarland)

The classifications of protected areas according to the IUCN are:

Category Ia: Strict Nature Reserve

- "Protected areas that are strictly set aside to protect biodiversity and also possibly geological/geomorphological features, where human visitation, use and impacts are strictly controlled and limited to ensure protection of the conservation values. Such protected areas can serve as indispensable reference areas for scientific research and monitoring."[24]

Category Ib: Wilderness Area

- "Protected areas that are usually large unmodified or slightly modified areas, retaining their natural character and influence, without permanent or significant human habitation, which are protected and managed so as to preserve their natural condition."[25]

Category II: National Park

- "Large natural or near natural areas set aside to protect large-scale ecological processes, along with the complement of species and ecosystems characteristic of the area, which also provide a foundation for environmentally and culturally compatible spiritual, scientific, educational, recreational and visitor opportunities."[26]

Category III: Natural Monument or Feature

- "Protected areas set aside to protect a specific natural monument, which can be a landform, sea mount, submarine cavern, geological feature such as a cave or even a living feature such as an ancient grove. They are generally quite small protected areas and often have high visitor value."[27]

Category IV: Habitat/Species Management Area

- "Protected areas aiming to protect particular species or habitats and management reflects this priority. Many category IV protected areas will need regular, active interventions to address the requirements of particular species or to maintain habitats, but this is not a requirement of the category."[28]

Category V: Protected Landscape/Seascape

- "A protected area where the interaction of people and nature over time has produced an area of distinct character with significant ecological, biological, cultural and scenic value: and where safeguarding the integrity of this interaction is vital to protecting and sustaining the area and its associated nature conservation and other values."[29]

Category VI: Protected Area with Sustainable Use of Natural Resources

- "Protected areas that conserve ecosystems and habitats, together with associated cultural values and traditional natural resource management systems. They are generally large, with most of the area in a natural condition, where a proportion is under sustainable natural resource management and where low-level non-industrial use of natural resources compatible with nature conservation is seen as one of the main aims of the area."[30]

Other Resources on Tropical Rainforest Ecology

Amazon Tree Diversity Network

- http://atdn.myspecies.info/

BirdLife International's Data Zone

- http://datazone.birdlife.org/home

Conservation Evidence

- http://www.conservationevidence.com/

Critical Ecosystem Partnership Fund

- http://www.cepf.net/Pages/default.aspx

Encyclopedia of Earth

- http://eol.org/

Global Biodiversity Information Facility

- http://www.gbif.org/

Global Biodiversity Outlook

- https://www.cbd.int/gbo/

GlobAllomeTree

- http://www.globallometree.org/

Great Ape Survival Partnership's (GRASP) apeAPP

- http://www.apeapp.org/

Living Planet Index

- http://www.livingplanetindex.org/home/index

Millennium Ecosystem Assessment

- http://www.millenniumassessment.org/en/index.html

National Biodiversity Strategies and Action Plans' Forum

- http://nbsapforum.net/#

Network of Regional Governments for Sustainable Development's Biodiversity Learning Platform

- http://www.nrg4sd.org/biodiversity/regions-biodiversity-learning-platform/

The IUCN Red List of Threatened Species

- http://www.iucnredlist.org/

TRAFFIC, The Wildlife Trade Monitoring Network

- http://www.traffic.org/

Tropicos®

- http://www.tropicos.org/

United Nations Environment Programme's World Conservation Monitoring Centre

- https://www.unep-wcmc.org/

World Wildlife Fund's The Global Observation and Biodiversity Information Portal

- http://panda.maps.arcgis.com/home/index.html

World Wildlife Fund's Wildlife and Climate Change

- https://www.worldwildlife.org/initiatives/wildlife-and-climate-change

Yale School of Forestry & Environmental Studies' Global Forest Atlas

- http://globalforestatlas.yale.edu/about-global-forest-atlas

Notes

1. Wilson. *The Creation*. 24.
2. Martin. *On the Edge*. 135.
3. Martin. *On the Edge*. 137–138.
4. Maretti, Cláudio. "The Amazon: There is Hope!—If We All Do the Right Thing." In *On the Edge*, written by Claude Martin. Berkeley: Greystone Books, 2015. 204.
5. Martin. *On the Edge*. 137.
6. Martin. *On the Edge*. 138.
7. International Potato Center. "Potato Facts and Figures." Accessed March 22, 2017. http://cipotato.org/potato/facts/
8. The University of Georgia. "MAIZE Project—Genetic Diversity of Maize." Accessed March 22, 2017. http://maize.uga.edu/?loc=diversity
9. The Rice Association. "Varieties." Accessed March 22, 2017. http://www.riceassociation.org.uk/content/1/10/varieties.html
10. IUCN. "About: Donors and Partners." Accessed February 20, 2017. https://www.iucn.org/about/donors-and-partners
11. IUCN. "About: Donors and Partners." Accessed February 20, 2017. https://www.iucn.org/about/donors-and-partners
12. Martin. *On the Edge*. 140.
13. Royte, Elizabeth. *The Tapir's Morning Bath*. 266.
14. Bisaccio, Daniel J. of Brown University. Interviewed by Brian McFarland. July 2016.
15. Royte, Elizabeth. *The Tapir's Morning Bath*. 37.
16. Wilson. *The Creation*. 75.

17. Redford, K.H. "The Empty Forest." In *On the Edge*, written by Claude Martin. Berkeley: Greystone Books, 2015. 159.
18. Martin. *On the Edge*. 6.
19. Martin. *On the Edge*. 155.
20. Ginn. *Investing in Nature*. 173.
21. Roberts, J. Timmons and Nikki Demetria Thanos. *Trouble in Paradise*. 79–80.
22. Diamond, Jared. "The Island Dilemma: Lessons of Modern Biogeographic Studies of the Design of Natural Reserves." *Biological Conservation* (7) (1975). England: Applied Science Publishers Ltd. Accessed February 20, 2017. http://jareddiamond.org/Jared_Diamond/Further_Reading_files/Diamond%201975.pdf. 129–146.
23. FAO. "Mesoamerican Biological Corridor: Regional Initiative for the Promotion of Forest Conservation." Accessed December 12, 2016. http://www.fao.org/docrep/ARTICLE/WFC/XII/MS15-E.HTM
24. IUCN. "Protected Areas: Category Ia: Strict Nature Reserve." Accessed February 20, 2017. https://www.iucn.org/theme/protected-areas/about/protected-areas-categories/category-ia-strict-nature-reserve
25. IUCN. "Protected Areas: Category Ib: Wilderness Area." Accessed February 20, 2017. https://www.iucn.org/theme/protected-areas/about/protected-areas-categories/category-ib-wilderness-area
26. IUCN. "Protected Areas: Category II: National Park." Accessed February 20, 2017. https://www.iucn.org/theme/protected-areas/about/protected-areas-categories/category-ii-national-park
27. IUCN. "Protected Areas: Category III: Natural Monument or Feature." Accessed February 20, 2017. https://www.iucn.org/theme/protected-areas/about/protected-areas-categories/category-iii-natural-monument-or-feature
28. IUCN. "Protected Areas: Category IV: Habitat/Species Management Area." Accessed February 20, 2017. https://www.iucn.org/theme/protected-areas/about/protected-areas-categories/category-iv-habitatspecies-management-area
29. IUCN. "Protected Areas: Category V: Protected Landscape/Seascape." Accessed February 20, 2017. https://www.iucn.org/theme/protected-areas/about/protected-areas-categories/category-v-protected-land scapeseascape
30. IUCN. "Protected Areas: Category VI: Protected Area with Sustainable Use of Natural Resources." Accessed February 20, 2017. https://www.iucn.org/theme/protected-areas/about/protected-areas-categories/category-vi-protected-area-sustainable-use-natural-resources

4

Global Environmental Policy

"Policy changes happen through far-sighted governments, political will, and a functioning multilateral system to set the regulatory framework for sustainability."[1]
—Dr. Claude Martin

Introduction

Global environmental policy, particularly policy focused on tropical rainforest conservation, is primarily done through local and national policy and through international trade policy.

With respect to tropical rainforest conservation policy, it is important to recognize that:

> Obviously, targets alone do not do the trick. They are only effective if the accompanying conditions are conducive to change: public awareness of the threats to ecosystems; nongovernmental organization campaigning and publicity with statistical and photographic evidence on the drivers of deforestation and the related fate of forest-dependent people; financial support and favorable working relationships between governments and nongovernmental organizations; and most importantly, political will and determination.[2]

B.J. McFarland, *Conservation of Tropical Rainforests*, Palgrave Studies in Environmental Policy and Regulation, https://doi.org/10.1007/978-3-319-63236-0_4

Thus, policy matters, especially effective implementation of policy. Similarly, all politics are local and voting is critical, particularly in democracies. It should also be noted that many countries have conflicting policies that create perverse incentives. For instance in Belize, if land is transferred to private ownership, the landowner must "improve" the land and "improvement" means clearing the land.[3] A similar incentive happens in Brazil with local communities who do not have official title to the land, but must clear the land to demonstrate their productive use of the land and, thus, to receive rights (i.e., but not necessarily title) to the land.

It is also important to understand that there are two main legal systems, which are referred to as English common law and Napoleonic code (also referred to as civil law or Napoleonic law). Common law is "generally *uncodified*. This means that there is no comprehensive compilation of legal rules and statutes. While common law does rely on some scattered statutes, which are legislative decisions, it is largely based on *precedent*, meaning the judicial decisions that have already been made in similar cases."[4] In contrast, civil law "is *codified*. Countries with civil law systems have comprehensive, continuously updated legal codes that specify all matters capable of being brought before a court, the applicable procedure, and the appropriate punishment for each offense."[5] This said, tropical countries such as Brazil, the Democratic Republic of the Congo, and Indonesia follow the civil law system, while tropical countries such as Belize, India, and Jamaica are based off the common law system.

While the following policies will primarily focus on land use and climate change policies, timber policies are also integral to land use policy in many tropical countries. This said, how a country handles its extractive industry has a big impact on land. For instance, what operators do during the timber cut and what they do afterward has a lot to do with land conversion. With privately owned land, the owner is incentivized to protect land for timber and with a concession model, such as in Indonesia, the concession owner controls the boundaries. However, without long-term leases and without clear policies outlining the responsibilities for pre- and postharvest, the owners can leave behind severely degraded forests and road systems that, for instance, in Peru, can later be accessed by illegal gold miners.[6]

State and Local Policies

Designing and implementing policy is challenging, especially in developing countries which at times lack transparency, have increased levels of corruption, and lack adequate funding. Yet, there are promising developments with the level of participation in the Governors' Climate and Forests Task Force (GCF) (Fig. 4.1). The GCF is a:

> {...} unique subnational collaboration between 35 states and provinces {as of April 2017} from Brazil, Colombia, Indonesia, Ivory Coast, Mexico, Nigeria, Peru, Spain, and the United States. The GCF seeks to advance jurisdictional programs designed to promote low emissions rural development and reduced emissions from deforestation and land use (REDD+) and link these activities with emerging greenhouse gas (GHG) compliance regimes and other pay-for-performance opportunities. More than 25% of the world's tropical forests are in GCF states and provinces, including more than 75% of Brazil's and more than half of Indonesia's. The GCF includes states and provinces that are leading the way in building comprehensive, jurisdiction-wide approaches to low emissions development and REDD+ as well as the only jurisdiction in the world (California) that is considering provisions that would recognize offsets from REDD+ as part of its GHG compliance system.[8]

Fig. 4.1 GCF Task Force membership (Credit: GCF Task Force)[7]

Due to the slow progress of policy making within national and international forums, some subnational jurisdictions—such as many of the GCF Task Force members—have taken a leading role. For instance, the State of Acre, Brazil, has some of the most advanced subnational policies and institutions aimed at reducing tropical deforestation. The State passed the Sistema de Incentivos para Serviços Ambientais (SISA, which translates into the State System of Incentives for Environmental Services) on October 22, 2010.[9] The State of Acre also has set up the Institute of Climate Change and Environmental Services Regulation. The Brazilian State of Rondônia is also exploring a similar, subnational jurisdictional approach as Acre.[10]

In Brazil, there is also a unique state and federal-level policy that aims to counteract deforestation by blacklisting municipalities with high levels of deforestation (such as in Mato Grosso and Pará as a result of soy and cattle) and restricting their access to agricultural credit.[11] This, in turn, is meant to encourage the transition to a "Green Municipality."[12]

While no tropical rainforests exist in the US State of California, California is a member of the GCF Task Force, and the state has passed progressive legislation that may enable the financing of tropical rainforest conservation via carbon finance. Assembly Bill 32 (AB 32), known as the California Global Warming Solutions Act of 2006, allows for international sector-based offsets, and the first such offsets may come from forest conservation activities in the State of Acre, Brazil.[13] To date, some of the largest purchasers of compliance-grade forestry offsets for the California market include Chevron, Tesoro Refining & Marketing, and Calpine Energy Services.[14] California, along with Chiapas, Mexico, and the State of Acre, signed a Memorandum of Understanding (MOU) on November 16, 2010, to further harmonize their subnational programs. More specifically, the MOU cites:

> Recognizing further the importance of focusing on issues of common interest between the Parties, such as reducing greenhouse gas emissions in the forest sector by preserving standing forests and sequestering additional carbon through the restoration and reforestation of degraded lands and forest, and through improved forest management practices;
>
> Recognizing further that the Governors' Climate and Forests (GCF) Task Force is a unique subnational collaboration between 14 states and

> provinces {now 35 states and provinces} from the United States, Brazil, Indonesia, Nigeria, and Mexico that seeks to integrate Reducing Emissions from Deforestation and Forest Degradation (REDD) and other forest carbon activities into emerging greenhouse gas (GHG) compliance regimes in the United States and elsewhere. As such, the GCF represents an important foundation for identifying enhanced partnerships.
>
> ARTICLE 2 The Parties will coordinate efforts and promote collaboration for environmental management, scientific and technical investigation, and capacity building, through cooperative efforts focused particularly on:
>
> a. Reducing greenhouse gas emissions from deforestation and land degradation—otherwise known as "REDD"—and sequestration of additional carbon through the restoration and reforestation of degraded lands and forests, and through improved forest management practices.
>
> b. Developing recommendations together to ensure that forest-sector emissions reductions and sequestrations, from activities undertaken at the sub-national level, will be real, additional, quantifiable, permanent, verifiable and enforceable, and capable of being recognized in compliance mechanisms of each party's state.[15]

Although not specifically focused on tropical rainforest conservation, an excellent book covering a US perspective on grassroots campaigning for measuring public opinion (e.g., selecting a pollster, designing and interpreting your poll), designing a winning measure (e.g., defining goals, identifying land, and setting priorities, establishing fiscal safeguards, and drafting ballot language), and running a conservation campaign (e.g., designing a winning campaign strategy, securing endorsements, developing a paid media plan, and setting up your field organization), please see Kim Hopper and Ernest Cook's book entitled, *Conservation Finance Handbook: How Communities Are Paying for Parks and Land Conservation.*

Federal Policies

While policy at the local level is critical, tropical rainforests often transcend state or province borders—not to mention country borders—and thus, policy at the federal level is vital to the conservation of tropical rainforests at a landscape level. However, federal policy—and similarly on the local level—can often conflict between both promoting and slowing tropical rainforest deforestation and degradation. For example:

> Government subsidization of projects like road building has further fueled both timber booms and large-scale development. Another favorite policy of forest-rich countries is to promote agricultural development and ranching in previously forested areas, sometimes with government subsidies so deep that the enterprises would be essentially uneconomical without them. Cattle ranching in the Amazon is the most well known example in this regard.[16]

From 2005 to 2013, the country of Brazil made remarkable progress on slowing deforestation by an estimated 70% from its 10-year average from 1996 to 2005.[17] In 2004, Brazil passed The Action Plan for Prevention and Control of Deforestation in the Legal Amazon (Plano de Ação para Prevenção e Controle do Desmatamento na Amazônia Legal in Portuguese or PPCDAm), which was:

> a governmental response to the increasing deforestation rates in the Amazon—has as main objective to reduce the deforestation rates through a series of integrated actions of territorial and land tenure ordering, environmental monitoring and control, fostering sustainable production activities, involving partnerships between federal institutions, state governments, municipalities, civil society entities and the private sector. Currently, it is on its third implementation phase (2012–2015).[18]

Then in 2009, Brazil passed Law Number 12.187 entitled the National Policy on Climate Change which "{…} contained emissions reduction targets by sector, including a reduction in Amazonian deforestation of 80 percent by 2020."[19]

An additional federal-level policy in Brazil is the Forest Code and specifically the requirement of legal reserves on private properties. This said, private properties in the Legal Amazon—which includes the states of Acre, Amapá, Amazonas, Maranhão, Mato Grosso, Pará, Rondônia, Roraima, and Tocantins—are required to maintain 80% of their forests as a legal reserve. However, Brazil's recent recession of 2015–2016, and the reduced enforcement of the Forest Code in states such as Pará, is one cause of the recent increase in deforestation.

With respect to Indonesia, in May 2011 and in conjunction with financing from Norway, Indonesia "instituted a nationwide moratorium on new license areas ('concessions') for oil palm plantations, timber plantations, and logging activity on primary forests and peatlands {…}."[20] This moratorium would be extended in May 2013 for 2 years and then again in May 2015.

Although not the host of the world's largest tropical rainforests, Costa Rica has some of the most novel and advanced federal policies to counteract tropical deforestation. For example, "the country's first forest-incentive program, established in 1979, gave an income tax deduction to landowners who engaged in reforestation projects."[21]

In addition to Costa Rica, Panama and Colombia both have relatively new laws prohibiting the cutting of trees which were not huge priorities due to the wars in both countries. Honduras and Ecuador have robust forest conservation laws, but such laws are mainly focused on watershed management (i.e., focus is not so much on timber or forestry dynamics).[22]

It is important to note that policy matters both within countries that host tropical rainforests and export raw materials—such as Brazil, the Democratic Republic of the Congo, and Indonesia—and countries with large economies that import many products from these host countries like the United States, the European Union, and Japan.

The United States adopted the Lacey Act of 1900, and then in 2008, the US Congress "followed suit and passed an amendment to the Lacey Act of 1900, which prohibits trafficking in illegally harvested plants and wildlife."[23] The US Endangered Species Act of 1973 is also an important, complimentary piece of legislation. Similarly, the Australian Government passed the Illegal Logging Prohibition Act of 2012.[24]

To counteract tropical rainforest deforestation and degradation, and specifically illegal logging, action is required on the international level. While "both Indonesia and Brazil have demonstrated that political will and resolute government action against corruption can make a difference. {…} International publicity, nongovernmental organization investigations, and intergovernmental schemes against illegal logging are also effective."[25]

Bilateral, Multilateral, and Regional Policies

There are numerous bilateral, multilateral, and regional policies aimed at cooperatively addressing tropical rainforest deforestation and degradation. A challenge within these forums is that most, if not all, countries strongly hold onto national sovereignty. Yet, because rainforests often transcend national borders (i.e., Amazon Basin, the Congo Basin, and the island of Borneo), there is a need to work on a regional level to address tropical deforestation and degradation.

The most active country facilitating bilateral agreements to specifically address tropical rainforest deforestation and degradation has been Norway. In 2007, Prime Minister Jens Stoltenberg launched the Norwegian International Climate and Forest Initiative (NICFI). Since NICFI was launched, Norway has committed funding to numerous countries to assist them with slowing tropical rainforest deforestation and deforestation including the countries of Brazil, Guyana, Indonesia, and Tanzania.

There was also a bilateral Memorandum of Understanding (MOU) signed on March 3, 2010, between Brazil and the United States of America on "cooperation regarding climate change." Such an MOU specifically includes:

> New areas of cooperation would be added, including, but not limited to, the following areas: reducing emissions from deforestation and forest degradation (REDD+); and low-carbon development (…) To exchange experiences on strategies and domestic policies, including carbon markets, to address climate change.[26]

Germany, via its development bank KfW, and the United States through the US Agency for International Development (USAID) and specifically through the former Forest Carbon, Markets, and Communities (FCMC) Program, have also channeled billions of dollars to tropical rainforest countries.

There are also important multilateral initiatives as well. Norway, alongside the United States and the United Kingdom, collectively pledged USD$280 million in 2013 toward tropical rainforest conservation.[27]

On a regional level, there is important policy work being done across the Amazon, the Congo Basin, and Borneo. Within the Amazon region, the Amazon Cooperation Treaty Organization is operational after the Amazon Cooperation Treaty was signed in 1978 between Bolivia, Brazil, Colombia, Ecuador, Guyana, Peru, Suriname, and Venezuela.[28] Within the Congo Basin—consisting of the countries of Cameroon, Central African Republic, Democratic Republic of the Congo (DRC), Republic of the Congo, Equatorial Guinea, and Gabon—are several initiatives. For example:

> The Yaoundé Declaration resulted from a meeting of Central African heads of state in Yaoundé, Cameroon, in 1999. It led to the Central African Forest Commission (COMIFAC) and fostered the creation of a large cluster of protected areas in Cameroon, the Republic of the Congo, the Central Africa Republic, and Gabon. The Yaoundé Declaration helped address the historical underrepresentation of protected areas in West and Central Africa compared to other tropical rainforest regions: by 2003 they made up 14 percent of the moist tropical forest area, half the percentage reached in South America.[29]

To further encourage such regional collaboration within the Congo Basin:

> At the World Summit in Johannesburg the US Secretary of State, Colin Powell, made a pledge of USD 53 million and in fact launched the Congo Basin Forest Partnership, a facilitating platform that brings together some seventy partners, including African countries, donor agencies, international organizations, nongovernmental organizations, scientific institutions, and the private sector.[30]

In February 2007, the Declaration on the Heart of Borneo Initiative was signed by the governments of Brunei, Indonesia, and Malaysia.[31]

There have also been regional collaborations within large, importing countries. For example, "as a major timber importer for its own consumption as well as timber processing, China is currently piloting a timber legality scheme in cooperation with Australia, Japan, and Indonesia."[32] In addition, the European Union in 2005 adopted similar legislation to the US Lacey Act known as the Forest Law Enforcement, Governance and Trade (FLEGT) regulation. A cornerstone of this FLEGT regulation is the Voluntary Partnership Agreements (VPA), which is a licensing program aimed at stopping illegal timber imports from countries such as Cameroon and Indonesia.[33]

International Policies

There are numerous international treaties, conventions, forums, and protocols that either directly aim to address tropical rainforest deforestation and degradation or tangentially aim to address related aspects such as climate change, biodiversity, and human rights. This includes but is not limited to:

- The Rio Convention of 1992, which established the Convention on Biological Diversity (CBD) and the United Nations Framework Convention on Climate Change (UNFCCC);
- Convention on International Trade in Endangered Species of Wild Fauna and Flora (CITES);
- International Tropical Timber Organization;
- Ramsar Convention on Wetlands;
- Universal Declaration of Human Rights;
- United Nations Declaration on the Rights of Indigenous Peoples (UNDRIP);
- Convention on the Elimination of All Forms of Discrimination Against Women;
- International Labor Organization (ILO) Convention;
- World Trade Organization (WTO) and other international trade agreements such as the Asia-Pacific Economic Cooperation (APEC), the North American Free Trade Agreement (NAFTA), and Mercosur; and
- The UN Forum on Forests, which was established in October 2000 and has universal membership.

Despite good intentions, such international policies appear to be less effective. As described in *The Economy of Nature*:

> Although governments have negotiated a wide array of global and regional agreements to protect certain ecosystems from degradation and extinction—such as the Ramsar Conventions on Wetlands, the Convention on Biological Diversity, and the Convention on the Law of the Sea—these agreements are mostly weak, lacking the participation, resources and systems of incentives and enforcement they need to be effective.[34]

Yet,

> The expectations on governments to ensure the conservation of biodiversity are significant. In fact, the Convention of Biodiversity, signed by almost all of the world's countries (except, sadly, the United States), commits the world to setting aside a fully representative protected areas system covering at least 10 percent of the globe before 2015.[35]

Furthermore, nearly every country on Earth submitted Intended Nationally Determined Contributions (INDCs) prior to the UNFCCC Conference of the Parties (COP) 21 event in Paris. Almost 200 countries signed onto the Paris Agreement in December 2015, which now commits countries to carry out their Nationally Determined Contributions (NDCs). Embedded in the Paris Agreement is an allowance for Internationally Transferred Mitigation Outcomes (ITMOs), which appears to allow for international offset and REDD+:

> The use of internationally transferred mitigation outcomes to achieve nationally determined contributions under this Agreement shall be voluntary and authorized by participating Parties. {…}
>
> Recognizes the importance of adequate and predictable financial resources, including for results-based payments, as appropriate, for the implementation of policy approaches and positive incentives for reducing emissions from deforestation and forest degradation, and the role of conservation, sustainable management of forests and enhancement of forest carbon stocks; as well as alternative policy approaches, such as joint mitigation and adaptation approaches for the integral and sustainable management of

> forests; while reaffirming the importance of non-carbon benefits associated with such approaches; encouraging the coordination of support from, inter alia, public and private, bilateral and multilateral sources, such as the Green Climate Fund, and alternative sources in accordance with relevant decisions by the Conference of the Parties.[36]

Alongside COP21, 18 countries signed a declaration on international carbon markets. This group, which is being led by New Zealand, is also comprised of Australia, Canada, Chile, Colombia, Germany, Iceland, Indonesia, Italy, Japan, Mexico, Netherlands, Panama, Papua New Guinea, South Korea, Senegal, Ukraine, and the United States of America.[37]

Furthermore, the FAO states:

> Agriculture and land use, land-use change and forestry (LULUCF) are among the most frequently included sectors in countries' mitigation contributions (targets and/or actions).
>
> When considered together, 89 percent of countries cover agriculture and/or LULUCF. Including the countries that mention bioenergy as a mitigation strategy, this percentage increases to 92 percent.[38]

The threshold for the Paris Agreement to come into force was met on October 4, 2016, and then the Paris Agreement officially came into force 1 month later on November 4, 2016. Even prior to COP21, many countries were beginning to regulate greenhouse gas emissions. This includes:

- The European Union Emissions Trading Scheme (EU ETS), the world's largest carbon market valued at USD$40.7 billion in 2015.[39]
- The California-Quebec Emissions Trading Scheme is currently the world's second largest carbon market with possible further linkages to Washington State, Oregon, Ontario, and Acre (i.e., for REDD+ offsets).
- China's developing national carbon market currently with seven pilot provinces and cities (Beijing, Chongqing, Guangdong, Hunan, Shanghai, Shenzhen, and Tianjin) planned with possible connections to other markets in the future.
- South Korea's cap-and-trade system.

Four important tropical rainforest conservation initiatives developed by multilateral agencies are the Central African Forest Initiative (CAFI), the Global Environmental Facility (GEF), UN-REDD Programme, and the World Bank's Forest Carbon Partnership Facility (FCPF). A fifth multilateral fund, known as the Green Climate Fund, was recently established in 2010.[40]

Endorsed in 2014 at the United Nations Climate Summit, the New York Declaration on Forests is now a global representation of governments, business, and civil society's commitment to stopping tropical rainforest deforestation.[41]

The World Trade Organization (WTO) and international free trade agreements—such as the Central American Free Trade Agreement (CAFTA)—can provide "{...} opportunities for governments to work together on environmental protection and conservation, and private sector initiatives will become even more important in the coming years."[42]

Company Policies

In addition to government policies, it is important to recognize that many companies around the world are adopting internal company policies that are designed to reduce the company's impact on the environment. For example, Puma implemented an environmental profit-loss statement,[43] and Microsoft has implemented an internal carbon price.[44] More recently, there has been the emergence of zero deforestation supply chain commitments from many of the world's largest private and publicly traded companies including Burger King, Cargill, McDonald's, and Nestlé.[45] According to Chain Reaction Research:

> More than 200 companies self-reported to CDP in 2016 to its annual Global Forests Report analyzing corporate risks and opportunities linked to the four commodities—soy, timber, cattle, and palm oil—that drive a significant portion of tropical deforestation.
>
> Based on the disclosures from 187 companies analyzed in this report, on average 24 percent of corporate revenues depend on commodities linked to deforestation: cattle, palm oil, soy and timber. This makes these revenues

vulnerable to the risks in producing and sourcing commodities that are responsible for deforestation.

But only one in five companies assess risks associated with deforestation beyond a six-year horizon across all commodities. Fewer than half of these companies have evaluated how the quantity or quality of key forest-risk commodities could affect their organization's growth strategy over the medium-term. With such a large proportion of company revenues attached to these forest-risk commodities, companies must plan for the long-term to ensure the future sustainable supply of these commodities.

Meanwhile, more than 350 global corporations have made zero-deforestation commitments. Therefore, if potential supply constraints from both physical deforestation risks and zero-deforestation procurement policies are not addressed, cost of goods sold could spike, potentially impacting revenue, profits, and future firm-wide financial valuation.

72 percent of companies report that they have identified sufficient future sustainable supply across all forest risk commodities. Meanwhile, 81 percent of companies in the Agricultural Production sector, which sits at the top of global commodity supply chains, have experienced impacts related to forest-risk commodities that have resulted in substantive changes to operations, revenue or expenditure in the past five years. But, Fewer than half—only 42 percent—of companies have accessed the quantity or quality of key forest risk commodities on their corporate strategies over the next five or more years.

Finally, the report showed that 47 percent of the companies considered deforestation as part of their enterprise risk management and operational risk strategies.

Demand for these forest risk commodities is expected to grow. Zion Market Research forecasts 7.2 percent compound annual growth rate (CAGR) for palm oil from 2015 to 2021 to a total $93 billion. This would result in additional capital and long-term equity at risk, if the more than 200 corporate zero-deforestation palm oil commitments were implemented.[46]

Jurisdictions and companies are also beginning to implement Produce & Protect models. Produce & Protect models are facilitated by regional government policies that support public-private partnerships to source sustainable products while reducing deforestation. These policy frameworks source commodities from areas practicing sustainable agricultural production, along with forest protection and protection of livelihoods. Per Chain Reaction Research:

> On January 20, 2017, media reported that Unilever has signed a three-year Memorandum of Understanding with governments in Indonesia to drive sustainable sourcing of palm oil. The three-year agreement is between the company, the Government of Central Kalimantan and various other regions.
>
> The Consumer Goods Forum and Tropical Forest Alliance 2020 (TFA 2020)—a global partnership focusing on reducing deforestation from the production of palm oil, soy, beef and pulp & paper—have been promoting "Produce & Protect" investment models such as Unilever's for the past two years. Produce & Protect models source commodities from areas practicing sustainable agricultural production, forest protection and protection of livelihoods.
>
> Beef, soy, palm oil and pulp & paper production is worth $180 billion annually across the tropical forest regions. 45 percent of these commodities, worth $81 billion annually, are exported globally.
>
> As reported by TFA 2020, 415 companies globally have made at least one relevant commitment to eliminate deforestation from the production of these four commodities. In the palm oil supply chain, 59 percent of the companies have committed to commodity specific policies, followed by pulp & paper with 53 percent of companies, soy with 21 percent and beef with 12 percent.
>
> Initially, the pilot program will impact 600 smallholders on 1,400 ha. This is important as smallholders in the palm oil industry in Indonesia contribute to 40 percent of Indonesia's overall production yet average productivity is 32 percent lower than palm oil plantation companies.
>
> The financial benefit from this program to investors and smallholders alike is the ability to use longer-term contracts to de-risk investments in sustainable agriculture intensification. This is forecasted to result in both improved yields per ha, and improved ability by corporations to achieve their sustainable sourcing commitments.[47]

The following timeline is an overview of some important global environmental policies as well as the dates certain governmental institutions were established.

Dates of Global Environmental Policy

Nineteenth Century

- 1876: "{…} Switzerland introduced a strict ban on forest clearing through one of the world's first forest laws (the Forstpolizeigesetz), after years of catastrophic landslides and floods, particularly in alpine areas."[48]

Twentieth Century

- 1900: The United States' Lacey Act is passed. The Lacey Act was later amended in 2008.[49]
- 1965: The original Brazil Forest Code, entitled Law No. 4.771, is adopted.[50]
- 1972: The UN Conference on the Human Environment is held in Stockholm.[51]
- 1972: The Convention Concerning the Protection of the World Cultural and Natural Heritage is signed.[52]
- 1973: The US Endangered Species Act is passed.[53]
- 1973: Convention on International Trade in Endangered Species of Wild Fauna and Flora (CITES) is passed and enters into force on July 1, 1975.[54]
- 1978: The Amazon Cooperation Treaty (ACT) was signed.[55] The Treaty was "institutionally strengthened" in 1995 with a Permanent Secretariat and that decision would later enter into force in 1998.[56]
- 1979: "The country's first forest-incentive program {Costa Rica}, established in 1979, gave an income tax deduction to landowners who engaged in reforestation projects."[57]
- 1980: "The {Indian} Forest Conservation Act of 1980 contained provisions to reduce deforestation and land use change and set out an obligation for forest plantation to offset deforestation for infrastructure projects."[58]
- 1985: The Tropical Forestry Action Plan of 1985 is adopted.[59]
- 1987: World Commission on Environment and Development (WCED) is held, and the *Our Common Future*, also known as the Brundtland Report, is published.[60]

- 1990: US Clean Air Act Amendments are signed into law, establishing a market-based trading system (i.e., cap and trade) for SOX and NOX emissions.[61]
- 1992: The United Nations Conference on Environment and Development (UNCED, also known as the Rio "Earth Summit") is held:
  - Agenda 21 is adopted;
  - The Rio Declaration on Environment and Development is adopted;
  - The United Nations Convention on Biological Diversity is adopted;
  - The United Nations Framework Convention on Climate Change (UNFCCC) is adopted;[62]
  - The Statement of Forest Principles is adopted, which "eventually was followed by the United Nations Forum on Forests in 2000."[63]
- 1992: United Nations Commission on Sustainable Development (CSD) is created "to ensure effective follow-up of United Nations Conference on Environment and Development (UNCED), also known as the Earth Summit."[64]
- 1993: The Convention on Biological Diversity enters into force.[65]
- 1996: "Costa Rica earned an international reputation for reforestation when it introduced a unique piece of legislation, the 1996 Forestry Law, that promotes conservation through payments for environmental services, reforestation, and conservation programs."[66]
- 1997: The Kyoto Protocol was adopted in Kyoto, Japan.[67]
- 1998: The United States' Tropical Forest Conservation Act (TFCA) is established.[68]

Twenty-First Century

- 2000: Cartagena Protocol on Biosafety is adopted and entered into force in September 11, 2003.[69]
- 2002: World Summit on Sustainable Development in Johannesburg.[70]
- 2005: "The role of deforestation in climate change had not been taken into account until it re-emerged in 2005 at the United Nations

Framework Convention on Climate Change Conference of the Parties (COP 11) in Montreal."[71]

- 2005: The Kyoto Protocol enters into force.[72]
- 2005: The European Union's Forest Law Enforcement, Governance and Trade (FLEGT) regulation is adopted.[73]
- 2006: Assembly Bill 32 (AB 32), the California Global Warming Solutions Act of 2006, is passed.[74]
- 2007: Bali Road Map—"The concept of REDD was taken up by environmental groups and some tropical countries, the latter informally organized into a Coalition for Rainforest Nations, and in 2007 at the United Nations Climate Change Conference in Bali, Indonesia, the Bali Road Map was adopted as a two-year plan to finalize a binding international agreement in December 2009 in Copenhagen."[75]
- 2008: In 2008, the "US Congress followed suit and passed an amendment to the Lacey Act of 1900, which prohibits trafficking in illegally harvested plants and wildlife."[76]
- 2009: In 2009, "Brazil launched a national climate change policy that contained emissions reduction targets by sector, including a reduction in Amazonian deforestation of 80 percent by 2020."[77]
- 2010: In May 2010, "Norway signed a letter of intent with Indonesia for a process of preparation, REDD+ readiness, and contributions for verified emissions."[78]
- 2010: The Mexican State of Chiapas, the Brazilian State of Acre, and the United States' State of California sign an MOU to collaborate on REDD.[79]
- 2011: Indonesia "instituted a nationwide moratorium on new license areas ('concessions') for oil palm plantations, timber plantations, and logging activity on primary forests and peat lands {…}."[80] This moratorium would be extended in May 2013 for 2 years and then again in May 2015.
- 2012: In 2012 "the Australian government also passed new legislation—the Illegal Logging Prohibition Act."[81]
- 2012: The latest revision of Brazil Forest Code, Law No. 12.651, is adopted.[82,83]
- 2012: The United Nations Conference on Sustainable Development—or Rio+20—took place in Rio de Janeiro, Brazil.[84]
- 2013: EU Timber Regulation came into force
- 2015: Paris Agreement of the UNFCCC is adopted.[85]

Dates When Government Institutions Were Established

Twentieth Century

- 1905: The United States Department of Agriculture's (USDA) Forest Service is founded.[86]
- 1920: League of Nations is created.[87]
- 1935: Natural Resources Conservation Service (NRCS) of the United States Department of Agriculture (USDA, formerly known as the Soil Conservation Service).[88]
- 1939: The International Institute of Tropical Forestry of the United States Department of Agriculture's Forest Service is established (then known as the Tropical Forest Experiment Station).[89]
- 1944: The Articles of Agreement of the International Bank for Reconstruction and Development (i.e., the World Bank) and the International Monetary Fund were drawn up and adopted; articles became effective on December 31, 1945.[90]
- 1945: Food and Agriculture Organization of the United Nations is founded.[91]
- 1945: United Nations (UN) officially comes into existence.[92]
- 1946: "The Division of Forestry and Forestry Products of the FAO had come into existence in May 1946 {…}."[93]
- 1948: Germany's KfW Bank is founded.[94]
- 1958: The US National Aeronautics and Space Administration (NASA) is established.[95]
- 1959: Inter-American Development Bank (IDB) is founded.[96]
- 1961: The Organisation for Economic Co-operation and Development (OECD) is established.[97]
- 1961: United States Agency for International Development (USAID).[98]
- 1964: African Development Bank (AfDB) becomes effective.[99]
- 1964: France's Office National des Forêts (ONF) is created. ONF International (ONFI) would later be established in 1997.[100]
- 1965: United Nations Development Programme is established. "UNDP is based on the merging of the United Nations Expanded

Programme of Technical Assistance, created in 1949, and the United Nations Special Fund, established in 1958."[101]

- 1961: Brazilian "Presidential decree creates the Grupo de Organização da Comissão Nacional de Atividades Espaciais (GOCNAE, The Group for the Organization of the National Space Activities Commission)," the precursor to INPE. The Instituto Nacional de Pesquisas Espaciais (INPE, The National Institute of Space Research) is later created in 1971.[102]
- 1966: Asian Development Bank (ADB) is opened.[103]
- 1972: United Nations Environment Programme (UNEP) is created by the UN General Assembly.[104]
- 1973: The World Heritage Trust is "established through UNESCO's 1973 Convention for the Protection of the World Cultural and Natural Heritage."[105]
- 1977: The International Fund for Agricultural Development (IFAD) is established as "a specialized agency of the United Nations."[106]
- 1986: International Tropical Timber Organization (ITTO) is established.[107]
- 1990: USDA National Agroforestry Center is created.[108]
- 1991: European Bank for Reconstruction and Development (EBRD) is opened.[109]
- 1991: Global Environmental Facility (GEF) is founded.[110]
- 1994: Uruguay Round of the General Agreement on Tariffs and Trade (GATT) concluded.[111]
- 1994: The World Trade Organization (WTO) is established.[112]

Twenty-First Century

- 2000: The UN Forum on Forests (UNFF) is established and now has universal membership.[113]
- 2000: UN Millennium Declaration (later to be known as the Millennium Development Goals or MDGs) is first adopted.[114] The MDGs are now known as the Sustainable Development Goals (SDGs), which were adopted in September 2015.[115]

- 2002: "At the World Summit in Johannesburg the US Secretary of State, Colin Powell, made a pledge of USD 53 million and in fact launched the Congo Basin Forest Partnership, a facilitating platform that brings together some seventy partners, including African countries, donor agencies, international organizations, nongovernmental organizations, scientific institutions, and the private sector."[116]
- 2004: World Bank's BioCarbon Fund is created.[117]
- 2005: Commission des Forêts d'Afrique Centrale (COMIFAC, Central African Forest Commission in English) is formally established after the "Treaty on the Conservation and Sustainable Management of Forest Ecosystems of Central Africa and Establishing the Central African Forest Commission" is signed by the Heads of State in Brazzaville.[118]
- 2007: Declaration on the Heart of Borneo Initiative signed by governments of Brunei, Indonesia, and Malaysia.[119]
- 2007: Norwegian Forest and Climate Initiative is launched.[120]
- 2007: World Bank's Forest Carbon Partnership Facility (FCPF) is launched.[121]
- 2008: UN-REDD Programme is founded.[122]
- 2009: Governors' Climate & Forests Task Force (GCF) is created.[123]
- 2011: USAID's Forest Carbon, Markets and Communities (FCMC) is launched.[124]
- 2012: Germany REDD Early Movers Programme is created.[125]

Other Resources on Global Environmental Policy

Climate Action Tracker's Tracking (I)NDCs

- http://climateactiontracker.org/indcs.html

Climate Change Litigation Databases

- http://climatecasechart.com/

- http://www.lse.ac.uk/GranthamInstitute/climate-change-laws-of-the-world/
- http://www.lse.ac.uk/GranthamInstitute/countries/

ECOLEX

- http://www.ecolex.org/

FAO's FAOLEX Database

- http://www.fao.org/faolex/en/

Forest Legality Initiative by the World Resources Institute

- http://www.forestlegality.org/

Legal Preparedness for REDD+

- https://dgroups.org/fao/law-for-redd-plus/

National Biodiversity Strategies and Action Plans' (NBSAP) Policy and Legislation Page

- http://nbsapforum.net/#categories/293

NDC Partnership's NDC Toolbox Navigator

- http://ndcpartnership.org/toolbox-navigator#tools

Planning for NDC Implementation: A Quick-State Guide

- http://www.cdkn.org/ndc-guide

The Little Book of Legal Frameworks for REDD+

- http://d6.globalcanopy.org/materials/little-book-legal-frameworks-redd

Trust for Public Land's LandVote Database

- https://tpl.quickbase.com/db/bbqna2qct?a=dbpage&pageID=8

United Nations' Audiovisual Library on International Law

- http://legal.un.org/avl/intro/introduction.html?tab=2
  Voigt, Christina (Ed.). *Research Handbook on REDD+ and International Law*. Massachusetts: Edward Elgar Publishing, Inc., 2016

WWF's Mapping REDD+: A Visual Guide to UNFCCC Decisions

- http://wwf.panda.org/what_we_do/footprint/forest_climate/forest_climate_publications/?292670/Mapping-REDD

Yale Center for Environmental Law and Policy

- http://envirocenter.yale.edu/

Notes

1. Martin. *On the Edge*. 165.
2. Martin. *On the Edge*. 187.
3. Waldon, Dr. Jeff. Interviewed by Brian McFarland. March 2017.
4. University of California at Berkeley's School of Law. "The Common Law and Civil Law Traditions." Accessed April 5, 2017. https://www.law.berkeley.edu/library/robbins/CommonLawCivilLawTraditions.html

5. University of California at Berkeley's School of Law. "The Common Law and Civil Law Traditions." Accessed April 5, 2017. https://www.law.berkeley.edu/library/robbins/CommonLawCivilLawTraditions.html
6. Waldon, Dr. Jeff. Interviewed by Brian McFarland. March 2017.
7. The Governors' Climate & Forests Task Force. "About: Membership." Accessed April 13, 2017. http://www.gcftaskforce.org/about
8. The Governors' Climate & Forests Task Force. "About: GCF Overview." Accessed April 13, 2017. http://www.gcftaskforce.org/about
9. State of Acre. "Unofficial Translation. State of Acre. Bill No. 2.308 of October 22. 2010." Accessed November 23, 2016. http://www.gcftaskforce.org/documents/Unofficial%20English%20Translation%20of%20Acre%20State%20Law%20on%20Environmental%20Services.pdf
10. Zwick, Steve. "Brazil's Rondônia State to Launch System of Payments for Ecosystem Services." *Ecosystem Marketplace*. Last modified January 14, 2015. http://www.ecosystemmarketplace.com/articles/brazils-rondonia-state-to-launch-system-of-payments-for-ecosystem-services/
11. Zwick, Steve and Ciro Calderon. "The Difficult Birth of Brazil's First "Green Municipality"." February 9, 2016. *Ecosystem Marketplace*. Accessed April 10, 2017. http://www.ecosystemmarketplace.com/articles/paragominas-the-green-revolution-that-almost-wasnt/
12. Zwick, Steve and Ciro Calderon. "Brazil's Green Municipalities: What Works? What Doesn't? Why?" March 6, 2016. *Ecosystem Marketplace*. Accessed April 10, 2017. http://www.ecosystemmarketplace.com/articles/brazils-green-municipalities-what-works-what-doesnt-why/
13. California Environmental Protection Agency. "International Sector-Based Offset Credits." Accessed December 7, 2016. https://www.arb.ca.gov/cc/capandtrade/sectorbasedoffsets/sectorbasedoffsets.htm
14. Ecosystem Marketplace. "State of Forest Carbon Finance 2016." Accessed February 14, 2017. http://www.forest-trends.org/documents/files/doc_5388.pdf. 8.
15. The State of Acre, the State of Chiapas, and the State of California. "Memorandum of Understanding on Environmental Cooperation between the State of Acre of the Federative Republic of Brazil, the State of Chiapas of the United Mexican States, and the State of California of the United States of America." http://www.gcftaskforce.org/documents/MOU_Acre_California_and_Chiapas.pdf
16. Speth, James Gustave and Peter M. Haas. *Global Environmental Governance*. 30.

17. Earth Innovation Institute. "Brazilian Agreement on Amazon Deforestation." Accessed December 12, 2016. http://earthinnovation.org/our-work/case-studies/amazon-deforestation/
18. Acre Climate Change and Regulation of Environmental Services Institute (IMC). "Jurisdictional, Subnational Program for Incentives for Environmental Services of Carbon of the State of Acre, Brazil—Acre ISA-Carbon Program." *VCS Project Database*. Last modified December 15, 2014. http://www.vcsprojectdatabase.org/services/publicViewServices/downloadDocumentById/17710. 87.
19. Elliott, Chris. "REDD+ Is a Little More Complex than 'Not Cutting Down a Tree'." In *On the Edge*, written by Claude Martin. Berkeley: Greystone Books, 2015. 241.
20. Busch, Jonah, Kalifi Ferretti-Gallon, Jens Engelmann, Max Wright, Kemen G. Austin, Fred Stolle, Svetlana Turubanova, Peter V. Potapov, Belinda Margono, Matthew C. Hansen and Alessandro Baccini. "Reductions in Emissions from Deforestation from Indonesia's Moratorium on New Oil Palm, Timber, and Logging Concessions." *Proceedings of the National Academy of Sciences of the United States of America*. Accessed November 29, 2016. https://www.ncbi.nlm.nih.gov/pmc/articles/PMC4321246/
21. Meyer, Shannon. "Working with Landowners to Provide Ecosystem Services: Costa Rica's Groundbreaking Experiment." In *Conservation Capital in the Americas*, edited by James N. Levitt. Massachusetts: Lincoln Institute of Land Policy, 2010. 210.
22. Eisenstadt, Keegan of ClearSky Climate Solutions. Interviewed by Brian McFarland. October 2016.
23. Martin. *On the Edge*. 82.
24. Australian Government's Department of Agriculture and Water Resources. "Illegal Logging." Accessed December 12, 2016. http://www.agriculture.gov.au/forestry/policies/illegal-logging
25. Martin. *On the Edge*. 79.
26. The Government of Brazil and the Government of the United States of America. "Memorandum of Understanding Between the Government of the Federative Republic of Brazil and the Government of the United States of America on Cooperation Regarding Climate Change." Accessed November 23, 2016. http://www.brazilcouncil.org/wp-content/uploads/2015/11/MOUonCooperationRegardingClimateChange-Mar032010.pdf
27. UK Government. "Press Release: Norway, UK and USA Come Together to Pledge Approximately $280 Million to Sustain the World's Forests."

Accessed December 12, 2016. https://www.gov.uk/government/news/norway-uk-and-usa-come-together-to-pledge-approximately-280-million-to-sustain-the-worlds-forests
28. Brazilian Ministry of Foreign Affairs. "Amazon Cooperation Treaty Organization—ACTO." Accessed November 23, 2016. http://www.itamaraty.gov.br/en/politica-externa/integracao-regional/6351-amazon-cooperation-treaty-organization-acto
29. Martin. *On the Edge*. 187.
30. Martin. *On the Edge*. 190.
31. World Wildlife Fund. "The Heart of Borneo Declaration." Accessed November 28, 2016. http://wwf.panda.org/what_we_do/where_we_work/borneo_forests/about_borneo_forests/declaration.cfm
32. Martin. *On the Edge*. 83.
33. European Forest Institute. "EU FLEGT Facility: What is a Voluntary Partnership Agreement?" Accessed December 12, 2016. http://www.euflegt.efi.int/what-is-vpa
34. Daily, Gretchen C. and Katherine Ellison. *The New Economy of Nature*. 7.
35. Ginn. *Investing in Nature*. 182.
36. UNFCCC. "Paris Agreement." Accessed August 19, 2016. https://unfccc.int/resource/docs/2015/cop21/eng/l09.pdf
37. Environmental Finance. "18 Countries Sign Declaration on International Carbon Markets." Accessed August 19, 2016. https://www.environmental-finance.com/content/news/18-countries-sign-declaration-on-international-carbon-markets.html
38. FAO. 2016. *The Agriculture Sectors in the Intended Nationally Determined Contributions: Analysis*, by Strohmaier, R., Rioux, J., Seggel, A., Meybeck, A., Bernoux, M., Salvatore, M., Miranda, J. and Agostini, A. Environment and Natural Resources Management Working Paper No. 62. Rome. http://www.fao.org/3/a-i5687e.pdf. xiv.
39. Carbon Pulse. "Global CO2 Trade Volume Down 19% in 2015, Value Up 9% -Point Carbon." Accessed August 19, 2016. http://carbon-pulse.com/14102/
40. Green Climate Fund. "GCF Features." Accessed April 4, 2017. http://www.greenclimate.fund/about-gcf/global-context#history
41. New York Declaration on Forests. "About the NYDF Assessment." Accessed December 12, 2016. http://forestdeclaration.org/about/
42. Simons, Paul E. "Foreword." In *Conservation Capital in the Americas*, edited by James N. Levitt. Cambridge, Massachusetts: Lincoln Institute of Land Policy, 2010. X.

43. Puma. "Environmental Profit and Loss Account." Accessed February 9, 2017.http://about.puma.com/en/sustainability/environment/environmental-profit-and-loss-account
44. Microsoft. "Internal Carbon Fee." Accessed February 9, 2017. Available: http://download.microsoft.com/download/0/A/B/0AB2FDD7-BDD9-4E23-AF6B-9417A8691CF5/Microsoft%20Carbon%20Fee%20Impact.pdf
45. Supply Change. "Profiles." Accessed February 9, 2017. http://supply-change.org/profiles/
46. Thoumi, Gabriel. "The Chain: Industry Report States Up To $900 Billion Revenue at Risk from Deforestation; Impacts Cost of Goods Sold." *Chain Reaction Research*. Last modified January 9, 2017. https://chainreactionresearch.com/2017/01/09/the-chain-industry-report-states-up-to-900-billion-revenue-at-risk-from-deforestation-impacts-cost-of-goods-sold/
47. Thoumi, Gabriel. "The Chain: Unilever Makes Produce & Protect Commitments, New Corporate Palm Oil Reporting Guidance Launched." *Chain Reaction Research*. Last modified January 24, 2017. https://chainreactionresearch.com/2017/01/24/the-chain-unilever-makes-produce-protect-commitments-marks-spencer-supports-new-corporate-palm-oil-reporting-guidance/
48. Martin. *On the Edge*. 124.
49. U.S. Fish & Wildlife Service. "Lacey Act." Accessed November 23, 2016. https://www.fws.gov/international/laws-treaties-agreements/us-conservation-laws/lacey-act.html
50. Presidency of the Republic. "Law No. 4.771." Accessed November 23, 2016. http://www.planalto.gov.br/ccivil_03/Leis/L4771.htm
51. United Nations. "United Nations Conference on the Human Environment (Stockholm Conference)." Accessed November 23, 2016. https://sustainabledevelopment.un.org/milestones/humanenvironment
52. United Nations Educational, Scientific and Cultural Organization. "Convention Concerning the Protection of the World Cultural and Natural Heritage." Accessed November 29, 2016. http://whc.unesco.org/en/conventiontext/
53. U.S. Fish & Wildlife Service. "A History of the Endangered Species Act of 1973." Accessed November 23, 2016. https://www.fws.gov/endangered/laws-policies/esa-history.html
54. CITES. "What is CITES?" Accessed November 29, 2016. https://cites.org/eng/disc/what.php

55. Brazilian Ministry of Foreign Affairs. "Amazon Cooperation Treaty Organization—ACTO." Accessed November 23, 2016. http://www.itamaraty.gov.br/en/politica-externa/integracao-regional/6351-amazon-cooperation-treaty-organization-acto
56. Amazon Cooperation Treaty Organization. "Amazon Cooperation Treaty." Accessed November 29, 2016. http://otca.info/portal/tratado-coop-amazonica.php?p=otca
57. Meyer, Shannon. "Markets for Ecosystem Services and Reclaiming the Great Dismal Swamp." In *Conservation Capital in the Americas*, edited by James N. Levitt. Massachusetts: Lincoln Institute of Land Policy, 2010. 210.
58. Martin. *On the Edge*. 128.
59. Columbia University. "Tropical Forestry Action Plan." Accessed November 29, 2016. http://www.ciesin.columbia.edu/docs/002-162/002-162.html
60. World Commission on Environment and Development. "Our Common Future." Accessed November 29, 2016. http://www.un-documents.net/our-common-future.pdf
61. US Environmental Protection Agency. "1990 Clean Air Act Amendment Summary." Accessed November 23, 2016. https://www.epa.gov/clean-air-act-overview/1990-clean-air-act-amendment-summary
62. United Nations. "UN Conference on Environment and Development (1992)." Accessed November 29, 2016. http://www.un.org/geninfo/bp/enviro.html
63. Martin. *On the Edge*. 181.
64. United Nations. "Commission on Sustainable Development (CSD)." Accessed November 29, 2016. https://sustainabledevelopment.un.org/csd.html
65. The Convention on Biological Diversity. "Introduction." Accessed November 29, 2016. https://www.cbd.int/intro/
66. Martin. *On the Edge*. 127.
67. United Nations Framework Convention on Climate Change. "Kyoto Protocol." Accessed November 23, 2016. http://unfccc.int/kyoto_protocol/items/2830.php
68. United States Agency for International Development. "Financing Forest Conservation: An Overview of the Tropical Forest Conservation Act." Accessed November 23, 2016. https://www.usaid.gov/biodiversity/TFCA

69. Convention on Biological Diversity. "The Cartagena Protocol on Biosafety." Accessed November 23, 2016. https://bch.cbd.int/protocol
70. United Nations. "World Summit on Sustainable Development." Accessed November 23, 2016. http://www.un.org/events/wssd/summaries/envdevj1.htm
71. Martin. *On the Edge*. 40.
72. United Nations Framework Convention on Climate Change. "Kyoto Protocol." Accessed November 23, 2016. http://unfccc.int/kyoto_protocol/items/2830.php
73. European Union. "FLEGT Voluntary Partnership Agreements (VPAs)." Accessed November 29, 2016. http://ec.europa.eu/environment/forests/flegt.htm
74. California Environmental Protection Agency. "Assembly Bill 32 Overview." Accessed November 23, 2016. https://www.arb.ca.gov/cc/ab32/ab32.htm
75. Elliott, Chris. "REDD+ Is a Little More Complex than 'Not Cutting Down a Tree'." In *On the Edge*, written by Claude Martin. Berkeley: Greystone Books, 2015. 235.
76. Martin. *On the Edge*. 82.
77. Elliot, Chris. "REDD+ Is a Little More Complex than 'Not Cutting Down a Tree'." In *On the Edge*, written by Claude Martin. Berkeley: Greystone Books, 2015. 241.
78. Ibid.
79. REDD Offset Working Group. "About ROW." Accessed November 23, 2016. http://stateredd.org/about-row/
80. Busch, Jonah, Kalifi Ferretti-Gallon, Jens Engelmann, Max Wright, Kemen G. Austin, Fred Stolle, Svetlana Turubanova, Peter V. Potapov, Belinda Margono, Matthew C. Hansen and Alessandro Baccini. "Reductions in Emissions from Deforestation from Indonesia's Moratorium on New Oil Palm, Timber, and Logging Concessions." *Proceedings of the National Academy of Sciences of the United States of America*. Accessed November 29, 2016. https://www.ncbi.nlm.nih.gov/pmc/articles/PMC4321246/
81. Martin. *On the Edge*. 83.
82. Presidency of the Republic. Civil House Cabinet Subcommittee for Legal Affairs. "Law No. 12.651." Accessed November 23, 2016. http://www.planalto.gov.br/ccivil_03/_Ato2011-2014/2012/Lei/L12651.htm
83. Machado, Frederico and Kate Anderson. "Brazil's New Forest Code: A Guide for Decision-Makers in Supply Chains and Governments."

WWF Brazil. 2016. http://assets.wwf.org.uk/downloads/wwf_brazils_new_forest_code_guide_1.pdf
84. United Nations. "United Nations Conference on Sustainable Development. Rio+20." Accessed November 23, 2016. https://sustainabledevelopment.un.org/rio20
85. United Nations Framework Convention on Climate Change. "Historic Paris Agreement on Climate Change." Accessed November 23, 2016. http://newsroom.unfccc.int/unfccc-newsroom/finale-cop21/
86. United States Department of Agriculture. "The U.S. Forest Service—An Overview." Accessed November 23, 2016. http://www.fs.fed.us/documents/USFS_An_Overview_0106MJS.pdf
87. Department of State's Office of the Historian. "The League of Nations, 1920." Accessed November 23, 2016. https://history.state.gov/milestones/1914-1920/league
88. United Stated Department of Agriculture. "History of NRCS." Accessed November 23, 2016. http://www.nrcs.usda.gov/wps/portal/nrcs/main/national/about/history/
89. International Institute of Tropical Forestry. "About the Institute." Accessed November 30, 2016. http://www.fs.usda.gov/main/iitf/about
90. World Bank Group. "World Bank Group Archivists' Chronology 1944–2013." Accessed November 23, 2016. http://pubdocs.worldbank.org/en/186241442500110286/PDF-World-Bank-Group-Archivists-Chronology-1944-2013.pdf
91. Food and Agriculture Organization of the United Nations. "History." November 23, 2016. http://www.fao.org/world-food-day/2016/history/en/
92. United Nations. "History of the United Nations." Accessed November 23, 2016. http://www.un.org/en/sections/history/history-united-nations/index.html
93. Martin, *On the Edge*, 7.
94. KfW. "KfW Turns 65." Last modified November 11, 2013. https://www.kfw.de/KfW-Group/Newsroom/Aktuelles/Pressemitteilungen/Pressemitteilungen-Details_171713.html
95. NASA. "History Home." Accessed November 28, 2016. http://history.nasa.gov/
96. Inter-American Development Bank. "History of the Inter-American Development Bank." Accessed November 23, 2016. http://www.iadb.org/en/about-us/history-of-the-inter-american-development-bank,5999.html

97. The Organisation for Economic Co-operation and Development. "History." Accessed November 23, 2016. http://www.oecd.org/about/history/
98. United States Agency for International Development. "USAID History." Accessed November 23, 2016. https://www.usaid.gov/who-we-are/usaid-history
99. African Development Bank. "History." Accessed November 23, 2016. http://www.afdb.org/en/about-us/corporate-information/history/
100. ONF International. "The ONF International Group." Accessed November 23, 2016. http://www.onfinternational.org/en/qui-sommes-nous/the-onf-international-group
101. United Nations Development Programme. "Frequently Asked Questions." Accessed November 23, 2016. http://www.undp.org/content/undp/en/home/operations/about_us/frequently_askedquestions.html#being
102. INPE. "History." Accessed November 28, 2016. http://www.inpe.br/ingles/institutional/about_inpe/history.php
103. Asian Development Bank. "Origins." Accessed November 23, 2016. http://www.adb.org/about/history
104. United Nations Environment Programme. "Milestones." Accessed November 23, 2016. http://www.unep.org/Documents.multilingual/Default.asp?DocumentID=287
105. Myers. *The Shrinking Ark*. 246.
106. International Fund for Agricultural Development. "Who We Are." Accessed March 21, 2017. https://www.ifad.org/who/overview
107. International Tropical Timber Organization. "About ITTO." Accessed November 23, 2016. http://www.itto.int/about_itto/
108. National Agroforestry Center. "About the NAC." Accessed November 23, 2016. http://nac.unl.edu/#about
109. European Bank for Reconstruction and Development. "History of the EBRD." Accessed November 23, 2016. http://www.ebrd.com/who-we-are/history-of-the-ebrd.html
110. World Bank Group. "The World Bank Group's Global Environment Facility Program." Last modified May 21, 2014. http://www.worldbank.org/en/topic/climatechange/brief/gef
111. World Trade Organization. "The Uruguay Round." Accessed November 23, 2016. https://www.wto.org/english/thewto_e/whatis_e/tif_e/fact5_e.htm
112. World Trade Organization. "20th Anniversary of the WTO." Accessed November 23, 2016. https://www.wto.org/english/thewto_e/20y_e/20y_e.htm

113. UN Economic and Social Development. "UN Forum on Forests." Accessed April 13, 2017. http://www.un.org/esa/forests/forum/index.html
114. UN Millennium Project. "What They Are." Accessed November 23, 2016. http://www.unmillenniumproject.org/goals/
115. Division for Sustainable Development. "Mission Statement." Accessed November 23, 2016. https://sustainabledevelopment.un.org/about
116. Martin, *On the Edge*, 190.
117. The World Bank's Carbon Finance Unit. "BioCarbon Fund." Accessed April 11, 2017. https://wbcarbonfinance.org/Router.cfm?Page=BioCF&FID=9708&ItemID=9708
118. COMIFAC. "Traité constitutif." Accessed November 28, 2016. http://www.comifac.org/en/node/28
119. World Wildlife Fund. "The Heart of Borneo Declaration." Accessed November 28, 2016. http://wwf.panda.org/what_we_do/where_we_work/borneo_forests/about_borneo_forests/declaration.cfm
120. Norwegian Agency for Development Cooperation. "Real-Time Evaluation of Norway's International Climate and Forest Initiative: Contribution to Measurement, Reporting and Verification." Accessed November 23, 2016. https://www.norad.no/globalassets/import-2162015-80434-am/www.norad.no-ny/filarkiv/vedlegg-til-publikasjoner/real-time-evaluation-of-norways-international-climate-and-forest-initiative-contribution-to-measurementreporting-and-verification.pdf
121. The REDD Desk. "Forest Carbon Partnership Facility." Accessed April 2, 2017. http://theredddesk.org/countries/actors/forest-carbon-partnership-facility
122. UN-REDD Programme. "About the UN-REDD Programme." Accessed November 23, 2016. http://www.un-redd.org/
123. Governors' Climate & Forests Task Force. "GCF Overview." Accessed November 23, 2016. http://www.gcftaskforce.org/about/gcf_overview/
124. United States Agency for International Development. "Forest Carbon, Markets and Communities (FCMC) Program: Semi-Annual Report (March–October 2011)." Accessed November 23, 2016. http://pdf.usaid.gov/pdf_docs/PA00JKG4.pdf
125. GIZ. "REDD Early Movers." Accessed November 23, 2016. https://www.giz.de/en/worldwide/33356.html

5

International Finance

"To bring a conservation dream to reality, a creative dealmaker may need to consider options that include revenues from working farms and forests; land sales associated with limited development opportunities; and the monetization of water, wildlife, and carbon-related ecosystem services."[1]
—Kim Elliman and Peter Howell

Introduction

According to the World Bank, the size of the world's Gross Domestic Product (GDP), in nominal terms, is approximately USD$73.5 trillion.[2] In 2015, the largest national economies in the world based on GDP were as follows (Table 5.1).[3]

This global GDP of USD$73.5 trillion is primarily derived from household consumption (57.1%), along with by government consumption (16.2%), fixed capital investments (25.4%), and investments in inventories (0.7%). While the services (62.6%) and industry sectors (30.4%) are the largest sectors contributing to GDP, followed by agriculture (6.4%), it is important to understand that the agricultural sector is still a very

B.J. McFarland, *Conservation of Tropical Rainforests*, Palgrave Studies in Environmental Policy and Regulation, https://doi.org/10.1007/978-3-319-63236-0_5

Table 5.1 Largest national economies in the world

	Country	GDP (trillion USD$)		Country	GDP (trillion USD$)
1	United States	$17.947	11	South Korea	$1.378
2	China	$10.866	12	Australia	$1.340
3	Japan	$4.123	13	Russia	$1.326
4	Germany	$3.356	14	Spain	$1.199
5	United Kingdom	$2.849	15	Mexico	$1.144
6	France	$2.422	16	Indonesia	$0.862
7	India	$2.074	17	Netherlands	$0.753
8	Italy	$1.815	18	Turkey	$0.718
9	Brazil	$1.775	19	Switzerland	$0.665
10	Canada	$1.551	20	Saudi Arabia	$0.646

Source: The World Bank

large employer (34.4%) when compared to industry (22.2%). The services sector, as of 2011, is estimated to be the largest employment sector (43.4%).[4]

With this context in mind, as of January 2016, "about $52 billion per year flows to conservation projects {not just tropical rainforest conservation projects}, the bulk of it is public and philanthropic funds. The best estimates are that $300 to $400 billion per year is needed to preserve healthy ecosystems on land and in the oceans, and with them the earth's natural capital stock of clean air, fresh water and species biodiversity."[5] Similarly, as stated by the former Commission on Climate and Tropical Forests, approximately USD$30 billion a year "will be required to reduce emissions from tropical deforestation by half in 2020."[6]

Going forward, "{…} to finance the preservation of the world's precious ecosystems will require $200 billion to $300 billion in additional capital, and private investment capital may be the main source of additional capital. Attracting such a level of private capital will require attractive risk-adjusted rates of return, in addition to clear and measurable conservation impacts."[7]

As reported in the *State of Private Investment in Conservation 2016*, approximately USD$8.2 billion was invested by the private sector between 2004 and 2015 in the following "three groups of conservation-oriented investments:

- Sustainable food and fiber production (e.g., forestry, agriculture, and fisheries);
- Habitat conservation (e.g., mitigation banking and forest carbon trading); and
- Water quality and quantity protection (e.g., watershed protection and water rights trading)."[8]

The additional USD$200 to USD$300 billion required to conserve the world's precious ecosystems, while a very large sum, "only" represents approximately 0.27–0.41% of the global USD$73.5 trillion economy. A small international currency trading tax (known as the Tobin tax), the implementation of a global carbon price, a USD$0.01 tax on each cup of coffee sold in the United States, or the scaling up of investable conservation deals could raise the necessary financing to conserve the last remaining tropical rainforests.

In contrast, the total global value of the food and agricultural industry—which is the largest contributor to tropical rainforest deforestation and degradation—is estimated at approximately USD$7.8 trillion as of 2015.[9]

Yet, consumers annually demonstrate their interest in animals and travel. For instance, the overall "travel and tourism industry is the largest business sector in the world, accounting for more than 230 million jobs and 10 percent of the global economy. The $6.5 trillion industry is one of the top five export earners in 80 percent of the world's countries and the top export earner in 60 of them."[10] The size of the United States' pet supplies industry was estimated at USD$62.75 billion in 2016.[11] Zoos and aquariums accredited by the Association of Zoos & Aquariums (AZA) "contributed more than $17.4 billion to the US economy in 2014."[12]

Between theater and home market sales, the gross revenue generated from the movies Rio 1 (2011)[13] and Rio 2 (2014)[14]—an animated movie based in Brazil which touched upon themes such as wildlife smuggling of parrots and tropical rainforest deforestation—topped more than USD$1.1 billion.

Before explaining the financial framework to be used to analyze the various conservation financing instruments, the following will briefly review some key terms.

Capital Markets

Capital markets, whether via the primary or secondary market, are where equities and bonds are traded in order to raise financing.

Equities and Bonds

In general, equities are shares in the stock issued by publicly traded companies or shares in a private company, whereas bonds are debt instruments. Bonds can be issued by governments or companies. Typically, a bond's principal is paid back at the end of the bond's maturity, and intermittent coupon payments are made to the investor(s).

Short-Term and Long-Term Debt

In addition to bonds, there are other forms of debt such as traditional bank loans and notes. Oftentimes, conservation organizations will utilize short-term loans, as opposed to long-term loans:

> Short-term loans can also pay for the up-front costs of a transaction, such as options, earnest money, or down payments, or to cover legal, appraisal, or planning costs. Restoration projects often have retention requirements (withholding of money until the project is completed), which hinder cash flow for the project.[15]

Long-term debt could be raised to finance education centers or offices,[16] along with establishing ecotourism operations.

Many conservation organizations have also established internal and external revolving loan funds, since it can be more challenging to take out a loan from a typical bank that may be unfamiliar with conservation.

> An internal fund, such as that of The Nature Conservancy (TNC), recirculates money within a large organization. OSI {Open Space Institute}, on the other hand, uses external revolving funds, which lend money to outside organizations seeking funding for conservation projects.[17]

Currencies

The value of a currency relative to another currency is constantly changing, and this can be particularly challenging in developing countries where currencies exhibit more volatility relative to currencies in developed countries such as the US dollar, Euro, or Japanese yen.

For instance, USD$1 million on March 23, 2012, was equivalent to R$1.82 million (Brazilian reais).[18] This same USD$1 million, 3 years later, on March 23, 2015, was equivalent to R$3.16 million (73.63% increase).[19] Similarly, €1 million on October 21, 2011, was equivalent to Rp12.29 billion (Indonesia rupiah).[20] Five years later, this same €1 million on October 21, 2016, was equivalent to Rp14.18 billion (15.38% increase).[21] Furthermore, £1 million on March 18, 2014, was equivalent to CDF1,515.27 (Congolese franc). Three years later, this same £1 million was equivalent to CDF1,641.01 (8.30% increase).

Currency movements, such as these examples, can impact investment decisions and directly impact conservation activities by making alternative land use activities, such as cattle ranching or timber extraction, more or less profitable.

Interest Rates

Interest rates are the rates charged to borrow money. Interest rates are usually pegged to the London Interbank Offered Rate (LIBOR) or to the US Prime Rate and then adjusted based on the borrower's creditworthiness. Due to increased risk, countries that host tropical rainforests and the companies that operate within these countries tend to have higher interest rates and thus a higher cost to borrow.

Project Finance

Project finance involves structuring the finance of a project where the project's cash flows are used to pay back investors and where there is no recourse against the project sponsor's corporate assets. Project finance is often used for large infrastructure projects.

Public Finance

Public finance is funding provided by a government, whether it be a town, city, municipality, or federal government. This public finance could be raised via taxation to pay for the establishment and maintenance of a national park or the issuance of a bond by a city to upgrade the public water supply and restoration of an upstream watershed.

Trade Finance

Trade finance is the funding provided for trade. In addition to direct financing for trade, there is also specialized contracting and insurance provided for such trade. Trade finance is often provided for the international trade of commodities such as cattle, soy, palm oil, timber, and pulp paper. This said, there are soft agricultural commodities (e.g., bananas, beef, coffee, corn, soy, and wheat), and there are hard commodities (e.g., diamonds, gold, and petroleum).

Financial Intermediaries

Financial intermediaries are market makers that connect borrowers with lenders. Such financial intermediaries include banks, traders, and brokers.

> Banks, especially local banks, are the most attractive commercial retail lenders. For land trusts, they have three big advantages: they are highly motivated to lend to people they know and trust, they are present in most midsize or larger communities, and they are often willing to take real estate (assuming it has sufficient value and meets certain other criteria) as collateral.[22]
>
> There are three potential problems with banks for land trusts: (1) because banks are commercial, profit-making ventures, they aren't always interested in lending on raw land, the liquidity of which might be less assured; (2) their lending rates can be expensive; and (3) unlike lenders with conservation as part of their mission, including certain foundations and conservation revolving funds, banks are not always as flexible and accommodating to land conservation borrowers.[23]

Cash Flows

Cash flows involve the sources, quantity, and timing of both revenue and expenses related to a particular project, program, or portfolio. Challenges can arise when there is a mismatch between cash outflows and cash inflows.

Conservation Finance

Conservation finance is a subset field of finance that involves deploying funds for the conservation and restoration of ecosystems. Conservation finance can come from public or private sources and can be structured in a variety of ways including project finance or through a debt offering. It is important to note that "some techniques work better for raising operational revenue, such as annual fundraising. Others are better for acquisition (e.g., capital campaigns) or stewardship (e.g., transfer fees)."[24]

Financial Frameworks

There are 3 financial frameworks that will be used to analyze each of the conservation financing instruments that will be discussed throughout the book. These 3 financial frameworks are the efficient frontier; the Global Association of Risk Professionals' Framework for Financial Risk Management; and the Risk, Return, Time (Horizon), Taxes, Liquidity, Legal and Unique (RRTTLLU) framework for Portfolio Management.

A common financial calculation that is conducted on investments is a basic review of expected revenue and expenses. More complex financial calculations such as Net Present Value (NPV), Internal Rate of Return (IRR), payback periods, and sensitivity analyses can also be performed.

These calculations can be important when making conservation investment decisions. For instance: If you need to raise USD$1 million in 5 years from now to launch a new protected area, with a 5% annual interest rate, how much do you need to raise each year starting at the end of year 1 until the end of year 5? The following assumes at the end of each year, you deposit an equal amount into a fund that returns 5% a year, compounded annually, and then you will liquidate the full USD$1 million starting day 1 of year 6. The answer is approximately USD$180,975. Similarly, if you need USD$10 million each year fixed for perpetuity to manage protected areas throughout the Congo Basin, what size of an endowment—with 7% annual returns—do you need to have? This example assumes you will not be raising additional funds. The answer is approximately USD$142,857,143.

Efficient Frontier

The efficient frontier is a conceptual framework to align the expected return of an investment with an appropriate level of risk. In contrast, an inefficient investment is an investment with a higher level of risk than the corresponding expected return warrants. Thus to make up for this deficiency, an investment can seek ways to reduce risk (e.g., through internally mitigating the risk or purchasing an insurance product) and/or increase the expected return (e.g., reducing costs, diversifying revenue, or through expanding sales) to reach the efficient frontier (Fig. 5.1).

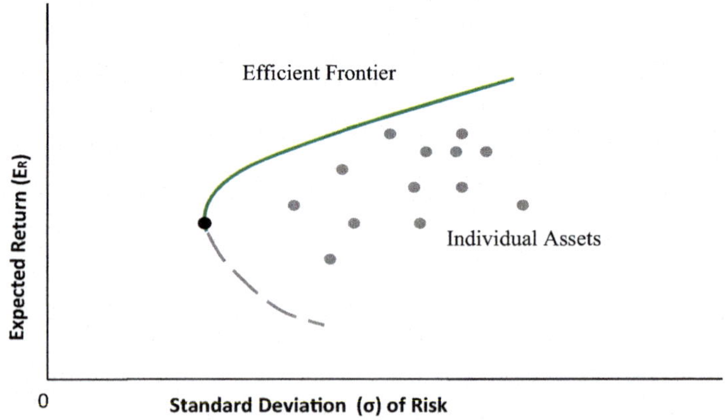

Fig. 5.1 Efficient frontier (Credit: Brian Joseph McFarland)

Financial Risk Management

The Global Association of Risk Professionals' (GARP) Framework for Financial Risk Managers (FRM) is one of the leading frameworks for identifying and managing risk. Risks are inserted into thematic categories. These categories are business, credit, reputation, liquidity, operational, strategic, market, and regulatory or legal risks.

This risk management framework allows analysts to assess financial instruments and their possible impacts over time. In addition, this assessment can inform decision-making for project design. Similar to the RRTTLLU framework (see below), this risk model allows for the integration of characteristics such as land use change, biodiversity loss, hydrological systems, and deforestation, along with human and labor rights into financial terms that may fit at times within financial models.

For example, the GARP thematic model shown below feeds into enterprise risk management tools that can be used to manage and model the overall risk of an organization. In this case, Environmental, Social and Governance (ESG) characteristics of an investment—including deforestation, human rights, and other risks—can be included within each risk category. These ESG risks often have direct operational risks which then result in indirect business, credit, reputation, liquidity, strategic, market, and regulatory or legal risks (Fig. 5.2).

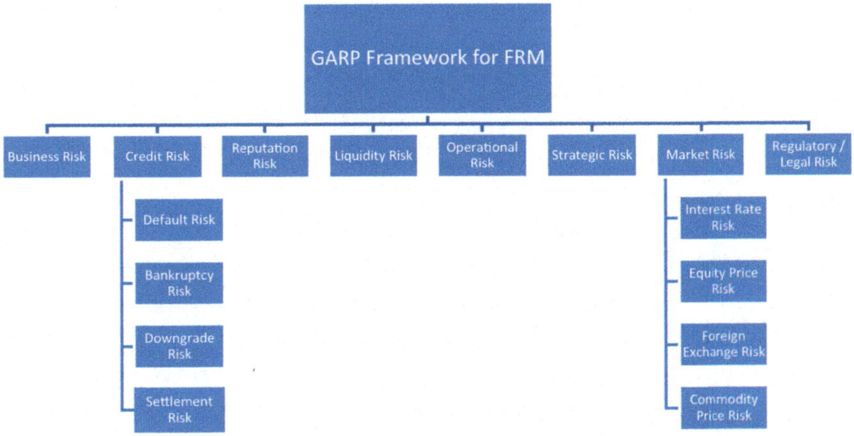

Fig. 5.2 GARP Framework for FRM (Credit: Brian Joseph McFarland)

Portfolio Management: Risk, Return, Time (Horizon), Tax, Liquidity, Legal, and Unique

To develop an Investment Policy Statement (IPS), the Risk, Return, Time (Horizon), Taxes, Liquidity, Legal and Unique (RRTTLLU) framework can be utilized.[25] This RRTTLLU complements the efficient frontier and the GARP Framework.

In the RRTTLLU framework, analysts review 5 constraints (time horizon, taxes, liquidity, legal, and unique) to inform both return and risk objectives. Conservation finance is informed by this RRTTLLU process of categorizing information, even though most in the field might not use this acronym. When the possible financial structure is described using risk and return objectives informed by the 5 constraints, it then fits inside the efficient frontier model. The following figure outlines the RRTTLLU process for a REDD+ offset project (Fig. 5.3)[26]:

Category	Metric	Example of Metric	Discussion
Return	Percentage	Internal rate of return	Describe required IRR
Risk	Percentage (can be based on standard deviation)	1. Likelihood of a default (lender, borrower, proponent, community, or purchaser); 2. Change in policy or law; 3. Environmental factors.	Effective risk management lowers cost of capital for project.
Time Horizon	Years	1. Duration of project; 2. Tenure of forest carbon offset; 3. Time between validation and verification of carbon offset project.	Financing needs to be managed to all time horizon criteria.
Taxes	Percentage	1. Tax assets and liabilities (for all counterparties); 2. Include income, social security, healthcare, value-added, property, deeds and carbon licensing.	Tax assets and liabilities impacts must be calculated.
Liquidity	Currency	1. Regular operating costs (e.g., MRV, salaries and profit-sharing); 2. Contingency; 3. One-time expenses (e.g., feasibility studies, baseline inventory, and land titling review).	Project viability includes sufficient income streams to cover liquidity constraints.
Legal	Contract	1. Emissions Reduction Purchase Agreement (ERPA); 2. Free Prior and Informed Consent (FPIC); 3. Local service and management contracts.	Includes clarity on rights and responsibilities, economic distribution models, schedules, and profit-sharing ladders, pathways for recourse, and description of process for settlement and clearing.
Unique	(No universal metric)	1. Community co-benefits; 2. Biodiversity factors; 3. Carbon buffer pool determination.	Discussion of co-benefits and unique risks presented.

Fig. 5.3 RRTTLLU for REDD+ projects (Credit: Gabriel Thoumi and John Waugh)

Other Resources on International Finance

Biodiversity Finance Initiative (BIOFIN) from UNDP

- http://www.biodiversityfinance.net/

Chain Reaction Research

- https://chainreactionresearch.com/

Earth Genome

- https://www.earthgenome.org/overview/

Global Footprint Network's Ecological Wealth of Nations

- http://www.footprintnetwork.org/content/documents/ecological_footprint_nations/

Global Green Economy Index 2016

- http://dualcitizeninc.com/GGEI-2016.pdf

Global Landscapes Forum: The Investment Case—A 2016 Donor and Partner Report

- http://www.landscapes.org/wp-content/uploads/docs/GLF-2016-London/GLF-London-Donor-and-Partner-Report-2016.pdf

Impact Alpha's Conservation Finance

- http://impactalpha.com/category/conservation/
- http://impactalpha.com/new-products-attract-investors-to-the-conservation-finance-marketplace/

Instrumentl

- https://www.instrumentl.com/

International Finance Corporation's Presentations and Factsheets

- http://www.ifc.org/wps/wcm/connect/corp_ext_content/ifc_external_corporate_site/about+ifc_new/ifc+governance/investor+relations/presentations+and+factsheets

The Forestry Hub

- http://match.thenode.co/

The Little Biodiversity Finance Book (3rd Edition)

- http://d6.globalcanopy.org/materials/little-biodiversity-finance-book

The Little Biodiversity Finance Book—Methodology Appendix

- http://d6.globalcanopy.org/materials/little-biodiversity-finance-book-methodology-appendix

The Little Climate Finance Book

- http://d6.globalcanopy.org/materials/little-climate-finance-book

The Little Forest Finance Book—14 Catalysts to Scale up Forest-Friendly Finance

- http://d6.globalcanopy.org/materials/little-forest-finance-book

The Nature Conservancy's Stewardship Calculator and Handbook

- http://www.conservationgateway.org/ConservationPlanning/ToolsData/Pages/stewardshipcalculator.aspx

Trust for Public Land, What Conservation Finance is Really About

- https://www.youtube.com/watch?v=PSVZsXEA8tw&feature=youtu.be

United Nations Environment Programme's Finance Initiative

- http://www.unepfi.org/

Yale University's Environmental Performance Index

- http://epi.yale.edu/data

Notes

1. Elliman, Kim and Peter Howell. "Anatomy of a Finance Program: Lessons from a Conservation Lender." In *Conservation Capital in the Americas*, edited by James N. Levitt. Massachusetts: Lincoln Institute of Land Policy, 2010. 157.
2. World Bank. "Gross Domestic Product 2015." Accessed November 23, 2016. http://databank.worldbank.org/data/download/GDP.pdf
3. Ibid.
4. The Central Intelligence Agency. "The World Factbook: The World: Economy." Accessed December 12, 2016. https://www.cia.gov/library/publications/the-world-factbook/geos/xx.html
5. Huwyler, Fabian, Jürg Käppeli, and John Tobin. "Conservation Finance: From Niche to Mainstream: The Building of an Institutional Asset

Class." Last modified January 2016. https://www.credit-suisse.com/media/assets/corporate/docs/about-us/responsibility/banking/conservation-finance-en.pdf. 3.

6. The Commission on Climate and Tropical Forests. "Protecting the Climate Forests: Why Reducing Tropical Deforestation is in America's Vital National Interest." October 2009. Accessed April 13, 2017. http://www.climateforestscommission.org/documents/cctf-report.pdf. 44.
7. Huwyler, Fabian, Jürg Käppeli, and John Tobin. "Conservation Finance: From Niche to Mainstream: The Building of an Institutional Asset Class." Last modified January 2016. https://www.credit-suisse.com/media/assets/corporate/docs/about-us/responsibility/banking/conservation-finance-en.pdf. 3.
8. Hamrick, Kelley. "State of Private Investment in Conservation 2016." *Ecosystem Marketplace*. December 2016. http://www.forest-trends.org/documents/files/doc_5474.pdf#. vii.
9. Plunkett Research, Inc. "Global Food Industry Statistics and Market Size Overview, Business and Industry Statistics." Accessed December 12, 2016. https://www.plunkettresearch.com/statistics/Industry-Statistics-Global-Food-Industry-Statistics-and-Market-Size-Overview/
10. Milder, Brian. "Conservation Finance in the Galápagos Islands: Capitalizing the 'Missing Middle'." In *Conservation Capital in the Americas*, edited by James N. Levitt. Massachusetts: Lincoln Institute of Land Policy, 2010. 110.
11. American Pet Products Association. "Pet Industry Market Size & Ownership Statistics." Accessed February 9, 2017. http://www.americanpetproducts.org/press_industrytrends.asp
12. Association of Zoos & Aquariums. "Zoo and Aquarium Statistics." Accessed February 9, 2017. https://www.aza.org/zoo-and-aquarium-statistics
13. The Numbers. "Rio 1 (2011)." Accessed February 9, 2017. http://www.the-numbers.com/movie/Rio#tab=more
14. The Numbers. "Rio 2 (2014)." Accessed February 9, 2017. http://www.the-numbers.com/movie/Rio-2#tab=more
15. Clark, Story. *A Field Guide to Conservation Finance*. 184.
16. Clark, Story. *A Field Guide to Conservation Finance*. 185.
17. Elliman, Kim and Peter Howell. "Anatomy of a Finance Program: Lessons from a Conservation Lender." In *Conservation Capital in the Americas*, edited by James N. Levitt. Massachusetts: Lincoln Institute of Land Policy, 2010. 148.

18. XE. "Current and Historical Rate Tables." Accessed November 30, 2016. http://www.xe.com/currencytables/?from=USD&date=2012-03-23
19. XE. "Current and Historical Rate Tables." Accessed November 30, 2016. http://www.xe.com/currencytables/?from=USD&date=2015-03-23
20. XE. "Current and Historical Rate Tables." Accessed November 30, 2016. http://www.xe.com/currencytables/?from=EUR&date=2011-10-21
21. XE. "Current and Historical Rate Tables." Accessed November 30, 2016. http://www.xe.com/currencytables/?from=EUR&date=2016-10-21
22. Clark, Story. *A Field Guide to Conservation Finance*. 191.
23. Clark, Story. *A Field Guide to Conservation Finance*. 192.
24. Clark, Story. *A Field Guide to Conservation Finance*. 94.
25. Vanguard Asset Management. "Categorise and Evaluate Data: The RRTTLLU Framework." Accessed December 12, 2016. https://www.vanguard.co.uk/documents/adv/literature/advice_process/categorise-and-evaluate-data.pdf
26. Thoumi, Gabriel and John Waugh. "5.4 Industry-Level Frameworks to Improve Access to REDD+ Financing." *ETFRN News* 54. December 2012. Accessed March 22, 2017. www.etfrn.org/file.php/57/5.4thoumi-waugh.pdf. 182.

6

The Origins and History of Conservation Finance

"The first initiatives to protect tropical forests were not driven by a concern for their biodiversity, but by the desire to establish a 'permanent forest estate' to preserve its timber value. In the nineteenth and early twentieth centuries, colonial governments introduced the first forest management legislation, and they established forest departments, at least in those countries whose timber supplies were important to the European market."[1]
—Dr. Claude Martin

Conservation finance, whether from public or private sources, is a subset field of finance that involves deploying funds for the conservation and restoration of ecosystems. The exact date when conservation finance for tropical rainforests began is a bit elusive.

In 1634, the Boston Commons was established and is considered "the oldest public open space in the United States. Indeed, it is very likely the oldest public open space in the English-speaking world."[2] The Bogd Khan Uul National Park in Mongolia was established "by the Mongolian government in 1778, {but} it was originally chartered by Ming Dynasty officials in the 1500s as an area to be kept off limits to extractive uses, protected for its beauty and sacred nature."[3]

B.J. McFarland, *Conservation of Tropical Rainforests*, Palgrave Studies in Environmental Policy and Regulation, https://doi.org/10.1007/978-3-319-63236-0_6

The first documented protected area, according to the World Database on Protected Areas, appears to be a nature park in Russia known as the Donskoj rybny zapovednik (Donskoe zapretnoe rybnoe prostranstvo) which was established in 1819.[4]

The first state park appears to be Yosemite National Park in California, United States, which was first designated as a state park in 1864,[5] while the first national park—aside from the Bogd Khan Uul National Park—appears to be Yellowstone National Park in the current US states of Wyoming, Montana, and Idaho in 1872.[6]

In 1891, the Trustees of Public Reservations (now the Trustees of Reservations, or TTOR) became the world's first regional land trust,[7] which established the world's first conservation easement in Boston, Massachusetts.[8]

It is difficult to discern when the first government international budgetary expenditures for conservation were allocated. However, it is likely this mechanism was first established in the late nineteenth or early twentieth century, when the world's first international conservation organizations were established. This includes the Society for the Preservation of the Wild Fauna of the Empire (now known as Fauna and Flora International or FFI), which was founded in 1903,[9] and the International Council for Bird Preservation (ICBP; the ICBP would rebrand as BirdLife International in 1993) which was established in 1922.[10] Waterton Glacier International Peace Park was established in 1932 between the United States and Canada and marked the world's first international peace park.[11]

Today in additional to more traditional national parks,

> {…} conservationists are experimenting with new ways to use private capital to protect land and water. They are developing markets for ecological services such as carbon, clean water, wetlands and more: creating revenue streams by sustainably harvesting timber; restoring and enhancing areas that add value and generate return from creational markets; developing local markets for locally grown products; finding conservation-minded buyers to purchase critical areas and protect them; creating conservation developments that truly preserve critical areas and allow very limited development; and teaming up with new and surprising partners from diverse sectors that are helping to capitalize conservation.[12]

Modern-day impact investing is comparable to socially responsible investing, which took "{…} off in the 1970s as part of grassroots campaigns to boycott firms in South Africa {as a result of Apartheid} as well as the nuclear weapons and tobacco industries."[13]

In the late 1970s, the concept of payments for ecosystem services (PES, referred to earlier as payments for environmental services) appears.[14]

Low-impact tourism likely began earlier, but in 1983, the term "ecotourism" is first coined by Mexican architect Héctor Ceballos-Lascuráin.[15]

Debt-for-nature swaps came along shortly thereafter with the first such swaps conducted in 1987.[16]

It appears that in 1991, the world's first major bioprospecting deal was struck between Merck & Co. and Costa Rica's Instituto Nacional de Biodiversidad (INBio).[17]

The sustainable, green procurement of locally produced commodities—particularly by Indigenous Peoples and forest-dependent communities—has existed for thousands of years. However, the origin of commercial, green procurement with independent certification standards likely started with the creation of the Forest Stewardship Council (FSC)[18] and the Sustainable Forestry Initiative (SFI) in 1994.[19]

The first widely recognized green bonds were issued in 2007 and 2008 by the World Bank[20] and European Investment Bank.[21]

In addition to the historical development of conservation financing instruments, it is also important to understand the historical development of other trends, such as population growth, cultural changes, migration of populations from rural to urban centers, changes in diets, and the emergence of technology which made measuring and monitoring tropical rainforests more possible.

- 1688–1715 (approximately): Dodo becomes extinct.[22]
- 1799: "When the great scholar and explorer Alexander von Humboldt embarked on his five-year expedition to Central and South America in 1799, accompanied by the botanist Aimé Bonpland, no European had penetrated the deep Amazonian forests before them."[23]
- 1800 (approximately): World population reaches 1 billion people.[24]
- 1848: Henry Walter Bates and Alfred Russel Wallace begin their exploration of the Brazilian Amazon.[25]
- 1859: Charles Darwin's *On the Origin of Species by Means of Natural Selection* is published.[26]

- 1869: "Westerners first learned of the Giant Panda in 1869 when French missionary Armand David, while in China, laid eyes on a distinctive black-and-white pelt and then bought a complete, dead specimen from local hunters."[27]
- 1898: "The term 'rainforest' goes back to the term 'tropischer Regenwald' first used by the German botanist A.F.W. Schimper in his Plant-Geography upon a Physiological Basis of 1898. But it was only in 1952 that P.W. Richards, in his momentous book The Tropical Rain Forest, made the term commonly known."[28]
- 1902: Mountain gorillas were discovered on "the ridges of the volcanic Virunga Mountains by German explorer Captain Robert von Beringe."[29]
- 1903–1990: Indonesia undertakes Transmigration Program.[30]
- 1913–1914: Theodore Roosevelt's Amazon expedition with Cândido Rondon.[31]
- 1914–1918: World War I.[32]
- 1918–1919: The "Spanish" influenza pandemic kills an estimated 50 million people.[33]
- 1930: World population reaches 2 billion people.[34]
- September 1, 1939, to September 2, 1945: World War II.[35]
- 1952: P.W. Richards wrote *The Tropical Rain Forest*.[36]
- 1959: World population reaches 3 billion people.[37]
- 1960s: Many new countries, particularly in Africa, gain independence including the DRC in 1960 and Cameroon in 1960.[38]
- 1960: Jane Goodall first discovers the use of tools among chimpanzees.[39]
- 1962: Rachel Carson's *Silent Spring* is first published.[40]
- 1967: Robert H. MacArthur and Edward O. Wilson first published *The Theory of Island Biogeography*.[41]
- 1968: The "term 'biological diversity' was coined by the conservation biologist Raymond F. Dasmann."[42]
- 1968: "While studying pollution control in the Great Lakes, University of Toronto economist John Dales hit on a way for the costs to be paid with minimal government intervention, by using tradable permits or allowances."[43]
- 1973–1974: Oil embargoes

- 1970: First Earth Day (April 22) is founded by US Senator Gaylord Nelson.[44]
- 1970: The Trans-Amazonia Highway has begun.[45]
- 1971: Dr. Seuss' *The Lorax* is first published.[46]
- 1972: *The Limits to Growth* is first published.[47]
- 1972–1973: The first "imagery from NASA Landsat satellites had already become available in 1972–1973)."[48]
- 1974: World population reaches 4 billion people.[49]
- 1976: The "term 'tropical moist forest' was introduced by Adrian Sommer as he attempted to assess the extent of these forests."[50]
- 1978: The "FAO and the United Nations Environment Programme (UNEP) started working on the Tropical Forest Resources Assessment Project {…}."[51]
- 1987: *Our Common Future* is published.[52]
- 1987: World population reaches 5 billion people.[53]
- 1988: The "{…} Brazilian government initiated the Legal Amazon Deforestation Monitoring Project (PRODES, Projeto de Monitoramento do Desflorestamento na Amazônia Legal), the world's most extensive national forest monitoring program."[54]
- 1990s: "Soy expansion in South America started in the early 1990s with the rapid increase of meat consumption and the demand for protein-rich feed for chickens, pigs, and cattle. Fish meal, an important cattle feed, was no longer available after the collapse of the Peruvian anchovy fisheries."[55]
- 1996: Jonathan Harr's *A Civil Action* is first published.[56]
- 1998: "When the *Washington Post* staff writer Joby Warrick published an article in 1998 under the title 'Mass extinction underway, majority of biologists say,' it triggered a huge wave of reactions. To date, the Mysterium.com website has accumulated over 300 links to authoritative reports and articles with evidence that a mass extinction is beginning."[57]
- 1999: World population reaches 6 billion people.[58]
- 2006: An Inconvenient Truth is released.[59]
- 2006: Sir Nicholas Stern's *Stern Review on the Economics of Climate Change* is released.[60]
- 2009: Research on the nine planetary boundaries is first presented: (1) stratospheric ozone depletion, (2) loss of biosphere integrity (func-

tional and genetic diversity), (3) chemical pollution and the release of novel entities, (4) climate change, (5) ocean acidification, (6) freshwater consumption and the global hydrological cycle, (7) land system change, (8) biogeochemical flows including nitrogen and phosphorus, and (9) atmospheric aerosol loading.[61]

- 2011: World population reaches 7 billion people.[62]
- 2012: Natural Capital Declaration launched at Rio+20 conference.[63]
- 2014: Years of Living Dangerously, a documentary event series, is released.[64]
- 2015: Pope Francis' Encyclical on the Environment is presented.[65]
- 2017: An Inconvenient Sequel: Truth to Power is released.[66]

Other Resources on the Origins and History of Conservation Finance

Conservation Finance Alliance

- http://www.conservationfinancealliance.org/ (new website)
- http://www.conservationfinance.org/ (old website)

Conservation Finance Network

- http://www.conservationfinancenetwork.org/

Conservation Finance Network's Conservation Finance 101

- http://www.conservationfinancenetwork.org/conservation-finance-101

Conservation Finance Network's Conservation Finance Video Playlists

- http://www.conservationfinancenetwork.org/2016/01/12/explore-our-conservation-finance-video-playlists

Conservation International's What on Earth is 'Conservation Finance?'

- http://blog.conservation.org/2016/12/what-is-conservation-finance/

E&E News' HOTCO_2MMODITY

- http://www.eenews.net/special_reports/hot_commodity

TEDxBeaconStreet: Innovate to Meet the Challenge of Conservation by Jim Levitt

- http://www.tedxbeaconstreet.com/jim-levitt/

Notes

1. Martin. *On the Edge*. 177.
2. Levitt. *From Walden to Wall Street*. 4.
3. The Nature Conservancy. "The World's Oldest National Park: Ghosts of Monks and Red Deer." Last modified November 10, 2009. http://blog.nature.org/conservancy/2009/11/10/worlds-oldest-national-park-mongolia-nature-charles-bedford-bogdkhan/
4. IUCN and UNEP-WCMC. "The World Database on Protected Areas (WDPA) [Online]." December 13, 2016. Cambridge, UK: UNEP-WCMC, 2016. www.protectedplanet.net
5. California Department of Parks and Recreation. "Yosemite 'State Park'." Accessed November 23, 2016. http://www.150.parks.ca.gov/?page_id=27539
6. National Park Service. "Yellowstone National Park: Frequently Asked Questions: History." Accessed November 23, 2016. https://www.nps.gov/yell/learn/historyculture/history-faqs.htm
7. Foster, Dave. "Meeting the Conservation Challenge in New England." In *Conservation Capital in the Americas*, edited by James N. Levitt. Massachusetts: Lincoln Institute of Land Policy, 2010. 20–21.

8. The Nature Conservancy. "Easements 101." Accessed December 13, 2016. https://www.nature.org/about-us/private-lands-conservation/conservation-easements/conservation-easements-101.pdf
9. Fauna & Flora International. "110 Years of Fauna & Flora International." Accessed November 23, 2016. http://www.fauna-flora.org/timeline/
10. BirdLife International. "Our History." Accessed November 23, 2016. http://www.birdlife.org/worldwide/partnership/our-history
11. UNESCO. "Waterton Glacier International Peace Park." Accessed December 13, 2016. http://whc.unesco.org/en/list/354
12. Clark, Story. *A Field Guide to Conservation Finance*. 331.
13. Daily, Gretchen C. and Katherine Ellison. *The New Economy of Nature*. 116.
14. Gómez-Baggethun, Erik, Rudolf de Groot, Pedro L. Lomas and Carlos Montes. "The History of Ecosystem Services in Economic Theory and Practice: From Early Notions to Markets and Payment Schemes." *Ecological Economics*. 69(6). April 2010. 1209–1218. doi:10.1016/j.ecolecon.2009.11.007; https://www.researchgate.net/publication/46490541_The_History_of_Ecosystem_Services_in_Economic_Theory_and_Practice_From_Early_Notions_to_Markets_and_Payment_Schemes
15. EcoClub. "International Ecotourism Monthly." Year 7. Issue 85. Accessed December 14, 2016. http://ecoclub.com/news/085.pdf. 2.
16. World Wildlife Fund. *Guide to Conservation Finance*. 41.
17. The North American Congress on Latin America. "Bioprospecting and Biopiracy in the Americas." Accessed December 13, 2016. http://nacla.org/news/bioprospecting-and-biopiracy-americas
18. Forest Stewardship Council. "Our History." Accessed November 23, 2016. https://ic.fsc.org/en/about-fsc/our-history
19. Sustainable Forestry Initiative. "Background." Accessed November 23, 2016. http://www.sfiprogram.org/about-us/basics-of-sfi/
20. The World Bank. "World Bank Green Bonds: Green Bond Issuances to Date." Accessed December 13, 2016. http://treasury.worldbank.org/cmd/htm/GreenBondIssuancesToDate.html
21. Climate Bonds Initiative. "History." Accessed December 13, 2016. https://www.climatebonds.net/market/history
22. Discover Wildlife. "The Greatest Wildlife Discoveries of All Time." Last modified February 20, 2012. http://www.discoverwildlife.com/animals/10-greatest-wildlife-discoveries-all-time?page=2
23. Martin. *On the Edge*. xx.
24. Worldometers. "World Population." Accessed November 23, 2016. http://www.worldometers.info/world-population/#pastfuture

25. Encyclopedia.com "Bates, Henry Walter." Accessed August 28, 2017. http://www.encyclopedia.com/people/literature-and-arts/english-literature-20th-cent-present-biographies/henry-walter-bates
26. HISTORY.com. "This Day in History: 1859: Origin of Species is Published." Accessed March 27, 2017. http://www.history.com/this-day-in-history/origin-of-species-is-published
27. National Geographic. "Who Discovered the Panda?" Last modified August 2016. http://www.nationalgeographic.com/magazine/2016/08/explore-panda-mania-history/
28. Martin. *On the Edge*. 10.
29. The Mountain Gorilla Conservation Fund. "Discovery." Accessed February 9, 2017. http://www.gorilla100.com/30-Discovery.html
30. The World Bank Group. "Transmigration in Indonesia." Accessed March 27, 2017. http://lnweb90.worldbank.org/oed/oeddoclib.nsf/DocUNIDViewForJavaSearch/4B8B0E01445D8351852567F5005D87B8
31. Almanac of Theodore Roosevelt. "Roosevelt—Rondon Scientific Expedition." Accessed November 23, 2016. http://www.theodore-roosevelt.com/trbrazil.html
32. Graham, John and Royde Smith. "World War I." *Encyclopedia Britannica*. Accessed November 29, 2016. https://www.britannica.com/event/World-War-I
33. Centers for Disease Control. "1918 Influenza: the Mother of All Pandemics." Accessed February 9, 2017. https://wwwnc.cdc.gov/eid/article/12/1/05-0979_article
34. Worldometers. "World Population." Accessed November 23, 2016. http://www.worldometers.info/world-population/#pastfuture
35. Graham, John and Royde Smith. "World War II." *Encyclopedia Britannica*. Accessed November 29, 2016. https://www.britannica.com/event/World-War-II
36. Cambridge University Press. "The Tropical Rain Forest." Accessed November 23, 2016. http://www.cambridge.org/us/academic/subjects/life-sciences/plant-science/tropical-rain-forest-ecological-study-2nd-edition
37. Worldometers. "World Population." Accessed November 23, 2016. http://www.worldometers.info/world-population/#pastfuture
38. About, Inc. "Independence or Birthday for Every Country." Accessed December 12, 2016. http://geography.about.com/od/lists/a/officiallist.htm
39. Jane Goodall Institute of Canada. "About Chimpanzees: Chimp Behavior." Accessed November 23, 2016. http://www.janegoodall.ca/about-chimp-behaviour-tool-use.php

40. RachelCarson.org. "Silent Spring." Accessed November 23, 2016. http://www.rachelcarson.org/SilentSpring.aspx
41. Princeton University Press. "The Theory of Island Biogeography." Accessed November 23, 2016. http://press.princeton.edu/titles/7051.html
42. Martin. *On the Edge*. 135
43. Smithsonian Magazine. "The Political History of Cap and Trade." Accessed November 23, 2016. http://www.smithsonianmag.com/air/the-political-history-of-cap-and-trade-34711212/?no-ist
44. Earth Day Network. "The History of Earth Day." Accessed November 23, 2016. http://www.earthday.org/about/the-history-of-earth-day/
45. Myers. *The Sinking Ark*. 138.
46. Random House. "Seuss Biography." Accessed November 23, 2016. http://m.seussville.com/biography.html
47. Club of Rome. "The Limits to Growth." Accessed November 23, 2016. http://www.clubofrome.org/report/the-limits-to-growth/
48. Martin. *On the Edge*. 13.
49. Worldometers. "World Population." Accessed November 23, 2016. http://www.worldometers.info/world-population/#pastfuture
50. Martin. *On the Edge*. 10.
51. Martin. *On the Edge*. 13.
52. Council on Foreign Relations. "Report of the World Commission on Environment and Development: Our Common Future (Brundtland Report)." Accessed November 23, 2016. http://www.cfr.org/economic-development/report-world-commission-environment-development-our-common-future-brundtland-report/p26349
53. Worldometers. "World Population." Accessed November 23, 2016. http://www.worldometers.info/world-population/#pastfuture
54. Martin. *On the Edge*. 33.
55. Martin. *On the Edge*. 97.
56. Penguin Random House. "A Civil Action." Accessed November 23, 2016. http://www.penguinrandomhouse.com/books/76093/a-civil-action-by-jonathan-harr/9780679772675/
57. Martin. *On the Edge*. 263.
58. Worldometers. "World Population." Accessed November 23, 2016. http://www.worldometers.info/world-population/#pastfuture
59. IMDb. "An Inconvenient Truth (2006)." Accessed March 29, 2017. http://www.imdb.com/title/tt0497116/

60. Cambridge University Press. "The Economics of Climate Change: The Stern Review." Accessed March 30, 2017. http://www.cambridge.org/catalogue/catalogue.asp?isbn=9780521700801
61. Stockholm Resilience Centre. "About the Research." Accessed April 11, 2017. http://www.stockholmresilience.org/research/planetary-boundaries/planetary-boundaries/about-the-research.html
62. Worldometers. "World Population." Accessed November 23, 2016. http://www.worldometers.info/world-population/#pastfuture
63. Natural Capital Finance Alliance. "The NCD and NCFA Story." Accessed December 13, 2016. http://www.naturalcapitalfinancealliance.org/about-the-natural-capital-declaration/
64. IMDb. "Years of Living Dangerously." Accessed March 29, 2017. http://www.imdb.com/title/tt2963070/?ref_=fn_al_tt_1
65. Vatican. "Laudato si'." Accessed November 23, 2016. http://w2.vatican.va/content/dam/francesco/pdf/encyclicals/documents/papa-francesco_20150524_enciclica-laudato-si_en.pdf
66. IMDb. "An Inconvenient Sequel: Truth to Power." Accessed March 29, 2017. http://www.imdb.com/title/tt6322922/

7

Government Domestic Budgetary Expenditures

"Government revenue allocations can be an important source of funding for biodiversity conservation. However, in many developing countries, where limited government budgets are needed to support basic human needs, direct government support of conservation is rare or inadequate. Fluctuations in international markets and poor economic conditions cause government revenues to vary year by year, resulting in insufficient and uncertain funding for conservation. By earmarking revenues collected through various fiscal instruments, governments can stabilize and even increase conservation allocations. Examples of such mechanisms include earmarked government taxes on tourism and on commodities such as gasoline, structured debt relief earmarked to conservation, and government bonds."[1]

—World Wildlife Fund

Introduction

Government domestic budgetary expenditures—whether coming from city, provincial, state, or national governments—are a financing mechanism for the conservation of tropical rainforests within local, state, and national parks, along with state and national forests. There are also

B.J. McFarland, *Conservation of Tropical Rainforests*, Palgrave Studies in Environmental Policy and Regulation, https://doi.org/10.1007/978-3-319-63236-0_7

extractive reserves, Indigenous Peoples' reserves, biosphere reserves, and ecological reserves which could be financed through these domestic budgetary expenditures.

Historical Overview

In 1634, the Boston Commons was established and is considered "the oldest public open space in the United States. Indeed, it is very likely the oldest public open space in the English-speaking world."[2] The Bogd Khan Uul National Park in Mongolia was established "by the Mongolian government in 1778, {but} it was originally chartered by Ming Dynasty officials in the 1500s as an area to be kept off limits to extractive uses, protected for its beauty and sacred nature."[3]

The first documented protected area, according to the World Database on Protected Areas, appears to be a nature park in Russia known as the Donskoj rybny zapovednik (Donskoe zapretnoe rybnoe prostranstvo) which was established in 1819.[4]

The first state park appears to be Yosemite National Park in California, United States, which was first designated as a state park in 1864. As explained by the California Department of Parks and Recreation:

> The State of California acquired the Yosemite Valley and the Mariposa Big Trees Grove Grant in 1864 due to the foresight of early preservationists like Mrs. Jessie Benton Fremont, Captain Israel Ward Raymond and California Senator John Conness. They initiated the theory of setting aside areas purely for the purpose of preservation and the enjoyment of the citizenry.
>
> When the junior California Senator Conness introduced the bill in the Senate on March 28, 1864, in beseeching his colleagues to vote, he stated that the bill's sole objective was the "preservation of the Yosemite Valley and the Big Tree Grove ... for the benefit of mankind." The bill breezed through the 38th Congress and was signed by President Abraham Lincoln on June 30, 1864.
>
> This historic legislation was the beginning of not only the California State Park system, but also the park notion of today. {...} With popularity come expenses, and the state of California refused numerous requests by the board of commissioners for funding to support the grant. John Muir

mounted a campaign in favor of the federal government's reacquiring the grant to be administered as part of the larger Yosemite National Park, and he had powerful backers. A campaign ensued to keep the grant under state control.

Muir gained the support of President Theodore Roosevelt in his effort to return the grant to federal control. Recession of the grant was approved by Congress, and President Roosevelt signed the bill June 11, 1906. The Yosemite Grant was returned to the federal government to become part of Yosemite National Park.

Although Yosemite and the Mariposa Grove are no longer owned by the state, early preservationists' idea hatched a mindset in 1864 that spread throughout the world: we must preserve places for their inherent value to humans and the creatures and features that define those places.[5]

The first national park appears to be Yellowstone National Park in the current US states of Wyoming, Montana, and Idaho in 1872. As explained by National Park Service:

Some sources list Hot Springs in Arkansas as the first national park. Set aside in 1832, forty years before Yellowstone was established in 1872, it was actually the nation's oldest national reservation, set aside to preserve and distribute a utilitarian resource (hot water), much like our present national forests. In 1921, an act of Congress established Hot Springs as a national park.

Yosemite became a park before Yellowstone, but as a state park. Disappointed with the results 26 years later in 1890, Congress made Yosemite one of three additional national parks, along with Sequoia and General Grant, now part of Kings Canyon. Mount Rainier followed in 1899.

As an older state park, Yosemite did have a strong influence on the founding of Yellowstone in 1872 because Congress actually used language in the state park act as a model. It's entirely possible that Congress may have preferred to make Yellowstone a state park in the same fashion as Yosemite, had it not been for an accident of geography that put it within three territorial boundaries. Arguments between Wyoming and Montana territories that year resulted in a decision to federalize Yellowstone.[6]

Yellowstone National Park was essentially carved out from vast federal territories in 1872. This occurred prior to the creation of the National

Park Service, which was established in 1916, and at the time the Park was under the management of the US Department of Interior. The Interior had a small appropriation and faced pressures on Yellowstone National Park from poaching, grazing, and unlawful development. The Secretary of the Interior asked the Secretary of the Army for assistance, and US Army appropriations were used to finance, develop, and manage Yellowstone National Park from 1886 to 1918. The US Army had adequate funding and would end up building facilities, making roads, patrolling the Park, and keeping visitors safe. In 1916, the National Park Service was established and the US Army remained at Yellowstone National Park for 2 more years. The first park rangers at Yellowstone National Park learned from Army officers. Nevertheless, there was not a lot of money flowing for a land management agency and to complement federal appropriations, there was additional philanthropic support in the 1910s and 1920s from John D. Rockefeller, Jr.[7]

While there were several unverified forest reserves established in Sri Lanka from 1875 to 1896 and the "first protected areas in Africa, the Hluhluwe, Umfolozi, and St. Lucia reserves in South Africa, were established in 1895,"[8] it appears the first protected area (i.e., which was designated a "restricted zone") in a developing country was the Cordillera Huayhuash in the Andes Mountains of Peru in 1902.[9]

The first officially established protected area of a tropical rainforest appears to be the Chior Wildlife Reserve in Peninsular Malaysia, which was designated in 1903.[10] Throughout the early to mid-twentieth century, few protected areas were established for tropical rainforests. Likewise:

> It was only from the 1970s that priority was given to establishing protected areas of tropical rainforest. With some rare exceptions of tropical rainforest national parks having been established earlier (for example the Taman Negara National Park in Peninsular Malaysia in 1939, or the Salonga National Park in the Democratic Republic of the Congo in 1956), the majority of the protected areas in the rainforest biome were established after 1970. This includes some of the most famous ones: the Taï National Park in Côte d'Ivoire in 1972, the Gunung Leuser National Park in Sumatra in 1980, and the Jaú National Park in the Brazilian state of Amazonas, also in 1980. But these first rainforest national parks, which all still exist, were not more than samples of the biome and primarily of scientific interest.[11]

Among the largest tropical rainforest countries: the first protected area in Brazil was the Reserva Biológica Estadual de Guaratiba (the Guaratiba State Biological Reserve in English) established in 1914; the first protected area in the DRC was the Salonga National Park established in 1956 (i.e., there is a non-verified hunting area in the DRC known as the Luama-Katanga possibly designated in 1905); and the first protected area in Indonesia was the Gua Nglirip Nature Reserve established in 1919 (i.e., there is a non-verified Arca Domas Nature Reserve possibly designated in 1913).[12]

Other notable moments in the history of domestic expenditures for conservation include:

- 1939: The Taman Negara National Park in Peninsular Malaysia is established.[13]
- 1943: Gombe Game Reserve is established. The Gombe Stream Research Center, made famous by Jane Goodall, was established in 1965.[14] The Gombe Game Reserve would be upgraded to the Gombe National Park in 1968.[15]
- 1956: The Salonga National Park in the Democratic Republic of the Congo is established.[16]
- 1960s: "Until the 1960s, protected areas in the Amazon were mainly established with limited objectives to protect a specific feature, such as a mountain or a set of waterfalls or caves. In the 1970s a significant public opinion movement in favor of conservation and sustainable development emerged, but the growth of land under protection was slow until the late 1980s. The first large protected areas of more than half a million hectares were established in this period—a tendency that continued later on."[17]
- 1961: The oldest open space program in the United States is New Jersey's Green Acres Program.[18]
- 1962: "Although state and local governments have set aside land for parks and recreation for a very long time, the modern movement to protect open space can be traced to a little-known report, Outdoor Recreation for America, released in January 1962 by the Outdoor Recreation Resources Review Commission."[19]
- 1964: The "concept of using royalty payments from offshore oil and gas production to fund conservation initiatives has a long history.

Since enabling legislation was signed by President Lyndon Johnson in 1964, the Land and Water Conservation Fund (LWCF), largely funded from offshore oil and gas royalties paid to the U.S. government, has supported the acquisition of millions of acres of recreational landscapes and critical wildlife habitat."[20]

- 1972: The Taï National Park in Côte d'Ivoire is established.[21]
- 1979: Brazil "established a set of thirty parks and preserves based in part on recommendations of the refugia theory."[22]
- 1979: Yasuní National Park is established.[23]
- 1980: The Jaú National Park in the Brazilian State of Amazonas is established.[24]
- 1980: The Gunung Leuser National Park in Sumatra, Indonesia, is established.[25]
- 1989: "The Nature Conservancy, with support from the United States Congress, launched the Parks in Peril program in 1989, designed to protect twenty million hectares in Central and South America and the Caribbean. {…}."[26]
- 1990: The "United Nations list of protected areas included 669 sites in the lowland tropical moist forest biome, covering some sixty-six million hectares. This corresponded to only a little more than 5 percent of the world's tropical rainforests and raised concern that this could be the end of progress toward the 10 percent target set at the World Parks Congress in Bali in 1982."[27]
- 2004: "Around 2004, Peru accomplished a significant mosaic in the Alto Purús national park area. The Purus-Manu conservation corridor, now about ten million hectares in size, includes previously established protected areas. It is important for the protection of indigenous peoples, including some in voluntary isolation."[28]
- 2005: The "Apuí Mosaic in the Brazilian state of Amazonas was created in 2005. Together with the creation of the Juruena and Campos Amazônicos National Parks in 2006, and complementing a series of other protected areas, this complex forms part of the Southern Amazon Mosaic—an area of more than seven million hectares."[29]
- 2005: Brazil "made significant progress with the Terra do Meio Mosaic in Pará in 2005–2006, which covers more than eleven million hect-

ares. The Terra de Meio Mosaic is adjacent to indigenous territories, complementing the conservation of an area larger than twenty-five million hectares, a large part of which falls in the Xingu River basin."[30]

- 2006: A "landmark achievement was the creation of the Calha Norte mosaic of Pará in 2006, complementing large existing protected areas, in total an area of almost thirty million hectares. This mosaic connects to protected areas of the Guiana Shield countries and is likely to constitute the largest protected pristine tropical rainforest area of the world."[31]
- 2008: In "2008–09, the federal government and the state government of Amazonas put 8.4 million hectares under protection, joining it is a mosaic with other preexisting protected areas to mitigate impacts from the BR-319 highway that connects Porto Velho with Manaus."[32]

Mechanisms of Instrument

Governments, whether local, state, provincial, or federal, have a variety of mechanisms to raise funds for conservation. In general:

> Three main forms of public funds are used: general revenue, bonding, and dedicated funding. Funding from general revenue is the least predictable because it is subject to fluctuating annual budgetary decisions. Bonding, once authorized, is earmarked for conservation. Because it typically needs to be appropriated each year by the legislature, however, it can also be undependable. The surest form of public spending comes from dedicated funds, which accrue money generated by a specific income source and therefore cannot be diverted by state agencies as priorities shift.[33]

Examples of conservation financing mechanisms for states with dedicated funding sources include but are not limited to lotteries, sales and use taxes, general obligation bonds, property tax levies, real estate transfer taxes, impact fees, special assessment, business improvement districts, and revenue bonds. Furthermore, there are environmental license plate sales and environmental penalty monies (i.e., fines for oil spills, illegal logging, or poaching) that are collected, along with gas taxes and royalties from oil, gas, and/or mineral concessions. The following programs, for instance, exist in the United States:

- *Alaska*: duck stamp program and a king salmon stamp program;
- *Georgia*: hunting and fishing licenses;
- *Maine*: affinity card;
- *Kentucky*: tax checkoff (i.e., allows individuals to donate tax return);
- *Maryland*: agricultural transfer tax;
- *Texas*: sporting goods tax; and
- *Virginia*: motor vehicle registration fee.[34]

Although applicable to the United States and not specifically to the conservation of tropical rainforests, the following are key mechanisms utilized by state governments throughout the United States:

> In many respects, conservation finance on a significant scale—in the sense of state and local government spending to acquire or protect parks and open space—dates back to the 1960s. That was the decade when New Jersey first offered a statewide "Green Acres" conservation finance bond to its voters for approval (1961) and when the country's first dedicated tax at the local government for land conservation was instituted in the city of Boulder, Colorado (1967). Conservation finance gathered steam in the following years. Many states launched major conservation initiatives. Some notable examples include Maryland's Program Open Space, funded by a real estate transfer tax (1969); Michigan's Land Trust Fund, backed by revenues from oil and gas leases on state lands (1976); the creation of a lottery in Colorado (1980); California's $776 million Proposition 70 general obligation bond (1988); a ten-year commitment by the Florida legislature to bond $300 million per year (1990); and a thirty-year dedication of $98 million in sales taxes per year to create the Garden State Preservation Trust in New Jersey (1998).[35]

On a more local level, "ballot measures may be referred by the legislative body (termed a referendum) or placed on the ballot by citizen petition (termed an initiative). Some measures are advisory in nature, others create statutory obligations, and yet others may actually amend government charters."[36]

An example of a well-financed protected area from government domestic expenditures is Yellowstone National Park, which today utilizes a diversity of sources to finance its operations. This includes an operating budget provided by the US Congress, along with charging visitor fees, receiving donations, and via concessions. With an annual budget of approximately USD$70 million, about half or approximately USD$34 million is designated by the US Congress to cover the operating expenses associated with day-to-day expenses, rangers on the ground, law enforcement, utility bills, and maintenance. In addition to this USD$34 million, within the appropriation budget, are line items for significant deferred maintenance and construction projects. In addition to appropriations from the US Congress, there are also funds provided to Yellowstone National Park from the US Department of Transportation to finance road construction and maintenance. Collectively, funds from the US Department of Transportation and the US Congress cover 60% of its budget. The remaining 40% of the budget is financed through entrance fees, concession franchise fees, and user fees (e.g., fishing license), along with through donor support. For instance, there is the Yellowstone Park Foundation that raises money for special projects within Yellowstone National Park, and there is also the Yellowstone Association which is a private nonprofit that operates bookstores and educational programs. A portion of the sales generated by the Yellowstone Association are provided to Yellowstone National Park.[37,38]

This situation is not typical around the world. While state and national parks are often provided funds from domestic governments, some parks do not have concession stands, while others do not collect visitor fees (i.e., see the Yasuní National Park case study). Furthermore, in many developing countries with other pressing development needs (i.e., education, healthcare, and energy access) and relatively young park services (i.e., as opposed to the US Park Service, which was established in 1916), park managers within the state, provincial, or country government can often struggle for identity and funding. For instance, there is pressure in places such as Mexico, Central America, and China where there is a commercialization of nature approach and conservation needs to finance itself, primarily through tourism.

Size of Instrument

As reported in the Protected Planet Report 2014, the world's terrestrial and inland water protected areas covered approximately 20.6 million square kilometers (2.06 billion hectares or 5.1 billon acres), representing 15.4% of the world's terrestrial and inland landscape.[39]

This analysis included 197,368 sites and included:

> {...} all sites designated at a national level (e.g. national parks), under regional agreements (e.g. Natura 2000 network) and under international conventions and agreements (e.g. natural World Heritage sites). The UNESCO Man and Biosphere Reserves were not included in the calculations, as many of their buffer areas do not meet the IUCN protected area definition. Proposed protected areas and protected areas recorded as points without a reported area were also excluded. In addition, all overlaps between different designation types were removed from the calculations to avoid double counting.[40]

These sites did not include privately owned protected areas and do not say anything about how well the protected areas are being managed. Another data point is that there are approximately 669 biosphere reserves in 120 countries.[41] The two locations with the highest percentage of protected areas are Central America and South America with 28.2% and 25%, respectively, of their terrestrial and inland areas incorporated under a protected area.[42] Although a bit outdated, in 2004, the annual national park budgets of developing countries were estimated to be approximately USD$1.6 billion.[43]

Introduction to Case Studies

The following case studies will examine Yasuní National Park in Northeastern Ecuador and the Salonga National Park in the Democratic Republic of the Congo. The Yasuní National Park is almost exclusively financed by the Ecuador Government, and the Salonga National Park, with limited domestic budgetary expenditures, has relatively recently started receiving international funds. Both of these Parks boast a tremendous amount of biodiversity, and in fact, Yasuní National Park has among the highest rates of biodiversity in the world. Both Parks have local communities living in, and around, the Parks, and although these Parks are remote, they are starting to experience deforestation as a result of oil exploration and poaching pressures.

Case Study #1: Yasuní National Park in Ecuador

Introduction

Yasuní National Park, located in Northeastern Ecuador on the border of Peru, is one of the most biologically diverse places on Earth. On a landscape level, "the area is: one of the two richest in the world for amphibian species, the second richest known to date for reptiles, within the top nine richest centers for vascular plants (and the top center for trees and shrubs), among the richest lowland areas for birds, high in mammal richness (particularly for bats), and very rich in fish species."[44] Similarly, the Smithsonian Tropical Research Institute "recorded 644 tree species in a single hectare of the Yasuní Reserve in the Ecuadorian Amazon, about as many as in all of North America! This is presumably the highest tree species diversity ever recorded. The Yasuní Reserve is also home to the only two uncontacted indigenous tribes in Ecuador, the Tagaeri and the Taromenane."[45]

This particular region, which includes the Ecuador Basin and a portion of neighboring Peru, is a biological hotspot with a tremendous diversity of life. In fact, there are new species regularly being added to the Library of Life. Such high rates of biodiversity in the western Amazon could be a result of its proximity to the foothills of the Andes and due to the glaciation periods and the refugia theory. Essentially, during the glaciation periods when the temperatures changed, tropical forests started turning into savannas, while some places remained and kept their characteristics which created islands of tropical rainforests and refugia. One of the biggest refugia in the Amazon was this location, and likewise, there is more diversification.[46]

Yasuní National Park was established in November 20, 1979, and then in "1989 Yasuní National Park was made a World Biosphere Reserve as part of UNESCO's Man and the Biosphere (MAB) Programme of UNESCO. As a consequence of this declaration, the park's management must comply with the Seville Strategy for Biosphere Reserves, adopted at the International Conference on Biosphere Reserves held in Seville, Spain in March 1995."[47]

At the time, there was no Ministry of Environment, and instead, the Park was under the management of the Ministry of Agriculture. Today, Yasuní National Park is approximately 982,000 hectares (2.4 million acres).[48] The location is relatively remote and does not see many tourists. There are several research stations including Tiputini Biodiversity Station, which is a collaboration between Boston University and the University of San Francisco Quito,[49] and the Yasuní Research Station, which is administered by the School of Biology at the Pontificia Universidad Católica del Ecuador (the Pontifical Catholic University of Ecuador in English) (Fig. 7.1).[50]

Yasuní National Park is owned by the Ecuador Government, and there are approximately 200 people living in isolation within the Park.

Identify the Problem

There are several problems facing Yasuní National Park including oil exploration and road construction, a relatively small operating budget, and illegal logging.

Prior to Yasuní being designated a national park, the area was very remote with only about 5% of Ecuador's population living in the Amazon Basin. Over time, access to this region increased mainly due to roads being built to access the country's petroleum resources. Likewise, the roads from Western Ecuador over the Andes Mountains to Eastern Ecuador were specifically created for oil exploration and extraction.

Ecuador's experience is different than Brazil's experience. Ecuador created roads for oil access and then people settled in afterward. In contrast, Brazil initially created roads to settle people (i.e., primarily due to the landless movement) and then cattle ranches were established.

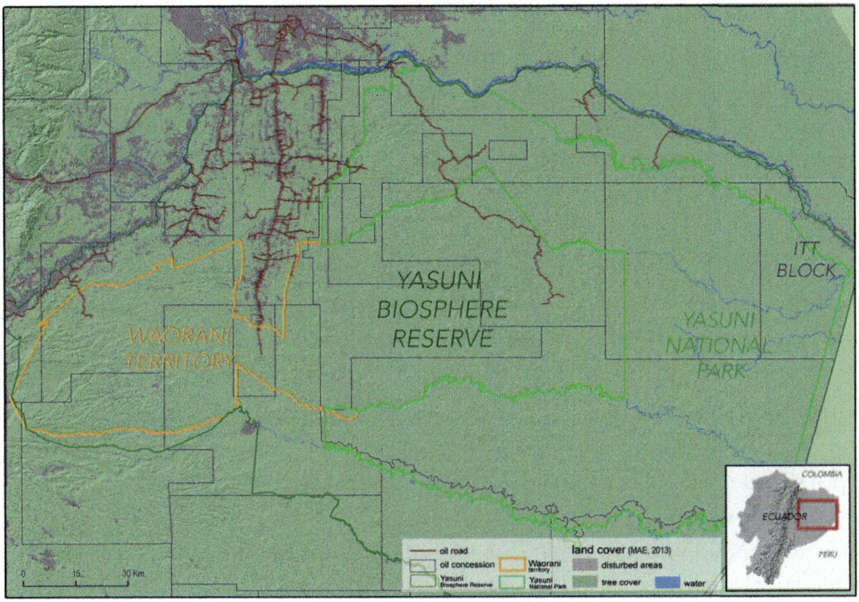

Fig. 7.1 Map of Yasuní National Park (Credit: Leo Zurita-Arthos, Universidad San Francisco de Quito) (The University of San Francisco Quito in English)

It is important to note that oil exploration and extraction in the Ecuadorian Amazon have a relatively small direct impact on tropical rainforest deforestation and degradation. The essential problems are the roads (which are known in Ecuador as "ecological access") and the paths cut for pipelines. With this increased access, local communities are deforesting more while also practicing illegal hunting and logging. Even local Indigenous Peoples, who do not want to deforest, hunt wildlife to the brink of local extinction.

Ecuador, which is a member of the Organization of the Petroleum Exporting Countries (OPEC), first began producing oil in 1921.[51] Several decades later, Texaco, as one of the original foreign gas and oil companies operating in Ecuador, first discovered oil in 1967.[52] In Ecuador, resources under the ground belong to the government and not to the owner of the land. Upon discovering oil near Yasuní National Park, Empresa Pública de Hidrocarburos del Ecuador (EP Petroecuador) took over the concession, and it slowly became Texaco's job to train local Ecuadorians. Due in part to corruption, EP Petroecuador claimed the site remediation was satisfactory and the concession was being returned in the same condition, but then Texaco left and the local communities saw the situation differently. As a result of this, there was a USD$8.6 billion lawsuit brought against Texaco, and this has been cited as an example of what can never be allowed to happen at Yasuní National Park.[53]

This said, oil was discovered in the region and the oil blocks were set before Yasuní National Park was officially established. Approximately 60 to 70% of the Park area has overlapping claims, and because the Park did not expropriate the lands, these overlapping claims continue. These oil blocks include, but are not limited to: Block 16 and the Ishpingo-Tambococha-Tiputini (ITT) Block within Yasuní National Park. Under the ITT Block are proven reserves of 944 million barrels of crude oil with a potential 1.53 billion barrels of reserve.[54] According to Amazon Watch:

> After the 2008 default on USD$3.2 billion in Brady Bonds to underwrite a major expansion in public sector spending and to keep its dollarised economy running, Ecuador found a willing partner in Beijing, which, in the midst of China's "Great Opening," sought investment and natural resources beyond its border.
>
> Since 2010, the China Development Bank and China Export-Import Bank have provided eleven loans to Ecuador totaling USD$15.2 billion in overseas foreign investment, largely for energy, infrastructure, and transportation projects. China is now Ecuador's largest creditor, providing around 60% of the government's finance. Ecuador's debt to China now stands at 38.7% of GDP. Many of these loan deals are to be paid by the sale of oil or fuel from Ecuador's state-run Petroecuador to PetroChina International Co, which commit much Ecuadorian crude to Beijing until 2024.
>
> These loans are driving a new oil boom in the Amazon rainforests of Ecuador. While it costs roughly USD$39 to extract a barrel of the heavy, viscous Amazonian oil, its market value averaged USD$35 for much of 2016. The 2016 collapse of crude prices forced the country to drill at a loss or to merely break even to keep production up to pay China. At current market prices, Ecuador is now forced to provide China with almost two barrels of oil to cover loan obligations compared to every one barrel just a year ago.
>
> Meeting contractual loan payments with oil is a major driver behind Ecuador's effort to open up new, pristine Amazon indigenous rainforest territory to oil drilling. Contracts were inked in 2015 for two new oil con-

cessions in a remote area of Ecuador's southern Amazon with Andes Petroleum, a subsidiary of China National Petroleum Company (CNPC), and SINOPEC, a move to ensure the production of oil with which Ecuador can pay the company's sole shareholder.[55]

Another problem is that up to 2015, Yasuní National Park only had approximately USD$1 million for its operating budget. There are no specific fees from the research stations that are payable to Yasuní National Park. Rather, a special price is given to researchers, and they have to pay for a research permit to the Ecuador Government. Historically, there was a fee of approximately USD$20 to enter the park, but now access is free.[56]

Illegal logging is also an issue. There are areas in Ecuador where loggers will cross over from neighboring Peru or Colombia and then smuggle the wood out of Ecuador. These illegal loggers will sometimes get a legal license to harvest wood, then go into the forest to do selective logging of illegal wood, and then use the legal license to "camouflage" the illegal wood. Illegal hunting, also along the roads, is becoming unsustainable.[57]

Additional problems include when Yasuní National Park was originally created, there was no acknowledgment of the Indigenous Peoples who lived within the Park. In the late 1980s, all Indigenous Peoples' areas were recognized. The Indigenous Peoples known as the Taromenane have part of their land in the Park and part of their land outside of the Park. The Taromenane have the right to live and hunt but must follow the Park laws on their lands within the Park.[58]

A further problem is that the head of Yasuní National Park recently resigned and, as of October 2016, had yet to be replaced. Another somewhat similar problem is that the Ministry of Environment wants to generate their own data (i.e., such as for Block 21), and although deforestation is relatively minute, the Ministry does not even have annual reports about deforestation in the park.[59]

Why the Problem Is Important

The problem of increasing access to Yasuní National Park and the ensuing tropical rainforest deforestation is that one of the world's most biologically diverse places is threatened. This increased access, in part, has resulted in the fact that "Yasuní is home to a considerable number of globally threatened species {...}. These include 13 documented vertebrate species and an estimated 56 plant species (28 documented in the park, with another 28 expected)."[60] Furthermore, "Yasuní does in fact harbor a considerable number of regional endemics. It has 43 documented vertebrates and an estimated 220–720 plants {...}."[61]

The Government of Ecuador cares about these issues, but not as much as it cares about generating international funds. Likewise, Yasuní National Park cannot stop the oil drilling because oil generates too much revenue—especially compared to any potential revenue from the Park such as visitor fees from research institutions and ecotourism operators, or from locally produced goods such as organic cocoa or sustainable fish farms.[62]

As David Romo Vallejo, from Universidad San Francisco de Quito (the University of San Francisco Quito in English), says, "If we are unable to protect this special place, how can we do it anywhere else?"[63]

How Problem Was Identified

The problems facing Yasuní National Park were identified via several methods. The problem of tropical rainforest deforestation and degradation is identified via satellite imagery and ground-truthing. Numerous nongovernmental organizations such as the Wildlife Conservation Society (WCS) and academic institutions such as the University of San Francisco Quito are conducting private research and/or assisting with monitoring efforts. For example, WCS is involved with monitoring poaching and illegal logging.[64]

Effectiveness of Process for Identifying Problem

The effectiveness of the process for identifying problems in and around Yasuní National Park is a bit mixed. The government is doing a relatively good job at monitoring tropical rainforest deforestation.

However, the same cannot be said with respect to poaching or illegal logging.

Steps Taken to Address the Problem

Private research is the primary step being taken to address the problem.

In terms of tropical rainforest deforestation, the Government of Ecuador engaged with climate negotiations centered on Reducing Emissions from Deforestation and forest Degradation (REDD+). Ecuador was trying to become an example with its Socio Bosque program that would have put money toward forest conservation, and more specifically, the Ecuador Government would have paid people to not clear forests. Ecuador was trying to prove the concept works and wanted to receive money for its REDD+ work. However, at that time, Ecuador did not have a baseline of deforestation. The Ecuador Government now has a baseline after hiring an Ecuadorian researcher who assisted with developing the baseline and helped reclassify ecosystems.[65]

Results

To date, some of the results of Yasuní National Park include:

- Yasuní was designated a national park in 1979 and, then in 1989, made a World Biosphere Reserve;
- The Park has conserved nearly 1 million hectares; and
- A tremendous amount of research has been generated on the Park's biodiversity.

Challenges and How They Were Met

There are several challenges facing Yasuní National Park, which have yet to be addressed. This includes that current reality that Ecuador's national economy is struggling with a deep recession due to the drop in oil prices and that no trust fund has yet been established for Yasuní National Park.

During the early days of Yasuní National Park, there was no established budget for the park—rather, there was simply a map depicting the general borders of the park. Up to 2015, Yasuní National Park had approximately USD$1 million for its operating budget. While these funds were in addition to salaries of people working at the Park, the funds are still insufficient. This is because there is nearly 1 million hectares of the Park to be managed, plus all the necessary equipment such as boats, vehicles, gas, office supplies, and technology. In addition, the salaries are often quite low for employees. This said, the USD$1 million operating budget for nearly 1 million hectares translates into only about USD$1 per hectare.

National parks in Ecuador, including Yasuní National Park, as well as national parks around the world need a management plan and an operating budget. The annual operating budget at Yasuní National Park should be approximately USD$5 to USD$10 million to provide some level of financial stability and to enable half the things required to manage the area. The German Government was willing to put up €32 million over 5 years, which is approximately USD$33.6 million to USD$44.6 million (i.e., depending on the lowest and highest exchange rate over a 5-year period from December 3, 2016).[66] This initial trust fund never materialized, in part, because the German Government requested more transparency and wanted to converse directly with local communities to gather their perceptions. In contrast, Ecuador interpreted this as a violation of national sovereignty and wanted the funds to go through the Ministry of Environment, as opposed to a third-party NGO. Germany appears to remain committed, but is focused on more of a regional approach and not specifically on Yasuní National Park.[67]

Beyond Results

As of right now, the results of Yasuní National Park are not sustainable, nor is the level of financing adequate. Yasuní National Park is barely moving from a paper park concept, and the Park is not taking advantage of all the countries who want to help. It appears as though government will is lacking or the directionality of that will is in the wrong place. Likewise, discourse is different than action.

Lessons Learned

Some of the lessons learned from the experiences at Yasuní National Park include:

- Despite drilling in the Ishpingo-Tambococha-Tiputini (ITT), Yasuní National Park is not gone. The drilling will create damage, but not so much that it will cause the Park to collapse. However, international monitoring and environmental activists will play an important role in the future of the Park.
- There is a need to work right away with local peoples and provide them with a very good transition that allows them to be a part of society. More specifically, this includes offering a twenty-first-century education to local children to help them get into college. This in turn will help them become future engineers and programmers and help to change the development dynamics of their local region.
- Ecuador's famous Galapagos Islands have lots of funding, and there are many lessons to be learned from the Galapagos Islands. For instance, research and conservation made the Galapagos Islands, not the other way around.
- Best-in-class protections need to be put in place for sound control of oil exploration, particularly by helping biologists and the government monitor the environmental health in Yasuní National Park.[68]

Other Resources on Yasuní National Park

Amazon Watch

- http://amazonwatch.org/

Boston University and Universidad San Francisco de Quito's (University of San Francisco, Quito) Tiputini Biodiversity Station

- https://www.bu.edu/cecb/tiputini/
- http://www.usfq.edu.ec/programas_academicos/Tiputini/Paginas/About-us.aspx

Center for Tropical Forest Science's Yasuní Page

- http://www.ctfs.si.edu/site/Yasuni

Multi-Partner Trust Fund Office's Ecuador Yasuní ITT Trust Fund

- http://mptf.undp.org/yasuni

Napo Wildlife Center

- http://www.ecoecuador.org/

Case Study 2: Salonga National Park in the Democratic Republic of the Congo

Introduction

Located in the Democratic Republic of the Congo (DRC) and covering approximately 3.6 million hectares (approximately 8.9 million acres), Salonga National Park is the largest tropical rainforest reserve in Africa and one of the largest such reserves in the world (Fig. 7.2).[69]

The Park, which was established in 1956, is divided among a Northern section and a Southern section, with both sections approximately the same size.

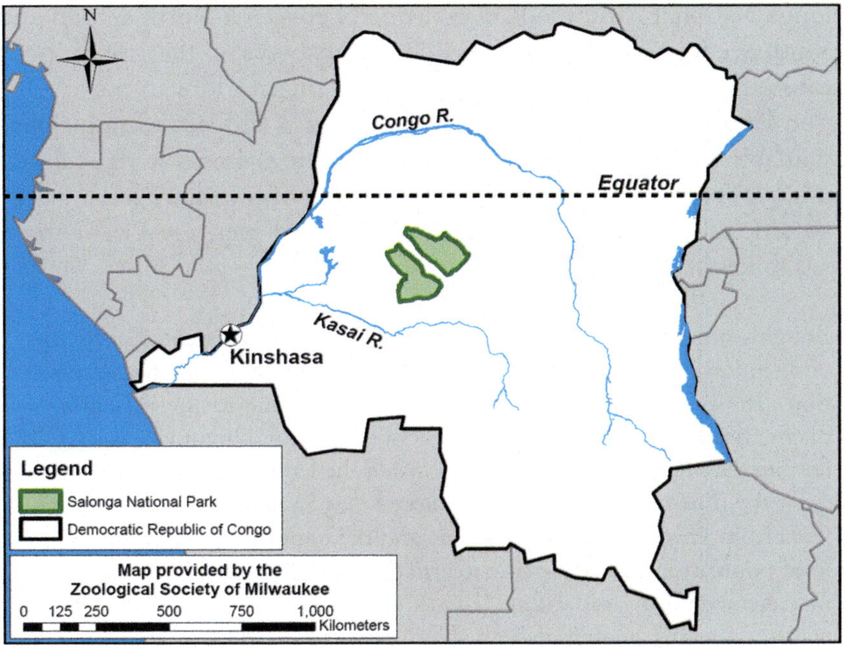

Fig. 7.2 Map of Salonga National Park (Credit: Zoological Society of Milwaukee)

Identify the Problem

There are several problems facing the Salonga National Park. The DRC experienced a horrific civil war from approximately 1997 to 2003 which has been termed "Africa's First World War," as the conflict spread to neighboring countries. The conflict, particularly in Eastern DRC, continues today.[70] As a result, the DRC is a very challenging place to work. The DRC, for instance, ranks #184 in the world for overall ease of doing business, according to the World Bank. Additional subcategories have the DRC ranked #156 for registering property, #162 for protecting minority investors, #171 for enforcing contracts, and #188 for trading across borders.[71] Similarly, Transparency International's Corruption Perceptions Index 2016 has the DRC tied for #156.[72]

Land tenure is another challenging problem in the DRC, and such land tenure depends on the region and the province. With respect to

Salonga National Park, the Park is divided between a North section and a South section. There was a corridor setup between the two sections because there were too many people within this corridor to move. One reason land tenure is a challenge is because it is difficult to implement sustainable fishery projects or help communities improve agricultural techniques to increase their yield if people do not own their land.

Additional problems, according to the World Heritage Convention, are as follows:

> Roughly one third of the southern sector of the Park is occupied by groups of pygmies and a part of this occupied land is claimed by the local population. The boundaries of the property are intact due to the existence of major rivers that form recognized, precise and natural boundaries and this despite the presence of some villages inside the Park.
>
> At the time of inscription, it was noted that Salonga National Park suffered from pressures such as poaching and the removal of vegetation by the local populations. A management structure, sufficient qualified staff and a management plan are lacking. The future of the Park cannot be assured without a strengthening of both the management structure and available financial means.
>
> Among the management problems requiring long-term attention are poaching using traditional methods, and more recently by the military with modern war weapons; pressure and human occupation by the Yaelima in the southern part and by the Kitawalistes in the northern area (with accompanying impacts, such as fire, deforestation for the sowing of food crops, logging for heating purposes, honey gathering and the building of pirogues); dispute of the Park boundaries by populations in certain areas; commercial traffic in bush meat; forestry exploitation by individuals in the southern part; and pollution of the Park waters with toxic products used for illicit fishing.[73]

It is important to note that despite the lack of infrastructure and the lack of financial resources, along with the presence of local settlements and poaching, deforestation is currently not a significant problem at Salonga National Park. This is in part due to the relatively low population density in the North and South sections and due to the Park's inaccessibility from major DRC cities such as Kinshasa (the nation's capital),

Lubumbashi, and Mbuji-Mayi. This inaccessibility is helpful for the preservation of the Park, but creates challenges for access by researchers, conservationists, and tourists. This said, site visit logistics are challenging, as there is little infrastructure (e.g., few passable roads).[74]

However it is important to note that deforestation could become a problem. For example, there are large fishing and poaching camps in the region known as Dare Dare (i.e., a region in the Southeast of the Watsi Kengo Sector is known as "Dare Dare," meaning "in/out" or "quick quick," and refers to the quick exploitation of a natural resource). There are also signs of permanent settlements (e.g., houses and large manioc fields) starting to appear in the Southeast portion of the North section from the Kasai Province. These smaller, permanent settlements—whether intentionally or unintentionally—create an avenue of support (e.g., provide food supplies such as manioc) for the larger, more organized poachers. This said, the local African forest elephant (*Loxodonta cyclotis*) population is becoming absolutely devastated by poachers, particularly in the Dare Dare region.[75] These threatened areas would be the place to specifically provide investments for protection and community development projects.

Historically, Salonga National Park received very little—if any—financial support from the DRC Government and has basically been neglected since the 1970s. Nowadays, the financing is almost entirely donor funding. The Central Africa Regional Program for the Environment (CARPE) is a USAID program that helps to finance landscapes within the DRC, such as the Salonga-Lukenie-Sankuru landscape which includes the Salonga National Park. One of the main problems is that there are few, if any, local people who have done protected area management. For example, there is a lack of business planning, and operating costs for protected area management are significantly underestimated.[76]

Why the Problem Is Important

The problems facing Salonga National Park are important because the Park is home to a tremendous amount of biodiversity. According to the World Wildlife Fund (WWF), the Salonga National Park is home to an estimated 51 species of mammals, 129 species of fish, and at least 223 species of birds. This includes providing refuge to African forest elephants (*Loxodonta cyclotis*) and possibly 40% of the world's bonobo (*Pan paniscus*)—also known as pygmy chimpanzee—population.[77] The African forest elephants are also key seed dispersers, and if lost, the composition of the local forests will be permanently altered.[78]

How Problem Was Identified

Over the last 10 years, satellite imagery has been utilized to identify the problems related to deforestation and the expansion of local settlements. In addition, several NGOs have established survey teams to determine the abundance and distribution of wildlife populations, as well as illegal hunting signs.[79]

Effectiveness of Process for Identifying Problem

The process of identifying the aforementioned problems has been effective. This said, there are many committed NGOs such as Oxfam, WWF, the Wildlife Conservation Society (WCS), and the Zoological Society of Milwaukee (ZSM) that have been working in the region for many years.

Steps Taken to Address the Problem

There are several different steps that have been taken to address the myriad of problems facing Salonga National Park. Originally, the Park was zoned for areas that contained the lowest population densities. Within the corridor between the North and South sections, along with throughout the remaining buffer zones, there are "soft development programs" being implemented by groups such as WWF and Oxfam. In addition, WWF, WCS, and the ZSM are working on programs within the Park.

An agreement between the DRC Government's park authority Institut Congolais pour la Conservation de la Nature (ICCN; Congolese Institute for the Conservation of Nature) and WWF was signed in August 2015, which officially provides for the co-management of Salonga National Park by WWF and ICCN.[80] A Salonga Financing Agreement was later signed in December 2016 which provides €17.3 million (approximately USD$18.3 million) from the European Union (EU), which specifically provides financing for the Park and surrounding areas.[81]

Furthermore, Dr. Allard Blom of WWF is helping to write business plans and financial spreadsheets for the Park and has published his work on managing protected areas in the DRC and their associated costs.[82]

Results

Most of the results at Salonga National Park have been organizational results, and this includes:

- Wildlife surveys have been done throughout the whole park from approximately 2002 to 2003, and it is currently being resurveyed. As explained by Dr. Blom of WWF, "you have to know your resource base before you can manage it."[83]
- Building infrastructure (e.g., housing and landing strips);
- Organizing and training staff (e.g., make an inventory of state employees and start retiring lots of state employees); and
- Putting in place the Spatial Monitoring and Reporting Tool (SMART) management system to track patrols. This includes mapping out patrols with GPS and regularly sending out patrols.[84]

Additional high-level results for Salonga National Park include, but are not limited to, the Park being established as a National Park and being designated as World Heritage Site.

Challenges and How They Were Met

One of the challenges facing Salonga National Park is the fact that the Park is enormous and there is a logistical challenge to visit and travel throughout the Park. This logistical challenge requires additional planning and significant costs.[85]

Another challenge is training and capacity building. Training and capacity building are doable and it is a work in progress, but it takes a lot of time and effort. In villages bordering the Park, the illiteracy rate is nearly 80%, and there are basically 2 generations with little formal education and without specialized skill sets. In addition, building capacity for park personnel needs to be done all the way up from the park guards to the park administrators.[86]

A related problem to training and capacity building is the challenge of patrols. In the past, there were very few organized patrols. Prior to the late 1990s and early 2000s, the Park received very little financial support and was basically just a paper park. Now there is more funding and organization is taking place at the Park. However, if the patrols arrest someone, there is a chance the person will be let free later at the level of the judicial process. This presents a major setback for the park personnel.[87]

The biggest challenge facing the Park is getting long-term financing in place. USAID, particularly through the CARPE mechanism, has helped, and for the first time, NGOs have been able to establish a baseline and measure results.[88] The Park will likely be financially dependent on major donors, such as the EU's Salonga Financing Agreement, for at least 20 years. There is a possibility to have bonobo tourism in the distant future, but there will likely be no significant tourism in the next 5 years and possibly not for the next 10 years.[89]

Furthermore, there are ongoing challenges associated with poverty and poaching, along with the lack of exposure as few people have heard of Salonga National Park despite its large area and importance for both biodiversity and the local communities.

Beyond Results

It is too early to tell whether the results at Salonga National Park are sustainable. The Park's management structure between Institut Congolais pour la Conservation de la Nature (ICCN; Congolese Institute for the Conservation of Nature) and WWF was recently signed in August 2015,[90] and the Salonga Financing Agreement was later signed in December

2016.[91] There is now serious money flowing to the Park and a national trust fund is being set up that might potentially progress with CARPE, or a separate trust fund specifically for the Park might be created.

Lessons Learned

Some of the lessons learned at Salonga National Park include:

- There is not just one fix. You need to simultaneously do several activities; if not, you will not be effective.
- One must lessen the need for poaching or find alternatives to poaching. At the same time, you need to reinforce your law enforcement capacity.
- Guardians of wildlife and protected areas must be held accountable, but often wildlife is still exploited, especially for ivory and the pet trade.[92]

Other Resources for Salonga National Park

African Biodiversity Conservation Group

- http://www.abcg.org/index

African World Heritage Sites

- http://www.africanworldheritagesites.org/natural-places/forests/salonga-national-park-dr-congo.html

Bonobo & Congo Biodiversity Initiative, A Project of the Zoological Society of Milwaukee

- http://www.bonoboconservation.com/where/salonga.php
- http://www.zoosociety.org/Podcasts/

Central Africa Regional Program for the Environment (CARPE)

- http://carpe.umd.edu/about/index.php/?tab=0

Spatial Monitoring and Reporting Tool (SMART)

- http://smartconservationtools.org/

Return

To generate revenue, or additional revenues, from government-owned protected areas, there are several options. First, the protected area could start charging a visitation fee, if no such fee exists (i.e., as is the case with Yasuní National Park). However, this visitation fee might not be significant if the protected area is very remote with few visitors (i.e., as is the case with Salonga National Park).

Second, the protected area could offer differentiated pricing. This might include a higher price for foreign visitors and a lower, or no cost, option for local tourists. In addition, the protected area could offer season passes or look to lease concession stands.

Another important consideration is that "federal and state governments cannot meet all local conservation needs. Therefore, states need to provide local governments with the legal authority to tax and dedicate revenues for land conservation (using property taxes, sales taxes, transfer taxes, bonding authority, and so on)."[93]

Whether it is assessing visitor and researcher fees in addition to using the government's general funds, diversifying revenue sources can help contribute to financial sustainability. For instance, the US State of "Minnesota has embraced a unique blend of funding sources to support its land conservation efforts: lottery revenue, cigarette taxes, general obligation bonds, and fees from environmental license plates."[94]

Risk

There are a variety of risks associated with a government's domestic expenditures for conservation initiatives.

Business Risk

One of the most significant business risks for a protected area that is primarily financed through government domestic expenditures is whether the same, or greater, amount of funds will be appropriated in the future. In addition, it is often difficult to predict the level of demand for the protected area's products and services, which could include the number of visitors paying an entrance or user fee or the number of future purchases at the concession stands.

Strategic Risk

There are numerous strategic risks associated with government domestic expenditures to local protected areas. Such strategic risks to establishing, maintaining, and expanding such protected areas include:

- Where to locate the local, state, or national park.
- The size of the protected area.
- Deciding on what recreational activities will be/are allowed. For example, is access to the protected area strictly forbidden to visitors, except qualified researchers? If tourists are allowed, how many are allowed, are they allowed to stay overnight, and are fires allowed? Is recreational hunting and/or fishing allowed, or just for subsistence purposes?
- What strategic partners to work with, whether its university researchers, nongovernmental organizations, and/or multinational companies.

Reputation Risk

Paper parks, where parks are merely designated areas without the proper management and funding to preserve the area, pose possibly the greatest reputational risk. This risk could be relevant to the host state or federal government, the responsible ministry or department, along with the park management. Illegal activities, whether it is being permitted or not addressed, pose reputational risks as well. Such activities could include illegal logging, poaching, and/or cultivation of illicit drugs.

Liquidity Risk

Government domestic expenditures—particularly to state or national parks—face a unique liquidity risk. This is because presumably the state or national park is illiquid and the manager (i.e., National Park Service and/or Ministry of Environment) of the protected area is unable to sell their holdings. Because they are not able to sell their holdings, they need to have a contingency plan if there are funding shortages. There can also be a liquidity risk presented if government budgeting comes after the protected areas run out of funds and are, thus, unable to make payroll or upgrade its facilities.

Operational Risk

Operational risks facing local, state, and national parks are oftentimes similar as those operational risks experienced by other businesses. This includes management capabilities, staff morale, employee retention, adequate training, and promotional opportunities. Unique operational risks depend on the structure of the government domestic expenditure. For instance:

- Are the concessions (e.g., gift shops and/or visitor centers) subcontracted out to another entity?
- Natural disasters such as hurricanes, tsunamis, and infestations of pests or disease.

Other operational risks include technology risk (e.g., adequacy of monitoring technology, whether it is the use and classification of remote sensing technology or on-the-ground wildlife cameras) and the proper use of internal controls to minimize fraud (e.g., are the collections of visitor fees at a remote entrance being properly recorded and submitted to management).

Legal and Regulatory Risk

There are lots of legal and regulatory risks, particularly surrounding traditional uses within and adjacent to the protected area. For example:

- From the State of Alaska in the United States to the Yasuní National Park in Ecuador, there is subsistence hunting by local communities.
- In Grand Teton National Park in the State of Wyoming in the United States, there is grazing rights that are legally granted.

As Steven Iobst, the recently retired deputy superintendent of Yellowstone National Park, said, "wildlife in the park, become livestock or pests outside of the park."[95] This said, wildlife can often become transboundary legal issues as wildlife—from bison, to jaguars, to elephants—cross over state and country borders.

Additional complications arise if there are overlapping concessions granted for energy (e.g., reservoirs, dams, or oil blocks as is the case for Yasuní National Park), agriculture (including palm oil and soy), logging, or mineral rights.

There can also be competing interests from local, state, or federal representatives who propose to divert funds from the protected area to other programs. Similarly, governments can decide to abolish protected areas. This happened, for instance, when the Brazilian President Michel Temer issued a decree in August 2017 to abolish a 46,000 square kilometer (4.6 million hectares) protected area known as the National Reserve of Copper and Associates (Renca) in order to allow mining operations. While it is not yet known how much deforestation will occur and whether remaining protections at Renca are sufficient, this exemplifies a major potential legal and regulatory risk.[i][i] BBC News. "Brazil opens vast Amazon reserve to mining." August 24, 2017. www.bbc.com/news/world-latin-america-41033228

Furthermore, there can be legal and regulatory risks associated with Free, Prior, and Informed Consent (FPIC) if a protected area does not equitably work with local communities. Such risks could arise, for instance, if the protected area is expanding its boundaries at the expense of local communities or if the protected area restricts access to the protected area (i.e., allows for little-to-no harvesting of wood or game by the local communities).

Credit Risk

Credit risk includes default risk, bankruptcy risk, downgrade risk, and settlement risk.

When a government—whether it is a city, state, or federal government—faces a credit risk such as a potential default or downgrade in its investment grade, then it becomes more expensive for the government to borrow. With increased costs of borrowing, governments may have a smaller domestic budget, and thus, its protected areas may experience a constricted budget.

Market Risk

Market risk includes interest rate risk, equity price risk, foreign exchange risk, and commodity price risk.

Interest rate risk could threaten a domestic government's ability to borrow funds if interest rates rise and make it prohibitively expensive.

Equity price risk is not as relevant for a domestic government, but could threaten a protected area if significant private companies are devalued and no longer operate tours to a protected area or close down nearby hotels (i.e., thereby decreasing the user and visitor fees).

Foreign exchange risk may make it more or less difficult for international visitors (e.g., researchers, tourists, and/or students) to visit the protected area.

Commodity price risk can have a direct or indirect impact on government domestic expenditures to local protected areas. Considering the case study on Yasuní National Park, if oil prices increase, then it becomes more likely that drilling for oil will continue and potentially increase. Similarly, an increase in the price of tropical hardwoods could increase the amount of illegal logging taking place in and around Yasuní National Park. Furthermore, if commodity prices increase for cattle and/or palm oil, then land use activities around the protected area such as cattle ranching and commercial agriculture may increase and threaten the buffer zone.

Risk, Return, Time (Horizon), Taxes, Liquidity, Legal and Unique (RRTTLLU)

Risk and Return

Please see above for the risk and return associated with government domestic budgetary expenditures. It is important to note that many companies put pressure on parks leading to tropical deforestation and degradation. These companies obtain financing from banks and exchanges. These banks and exchanges may need to be held accountable for their financing of illegal land grabbing and ensuing tropical deforestation (i.e., essentially financing the stealing of other peoples' property).

For example, as described by Chain Reaction Research, United Cacao deforested thousands of hectares in Peru to which they did not have the legal title, but rather were Indigenous Peoples' traditional land.[96] In other words, when United Cacao raised USD$10 million in the December 2014 initial public offering (IPO) on the London Stock Exchange's Alternative Investment Market, should the exchange not have risked its own reputation for allowing United Cacao to raise funding for potentially illegal activities on the exchange? As a result, United Cacao—and its other companies—Cacao del Perú Norte S.A.C., Plantaciones de Ucayali S.A.C., and Plantaciones de Pucallpa S.A.C. should not have been allowed to use the London Stock Exchange to finance their potentially illegal deforestation activities, according to the Peruvian Government.[97] Furthermore, United Cacao likewise traded its debt on the NEX Exchange in London. In other words, the exchanges' reputation suffered from their client's actions.

In another similar case, according to Greenpeace, the Hongkong and Shanghai Banking Corporation (HSBC) financed billions of dollars of deforestation in Southeast Asia, and much of it possibly illegal. In addition, some of this alleged deforestation took place on community lands, high conservation value (HCV) lands, and Ramsar wetlands.[98] This created a significant problem for HSBC when 30,000 of its clients threatened to switch banks because of these activities allowed by HSBC.[99]

In both cases, the clients of the London Stock Exchange and HSBC threatened and destroyed primary forests—often illegally—using their ability to finance their activities via the bank and the exchange. This shows that without sufficient due diligence, financial institutions may assume the risk of ill-advised financial deals.

Time Horizon

The time horizon for government domestic budgetary expenditures, particularly investments into national parks, is long term. In fact, the time horizon could be in perpetuity with ongoing investments into park infrastructure, monitoring, and staff. Shorter time horizons may exist for smaller investments such as investments into the creation of concessions. As such, long-term budgeting needs to be driven by ways to frame out both financing in perpetuity alongside immediate short-term needs to strengthen protection of the park (e.g., paying park rangers their salaries, expanding staff, offering training, and monitoring for deforestation).

Taxes

There are likely limited, if any, taxes due to the government if the government is the sole provider of investments. However, there could be sales tax levied on concessions, employment taxes, and/or some form of tourism tax for visitors (e.g., if visitors stay at onsite facilities). One incentive to attract concessioners would be to offer a reduced tax rate for their services rendered.

Liquidity

Liquidity might exist with concessions or other onsite facilities (e.g., tourist attraction or lodging) that could be transferred to another owner. However, the actual underlying protected area is likely to be illiquid unless the federal government can transfer the protected area to a state or province (i.e., or vice versa).

Legal

Legal considerations associated with government domestic budgetary expenditures will vary from country to country. Again, legal risks may arise as a result of improper Free, Prior, and Informed Consent (FPIC) when deciding what activities can take place in the protected area and in whose name is access granted. In addition, legal risks may arise when offering (or not offering) employment opportunities to local communities or when expanding the protected area's boundaries.

Unique

The unique aspects for financing protected areas are that their time horizon is in perpetuity. In addition, their revenue streams are often limited, and their activities must finance measures to prevent exploitation.

The drivers of deforestation in these protected areas (e.g., Yasuní National Park) are often financial policies at banks and exchanges halfway around the world. This means that communication between the communities in and near the protected areas and financiers and regulators is complicated and often nonexistent. As a result, protected areas might benefit by building into their budgeting some cash flows that fund financial sector engagement as a tool to protect the protected areas, similar to the United Cacao and HSBC case. Likewise, protected areas might benefit from asking for financial institutions to support monitoring global financial flows that drive deforestation and degradation in and around each protected area.

Another unique aspect of protected areas is that flora, and especially fauna, migrates, and this migration may take place to an area outside of the protected area. Thus, a protected area may have been designed to protect African savanna elephants (*Loxodonta africana*) and attracts paying tourists to specifically see these elephants, but then the elephants migrate out of the protected area during a particular period of the year.

Policy Analysis

Defining the Problem

The essential problem is although deforestation generally occurs at a higher rate outside of protected areas, than within areas officially designated for protection, this is not always the case. Some government-sanctioned protected areas (e.g., state forests or national parks) are poorly funded, and due to a lack of infrastructure, lack of adequate staff, and low salaries, along with limited training, some of these protected areas are experiencing higher rates of deforestation.

Furthermore, some parks are so poorly funded that they are considered paper parks. Even if a protected area is well financed by government domestic expenditures with well-paid staff, the regional land use surrounding the protected area can have a significant impact on the park.

The cause of this deforestation depends on the given country, but is largely attributed to cattle ranching along with commercial and subsistence agriculture. The problems for implementing a policy to commit domestic budgetary expenditures to protected areas include:

- Small domestic budgets with competing domestic interests such as healthcare, energy access, and education.
- Competing land uses (e.g., agriculture, mining, and cattle ranching) and the need for foreign exchange.
- Squandered resources due to corruption (e.g., informal or formal bribing).

Establishing Goals

The Aichi Biodiversity Targets of the Convention on Biological Diversity (CBD) are an important collection of 20 targets designed to outline and establish international conservation goals. More specifically, Target #11 states:

> By 2020, at least 17 per cent of terrestrial and inland water, and 10 per cent of coastal and marine areas, especially areas of particular importance for biodiversity and ecosystem services, are conserved through effectively and equitably managed, ecologically representative and well connected systems of protected areas and other effective area-based conservation measures, and integrated into the wider landscapes and seascapes.[100]

Similarly, the Sustainable Development Goals (SDGs) have the overarching goal of Target #15 to "protect, restore and promote sustainable use of terrestrial ecosystems, sustainably manage forests, combat desertification, and halt and reverse land degradation and halt biodiversity loss."[101]

Thus, the goal is for countries—particularly tropical rainforest countries—to establish a significant (as defined, at a minimum, by the Aichi Targets) and representative protected area for each biome and ensure that such a protected area is well financed, conserves biodiversity, and benefits local communities and that its staff is well paid.

Selecting a Policy

To select a policy, further analysis could be done in several manners. First, one could look at which countries have established the highest percentage of protected areas relative to land area and examine the specific public policies and the specific financing mechanisms that enabled this establishment of protected areas. However, this analysis can be misleading. For instance, on a national level as of 2014, the two rainforest countries with the highest percentage of their land area devoted to protected areas were New Caledonia (54.3%) and Venezuela (53.9%). In contrast, Brazil had 28.4%, the DRC had 12.1%, and Indonesia had 14.7%, while the United States had 13.9%, Germany had 37.4%, and Japan had 19.4%.[102]

Venezuela has established several large national parks including the Canaima, Jaua-Sarisariñama, Parima Tapirapecó National Park (i.e., which appears to provide ecological connectivity to Brazil's Amazonas

National Forest or Floresta Nacional do Amazonas), and Serrania de la Neblina National Park (i.e., which appears to provide ecological connectivity to Brazil's Pico da Neblina National Park or Parque Nacional do Pico da Neblina). However, with Venezuela's recent economic, humanitarian, and political crises, it is unlikely such protected areas are adequately financed by the government's domestic budgetary allocations.[103] This said, the higher percentage of land area set aside for protection does not necessarily translate into quality protected areas as many of these areas might be considered "paper parks."

On the surface, Brazil, Costa Rica, Ecuador, and India appear to have some of the better domestic policies to encourage the establishment of protected areas. For instance, Brazil has one of the more ambitious Aichi Target #11 among rainforest countries. Thus:

> By 2020, at least 30% of the Amazon, 17% of each of the other terrestrial biomes, and 10% of the marine and coastal areas, especially areas of particular importance for biodiversity and ecosystem services, are conserved through protected areas foreseen under the SNUC Law and other categories of officially protected areas such as Permanent Protection Areas, legal reserves, and indigenous lands with native vegetation, ensuring and respecting the demarcation, regularization, and effective and equitable management, so as to ensure ecological interconnection, integration and representation in broader landscapes and seascapes.[104]

There are numerous policies in Brazil and a "more comprehensive system of environmental legislation has been developed. In 2009, the Ministry of Environment updated its inventory of national environmental legislation, identifying that there were 550 legal instruments that related to implementing the global biodiversity targets."[105] Brazil's revised Forest Code, Law Number 12.651, was passed on May 25, 2012, and although not specifically targeting the creation of protected areas and not necessarily connected to domestic budgetary allocations, the law requires that 80% of a private property in the Amazon biome shall remain a forested, legal reserve.[106]

Costa Rica, through a wide variety of policies and in conjunction with establishing national parks, has transitioned its economy toward ecotourism and away from an agricultural base.

Furthermore, Costa Rica is also seeking carbon neutrality. As explained by the REDD Desk:

> The cornerstone of environmental legislation in Costa Rica is the General Environmental Law (Law 7554) of 1995, which created key institutions including the Ministry of Environment and Energy (MINAE) and the National Environmental Council. It also sets out the categories of the national system of Protected Wildlife Areas (ASP). It was complemented three years later by the Biodiversity Law (Law 7788) of 1998, which created the National System of Conservation Areas (SINAC) by merging the National Park Service, the Forestry Department and the Wildlife Service.[107]

Ecuador also had an ambitious Aichi Target, where "by 2017, Ecuador has increased the proportion of the continental territory under conservation or environmental management to 35.9%."[108] Specific Ecuador policies include in "2008 the Ecuadorian people approved a Constitution that was the first in the world to recognize the Rights of Nature,"[109] and particularly the 2002 Forestry and Conservation of Natural Areas and Wildlife Law.

India has recently developed one of the most innovative and promising public policies to encourage restoration of forests and a reduction in deforestation. This said, India's federal policy in a broad sense is valuing the country's natural capital and specifically forest cover. Through the federal government's authority to transfer payments to individual states, there is now a forest cover incentive within the "horizontal devolution formula." Essentially each state will receive a forest cover factor, and if the state increases or decreases forest cover, then the state's money from the federal government shall increase or decrease.[110] India also released in 2016 a draft National Forest Policy that states "Environmental cess, green tax, carbon tax etc. may be levied on certain products and services for facilitating ecologically responsible behaviour, garnering citizen's contribution and supplementing financial resources."[111]

Implementing a Policy

Authority for implementing these public policies and allocating government domestic expenditures can reside at the local, state or provincial, and federal levels.

On a national level, Brazil's Ministry of Environmental Affairs is responsible for much of the country's environmental laws,[112] with some shared responsibilities (e.g., at national forests) delegated to its Brazil Forest Service.[113] In addition, enforcement of Brazil's environmental laws, particularly with respect to illegal deforestation, is primarily done by the Instituto Brasileiro do Meio Ambiente e dos Recursos Naturais Renováveis (IBAMA; Brazilian Institute of the Environment and Renewable Natural Resources).[114]

Ecuador's Ministry of Environment appears to have numerous undersecretaries that have at least some degree of responsibility for protected areas including the National Department for Biodiversity, the National Forestry Department, and the National Department for Environmental Services and Conservation Tools.[115]

Costa Rica's Ministry of Environment and Energy is the implementing entity, and specifically its institution called the Sistema Nacional de Áreas de Conservación (SINAC; the National Conservation Areas System), "which promotes sustainable forestry management and regulates national protected areas as well as timber harvesting on private land."[116]

India's federal government and state governments will primarily implement the tax transfer. In addition, the Ministry of Environment, Forest and Climate Change will manage the revised National Forest Policy.[117]

Evaluating the Policy

Despite the political and economic crisis in Brazil (see Chap. 8 for evaluation of Brazil's Forest Code in the case study entitled Brazil's Forest Code and Legal Reserve Credits), Brazil is the:

> {…} country that created the greater number and extension of protected areas in the world in the 2006–2010 period. By mid-2010, 27.10% of the Brazilian Amazon biome was officially protected, as were 7.33% of the Caatinga; 8.43% of the Cerrado; 8.99% of the Atlantic Forest; 4.79% of the Pantanal; 3.50% of the Pampas; and 3.14% of the Brazilian Coastal and Marine zone (including the territorial sea and the Exclusive Economic Zone).[118]

Furthermore, the Amazon Region Protected Areas Program (ARPA; see ARPA case study in Chap. 9) and the Amazon Fund (see case study Norway's International Climate and Forest Initiative in Chap. 9) have helped contribute to the financial sustainability of Brazil's protected areas. Brazil's deforestation rate has recently increased, in part due to reduced enforcement of the Forest Code, but Brazil's dramatic reduction in deforestation from 2005 to 2013 relative to its 10-year average from 1996 to 2005 should not be understated.[119]

Ecuador's policies, although with an ambitious Aichi Target, seem to be less effective. For instance, the case study on Yasuní National Park demonstrates the ongoing challenge of insufficient funding, overlapping land use, and deforestation—despite being one of the most biologically diverse places on Earth. Furthermore, "in early 2013, Ecuador was one of 12 countries targeted by an Interpol operation to crack down on the illegal timber trade in Central and South America."[120]

Costa Rica's collection of policies has enabled the country to reverse its deforestation trend and become a global leader in ecotourism (see case study on Rara Avis Rainforest Lodge & Reserve in Chap. 12), along with successfully completing several unique financing arrangements such as debt-for-nature swaps (see case study Área de Conservación Guanacaste or Guanacaste Conservation Area in Chap. 13) and bioprospecting (see case study Merck & Co. and Costa Rica's INBio in Chap. 14).

India's policy is still relatively new, but has the potential to be a significant public policy development. This said, the size of the government's annual, domestic expenditure to individual states to incentivize and reward forest conservation is estimated at approximately USD$6 billion.[121]

Future Outlook for Instrument

There are elements that are within the control of the domestic governments and elements that are out of their control. As previously mentioned, governments in developing countries often own the majority of the country's land, as opposed to the private sector. This creates a potential opportunity for the expansion of existing national parks and the creation of additional protected areas. However, many governments are facing budget constraints, and domestic budgetary allocations to protected areas provide few opportunities for the private sector. Nevertheless, domestic budgetary allocations will remain a critically important source of financing for the conservation of tropical rainforest for the foreseeable future.

Other Resources on Government Domestic Budgetary Expenditures

Conservation International's Global Conservation Fund

- http://www.conservation.org/projects/Pages/global-conservation-fund.aspx

ENVIS Centre on Wildlife & Protected Areas' National Parks of India

- http://wiienvis.nic.in/Database/npa_8231.aspx

Parks Watch (Including Profiles on Select Parks in Central and South America)

- http://www.parkswatch.org/

Protected Planet, Including World Database on Protected Areas and Protected Areas Management Effectiveness

- https://www.protectedplanet.net/

Smithsonian Tropical Research Institute's Center for Tropical Forest Science and Forest Global Earth Observatories

- http://www.ctfs.si.edu/group/About/

United States Department of Agriculture's Conservation Reserve Program

- https://www.fsa.usda.gov/programs-and-services/conservation-programs/conservation-reserve-program/index

United States Department of Agriculture's Conservation Stewardship Program

- https://www.nrcs.usda.gov/wps/portal/nrcs/detailfull/national/programs/financial/csp/?cid=nrcs143_008316

U.S. Fish & Wildlife Service's Cooperative Endangered Species Conservation Fund

- https://www.fws.gov/endangered/grants/

U.S. Fish & Wildlife Service's Great Ape Conservation Fund

- https://www.fws.gov/international/wildlife-without-borders/great-ape-conservation-fund.html

U.S. Fish & Wildlife Service's Migratory Bird Conservation Commission and Fund

- https://www.fws.gov/refuges/realty/mbcc.html

Notes

1. World Wildlife Fund. *Guide to Conservation Finance*. 41.
2. Levitt. *From Walden to Wall Street*. 4.
3. The Nature Conservancy. "The World's Oldest National Park: Ghosts of Monks and Red Deer." Last modified November 10, 2009. http://blog.

nature.org/conservancy/2009/11/10/worlds-oldest-national-park-mongolia-nature-charles-bedford-bogdkhan/
4. IUCN and UNEP-WCMC. "The World Database on Protected Areas (WDPA) [Online]." Accessed December 13, 2016. Cambridge, UK: UNEP-WCMC, 2016. www.protectedplanet.net
5. California Department of Parks and Recreation. "Yosemite 'State Park'." Accessed November 23, 2016. http://www.150.parks.ca.gov/?page_id=27539
6. National Park Service. "Yellowstone National Park: Frequently Asked Questions: History." Accessed November 23, 2016. https://www.nps.gov/yell/learn/historyculture/history-faqs.htm
7. Iobst, Steve. Interviewed by Brian McFarland. September 2016.
8. Martin. *On the Edge*. 179.
9. IUCN and UNEP-WCMC. "The World Database on Protected Areas (WDPA) [Online]." Accessed December 13, 2016. Cambridge, UK: UNEP-WCMC. 2016. www.protectedplanet.net
10. Ibid.
11. Martin. *On the Edge*. 179.
12. IUCN and UNEP-WCMC. "The World Database on Protected Areas (WDPA) [Online]." Accessed December 13, 2016. Cambridge, UK: UNEP-WCMC. 2016. www.protectedplanet.net
13. Martin. *On the Edge*. 179.
14. Jane Goodall Institute. "Gombe Stream Research Center." Accessed November 29, 2016. http://www.janegoodall.org/what-we-do/protecting-great-apes/gombe/
15. Tanzania National Parks. "Gombe National Park." Accessed November 29, 2016. http://www.tanzaniaparks.go.tz/index.php?option=com_content&view=article&id=28&Itemid=186
16. Martin. *On the Edge*. 179.
17. Maretti, Cláudio. "The Amazon: There is Hope!—If We All Do the Right Thing." In *On the Edge*, written by Claude Martin. Berkeley: Greystone Books, 2015. 208.
18. McQueen, Mike and Ed McMahon. *Land Conservation Financing*. 31–33.
19. McQueen, Mike and Ed McMahon. *Land Conservation Financing*. 2.
20. More, Jeffrey T. "The Gray and the Green: The Built Infrastructure and Conservation Investment." In *From Walden to Wall Street*, edited by James Levitt. Massachusetts: Lincoln Institute of Land Policy, 2005. 179.

21. Martin. *On the Edge*. 179.
22. Roberts, J. Timmons and Nikki Demetria Thanos. *Trouble in Paradise*. 133.
23. SOS Yasuní. "Yasuní National Park." Accessed December 3, 2016. http://www.sosyasuni.org/en/index.php?option=com_content&task=view&id=36&Itemid=26
24. Martin. *On the Edge*. 179.
25. Martin. *On the Edge*. 179.
26. Martin. *On the Edge*. 180.
27. Martin. *On the Edge*. 182.
28. Maretti, Cláudio. "The Amazon: There is Hope!—If We All Do the Right Thing." In *On the Edge*, written by Claude Martin. Berkeley: Greystone Books, 2015. 209–210.
29. Ibid.
30. Ibid.
31. Ibid.
32. Ibid.
33. Elliman, Kim and Peter Howell. "Anatomy of a Finance Program: Lessons from a Conservation Lender." In *Conservation Capital in the Americas*, edited by James N. Levitt. Massachusetts: Lincoln Institute of Land Policy, 2010. 150–152.
34. McQueen, Mike and Ed McMahon. *Land Conservation Financing*. Appendix A.
35. Cook, Ernest and Matt Zieper. "State and Local Government Funding of Land Conservation: What Is the Full Potential?" In *From Walden to Wall Street*, edited by James Levitt. Massachusetts: Lincoln Institute of Land Policy, 2005. 52.
36. Hopper and Cook, *Conservation Finance Handbook*. 15.
37. National Park Service. "Budget Justifications and Performance Information: Fiscal Year 2017." Accessed December 13, 2016. https://www.nps.gov/aboutus/upload/FY17-NPS-Greenbook-for-website.pdf
38. Iobst, Steve. Interviewed by Brian McFarland. September 2016.
39. Juffe-Bignoli, D., N.D. Burgess, H. Bingham, E.M.S. Belle, M.G. de Lima, M. Deguignet, B. Bertzky, A.N. Milam, J. Martinez-Lopez, E. Lewis, A. Eassom, S. Wicander, J. Geldmann, A. van Soesbergen, A.P. Arnell, B. O'Connor, S. Park, Y.N. Shi, F.S. Danks, B. MacSharry and N. Kingston. "Protected Planet Report 2014." Cambridge, UK: UNEP-WCMC, 2014. 7–14. http://wdpa.s3.amazonaws.com/WPC2014/protected_planet_report.pdf

40. Juffe-Bignoli, D., N.D. Burgess, H. Bingham, E.M.S. Belle, M.G. de Lima, M. Deguignet, B. Bertzky, A.N. Milam, J. Martinez-Lopez, E. Lewis, A. Eassom, S. Wicander, J. Geldmann, A. van Soesbergen, A.P. Arnell, B. O'Connor, S. Park, Y.N. Shi, F.S. Danks, B. MacSharry and N. Kingston. "Protected Planet Report 2014." Cambridge, UK: UNEP-WCMC, 2014. 8. http://wdpa.s3.amazonaws.com/WPC2014/protected_planet_report.pdf
41. UNESCO. "Biosphere Reserves—Learning Sites for Sustainable Development." Accessed December 13, 2016. http://www.unesco.org/new/en/natural-sciences/environment/ecological-sciences/biosphere-reserves/
42. Juffe-Bignoli, D., N.D. Burgess, H. Bingham, E.M.S. Belle, M.G. de Lima, M. Deguignet, B. Bertzky, A.N. Milam, J. Martinez-Lopez, E. Lewis, A. Eassom, S. Wicander, J. Geldmann, A. van Soesbergen, A.P. Arnell, B. O'Connor, S. Park, Y.N. Shi, F.S. Danks, B. MacSharry and N. Kingston. "Protected Planet Report 2014." Cambridge, UK: UNEP-WCMC, 2014. 7–14. http://wdpa.s3.amazonaws.com/WPC2014/protected_planet_report.pdf
43. Molnar, Augusta, Sara J. Scherr and Arvind Khare. "Who Conserves the World's Forests? A New Assessment of Conservation and Investment Trends." *Forest Trends and Ecoagriculture Partners*. 2004. Washington, DC. http://www.forest-trends.org/documents/files/doc_163.pdf
44. Bass, Margaret S. and et al. "Global Conservation Significance of Ecuador's Yasuní National Park." *PLoS ONE*. 5(1). 2010. 7. Accessed December 6, 2016. http://journals.plos.org/plosone/article?id=10.1371/journal.pone.0008767
45. Martin. *On the Edge*. 138.
46. Vallejo, David Romo from the University of San Francisco Quito. Interviewed by Brian McFarland. October 2016.
47. SOS Yasuní. "Yasuní National Park." Accessed December 3, 2016. http://www.sosyasuni.org/en/index.php?option=com_content&task=view&id=36&Itemid=26
48. UNESCO MAB Programme. "Biosphere Reserve Information: Yasuní." Accessed December 3, 2016. http://www.unesco.org/mabdb/br/brdir/directory/biores.asp?mode=gen&code=ECU+02
49. Boston University's Center for Ecology & Conservation Biology. "Tiputini Biodiversity Station." Accessed December 3, 2016. https://www.bu.edu/cecb/tiputini/

50. The Pontificia Universidad Católica del Ecuador. "Inicio." Accessed December 3, 2016. http://www.yasuni.ec/inicio/
51. OPEC. "Ecuador Facts and Figures." Accessed December 3, 2016. http://www.opec.org/opec_web/en/about_us/148.htm
52. PBS. "Extreme Oil: Ecuador: Economics." Accessed December 3, 2016. http://www.pbs.org/wnet/extremeoil/journey/ecuador2.html
53. Fortune Magazine. "Chevron Wins Major Ruling in $8.6 Billion Ecuador Pollution Case." Last modified August 9, 2016. http://fortune.com/2016/08/09/chevron-wins-major-ruling-in-8-6-billion-ecuador-pollution-case/
54. Larrea, Carlos, Natalia Greene, Laura Rival, Elisa Sevilla and Lavinia Warnars. "Yasuni-ITT Initiative: Bid Idea from a Small Country." Accessed April 6, 2017. https://www.google.com/url?sa=t&rct=j&q=&esrc=s&source=web&cd=1&ved=0ahUKEwjSn9-9-Y7TAhVIzIMKHRLjAGcQFggaMAA&url=http%3A%2F%2Fmdtf.undp.org%2Fdocument%2Fdownload%2F4545&usg=AFQjCNFGeqlbBay2tqfwYhf6lyx1X1zWtw&bvm=bv.151426398,d.amc&cad=rja. 16.
55. Koenig, Kevin. "China's Amazon Footprint Gets Scant Attention in Ecuador's Election." *Amazon Watch*. Last modified February 23, 2017. http://amazonwatch.org/news/2017/0223-chinas-amazon-footprint-gets-scant-attention-in-ecuadors-election
56. Vallejo, David Romo. Interviewed by Brian McFarland. October 2016.
57. Ibid.
58. Ibid.
59. Ibid.
60. Bass, Margaret S. and et al. "Global Conservation Significance of Ecuador's Yasuní National Park." *PLoS ONE.* 5(1). January 2010. 8. Accessed December 6, 2016. http://journals.plos.org/plosone/article?id=10.1371/journal.pone.0008767
61. Bass, Margaret S. and et al. "Global Conservation Significance of Ecuador's Yasuní National Park." *PLoS ONE.* 5(1). January 2010. 11. Accessed December 6, 2016. http://journals.plos.org/plosone/article?id=10.1371/journal.pone.0008767
62. Vallejo, David Romo. Interviewed by Brian McFarland. October 2016.
63. Ibid.
64. Ibid.
65. Ibid.
66. XE. "XE Currency Charts: EUR to USD." Accessed December 3, 2016. http://www.xe.com/currencycharts/?from=EUR&to=USD&view=5Y

67. Vallejo, David Romo. Interviewed by Brian McFarland. October 2016.
68. Ibid.
69. UNESCO. "Salonga National Park." Accessed December 28, 2016. http://whc.unesco.org/en/list/280
70. BBC News. "DR Congo Country Profile." Accessed February 9, 2017. http://www.bbc.com/news/world-africa-13283212
71. World Bank Group. "Doing Business: Economy Rankings." Accessed January 12, 2017. http://www.doingbusiness.org/rankings
72. Transparency International. "Corruption Perceptions Index 2016." Accessed February 8, 2017. http://www.transparency.org/news/feature/corruption_perceptions_index_2016
73. World Heritage Convention. "State of Conservation: Salonga National Park (Democratic Republic of the Congo)." Accessed February 9, 2017. http://whc.unesco.org/en/soc/3368
74. Blom, Dr. Allard. Interviewed by Brian McFarland. January 2017.
75. Reinartz, Dr. Gay Edwards. Interviewed by Brian McFarland. January 2017.
76. Blom, Dr. Allard. Interviewed by Brian McFarland. January 2017.
77. World Wildlife Fund. "Salonga National Park." Accessed February 9, 2017. http://www.wwf-congobasin.org/where_we_work/priority_places/salonga_national_park/
78. Blom, Dr. Allard. Interviewed by Brian McFarland. January 2017.
79. Reinartz, Dr. Gay Edwards. Interviewed by Brian McFarland. January 2017.
80. WWF. "Salonga Programme." Accessed February 9, 2017. http://www.wwf-congobasin.org/where_we_work/democratic_republic_of_congo/salonga_programme/
81. WWF. "Salonga Financing Agreement Signature." Accessed February 24, 2017. http://wwf.panda.org/?289151/Salonga-Financing-Agreement-Signature
82. Blom, Allard. "An Estimate of the Costs of an Effective System of Protected Areas in the Niger Delta—Congo Basin Forest Region." *Biodiversity and Conservation*. 13. 2004. 2661–2678. https://www.cbd.int/financial/values/DRCongo-costprotectedareas.pdf
83. Blom, Dr. Allard. Interviewed by Brian McFarland. January 2017.
84. Ibid.
85. Reinartz, Dr. Gay Edwards. Interviewed by Brian McFarland. January 2017.

86. Ibid.
87. Ibid.
88. Ibid.
89. Blom, Dr. Allard. Interviewed by Brian McFarland. January 2017.
90. WWF. "Salonga Programme." Accessed February 9, 2017. http://www.wwf-congobasin.org/where_we_work/democratic_republic_of_congo/salonga_programme/
91. WWF. "Salonga Financing Agreement Signature." Accessed February 24, 2017. http://wwf.panda.org/?289151/Salonga-Financing-Agreement-Signature
92. Reinartz, Dr. Gay Edwards. Interviewed by Brian McFarland. January 2017.
93. Hopper and Cook, *Conservation Finance Handbook*. 13.
94. McQueen, Mike and Ed McMahon. *Land Conservation Financing*. 46.
95. Iobst, Steve. Interviewed by Brian McFarland. September 2016.
96. Environmental Investigation Agency. "Sign on Letter." May 2, 2016. Accessed April 11, 2017. https://content.eia-global.org/posts/documents/000/000/447/original/LSE_AIM_Sign_on_Letter_ENGLISH.pdf?1468864987. 2.
97. Thoumi, Gabriel. "The Chain: London Stock Exchange Suspends Trading of United Cacao." *Chain Reaction Research*. Last modified January 17, 2017. https://chainreactionresearch.com/2017/01/17/the-chain-london-stock-exchange-suspends-trading-of-united-cacao/
98. Greenpeace. "Dirty Bankers: How HSBC is Financing Forest Destruction for Palm Oil." January 2017. Accessed April 11, 2017. http://www.greenpeace.org/international/Global/international/publications/forests/2017/Greenpeace_DirtyBankers_final.pdf. 12–23.
99. Rahmawati, Annisa. "Revealed: HSBC is Funding Forest Destruction." *Greenpeace International*. Last modified January 17, 2017. http://www.greenpeace.org/international/en/news/Blogs/makingwaves/hsbc-indonesia-forest-destruction-palm-oil/blog/58508/
100. Convention on Biological Diversity. "Aichi Biodiversity Targets." Accessed April 8, 2017. https://www.cbd.int/sp/targets/
101. United Nations. "Sustainable Development Goal 15." Accessed April 8, 2017. https://sustainabledevelopment.un.org/sdg15
102. World Bank. "Terrestrial Protected Areas (% of Total Land Area)." Accessed February 20, 2017. http://data.worldbank.org/indicator/ER.LND.PTLD.ZS?year_high_desc=true

103. The New York Times. "Crisis Upon Crisis in Venezuela." March 29, 2017. Accessed April 8, 2017. https://www.nytimes.com/2017/03/29/opinion/crisis-upon-crisis-in-venezuela.html?_r=0
104. Convention on Biological Diversity. "Brazil—National Targets." Accessed March 29, 2017. https://www.cbd.int/countries/targets/?country=br
105. Convention on Biological Diversity. "Brazil—Country Profile." Accessed April 8, 2017. https://www.cbd.int/countries/profile/default.shtml?country=br#measures
106. Presidency of the Republic, Civil House Cabinet Subcommittee for Legal Affairs. "Law No. 12.651, of 25 May 2012." Accessed December 5, 2016. http://www.planalto.gov.br/ccivil_03/_ato2011-2014/2012/lei/l12651.htm
107. The REDD Desk. "Costa Rica: Legal Frameworks." Accessed April 13, 2017. http://theredddesk.org/countries/costa-rica/legal-frameworks
108. Convention on Biological Diversity. "Ecuador: National Targets." Accessed March 29, 2017. https://www.cbd.int/countries/targets/?country=ec
109. Forest Legality Initiative. "Risk Tool: Ecuador." Accessed April 8, 2017. http://www.forestlegality.org/risk-tool/country/ecuador
110. Busch, Jonah. "India's Big Climate Move." February 27, 2015. *Center for Global Development*. Accessed April 8, 2017. https://www.cgdev.org/blog/indias-big-climate-move
111. India Institute of Forest Management, Bhopal. "National Forest Policy, 2016: Empowered Communities, Healthy Ecosystems, Happy Nation." Last modified May 2016. http://www.moef.nic.in/sites/default/files/Draft%20National%20Forest%20Policy%2C%202016.pdf. 38.
112. Ministry of Environmental Affairs. "Florestas." Accessed April 8, 2017. http://www.mma.gov.br/florestas
113. Brazil Forest Service. "Home." Accessed April 8, 2017. http://www.florestal.gov.br/
114. IBAMA. "Home." Accessed April 8, 2017. http://www.ibama.gov.br/
115. The REDD Desk. "Ministry of the Environment (Ecuador)." Accessed April 8, 2017. http://theredddesk.org/countries/actors/ministry-environment-ecuador
116. The REDD Desk. "Ministry of Environment and Energy (Costa Rica)." Accessed April 8, 2017. http://theredddesk.org/countries/actors/ministry-environment-and-energy-costa-rica

117. India Institute of Forest Management, Bhopal. "National Forest Policy, 2016: Empowered Communities, Healthy Ecosystems, Happy Nation." Last modified May 2016. http://www.moef.nic.in/sites/default/files/Draft%20National%20Forest%20Policy%2C%202016.pdf. 38.
118. Convention on Biological Diversity. "Brazil—Country Profile." Accessed March 29, 2017. https://www.cbd.int/countries/profile/default.shtml?country=br#measures
119. Earth Innovation Institute. "Brazilian Agreement on Amazon Deforestation." Accessed December 12, 2016. http://earthinnovation.org/our-work/case-studies/amazon-deforestation/
120. Forest Legality Initiative. "Risk Tool: Ecuador." Accessed April 8, 2017. http://www.forestlegality.org/risk-tool/country/ecuador
121. Busch, Jonah. "India's Big Climate Move." February 27, 2015. *Center for Global Development*. Accessed April 8, 2017. https://www.cgdev.org/blog/indias-big-climate-move

8

Tax Deductions and Conservation Easements

"Conservation easements are one of the most powerful, effective tools available for the permanent conservation of private lands. Their use has successfully protected millions of acres of land while keeping it in private hands and generating significant public benefits."[1]
—The Nature Conservancy

Introduction

As opposed to the earlier and more traditional approach where governments raise conservation finance through taxation (e.g., such as via sales taxes or real estate transfer taxes), there is also the ability for local entities (i.e., governments and/or land trusts) to utilize a tax deduction to raise conservation finance. Oftentimes, this use of a tax deduction for conservation purposes is for the establishment of a conservation easement or to encourage donations to an eligible nonprofit organization which has a conservation-related mission.

For an excellent entrepreneurial resource on establishing and operating a land trust (i.e., positioning your organization, raising money, and

B.J. McFarland, *Conservation of Tropical Rainforests*, Palgrave Studies in Environmental Policy and Regulation, https://doi.org/10.1007/978-3-319-63236-0_8

borrowing money), please read Story Clark's book entitled, *A Field Guide to Conservation Finance*.

Historical Overview

In 1891, the Trustees of Public Reservations (now the Trustees of Reservations, or TTOR) became the world's first regional land trust,[2] which established the world's first conservation easement in Boston, Massachusetts.[3]

Other notable moments in the history of utilizing tax deductions and conservation easements include:

- 1961: The Nature Conservancy (TNC) structures its first conservation easement.[4]
- 1976: US Federal Law "allows landowners to begin reporting the value of an easement as a tax deduction."[5]
- 1970s: "{…} the donation of conservation easements to a land trust, yielding a federal charitable tax deduction, a practice pioneered in the state of Maine in the 1970s,"[6] particularly by Tom Cabot along with Peggy and David Rockefeller.[7]
- 1983: North Carolina "established the first state income tax credit for owners who donated conservation restrictions on their land."[8]
- 1983: The "concept of a statewide enabling authority that will preserve special community character had its original inspiration in the Nantucket Islands Land Bank (1983), the first program of its kind in the United States."[9]
- 1989: "Conservation-minded developers were likely the first to use transfer fees and to use them successfully. A developer in South Carolina created what may be the earliest programs, adapting them from the model of early, publicly levied real estate transfer taxes. (Some government agencies tax land transfers to generate revenue for public programs). In 1989, Jim Light and his partner, James Chaffin, of Chaffin/Light Associates, instituted a private transfer 'assessment' at their Spring Island luxury development in South Carolina. This 3000-acre development includes 1200 acres of nature preserves and open space with live oak forests, marsh, and waterways {…}."[10]

- 1990s: "The land trust movement truly turned national in scope in the 1990s. For the first time {…} land trusts protected open space in all fifty states, the District of Columbia, and Puerto Rico."[11]
- 1990s: In the late 1990s, "Colorado created one of the most generous conservation tax programs with an unusual twist. Landowners donating easements may sell the tax credits to individuals or corporations that can more readily use the tax credits."[12]
- 1990: TNC facilitates a conservation easement in Costa Rica, marking the first such easement in Latin America.[13]
- 2001: The Amazon Conservation Association (ACA) and Asociación para la Conservación de la Cuenca Amazónica (ACCA) established the Los Amigos Conservation Concession, "the world's first private conservation concession."[14]
- 2003–2006: The Nature Conservancy and its partner Defensores de la Naturaleza initiate talks and then later close in June 2006 on a conservation easement deal for two parcels known as the Naranjitos, which would become a part of Guatemala's Sierra del Lacandón National Park and the greater Maya Biosphere Reserve.[15]
- 2012: Brazil's revised Forest Code, Law Number 12.651, is passed on May 25, 2012.[16]
- 2014: The Yela Forest Conservation Easement becomes one of the first conservation easements outside of the Americas.[17]

Mechanisms of Instrument

Conservation easements are essentially "a perpetual deed restriction placed on a property to protect its ecological and open space values."[18] These perpetual deed restrictions are permanent legal restrictions placed on a given property. In the United States, the specific mechanism may involve the purchase and retirement or extinguishment of residential and/or commercial development rights for a given property. In addition, the property might be donated to a nonprofit organization such as a local land trust or a large conservation organization such as The Trust for Public Land (TPL) or The Nature Conservancy (TNC), or possibly donated to a local government (i.e., a town conservation commission) so the property can be maintained as conservation land. If the parcel and/or its development rights are donated to an eligible entity, then the landowner—in the United States—is entitled to federal and/or state tax incentives. The eligible entity is often then responsible for regularly monitoring the property to ensure the deeded restrictions are being followed and, if not, the entity can legally enforce the restrictions by using litigation, if necessary.

In the United States:

> The conservation easement has become a very popular and successful tool for three reasons. First, the landowner does not have to relinquish ownership of the property and can still live on it, sell it, or pass it on to heirs. Second, the donation of a conservation easement entitles the donating landowner to tax savings for charitable contributions. Finally, conservation easements do not necessarily prohibit {all} development.[19]

Easements that allow for some development are also known as limited development projects. A limited development project, for instance, could involve the establishment of a residential condominium project, in conjunction with the preservation via a conservation easement of an adjacent watershed and primary forest.

Furthermore, land trusts can manage their properties in numerous ways to encourage recreational use, public visibility, and to conserve the ecological value of the land:

> Some land trusts conduct field trips to visit protected properties. This is a particularly attractive opportunity for supporters if the properties are under easement and not regularly accessible to the public. Other land trusts create trail systems that encourage citizens to see and enjoy the landscape. Some sponsor workdays, fishing trips, hikes, mountain bike trips, and overnight camping trips; others bring in speakers, take on educational project such as inviting schools to use protected land, and hold events featuring stories by old-time landowners or discussions by biologists and ecologists.[20]

While the landowner may earn federal and/or state tax incentives, the purchaser of a conservation easement might raise funds through traditional fundraising channels and/or by acquiring bridge financing through local or regional revolving funds.

There are a variety of ways to structure a purchase and sale. For instance, "several legal instruments can give a buyer time to raise money. They include options, lease options, rights of first refusal, and certain structures of sales contracts. Low-cost options are the best because they buy time during which the buyer can decide whether to make the purchase but the seller cannot sell to another party."[21]

An historical advantage of land trusts and nonprofit organizations, such as the TPL, is that they were "nimbler and less burdened by the bureaucratic constraints than the government was and could offer tax incentives. TPL could often buy high-priority land or easements for less than their market value. Additionally, landowners usually preferred to negotiate land sales with private organizations rather than with governments."[22] Finally, in the United States, land trusts can insure their conservation easements via a variety of pooled insurance products.[23]

While conservation easements have been used in developed countries such as Australia (i.e., where they are called "conservation covenants"), Canada, and the United States, their use in countries with tropical rainforests is more limited. Such conservation easements have been used in Costa Rica, Guatemala, and Mexico,[24] but the difficult task is how to apply these conservation easements with a tax deduction benefit and local land trust model to countries such as Brazil, the Democratic Republic of the Congo, and Indonesia?

In addition to conservation easements, in some countries such as the United States, individuals and businesses can donate to eligible nonprofit organizations—including organizations with a mission related to tropical rainforest conservation—and receive a tax deduction.

Size of Instrument

According to the National Conservation Easement Database (NCED), there are more than 130,000 conservation easements in the United States which are protecting an estimated 40 million acres (16.2 million hectares).[25] Similarly, the 2015 National Land Trust Census Report states that in the United States:

- 56 million acres (22.7 million hectares) were conserved by state, local, and national land trusts as of year-end 2015.
- 6.25 million people visited land trust properties in 2015.

- 77% of the total acres owned and under easement were held by an accredited land trust.
- USD$2.18 billion was the amount in endowments and dedicated funding managed by state, local, and national land trusts.
- Land trusts had more than 8,000 staff, nearly 16,000 board members, more than 200,000 volunteers, and nearly 5 million members and financial supporters.[26]

The number of conservation easements in tropical rainforest countries is more difficult to obtain. The Nature Conservancy (TNC), as of 2014, had reportedly worked with 23 countries on international conservation easements. This figure included 8 easements (i.e., 7 in Guatemala and 1 in Costa Rica) held by TNC.[27]

With respect to the size of foundations and philanthropy, according to Steven Lawrence, Director of Research for the Foundation Center, the top 20 climate change foundations contributed approximately USD$798 million and made a total of 602 grants in 2008.[28] Some of the family and corporate foundations that have been involved in forest conservation include:

- David and Lucile Packard Foundation;
- Doris Duke Charitable Foundation;
- Ford Foundation;
- Gordon and Betty Moore Foundation; and
- Turner Foundation.

The Gordon and Betty Moore Foundation provided a "ten-year, USD$261 million grant to Conservation International and committed USD$40 million to its Andes-Amazon Initiative to help 2.7 million square miles {699 million hectares} in South America."[29] For an excellent resource written by Environment Services and commissioned by Gordon and Betty Moore Foundation, see *An Analysis of International Conservation Funding in the Amazon.*[30]

Introduction to Case Studies

The following case studies will look at Brazil's Forest Code and the creation of Legal Reserve Credits, along with the Naranjitos Conservation Easements in Guatemala and the Yela Forest Conservation Easement.

Brazil's Forest Code and the creation of Legal Reserve Credits are not a traditional conservation easement in the sense that the landowner receives a tax deduction. However, a landowner in Brazil may be eligible to effectively place a lien on their property and transfer the rights to the excess forest cover (as defined by the Forest Code) to another landowner who has a deficit in forest cover in exchange for compensation.

The Naranjitos Conservation Easements in Guatemala is a successful example of The Nature Conservancy, and its local partner La Fundación Defensores de la Naturaleza, structuring a conservation easement deal for 2 parcels known as the Naranjitos, which would later become a part of the Sierra del Lacandón National Park and the greater Maya Biosphere Reserve. The Maya Biosphere Reserve and the adjacent landscape known as the Selva Maya Initiative form the largest contiguous rainforest north of the Amazon.

The Yela Forest Conservation Easement case study is another successful example of utilizing a conservation easement to help conserve tropical forests. In fact, this was one of the first conservation easements structured outside of the Americas, the easement has a unique feature that provides for perpetual financing, and the easement is helping to protect the largest remaining ka (*Terminalia carolinensis*) forest in the world.

Case Study #1: Brazil's Forest Code and Legal Reserve Credits

Introduction

The Brazilian Forest Code (Forest Code) was initially established in 1965, but the Forest Code was not really enforced as a law for decades. During the past years, the improvement on the enforcement of the Forest Code led to a massive discussion of the law and its revision. Despite the reduction of requirements in the law, the recently approved Brazilian Forest Law Number 12.651 of May 25, 2012[31] (commonly referred to as the "new Forest Code"), still imposes a series of obligations among landowners and/or users that have the potential to generate positive social and environmental impacts on an unprecedented scale. The main requirements of the new Forest Code are that all rural properties in Brazil shall:

1. Be entered in the newly created Cadastro Ambiental Rural (CAR; Rural Environmental Registry), which is a georeferenced, digital registry system connected to satellite images that enables monitoring and mapping of land use in rural properties. Once complete, this system will provide an invaluable tool for land use planning and management and for the government to enforce compliance with this legislation. According to the new Forest Code, all rural properties will have to be in the CAR within a maximum of 2 years after publication of a government normative, after which they will be subjected to a series of legal sanctions and will lose the right to access agricultural finance (Fig. 8.1).
2. Maintain natural vegetation along water courses and steep slopes in order to protect water resources and prevent erosion. These areas are called Areas de Preservação Permanente (APPs; Areas of Permanent

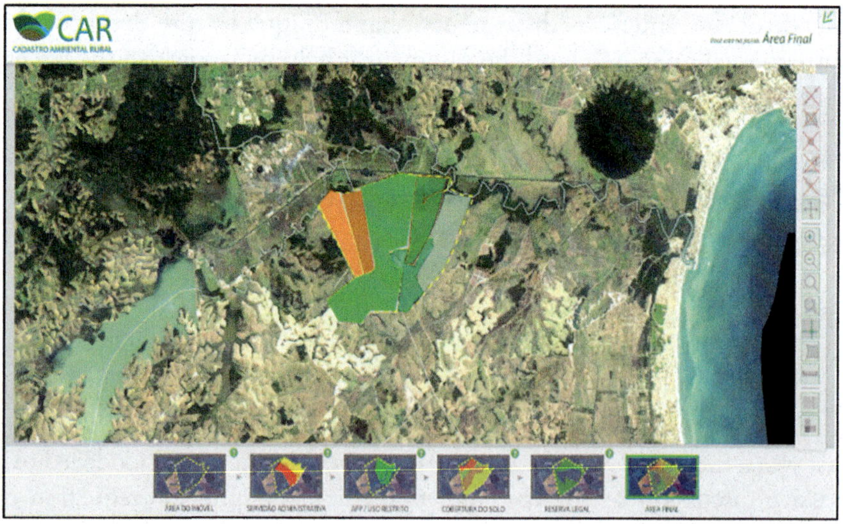

Fig. 8.1 Example of property listed in CAR (Credit: Pedro Moura Costa)

Preservation), and the size of the APPs varies depending on the width of the water course, the slope, and the size of the farm. A large number of rural properties, however, do not currently maintain sufficient APPs.

3. Maintain a certain amount of land under native vegetation (called "Forest Legal Reserves" or "Reserva Legal," in Portuguese). The size of Forest Legal Reserves can vary from 20% to 80% of rural properties, according to the type of vegetation (biome) and the region in which the properties are located. More specifically, properties with forest cover located in the Amazon Biome need to hold an 80% legal reserve, while properties in other biomes (e.g., Atlantic Forest, Cerrado Pampa, Caatinga, and the Pantanal) need to keep only 20% of native vegetation. An exception is made in the case of properties containing Cerrado vegetation located in the Amazon region, which need to maintain a 35% legal reserve. Properties that do not have a sufficient area of legal reserve can comply with the law in different ways such as by planting or regenerating areas within the rural property itself or through two offsetting mechanisms created in the Code: Cotas de Reserva Ambiental (CRAs; "Forest Reserve Credits" or "Environmental Reserve Credits") and the Compensação em Unidades de Conservação (UCs; Consolidation of Conservation Areas) by "purchasing privately held land inside areas subsequently designated as conservation units and donating them to the government (Fig. 8.2)."[32]

While not a traditional conservation easement as administered in the United States with an accompanying tax deduction, this mechanism still effectively places a lien on the landowner's property in exchange for compensation. A CRA represents 1 hectare (2.47 acres) of "forest legal reserve that is surplus to the amount required by law to be maintained in any given rural property."[33] In order to utilize the CRAs, the landowner's property must be in the same Brazilian state and be within the same biome, and the landowner's property must have been deforested prior to July 2008. The CRAs can be established for a definitive period of time or in perpetuity.

To help facilitate the transactions of Forest Reserve Credits, in 2011, Pedro Moura Costa and his brother Mauricio Moura Costa created Bolsa de Valores Ambientais Rio de Janeiro (BVRio, Rio de Janeiro

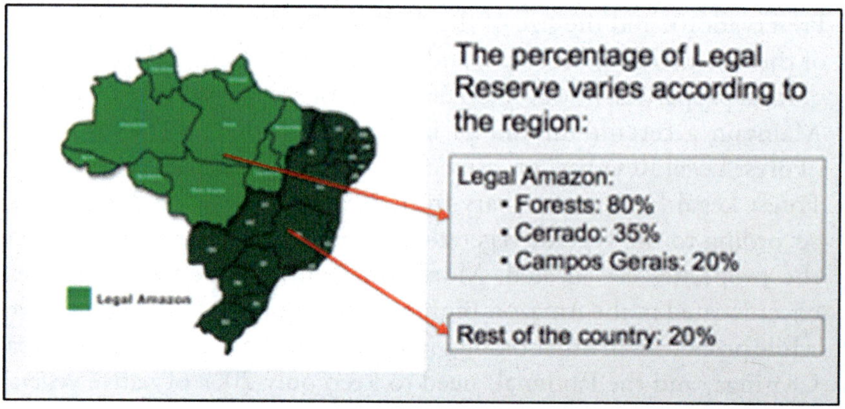

Fig. 8.2 Legal reserve requirements in Brazil (Credit: Pedro Moura Costa)

Environmental Exchange) and the BVRio Institute. BVRio operates the environmental exchange, and BVRio Institute develops policies and promotes the concept of market mechanisms to facilitate compliance with environmental laws through publications, policy papers, raising awareness, and training.[34]

CRAs are transacted on the BVRio exchange via a Contrato de Desenvolvimento e Venda de Cotas de Reserva Ambiental para Entrega Futura (CRAF; Contract of Development and Sale of Forest Reserve Credits).

The general process, which can slightly vary among states, for a landowner to have CRAs created for their property is as follows:

- Complete application form.
- Provide documentation identifying the property owner and provide documentation demonstrating landownership.
- Provide the Certificado de Cadastro de Imóvel Rural (CCIR; Certificate of Registration of Rural Property).
- Submit a topographic map prepared by a qualified professional, along with a description of the property and an explanation of how one would access the property.
- Declaration of up-to-date Imposto Territorial Rural (ITR; Rural Land Tax).[35]

Identify the Problem

One of the main problems that BVRio and the Forest Reserve Credits mechanism were designed to address is the fact that an estimated "{…} 4 million properties covering ca. 250 million hectares {618 million acres} don't have sufficient area of Legal Reserve. After matching those with surplus and deficit, it is estimated that there is still a net Legal Reserve 'deficit' of around 30 to 60 million hectares {74 million to 148 million acres}."[36] The CRA mechanism was designed to facilitate compliance with the Brazilian Forest Code.

Why the Problem Is Important

The problem is important to a variety of stakeholders. First, landowners need to legally comply with the new Brazilian Forest Code and seek cost-effective solutions. As explained by BVRio, "those who do not have sufficient area of Legal Reserve can comply with the law in different ways. One option is planting or regenerating areas within the rural property itself. However, this requires landowners to set aside areas currently used for other agricultural activities and engage in the expensive activity of forest restoration."[37]

Second, the government needs to promote compliance and will eventually need to enforce this compliance, due to the pressure they are receiving from companies and NGOs.

Third, the problem is important to companies—particularly large purchasers of Brazilian agricultural commodities such as cattle and soy—because of the pressure they are facing from stakeholders.

Fourth, the problem is important to NGOs—especially environmental NGOs—because non-compliance with the Brazilian Forest Code translates into deforestation, habitat fragmentation, and opportunities for forest restoration.

How Problem Was Identified

There are 3 pressure groups which have contributed to the identification of non-conformance with the new Brazilian Forest Code. These pressure groups are the Brazilian Government, NGOs, and companies (i.e., particularly large, purchasing companies).

Although the Brazilian Government has not toughened up on enforcement, the government has heavily invested in remote sensing technology and is equipped to identify non-compliance areas and threats to primary forest.[38]

In parallel, if the Brazilian Government does not enforce compliance with the Brazilian Forest Code, the system of remote sensing technology can be utilized by the NGO community to identify the ongoing problems of non-conformance. These NGOs include domestic organizations such as Imazon and IPAM, along with international organizations including WWF, TNC, and Conservation International.[39] In addition, an Observatório do Código Florestal (OCF; Forest Code Observatory) was established to monitor implementation with the Code.[40]

In addition, purchasing companies—particularly large buyers of Brazil's agricultural commodities such as cattle and soy—are making global commitments that the commodities they purchase are in compliance with local laws and do not contribute to deforestation.[41]

Effectiveness of Process for Identifying Problem

The process of the 3 pressure groups identifying the problem of non-conformance and the process of creating BVRio to provide solutions for compliance have been relatively effective. However, while the process has effectively identified the problem, the problem is still not resolved.

Steps Taken to Address the Problem

Several of the steps taken to address the problem include:

- Revision and passage of the new Forest Code, Law Number 12.651, on May 25, 2012.[42]
- Establishment of the Cadastro Ambiental Rural (CAR; Rural Environmental Registry).
- BVRio is established as an exchange to transact Forest Reserve Credits.

BVRio also published a practical guide for how buyers can detect non-compliance and, if detected, how to then demand compliance from suppliers with the new Brazilian Forest Code.[43]

Results

Although the passage of the new Forest Code and the establishment of BVRio are still relatively new, there are currently an estimated 5,000 participants representing 5 million hectares {12.4 million acres} of rural land on the BVRio exchange.[44] Furthermore, according to the federal government, "66% of the total area subject to registration was registered with the CAR by January 2016. Of the total of ca. 398 million hectares {983 million acres} that need to be included in the CAR, however, there are 135 million hectares {333 million acres} that still needs to be registered."[45]

Standardized contracts have been developed, and users are able to modify the price (R$ per hectare), size of the reserve in hectares (similar to the number of CRAs available for sale), and the lifetime of the CRAs (denoted in years).[46]

In addition, the trading platform has been established and officially launched. Users are now able to review:

- General location of the property, including both the State and the biome;
- Land tenure status;
- Map and description of property; and
- Whether the property is registered with the CAR and/or the land registry authorities.[47]

It is important to note that the implementation of the new Forest Code will result in significant benefits for the climate and tropical forests. For instance, BVRio Environmental Exchange has assembled a portfolio of 2 billion metric tonnes of carbon dioxide ($mtCO_2$) for an aggregate cost of approximately USD$250 million (an average of less than USD$0.13/$mtCO_2$), "based on the price that hundreds of individual farmers asked for maintaining their legal reserves."[48]

Ultimately, an innovative model has been created that continues to engage potential participants and has created a lot of awareness throughout the pre-operational stage. This model demonstrates that market mechanisms exist and that the model could be beneficial to facilitate compliance with the new Brazilian Forest Code.[49]

Challenges and How They Were Met

One of the major challenges is that there are still too few transactions and not enough liquidity for appropriate price discovery. As explained by BVRio, "at this point in time, however, CRAs are not issued in volumes large enough to allow the development of a spot market. Furthermore, many states in Brazil are still not fully equipped to enable the issuance of CRAs, delaying the process of creation of these instruments there."[50]

To address this challenge, BVRio developed specific contracts known as CRAFs to allow for the future delivery of CRAs.

Another major, related challenge was reforming the Forest Code and the opposition from a variety of stakeholders against the law. The new Forest Code, which was passed on May 25, 2012, still faces ongoing challenges of implementation and enforcement. To address this challenge, "assistance to producers registering with the CAR and use of the Compensation Mechanisms, as well as support to enforcement agencies and increase transparency are essential factors in making the Law achievable."[51]

Beyond Results

The results of the overall program, including the new Forest Code and BVRio, will be sustainable when things eventually get back on track for the country of Brazil. This said, the difficulty of implementing any new program is the initial stage. However, if there are no changes in the obligations of landowners under the new Forest Code, if Brazil's monitoring capacity continues, and if Brazil's enforcement of compliance with the new Forest Code improves, then the program should remain in place.[52]

Currently, a similar exchange for excess Forest Reserve Credits is not being implemented in any other tropical rainforest country, but the concept could certainly apply elsewhere.[53]

Lessons Learned

A few lessons learned by Dr. Pedro Moura Costa include:

The BVRio exchange for Forest Reserve Credits has seen less volume transacted than expected, not because the exchange is in the wrong place, but rather because the exchange came online at the wrong time. The political crisis in Brazil has resulted in weak enforcement of environmental laws, such as the new Forest Code. If the mechanism was geared toward a domestic market in the United States, the European Union, or China, there would likely be greater success.[54] This said, the exchange was launched at a time when Brazil was undergoing severe political and economic crises. Such crises include: the impeachment of now-former President Dilma Rousseff in August 2016[55]; the arrest of now-former lower house speaker Eduardo Cunha in October 2016[56]; numerous arrests and scandals related to corruption at some of Brazil's largest companies including Odebrecht, Andrade Gutierrez, and Petrobras[57]; and a multiyear, nationwide recession spanning at least 2015 and 2016.[58] Furthermore, Brazil's then Vice President Michel Temer under former President Dilma Rousseff, who is now the country's new president, is also facing allegations of fraud and corruption.[59] These events have contributed to a massive resistance from companies to comply with environmental laws, including the new Brazilian Forest Code.

Another related lesson learned is that it is possible BVRio overengineered the platform in the beginning. This said, BVRio was not expecting the crises in Brazil and thus, in contrast, was expecting the drive to compliance would happen faster than what is currently happening.[60]

Other Resources on Brazil's Forest Code and Legal Reserve Credits

ARPA Program

- http://programaarpa.gov.br/

BVRio: Environmental Exchange

- http://www.bvrio.com/login/login.do

BVRio Institute

- http://bvrio.org/

BVRio Institute Case Study: Using Big Data to Detect Illegality in the Tropical Timber Sector

- http://wp.bvrio.org/wp-content/uploads/2016/07/BVRio-Big-data-to-detect-timber-illegality.pdf?utm_source=Live%20Ingl%C3%AAs&utm_medium=Facebook&utm_term=Relat%C3%B3rio%20Big%20Data&utm_content=Relat%C3%B3rio%20ingl%C3%AAs%20apresentado%20no%20live%20com%20Pedro&utm_campaign=Live%20Biga%20Data

Cadastro Ambiental Rural (Rural Environmental Registry)

- Federal: http://www.car.gov.br/#/
- Minas Gerais: http://www.car.mg.gov.br/#/site
- Rio Grande do Sul: http://www.car.rs.gov.br/#/site

Observatório do Código Florestal (OCF; Forest Code Observatory in English)

- http://www.observatorioflorestal.org.br/

Case Study #2: Naranjitos Conservation Easements in Guatemala

Introduction

In 1990, the Guatemalan Government created the Sierra del Lacandón National Park, which is approximately 202,429 hectares (500,000 acres). However at the time of the Park's creation,

> {…} not all the land within its borders was publicly owned. Two parcels called Naranjitos, totaling nearly 77,000 acres {31,174 hectares}, occupied a central position within the park's nucleus. {…} On June 2nd, 2006, The Nature Conservancy and its partner Defensores de la Naturaleza closed on the purchase of these two key properties. This simple act was the culmination of research, planning, documentation, and negotiation dating back to initial talks with the owners in October 2003.[61]

The Naranjitos' incorporation into the Sierra del Lacandón National Park, alongside several other protected areas (e.g., Laguna del Tigre National Park, Mirador National Park, and Tikal National Park), buffer zones, and multi-use zones, collectively makes up the Maya Biosphere Reserve in the Petén Region of Northern Guatemala (Fig. 8.3).

The Maya Biosphere Reserve is approximately 2.1 million hectares (5.2 million acres) and is contiguous to Mexico's Montes Azules Biosphere Reserve and Mexico's Calakmul Biosphere Reserve.[62] With the addition of Belize, there was a regional program called the Conservation and Sustainable Use of the Selva Maya, which from 2010 to 2014 was "being

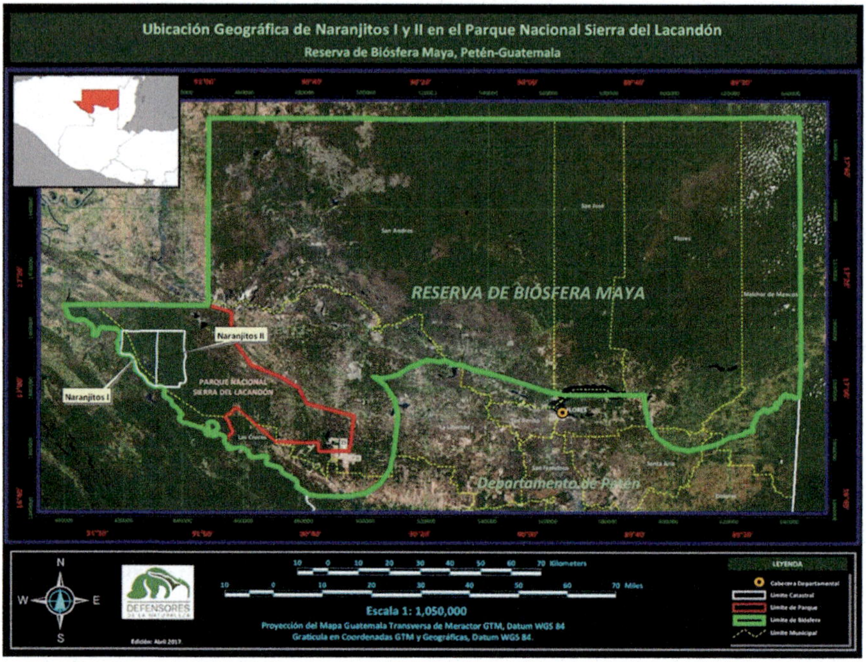

Fig. 8.3 Map of Naranjitos Conservation Easements in Guatemala (Credit: Defensores de la Naturaleza)

implemented by the Central American Commission on Environment and Development (Comisión Centroamericana de Ambiente y Desarrollo, CCAD) on behalf of GIZ (Deutsche Gesellschaft für Internationale Zusammenarbeit). It is funded by the German Federal Ministry of Economic Cooperation and Development (BMZ)."[63]

Fundación Defensores de la Naturaleza (FDN) continues to work with the Naranjitos parcel and the Sierra del Lacandón National Park, and helps to manage the Sierra de las Minas Biosphere Reserve, the Bocas del Polochic Wildlife Refuge, and the United Nations National Park.[64]

Identify the Problem

One of the major historical problems in the region is that Guatemala experienced 36 years of armed conflict. This conflict, after 100,000 people passed away and after a million more people were displaced, only came to an end with the 1996 Peace Accords.[65] During this conflict and particularly in the late 1980s and early 1990s, there was a significant increase in deforestation throughout the region as a result of illegal logging, hunting, and illegal settlement invasions within forests areas.[66]

Specific to the conservation easement, without the Naranjitos parcels, "the park was fragmented, easily invaded, poached and looted, making Lacandón's long-term conservation untenable."[67] This said, before FDN and TNC facilitated the conservation easement, there was potential problem that the Naranjitos parcels could potentially have been sold to loggers or cattle ranchers.[68]

The main threat today facing the region is the advance of the agricultural frontier that carries with it many problems including forest fires.[69] In addition, there are also ongoing problems associated with the future direction of the Maya Biosphere Reserve, particularly the Mirador National Park, and, for instance, whether bioprospecting and more ecotourism should be allowed. Likewise:

> Conservation of the Mirador region, its rainforest, and the ruins of ancient cities scattered throughout the area has been mired in controversy for more than a decade. Archaeologists, local communities, government officials, environmental organizations, and private-sector corporations have a range of opinions and proposals on how best to move forward with conservation and tourism. At its core, the issue boils down to how the area is used, who controls it, and who benefits.[70]

The problems, while complicated, will need to be equitably addressed to ensure the long-term sustainability of the Maya Biosphere Reserve.

Why the Problem Is Important

The problems associated with deforestation are important because the region includes dozens of important archeological sites of the ancient Mayan civilization, including the world famous Tikal National Park. In addition, the region "harbors tropical rainforest, old growth forests, seasonally inundated lowland forests, wetlands, rivers, pine savannas, and caves, and is home to endangered species such as the giant anteater, scarlet macaw, jaguar, tapir, jabiru stork, ocelot and spider and howler monkeys."[71]

For both the Guatemalan Government and FDN, the loss of forest cover and biodiversity is important to address. In addition, local communities have gradually become aware that conserving natural resources is important for future generations. This said, the local communities have also seen how climate change has impacted microclimates in their communities and are concerned about these developments.[72]

Furthermore, it is important to note that the Selva Maya Initiative, which includes the protected areas, buffer zones, and multi-use zones throughout Belize, Guatemala, and Mexico, forms the largest contiguous rainforest north of the Amazon.[73]

How Problem Was Identified

The problems in the region have been identified through constant monitoring and presence in the strategic areas by Resource Guards. These Guards are able to identify areas vulnerable to deforestation, depredation, and invasions and are then able to inform the authorities for follow-up action.[74]

Local NGOs, such as FDN, along with international NGOs including TNC, Rainforest Alliance, and Wildlife Conservation Society, also work in the region.

Effectiveness of Process for Identifying Problem

With respect to the Naranjitos easements, the Guatemalan Government was fully supportive of the acquisition of these lands within the Sierra del Lacandón National Park. Naranjitos I and II were previously owned by a family of cattle ranchers in Guatemala, and the family was satisfied with the conservation easement. In 1999, FDN and TNC purchased "Centro Campesino," which was an 8,000 hectare parcel also located in Sierra Lacandón; this was owned by a community years before the civil war came to the Sierra Lacandón area. TNC purchased the land and the Guatemalan Government bought the communities some land outside of the protected areas. The properties that were purchased from the communities involved holding several meetings over a 2-year period to negotiate ownership. In the end, the communities were totally satisfied with the conditions of the purchase. Consultation with communities was not necessary for the other parcel owned by the family of cattle ranchers.

Steps Taken to Address the Problem

Some of the main steps taken to address the problem include:

- Establishing an institutional presence in the protected areas as a strategy to address threats;
- Community work over many years has had a positive impact on the conservation of natural resources. For instance, FDN works with communities that are in the park so that their agricultural activities adopt good practices, which helps to eliminate the problems associated with the advance of the agricultural frontier, thus reducing forest fires.
- The land purchase strategy and conservation easements on the Naranjitos parcels have also been important for the conservation of strategic ecosystems. More specifically, this included negotiation of the land, purchase of the land, registration of the property, and then placement of the conservation easement.[75]

A few additional steps taken to address the problem include the establishment of the Sierra del Lacandón National Park and the establishment of the Maya Biosphere Reserve in 1990.[76]

Furthermore, the Tropical Agricultural Research and Higher Education Center (CATIE) "received a multi-year $1 million grant from the Inter-American Development Bank to implement a tri-national conservation project in the Maya Forest based on four strategies from the eco-regional plan."[77]

Results

One of the major results with respect to the Naranjitos parcels is that:

> In accordance with Conservancy policy, the land acquisition project includes not only $2.5 million for the purchase and transaction costs. In consideration of the size and management requirements of Naranjitos, another $2 million will be raised for short term stewardship and to establish an endowment to fund the area's long term conservation. The title for this property is held by our partner Defensores de la Naturaleza who will manage the property and oversee stewardship activities. A conservation easement ensures protection of this forest in perpetuity.[78]

According to FDN, the forest areas have been very well conserved, especially in areas owned by FDN and the community areas that have signed cooperation agreements with CONAP (Consejo Nacional de Areas Protegidas; The National Council for Protected Areas). In the Sierra del Lacandón National Park, more than 17,000 hectares (41,990 acres) of forest have been recovered over the last 4 years, there are more than 100,000 hectares (247,000 acres) in the "intangible area" that are well preserved, and deforestation has been reduced from 1,200 hectares (2,964 acres) per year to almost 500 hectares (1,235 acres) per year.[79]

In addition,

> Through the Certified Sustainable Products Alliance (CSPA), a three-year partnership between the Rainforest Alliance {RA} and USAID's Global Development Alliance, RA has been working to increase biodiversity conservation efforts and improve the livelihoods of farmers and workers in Central America and Mexico. Since its inception in 2004, the CSPA has enjoyed great success in bringing certified products to market by partnering with over 30 private sector partners, such as Kraft Foods, Nespresso, IKEA, Chiquita Brands International, Citigroup, and Gibson Musical Instruments who have made firm and significant commitments to purchasing sustainable products. {…} In Guatemala, more than 9,500,000 boxes of certified bananas have been sold, improving working conditions for more than 8,500 farm workers. More than $7,300,000 in gross sales of certified coffee has been generated in Guatemala, improving working conditions for more than 12,000 farm workers.[80]

Furthermore, there is monitoring of the populations of jaguars and other wildlife.[81]

Challenges and How They Were Met

Some ongoing challenges are poaching, illegal logging, and invasions into the area. These activities take place in remote areas of Lacandón and Sierra de las Minas, which makes access difficult and expensive for park rangers and police. Despite these challenges, FDN provides support for patrols of 4–10 people to access these remote areas.

Another ongoing challenge is that:

> The main point of contention with regard to the proposed Mirador wilderness area is that it includes lands that belong to three forest concessions: La Gloria, Uaxactún, and Carmelita. There are eleven active forest concessions in the multi-use zone of the Maya Biosphere Reserve. La Gloria and one other are managed by companies, and nine by communities, including Uaxactún, and Carmelita. Communities harvest, process, and market wood and a variety of non-timber forest products, including ornamental xate (*Chamaedorea elegans, C. ernesti-augusti, C. oblongata,* and *C. nerochlamys*) palm fronds and ramón (*Brosimum alicastrum*) tree seeds.[82]

Furthermore, the Maya Biosphere Reserve has insufficient funding. The main donor had been the Guatemalan Government through its Forest Incentives Program (PINFOR), which recently ended. Countries such as Germany and the United States are currently donors, and several NGOs including TNC and CI previously supported conservation in the area for many years, but no longer provide additional financial resources. To address this challenge, FDN hopes the sale of carbon offset credits from the GuateCarbon REDD+ Project will provide future financing for protection.[83] The GuateCarbon REDD+ Project has been validated to the voluntary carbon certification standards known as the Verified Carbon Standard (VCS) and the Climate, Community and Biodiversity Standard (CCBS), and the Project has been verified to the CCBS.[84]

Beyond Results

According to FDN, many communities have begun to manage their forests in a sustainable way, and it is expected that this will be replicated by other surrounding communities.[85] In addition, with respect to the Naranjitos parcels, the easements are in perpetuity and only low-impact tourist activities are allowed.

Lessons Learned

A few lessons learned by FDN are that the purchase of land for conservation has been a very successful strategy. This includes FDN ensuring the conservation of Guatemala's main ecosystems and biodiversity, along with important stocks of both germplasm and carbon. FDN continues to believe that the purchase of land—particularly the policy frameworks—must be strengthened, as long as it does not negatively affect local populations. This said, it is a great challenge for FDN to maintain these areas, and FDN believes that additional conservation financing mechanisms, such as payments for ecosystem services (e.g., water, carbon, biodiversity, and soils), can be coupled with conservation easements for the land's long-term conservation. It is also important to support local communities who know that these conservation areas bring many benefits in their sustainable development.[86]

Other Resources on Naranjitos Conservation Easements in Guatemala

Nature Defenders Foundation (Foundation Defensores de la Naturaleza)

- https://defensores.org.gt/en#conocenos
- https://defensores.org.gt/en/%C3%A1reas-protegidas/sierra-del-lacand%C3%B3n-national-park
- http://www.friendsfdn.org/sierra-del-lacandon-national-park

GuateCarbon Project Website

- http://guatecarbon.com/el-proyecto/

The Nature Conservancy

- https://www.nature.org/ourinitiatives/regions/centralamerica/guatemala/placesweprotect/sierra-del-lacandon-national-park-of-guatemala.xml?redirect=https-301

VCS Project Database: Reduced Emissions from Avoided Deforestation in the Multiple Use Zone of the Maya Biosphere Reserve in Guatemala (GuateCarbon)

- http://www.vcsprojectdatabase.org/#/project_details/1384

Wildlife Conservation Society: Best of the Wild: Wildlife Conservation Society and the Maya Biosphere Reserve

- http://fscdn.wcs.org/2015/07/24/7t8fl545da_MBR_prospectus.pdf

Case Study #3: Yela Forest Conservation Easement in Micronesia

Introduction

Despite its relatively small size at approximately 78 acres (31.6 hectares), the Yela Forest Conservation Easement on the island of Kosrae State in the Federated States of Micronesia (FSM) deserves special attention because in 2014, it became one of the first few conservation easements to be completed outside of the Americas (Fig. 8.4).[87]

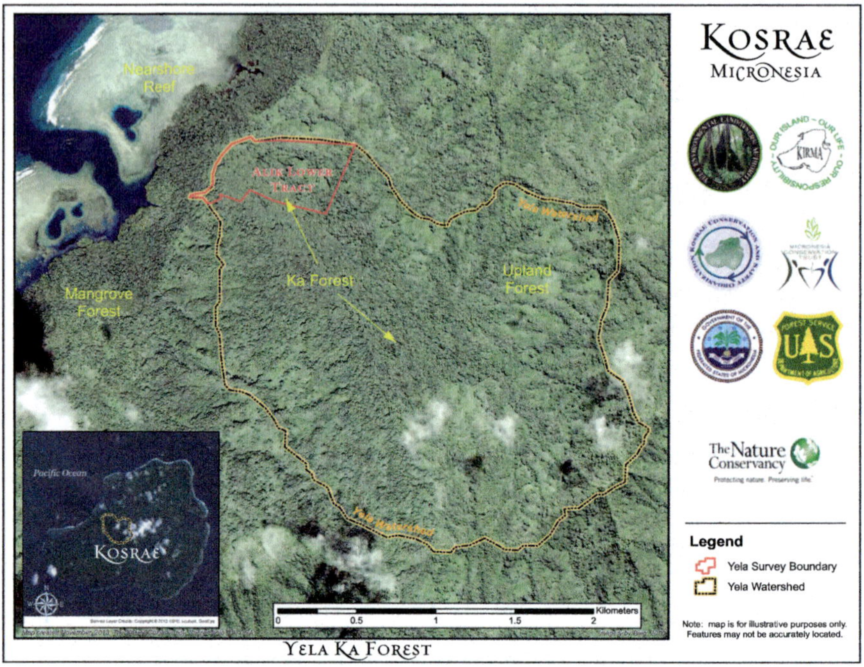

Fig. 8.4 Map of Yela Forest Conservation Easement (Credit: The Nature Conservancy)

The State of Kosrae, along with the neighboring island States of Yap, Chuuk, and Pohnpei, make up FSM. FSM is located in the Pacific Ocean near the Philippines, Indonesia, and Papua New Guinea.

Identify the Problem

The essential problem facing the land prior to the conservation easement was that a road, known as the Kosrae Circumferential Road Project, was proposed to be constructed through the land. This said, the "Alik family approached both The Nature Conservancy and the USDA Forest Service a decade ago {around 2004} when a road was proposed through the freshwater swamp forest which would likely lead to its demise through a change of hydrology and increased human access for development."[88]

Why the Problem Is Important

The reason the proposed Kosrae Circumferential Road Project was a problem is because the "entire forest, named Yela, stands at approximately 400 acres and is the largest remaining ka (*Terminalia carolinensis*) forest in the world."[89] In addition to several endemic plant species, the Yela Forest is home to the endemic ka trees and to the near threatened Micronesian imperial pigeon (*Ducula oceanica*).[90]

This rare freshwater swamp is also important to traditional livelihoods for the harvesting of eels, wild pigs, and "taro leaves for underground ovens or ums are gathered there."[91] Furthermore, "the forest provides freshwater fish from the Yela River and supports many plants used for traditional medicine. The trunk and immense buttresses of the ka tree were traditionally used to make canoes, and the nuts are an edible treat for children."[92]

It is also important to note that these freshwater swamp forests store significant amounts of carbon per acre both in the timber (i.e., large stature ka trees) and in the peat soils. Harvesting or drainage of this forest would have released this carbon stock into the atmosphere.[93]

Had the Kosrae Circumferential Road Project been constructed, the road would have disrupted the natural hydrology of the area, through diversion, impoundment, or drainage. The valley currently has undisturbed natural hydrology and ecosystems encompassing upland rainforest, freshwater swamp forest, and mangrove forest. Unfortunately, almost everywhere else in Southeast Asia and in the Pacific, such systems have been transformed by roads, conversion of mangroves to aquaculture, and/or conversion of freshwater swamp forests to taro, rice, or other agriculture/agroforestry land uses.[94]

The preservation of the Yela Forest is also important for the government to reach protected habitat goals established in the Micronesia Challenge. The Micronesia Challenge was signed in 2006 by the following five jurisdictions:

- The Commonwealth of the Northern Mariana Islands;
- The Federated States of Micronesia;

- The Republic of Palau;
- The Republic of the Marshall Islands; and
- The US Territory of Guam.

The goal of the Micronesia Challenge is to conserve 20% of terrestrial resources and 30% of marine resource by 2020.[95] This said, "{…} with the first easement in place it will offer a new tool for protection. The Yela effort will advance the terrestrial goal for the FSM and pave the way for other easement deals on Kosrae, in Micronesia, and across the Pacific and will do so for substantially less than it would cost those governments to purchase the land."[96]

How Problem Was Identified

The problem of the potential Kosrae Circumferential Road Project was identified by the landowners when they received a letter from the Kosrae State Government requesting the landowners to grant the government an easement for the construction of the road. The government had received grant funding, but the landowners wanted to conserve the area. It was at this time that the landowners approached TNC and the USDA Forest Service. TNC began exploring the conservation easement model with the government of Governor Robert Weilbacher.[97]

Effectiveness of Process for Identifying Problem

The process of identifying the problem of a potential road being constructed and then developing a conservation easement was very effective, despite the fact that the entire process took approximately 10 years. The landowners formed a nonprofit known as the Yela Environment Landowners Authority (Y.E.L.A.) and spent a lot of time educating their fellow islanders about the importance of the forest and how the forest can, among other things, bring ecotourists to the State of Kosrae to see its treasures such as the ka forest.[98]

Steps Taken to Address the Problem

There were numerous steps taken to address the problem. Such steps include:

- The landowners formed Y.E.L.A.
- Support was coordinated among several organizations including the Kosrae Island Resource Management Authority (KIRMA), the Micronesia Conservation Trust (MCT), FSM (including the FSM Department of Resources and Development) and the State of Kosrae, TNC, the Kosrae Conservation & Safety Organization (KCSO), the USDA Forest Service, and the David and Lucile Packard Foundation.
- The legal process of creating a conservation easement was undertaken and an endowment was structured with the MCT.
- The partners worked with the governors of Kosrae and received the legal opinion from the Attorney General.[99]

Results

There are numerous results from the Yela Forest Conservation Easement. The two most important results are one of, if not the, first official conservation easement outside of the Americas was established and a unique financing structure was arranged.

The Yela Forest Conservation Easement will preserve the world's largest remaining ka forest, protect an important watershed, and allow for traditional usage. Furthermore, the easement allows the Alik family to build a low-impact boardwalk and install some educational kiosks throughout the ka forest for visitors, as long as these small improvements are made using natural and biodegradable materials with no earthmoving equipment.[100] As of April 2017, the boardwalk and visitors hut are currently located outside the easement.

The easement is being managed by the KIRMA in collaboration with the landowners nonprofit Y.E.L.A. and indirectly by MCT (i.e., which manages the trust fund).[101] This said, KIRMA should conduct annual monitoring, but the wording of the conservation easement agreement allows for scheduled annual payments to proceed even if KIRMA is not timely with its monitoring. This provision is important to maintain trust and gives the landowner the benefit of the doubt, at least in the short-term.[102]

The unique financing structure involves the proceeds from the easement will be managed by the Micronesia Conservation Trust and invested by Morgan Stanley. More specifically:

> Instead of the person or persons who sign the sale agreement solely benefitting from the sale proceeds, as is often the case in the US, the Alik family has invested that income into a trust fund managed by the Micronesia Conservation Trust.
>
> Every year after a forest inspection {i.e., easement monitoring} shows the terms of the easement have been upheld, each of the families will receive interest income from the invested principal derived from the easement sale proceeds that will be close to an average salary on Kosrae. In addition a portion of the annual dividend will be reinvested in the principal to ensure the trust will grow with inflation.[103]

The cost of the easement was approximately USD$500,000, and upon closing, there was a small payout of approximately USD$25,000 to USD$30,000 to the landowners. The remaining balance was invested in perpetuity with the MCT. Thus, if the land is sold, the land is encumbered with the easement and the principal/endowment goes with the property. Another important feature of the conservation easement is that the landowners only receive a portion of the proceeds generated from their principal if they regularly pass the compliance monitoring.[104]

It is also important to note that 75% of the funding for the Yela Forest Conservation Easement was granted from the USDA Forest Service's Forest Legacy Program to KIRMA. This competitive Program is for a few, outstanding projects located in the United States and affiliated islands. Of the Legacy Projects, the Yela Forest Conservation Easement Project is outstanding for its commitment to education and accessibility, including school groups and ecotourists. Like many donors and programs, Forest Legacy does not offer 100% financing. The remaining 25% match for the

Yela Forest Conservation Easement was granted by the David and Lucile Packard Foundation to MCT and facilitated by TNC. The easement now belongs to KIRMA and the endowment is managed by MCT.[105]

Challenges and How They Were Met

The 3 big challenges were getting a legal opinion, getting title, and working through cultural differences. Although Micronesia is based off the US legal system, there was a challenge regarding whether conservation easements were legal. To overcome this challenge, the Governor of Kosrae requested the Attorney General to give a legal opinion on whether a conservation easement was legally allowed and the opinion found in favor of granting easements, as long as traditional use is allowed.[106]

With respect to getting title, many lands in Micronesia are not officially titled. More specifically, there were 10 families that owned the property, but they did not technically have title. The families had to go to Land Court and then had to publicly post that the families were requesting title and to request that any counterclaims come forward. One person did come forward and after a year of court proceedings, the families received title to 78 acres (down from approximately 84 acres).[107]

Lastly, there was a challenge working through cultural differences. For instance, there were long discussions about buying and selling land, along with discussion of how appraisals work and how you value land. This said, the USDA Forest Service needs to have an appraisal before it is able to provide a grant. Furthermore, conservation easement legal documents are long and detailed. Thus, it was a challenge to work through the documents with 10 families, especially since the easement would be placed in perpetuity and transcend generations. It was also a unique concept to explain that TNC and its partners were not actually buying the land but rather were buying out the development rights.[108]

Beyond Results

The Yela Forest Conservation Easement is sustainable as the easement will last in perpetuity. In addition, the landowners are happy about the easement, the government supports more conservation easements, and the concept is starting to spread via education to neighboring islands.

There is another potential conservation easement to be placed on an adjacent property to the Yela Forest Conservation Easement to further protect the watershed, and the funding for this second conservation easement is in place. In addition, there is funding pending in the Fiscal Year 2017 President's Budget for 75% of the cost for a third, larger parcel at the mouth of the valley, beside the Alik tract. This parcel was threatened by a proposed water development project, as well as by the potential agroforestry clearing or farm roads entering from the West.

The match for the second easement was already appropriated by the FSM legislature, which shows both local and national commitments. The match for the potentially funded third easement may come via a landowner donation (i.e., most common in the United States, but in the United States, the landowner can receive a tax donation) or from other revenue sources which may potentially include the voluntary carbon market. It is important to note that these 3 parcels (i.e., the Yela Forest Conservation Easement and the other 2 potential easements) have a different match.[109]

Furthermore, there is at least one large landowner on the island of Pohnpei who is interested in conservation easements as a result of the Yela Forest Conservation Easement.[110]

Lessons Learned

A few lessons learned by Michael Conner of The Nature Conservancy include:

- It is a good idea to get the people that are interested in protecting the land to immediately start securing title. This took a long time for the Yela Forest Conservation Easement, and in hindsight, this could have been done at the same time as securing the legal opinion of the Attorney General.
- Now that the first conservation easement is completed, future conservation easements should be easier. Future landowners will have the

security that their easement is not the first one. In addition, no one feels shorted from the Yela Forest Conservation Easement and there is a high level of trust.
- The conservation easement is an especially good model in Micronesia. This is because there is traditional landownership, and in more remote areas, the conservation easement model allows people to use their land for traditional livelihoods, the land stays in their family, and the land is protected forever.[111]

Other Resources on Yela Forest Conservation Easement

Federated States of Micronesia, Kosrae Page

- http://www.visit-micronesia.fm/state/k_top.html

Kosrae Conservation and Safety Organization

- http://kosraeconservation.blogspot.com/

Micronesia Challenge

- http://themicronesiachallenge.blogspot.com/

Micronesia Conservation Trust

- http://www.ourmicronesia.org/

USDA Forest Service's Forest Legacy Program

- https://www.fs.fed.us/cooperativeforestry/programs/loa/flp.shtml

Return

There are numerous returns—both financial and non-financial—from conservation easements. The landowner, at least in the United States, can generate state and/or federal tax savings by donating their land to an eligible nonprofit (e.g., local land trust, TPL, and/or TNC) or by donating the conservation easement. Alternatively, the landowner may generate a financial return by undertaking a limited development project. In the United States, these tax credits—depending on the state—can be transferred to a third party for a fee. These third parties are likely in need of a tax credit and, as such, may choose to purchase these tax credits at a discount.

In Brazil, a landowner may be able to generate a financial return by selling their excess legal reserve in exchange for a conservation easement placed on their property.

From a local government's perspective with respect to traditional conservation easements, there is still the opportunity—albeit at a lower rate—to collect property taxes. It is also possible that a town's property values will increase due to the presence of open space conserved via conservation easements and that this increase in property values will result in more revenue for the local government.

Additional opportunities for a financial return include the allowance for hunting and tourism.

Another source of revenue, although not necessarily via the reduction in taxes, is through the use of transfer fees. As explained by Story Clark, a transfer fee is essentially a "voluntary agreement between a landowner and a land trust that the landowner will pay a percentage of the gross sale price of a specific parcel of land to a land trust when the parcel is sold."[112]

Three ways to structure a transfer fee are as follows:

> In the first, land trusts negotiate transfer fees with individual landowners who are giving conservation easements on their property. In the second application, a small handful of conservation-minded developers fund the stewardship of the open space in their developments, associated conservation programs, or new land conservation with transfer fees on lot sales. {...} A third application of the fee has yet to receive much use, if any. A landowner can write the transfer obligation into the deed simply as a way of making a future gift to a land trust.[113]

There are also numerous non-financial returns. For example, with the onetime cost of purchasing an easement (i.e., along with the smaller, ongoing costs of monitoring), the ecosystem services provided by the protected area may include the protection of an important watershed and supply of drinking water, erosion protection, flood storage, and habitat for pollinators and seed dispersers. Conservation easements can also provide an educational opportunity for schools and recreation opportunities for local communities.

Furthermore, food security can be enhanced through preserving prime agricultural lands and preventing its conversion to an alternative land use such as residential or commercial development.

Similar to purchasing a house, there are many costs associated with purchasing a conservation easement. This includes, but is not limited to, the costs of:

- Obtaining a credit report;
- Purchasing title insurance;
- Appraisal of the land value;
- Legal expenses, including title review and environmental quality assessment (e.g., presence of hazardous materials or invasive species, water quality issues, and/or soil quality); and
- Site visits.

Risk

There are a variety of risks associated with identifying areas for possible acquisition via a conservation easement, along with structuring the purchase and maintaining the easement in perpetuity.

Business Risk

There are numerous business risks associated with purchasing, maintaining, and monitoring a conservation easement. These business risks include:

- Identifying, and possibly managing, the presence of hazardous materials;
- Presence, or later identification, of infestations of invasive species and/or disease; and
- Raising sufficient funds in time at a competitive rate.

Conservation easement insurance is a mechanism to mitigate business risk.

Strategic Risk

When selecting a particular tract of land for a possible conservation easement, some of the strategic risks that should be assessed include:

- The extent of restoration work to be done on the property. This could include little-to-no restoration work to more extensive work with respect to planting mixed, native species and recreating streams.
- Availability of additional, adjacent parcels to scale up the impact. For instance, in New England, many privately owned parcels are relatively small, and thus, it is difficult to acquire large tracts of contiguous land in order to preserve a landscape.
- Intergenerational equity, which takes place when the next generation assumes ownership and then may possibly seek to unwind or ignore the conservation easement.
- Managing the relatively few cash inflows (e.g., onetime, upfront payment for the easement or a once-in-30-year timber harvest) with the relatively more frequent cash outflows (e.g., monitoring conservation easement, property taxes, and site maintenance).

Reputation Risk

On such reputational risk occurred in 2003, when {…} "The Washington Post, in a series of stinging investigative articles directed at TNC, made allegations of insider sales with trustees and other closely affiliated people, suggesting conflicts of interest in evaluating easement gifts and sales."[114]

Liquidity Risk

Conservation easements, due to their perpetual existence, can pose a unique liquidity risk to the easement holder because the owner of the easement is unable to remove the easement and sell the land at a presumably higher value. However, the property owner can still sell the land that is encumbered with the easement, or the property owner may be able to sell, depending on the state and/or federal laws, the tax credit to an individual or company with a tax liability.

Operational Risk

Operational risks for conservation easements and tax deductions include insufficient internal controls (e.g., poor record keeping of contracting obligations associated with an easement), incompetent management, and fraud (e.g., falsifying information). Furthermore, there is significant operational risk that is associated with easements not being properly monitored. (i.e., because the easement has been forgotten about or due to a lack of will from the easement holder to undertake monitoring).

Legal and Regulatory Risk

Several types of legal and regulatory risk exist for conservation easements. This includes:

- The slight chance of eminent domain taking the land, even after the land has been encumbered with a perpetual conservation easement.
- Political opposition to conservation in favor of more residential and/or commercial development.
- Situations where the current, or a future, landowner violates the legal and regulatory requirements embedded in the conservation easement. This could result in an expensive and time-consuming litigation between the landowner and the land trust. Such violations may be intentional, or could be unintentional, as a result of a confusing or poorly translated contract.

- The conservation easements not being monitored (i.e., because the easement has been forgotten about or due to a lack of will from the easement holder to undertake monitoring).

Perhaps the greatest legal and regulatory risk is adapting the conservation easement model from the United States and applying the concept to countries with tropical rainforests. For instance, "{…} TNC and other conservation organizations have worked to adapt the legal framework that allows for conservation easements to work within the Napoleonic legal system that prevails in Chile and most of Latin America. Easement templates have been drafted and tested in several Latin American countries, including Chile. Adapting English common law legal instruments to Napoleonic law poses a significant challenge, however."[115]

Although a rare occurrence, there is the possibility that the agency or department responsible for taxation, such as the US Internal Revenue Service (IRS), may not recognize the tax deduction. This could result from a fraudulent land trust, from a land trust that loses its nonprofit status, or as a result of an indefensible valuation of the land (i.e., to increase the tax deductibility).

Furthermore, there is a risk associated with what is, or is not, included in the easement such as whether or not carbon rights are contractually transferred from the property owner to the easement holder and/or who assumes the responsibilities for force majeure as outlined in the conservation easement.

Credit Risk

Credit risk includes default risk, bankruptcy risk, downgrade risk, and settlement risk. To properly value a conservation easement and to mitigate the risk of a default or bankruptcy, Story Clark suggests that "{…} land trusts should encourage their donors to retain state-certified, respected appraisers who are knowledgeable in this area of valuation and to follow Uniform Standards of Professional Appraisal Practice. However, in many parts of the country, such appraisers are hard to come by."[116] This said, a potential credit risk is that if it is tough to appraise a property in some parts of the United States, it is presumably much more difficult to appraise the appropriate purchase price for a conservation easement on lands adjacent to a remote national park in Belize, Indonesia, or Peru.

An example of settlement risk is if the property owner transfers the tax credit and/or the conservation easement to another party, but the other party does not pay for the tax credit and/or conservation easement.

Market Risk

Market risk includes interest rate risk, equity price risk, foreign exchange risk, and commodity price risk. Interest rate risk and foreign exchange risk could make the purchase of land prohibitively expensive and simultaneously make the tax deduction insignificant. Similarly, commodity price risks could increase the value of land beyond the affordability for a local land trust. Equity price risk is less relevant to the purchase of a conservation easement for tax deductions.

Risk, Return, Time (Horizon), Taxes, Liquidity, Legal and Unique (RRTTLLU)

Risk and Return

Please see above for the risk and return associated with tax deductions and conservation easements.

Time Horizon

The time horizon for tax deductions and conservation easements will likely vary. Thus, the requirement of maintaining a conservation easement (e.g., maintaining trails, monitoring for deforestation and degradation, and/or removing invasive species) is a long-term commitment that may exist in perpetuity. In contrast, the tax deductions may exist over a shorter time frame than the length of the conservation easement.

As outlined in the case study of Brazil's new Forest Code and the transfer of Forest Reserve Credits (CRAs in Portuguese), there is a variable time horizon with the CRAs. These CRAs could, depending on the structure of the contract, exist in perpetuity or cease to exist after a finite period of time.

Taxes

There are likely reduced taxes (e.g., inheritance taxes, sales taxes, income taxes, and particularly reduced property taxes) in exchange for establishing a conservation easement. In addition, if a business is undertaking the conservation easement, then there could be a reduction in business-related taxes. However, there could be various fines, penalties, and back due taxes reinstated as a result of failing to maintain the conservation easement.

Liquidity

Properties with conservation easements may be less liquid than comparable properties. Yet, there still might be opportunities for the property to be sold or transferred to another entity such as a local government, a nonprofit organization (e.g., local land trust or international conservation organization such as TNC or TPL), or to an individual conservation buyer.

Legal

Legal considerations associated with tax deductions and conservation easements will vary from country to country. In the United States, for instance, these legal considerations for conservation easements and tax deductions will even vary from state to state. The key legal issues will be spelled out in the conservation easement contract, and the legal risk is that those legal obligations are not abided by or not transferred with the property. In addition, neighboring landowners or other stakeholders may violate the terms of the conservation easement (e.g., by harvesting timber or building roads).

Unique

Conservation easements may be challenged when legally reviewed by succeeding generations. Thus, as a land title encumbered with a conservation easement is passed from generation to generation, future generations may challenge the legal viability of a previous generation's conservation easement. Similarly, there can be a lack of understanding of what exactly perpetuity means in easement terms. It is very important to spend a lot of time on this topic, especially in poor rural areas, because the easement buyer may offer a relatively lot of money and people do not really care about the details when faced with an upfront, lump sum of money. To mitigate this risk, the decision must be the communities' decision and not what a government or NGO wants. Also, the communities have a desire to conserve the lands in perpetuity; if they are in it for the money, and not conservation, then a risk is presented.[117]

Conservation easements may also face challenges when neighboring property holders or governments seek to use the property that is encumbered with the easement. In many cases, the conservation easement monitoring funds—usually assigned to a third-party NGO—become a liability in perpetuity. This means that funds are safe kept, or should be kept, at a financial institution to fund monitoring in perpetuity. Yet these funds may not be sufficient when a well-funded adversary—such as a neighboring property holder or government—seeks to use eminent domain or other legal tools to negatively impact the legal strength of the conservation easement.

Policy Analysis

Defining the Problem

Two problems associated with tax deductions and conservation easements are the fact that there is a low level of private ownership in many of the countries that host tropical rainforests and the fact that few, if any, conservation easements have been structured in these countries. As previously mentioned, while forests are largely privately owned in countries such as the United States, Japan, and the United Kingdom, "today, central governments own almost 80 percent of the remaining intact forests in developing countries."[118] For instance:

- United States: 179 million hectares (ha) privately owned versus 130 million ha publicly owned;
- Japan: 14.8 million ha privately owned versus 10.2 million ha publicly owned;
- United Kingdom: 2.2 million ha privately owned versus 868,000 ha publicly owned;
- Brazil: 102 million ha privately owned versus 308 million ha publicly owned;
- Indonesia: 12 million ha privately owned versus 82 million ha publicly owned;
- Democratic Republic of the Congo: 0 ha reported as privately owned versus 154 million ha publicly owned[119]

This said, it is challenging to give landowners a tax deduction in Indonesia with its concession model. In addition, countries without a strong rule of law (e.g., the Democratic Republic of the Congo), countries with significant overlapping title claims (e.g., Brazil), and countries with a Napoleonic code of law as opposed to English common law (i.e., which has been the system of law used most often for conservation easements) present a unique challenge for governments seeking to establish public policies enabling conservation easements. For example:

- How would countries such as the Democratic Republic of the Congo and Indonesia, which follow the civil law system, grant conservation easements?
- To achieve a tax deduction, will the country's tax code need to be revised?
- Would there be some type of clawback provision if the terms of the conservation easement were violated?
- How would the terms of the conservation easement be enforced, especially if the proposed conservation easement is in a remote part of the country?
- Can local communities and/or Indigenous Peoples participate in conservation easements?

Establishing Goals

From a government's perspective with a goal of promoting forest conservation, one option—among the diversity of options—is to develop public policies that enable private landowners to receive a financial incentive (i.e., possibly through a tax deduction for a conservation easement) to conserve their privately owned forests.

Selecting a Policy

Although not specifically targeting the creation of conservation easements and not necessarily connected to a tax deduction, Brazil's new Forest Code requires a legal reserve of 80% on a private rural property in the Amazon Biome. The ability to trade Forest Reserve Credits is possibly a very powerful public policy that can facilitate compliance with the new Forest Code and simultaneously reduce tropical rainforest deforestation on a massive scale.

In Micronesia, the legal opinion of the Attorney General served as a cornerstone for the Memorandum of Understanding signed between the Kosrae Conservation & Safety Organization (KCSO), the Kosrae Island Resource Management Authority (KIRMA), and the nonprofit Y.E.L.A. which was established by the landowners.

Guatemala does not have specific policies about conservation easements. The most common use is "easement of passage," and it is defined in the national laws and it has national applicability.[120]

Implementing a Policy

In Brazil, both state governments and the federal government have responsibilities related to implementing the new Forest Code and particularly the registration of rural properties with the Cadastro Ambiental Rural (CAR; Rural Environmental Registry in English).

In Micronesia, the management of the endowment, along with the management of the easement and assurance of its effective implementation, will be conducted by several entities including KIRMA, the Micronesia Conservation Trust, and Y.E.L.A.[121]

If Guatemala had a conservation easement policy, the Consejo Nacional de Áreas Protegidas (CONAP; National Council for Protected Areas) would be the entity in charge of implementing the policy.[122]

Evaluating the Policy

The new Brazilian Forest Code, in part due to the country's political and economic crises, has not been effectively enforced. With better enforcement of the new Forest Code, there will be a greater need to comply with the legal reserve requirement. If these developments coincide with the registration of rural properties in the CAR, this should facilitate more transactions of Forest Reserve Credits. Taken together, the new Brazilian Forest Code could have a significant impact on the conservation of tropical rainforests.

In Micronesia, the legal opinion of the Attorney General not only allowed for the Yela Forest Conservation Easement in Kosrae but has now opened up the possibility of additional conservation easements on Kosrae and on other islands in Micronesia. The legal option, thus, seems to be very effective at facilitating more conservation easements in the region.

Despite Guatemala's lack of a specific, national policy on conservation easements, the conservation easements at Lacandón and Sierra de las Minas National Parks have been effective for conservation purposes.[123]

Future Outlook for Instrument

A potentially useful tool for conservation easements would be to establish a comparable website to the Database of State Incentives for Renewables & Efficiency (DSIRE).[124] Although no longer active, it appears a similar site was originally created by the Sonoran Institute in Tucson, Arizona, called Conservation Assistance Tools. Such a site could compile all the local, state, provincial, and/or federal incentives to purchase and manage conservation easements, and this could spur the future development of more conservation easements.

In addition, the creation of future conservation easements should assess the policy frameworks concerning private landownership and examine how conservation easements can be legally structured. For instance, while not targeting tropical rainforests, there has been work by The Nature Conservancy in Chile with conservation easements. In addition to Chile, there are similar large private lands in Uruguay, which has rainforests, and there are family forest owner associations in Europe, which is a model that could possibly be adapted to rainforest countries.[125]

Another angle to explore for conservation easements is the growth of philanthropy, particularly among the high-net-worth (HNW) and ultra-high-net-worth (UHNW) citizens of countries such as China.[126]

The future of conservation easements could start to look at alternative sites that are not privately owned in the traditional sense. For example, even if you do not own the land, but you have a long-term lease such as a forestry concession, there could be an opportunity for conservation easements. In Canada, for instance, there is a lot of environmental pressure to stop harvesting in the Northern Boreal Forests. This harvesting is economically marginal for lumber companies, and with climate change making it riskier to operate there, this opens up incentives for conservation easements to be structured to buy back a part of the timber license.[127] Perhaps the same mechanism could eventually be applied to countries such as the DRC and Indonesia.

Lastly, from the perspective of the financial sector, land trusts that purchase and manage conservation easements might present unique partnership opportunities in the future. Thus, land trusts are possibly under-banked and may lack a universal banker/investor across their geographic territories, and many land trusts have HNW directors on their board. Most land trusts have at least some institutional savings (i.e., investment of endowment) and will have demand for additional legal and financial services such as making payroll, short-term cash flow needs, and structuring both the land trust and the easements, along with tax support and assistance with insurance products.

Other Resources on Tax Deductions and Conservation Easements

Agricultural Conservation Easement Program of the US Department of Agriculture

- https://www.nrcs.usda.gov/wps/portal/nrcs/main/national/programs/easements/acep/

Conservation Almanac

- http://www.conservationalmanac.org/secure/

Conservation Resource Center's Tax Credit Exchange

- http://taxcreditexchange.com/

Foundation Center's GrantCraft

- http://www.grantcraft.org/

LandScope America

- http://www.landscope.org/

Land Trust Standards and Practices from the Land Trust Alliance

- http://www.landtrustalliance.org/topics/land-trust-standards-and-practices

Maine Coast Heritage Trust (MCHT), "which is one of the most user-friendly conservation revolving loan funds in the United States."[128]

- http://www.mcht.org/

National Conservation Easement Database

- http://conservationeasement.us/

The Conservation Registry

- http://www.conservationregistry.org/

The Cyrus R. Vance Center for International Justice, which looked into tax incentives for Chile to structure tax deductible conservation easements.

- http://www.vancecenter.org/vancecenter/

The Learning Center from the Land Trust Alliance

- http://learningcenter.lta.org/login

The Nature Conservancy's Conservation Buyer Program

- http://www.nature.org/about-us/private-lands-conservation/conservation-buyer/properties/index.htm
- http://www.nature.org/about-us/private-lands-conservation/conservation-easements/all-about-conservation-easements.xml

US Forest Service's Forest Legacy Program

- http://www.fs.fed.us/spf/coop/programs/loa/flp.shtml

Notes

1. The Nature Conservancy. "About Us: Private Lands Conservation." Accessed April 29, 2017. https://www.nature.org/about-us/private-lands-conservation/index.htm
2. Foster, Dave. "Meeting the Conservation Challenge in New England." In *Conservation Capital in the Americas*, edited by James N. Levitt. Massachusetts: Lincoln Institute of Land Policy, 2010. 20–21.
3. The Nature Conservancy. "Easements 101." Accessed December 13, 2016. https://www.nature.org/about-us/private-lands-conservation/conservation-easements/conservation-easements-101.pdf
4. Ibid.
5. Ibid.
6. Levitt, James N. *Conservation Capital in the Americas*. 3.
7. Levitt, James N. *Conservation Capital in the Americas*. 27.
8. Hocker, Philip M. "Transferable State Tax Credits as a Land Conservation Incentive." In *From Walden to Wall Street*, edited by James Levitt. Massachusetts: Lincoln Institute of Land Policy, 2005. 126.
9. Levitt (Ed.). *Conservation Capital in the Americas*. 31.
10. Clark, Story. *A Field Guide to Conservation Finance*. 164.
11. McQueen, Mike and Ed McMahon. *Land Conservation Financing*. 15.
12. Ginn. *Investing in Nature*. 124.
13. The Nature Conservancy. "Easements 101." Accessed December 13, 2016. https://www.nature.org/about-us/private-lands-conservation/conservation-easements/conservation-easements-101.pdf
14. Amazon Conservation Association. "Conservation Concessions." Accessed December 13, 2016. http://www.amazonconservation.org/ourwork/conservation.html
15. World Environment Center, The Nature Conservancy, and Rainforest Alliance. "Selva Maya Project. A Capacity-Building Alliance to Conserve the Maya Forest of Mexico, Guatemala and Belize: A Report to Pfizer." February 2007. Accessed April 11, 2017. http://www.wec.org/programs-initiatives/Selva%20Maya%20Final%20Report%202007.pdf. 15.

16. Presidency of the Republic, Civil House Cabinet Subcommittee for Legal Affairs. "Law No. 12.651, of 25 May 2012." Accessed December 5, 2016. http://www.planalto.gov.br/ccivil_03/_ato2011-2014/2012/lei/l12651.htm
17. Conner, Mile. Interviewed by Brian McFarland. April 2017.
18. Meyer, Shannon. "Markets for Ecosystem Services and Reclaiming the Great Dismal Swamp." In *Conservation Capital in the Americas*, edited by James N. Levitt. Massachusetts: Lincoln Institute of Land Policy, 2010. 199.
19. Tepper, Henry and Victoria Alonso. "The Private Lands Conservation Initiative in Chile." In *Conservation Capital in the Americas*, edited by James Levitt. Cambridge, Massachusetts: Lincoln Institute of Land Policy, 2010. 57.
20. Clark, Story. *A Field Guide to Conservation Finance*. 27.
21. Clark, Story. *A Field Guide to Conservation Finance*. 55.
22. Clark, Story. *A Field Guide to Conservation Finance*. 184.
23. Land Trust Alliance. "Insurance: Insurance for Land Trusts." Accessed March 29, 2017. http://www.landtrustalliance.org/topics/insurance
24. Tabas, Philip. "New Developments in Conservation Easements." *American College of Environmental Lawyers*. Last modified June 16, 2014. http://www.acoel.org/post/2014/06/16/New-Developments-in-Conservation-Easements.aspx
25. National Conservation Easement Database. "NCED at a Glance." Accessed December 8, 2016. http://conservationeasement.us/
26. Land Trust Alliance. "2015 National Land Trust Census Report: Our Common Ground and Collective Impact." Accessed March 29, 2017. http://s3.amazonaws.com/landtrustalliance.org/2015NationalLandTrustCensusReport.pdf. 3 and 10.
27. The Nature Conservancy. "Easements 101." Accessed April 4, 2017. https://www.nature.org/about-us/private-lands-conservation/conservation-easements/conservation-easements-101.pdf
28. Lawrence, Steven. "Climate Change: The U.S. Foundation Response." Last modified February 2010. http://foundationcenter.org/gainknowledge/research/pdf/researchadvisory_climate.pdf
29. McQueen, Mike and Ed McMahon. *Land Conservation Financing*. 126.
30. Castro de la Mata, Gonzalo and Sait Riega-Campos. "An Analysis of International Conservation Funding in the Amazon." March 2014. http://www.vale.com/brasil/PT/aboutvale/news/Documents/Amazon-Conservation-Funding-Analysis-Publication-2014.pdf

31. Presidency of the Republic, Civil House Cabinet Subcommittee for Legal Affairs. "Law No. 12.651, of 25 May 2012." Accessed December 5, 2016. http://www.planalto.gov.br/ccivil_03/_ato2011-2014/2012/lei/l12651.htm
32. BVRio. "Climate Value for Money: 2 GtCO2 Storage for U$ 250 million: Forest Legal Reserve Credits in Brazil." Last modified June 28, 2016. http://www.bvrio.org/2016/06/28/climate-value-for-money-2-gtco2-storage-for-u-250-million-forest-legal-reserve-credits-in-brazil-2/
33. BVRio. "Introduction to the Market of Legal Reserve Credits." Last modified September 20, 2012. 3.
34. Moura Costa, Pedro of BVRio. Interviewed by Brian McFarland. March 2017.
35. BVRio. "Introduction to the Market of Legal Reserve Credits." Last modified September 20, 2012. 8.
36. BVRio. "Introduction to the Market of Legal Reserve Credits." Last modified September 20, 2012. 2.
37. BVRio. "Introduction to the Market of Legal Reserve Credits." Last modified September 20, 2012. 3.
38. Moura Costa, Pedro of BVRio. Interviewed by Brian McFarland. March 2017.
39. Ibid.
40. Forest Code Observatory. "Home." Accessed March 22, 2017. http://www.observatorioflorestal.org.br/
41. Moura Costa, Pedro of BVRio. Interviewed by Brian McFarland. March 2017.
42. Presidency of the Republic, Civil House Cabinet Subcommittee for Legal Affairs. "Law No. 12.651, of 25 May 2012." Accessed December 5, 2016. http://www.planalto.gov.br/ccivil_03/_ato2011-2014/2012/lei/l12651.htm
43. BVRio. "Guia prático para a análise do atendimento ao Código Florestal." 2017. Accessed March 22, 2017. http://www.bvrio.org/wp-content/uploads/2017/03/BVRio-Proforest-Guia-Co%CC%81digo-Florestal-Observatorio-CF.pdf?_ga=1.58364847.483755547.1487799189
44. Moura Costa, Pedro of BVRio. Interviewed by Brian McFarland. March 2017.
45. BVRio. "Diagnostic, Opportunities and Challenges of Implementing the New Brazilian Forest Code." Last modified March 16, 2016. http://www.bvrio.org/2016/03/16/diagnostic-opportunities-and-challenges-of-implementing-the-new-brazilian-forest-code/

46. BVRio. "Introduction to the Market of Legal Reserve Credits." Last modified September 20, 2012. 4.
47. BVRio. "Introduction to the Market of Legal Reserve Credits." Last modified September 20, 2012. 5.
48. BVRio. "Climate Value for Money: 2 GtCO2 Storage for U$ 250 million: Forest Legal Reserve Credits in Brazil." Last modified June 28, 2016. http://www.bvrio.org/2016/06/28/climate-value-for-money-2-gtco2-storage-for-u-250-million-forest-legal-reserve-credits-in-brazil-2/
49. Moura Costa, Pedro of BVRio. Interviewed by Brian McFarland. March 2017.
50. BVRio. "Introduction to the Market of Legal Reserve Credits." Last modified September 20, 2012. 3.
51. BVRio. "Climate Value for Money: 2 GtCO2 Storage for U$ 250 Million: Forest Legal Reserve Credits in Brazil." Last modified June 28, 2016. http://www.bvrio.org/2016/06/28/climate-value-for-money-2-gtco2-storage-for-u-250-million-forest-legal-reserve-credits-in-brazil-2/
52. Moura Costa, Pedro of BVRio. Interviewed by Brian McFarland. March 2017.
53. Ibid.
54. Ibid.
55. Romero, Simon. "Dilma Rousseff is Ousted as Brazil's President in Impeachment Vote." Last modified August 31, 2016. *The New York Times*. https://www.nytimes.com/2016/09/01/world/americas/brazil-dilma-rousseff-impeached-removed-president.html?_r=0
56. The Guardian. "Brazilian Politician Who Led Rousseff Impeachment Arrested on Corruption Charges." Last modified October 16, 2016. https://www.theguardian.com/world/2016/oct/19/eduardo-cunha-arrested-corruption-charges-brazil
57. The Guardian. "Brazil Police Arrest Building Executives over Petrobras Bribes Scandal." Last modified June 19, 2015. https://www.theguardian.com/world/2015/jun/19/brazilian-police-arrest-building-executives-petrobras-bribes-scandal
58. Gillespie, Patrick. "Brazil's Worst Recession: 8 Consecutive Quarters of Contraction." Last Modified March 7, 2017. *CNN Money*. http://money.cnn.com/2017/03/07/news/economy/brazil-gdp-2016/
59. Romero, Simon. "Michel Temer Government in Brazil Reels as Dozens Face New Graft Investigations." April 11, 2016. *The New York Times*. Accessed April 13, 2017. https://www.nytimes.com/2017/04/11/world/

americas/brazil-michel-temer-investigation-petrobras-odebrecht.html?mabReward=A1&recp=0&action=click&pgtype=Homepage&region=CColumn&module=Recommendation&src=rechp&WT.nav=RecEngine&_r=0
60. Moura Costa, Pedro of BVRio. Interviewed by Brian McFarland. March 2017.
61. World Environment Center, The Nature Conservancy, and Rainforest Alliance. "Selva Maya Project. A Capacity-Building Alliance to Conserve the Maya Forest of Mexico, Guatemala and Belize: A Report to Pfizer." February 2007. Accessed April 11, 2017. http://www.wec.org/programs-initiatives/Selva%20Maya%20Final%20Report%202007.pdf. 15.
62. UNESCO. "MAB Biosphere Reserves Directory: Biosphere Reserve Information: Guatemala: Maya." Accessed April 11, 2017. http://www.unesco.org/mabdb/br/brdir/directory/biores.asp?code=GUA+01&mode=all
63. The REDD Desk. "Conservation and Sustainable Use of the Selva Maya." Accessed April 12, 2017. http://theredddesk.org/countries/initiatives/conservation-and-sustainable-use-selva-maya
64. FDN. "Protected Areas." Accessed April 12, 2017. https://defensores.org.gt/en#areasProtegidas
65. Rohter, Larry. "Guatemalans Formally End 36-Year Civil War, Central America's Longest and Deadliest." December 30, 1996. *New York Times*. Accessed April 12, 2017. http://www.nytimes.com/1996/12/30/world/guatemalans-formally-end-36-year-civil-war-central-america-s-longest-deadliest.html
66. Marquez, Javier and Oscar Nuñez. Interviewed by Brian McFarland. April 2017.
67. World Environment Center, The Nature Conservancy, and Rainforest Alliance. "Selva Maya Project. A Capacity-Building Alliance to Conserve the Maya Forest of Mexico, Guatemala and Belize: A Report to Pfizer." February 2007. Accessed April 11, 2017. http://www.wec.org/programs-initiatives/Selva%20Maya%20Final%20Report%202007.pdf. 15.
68. Marquez, Javier and Oscar Nuñez. Interviewed by Brian McFarland. April 2017.
69. Ibid.
70. Cuffe, Sandra. "Controversial Park Plans in Guatemala's Maya Biosphere Reserve." June 17, 2016. *Mongabay*. Accessed April 13, 2017. https://news.mongabay.com/2016/06/controversial-park-plans-in-guatemalas-maya-biosphere-reserve/

71. World Environment Center, The Nature Conservancy, and Rainforest Alliance. "Selva Maya Project. A Capacity-Building Alliance to Conserve the Maya Forest of Mexico, Guatemala and Belize: A Report to Pfizer." February 2007. Accessed April 11, 2017. http://www.wec.org/programs-initiatives/Selva%20Maya%20Final%20Report%202007.pdf. 2.
72. Marquez, Javier and Oscar Nuñez. Interviewed by Brian McFarland. April 2017.
73. The REDD Desk. "Conservation and Sustainable Use of the Selva Maya." Accessed April 12, 2017. http://theredddesk.org/countries/initiatives/conservation-and-sustainable-use-selva-maya
74. Marquez, Javier and Oscar Nuñez. Interviewed by Brian McFarland. April 2017.
75. Ibid.
76. UNESCO. "MAB Biosphere Reserves Directory: Biosphere Reserve Information: Guatemala: Maya." Accessed April 11, 2017. http://www.unesco.org/mabdb/br/brdir/directory/biores.asp?code=GUA+01&mode=all
77. World Environment Center, The Nature Conservancy, and Rainforest Alliance. "Selva Maya Project. A Capacity-Building Alliance to Conserve the Maya Forest of Mexico, Guatemala and Belize: A Report to Pfizer." February 2007. Accessed April 11, 2017. http://www.wec.org/programs-initiatives/Selva%20Maya%20Final%20Report%202007.pdf. 4.
78. World Environment Center, The Nature Conservancy, and Rainforest Alliance. "Selva Maya Project. A Capacity-Building Alliance to Conserve the Maya Forest of Mexico, Guatemala and Belize: A Report to Pfizer." February 2007. Accessed April 11, 2017. http://www.wec.org/programs-initiatives/Selva%20Maya%20Final%20Report%202007.pdf. 15.
79. Marquez, Javier and Oscar Nuñez. Interviewed by Brian McFarland. April 2017.
80. World Environment Center, The Nature Conservancy, and Rainforest Alliance. "Selva Maya Project. A Capacity-Building Alliance to Conserve the Maya Forest of Mexico, Guatemala and Belize: A Report to Pfizer." February 2007. Accessed April 11, 2017. http://www.wec.org/programs-initiatives/Selva%20Maya%20Final%20Report%202007.pdf. 4.
81. Marquez, Javier and Oscar Nuñez. Interviewed by Brian McFarland. April 2017.
82. Cuffe, Sandra. "Controversial Park Plans in Guatemala's Maya Biosphere Reserve." June 17, 2016. *Mongabay*. Accessed April 13, 2017. https://

news.mongabay.com/2016/06/controversial-park-plans-in-guatemalas-maya-biosphere-reserve/

83. Marquez, Javier and Oscar Nuñez. Interviewed by Brian McFarland. April 2017.
84. VCS Project Database. "Reduced Emissions from Avoided Deforestation in the Multiple Use Zone of the Maya Biosphere Reserve in Guatemala (GuateCarbon), Guatemala." Accessed April 14, 2017. http://www.vcsprojectdatabase.org/#/project_details/1384
85. Marquez, Javier and Oscar Nuñez. Interviewed by Brian McFarland. April 2017.
86. Ibid.
87. Conner, Mile. Interviewed by Brian McFarland. April 2017.
88. Conner, Mike. "Protecting the Yela Forest: A New & Improved Conservation Tool for Micronesia." *The Nature Conservancy's Conservancy Talk*. Last modified March 31, 2014. http://blog.nature.org/conservancy/2014/03/31/protecting-the-yela-forest-a-new-and-improved-conservation-tool-for-micronesia/
89. Conner, Mike. "Protecting the Yela Forest: A New & Improved Conservation Tool for Micronesia." *The Nature Conservancy's Conservancy Talk*. Last modified March 31, 2014. http://blog.nature.org/conservancy/2014/03/31/protecting-the-yela-forest-a-new-and-improved-conservation-tool-for-micronesia/
90. IUCN Red List of Threatened Species. "Ducula Oceanica." Accessed April 3, 2017. http://www.iucnredlist.org/details/22691663/0
91. Conner, Mike. "Protecting the Yela Forest: A New & Improved Conservation Tool for Micronesia." *The Nature Conservancy's Conservancy Talk*. Last modified March 31, 2014. http://blog.nature.org/conservancy/2014/03/31/protecting-the-yela-forest-a-new-and-improved-conservation-tool-for-micronesia/
92. Leberer, Trina. "Protecting the World's Last Ka Forest." *TreeHugger*. Last modified May 13, 2014. http://www.treehugger.com/endangered-species/protecting-worlds-last-ka-forest.html
93. Friday, Kathleen. Interviewed by Brian McFarland. April 2017.
94. Ibid.
95. The Nature Conservancy. "Micronesia: How We Work: The Micronesia Challenge." Accessed April 3, 2017. https://www.nature.org/ourinitiatives/regions/asiaandthepacific/micronesia/howwework/index.htm?redirect=https-301

96. Conner, Mike. "Protecting the Yela Forest: A New & Improved Conservation Tool for Micronesia." *The Nature Conservancy's Conservancy Talk*. Last modified March 31, 2014. http://blog.nature.org/conservancy/2014/03/31/protecting-the-yela-forest-a-new-and-improved-conservation-tool-for-micronesia/
97. Conner, Mile. Interviewed by Brian McFarland. April 2017.
98. Ibid.
99. Ibid.
100. Friday, Kathleen. Interviewed by Brian McFarland. April 2017.
101. Conner, Mile. Interviewed by Brian McFarland. April 2017.
102. Friday, Kathleen. Interviewed by Brian McFarland. April 2017.
103. Conner, Mike. "Protecting the Yela Forest: A New & Improved Conservation Tool for Micronesia." *The Nature Conservancy's Conservancy Talk*. Last modified March 31, 2014. http://blog.nature.org/conservancy/2014/03/31/protecting-the-yela-forest-a-new-and-improved-conservation-tool-for-micronesia/
104. Conner, Mile. Interviewed by Brian McFarland. April 2017.
105. Friday, Kathleen. Interviewed by Brian McFarland. April 2017.
106. Conner, Mile. Interviewed by Brian McFarland. April 2017.
107. Ibid.
108. Ibid.
109. Friday, Kathleen. Interviewed by Brian McFarland. April 2017.
110. Conner, Mile. Interviewed by Brian McFarland. April 2017.
111. Ibid.
112. Clark, Story. *A Field Guide to Conservation Finance*. 162.
113. Clark, Story. *A Field Guide to Conservation Finance*. 164.
114. Ginn. *Investing in Nature*. 126.
115. Tepper, Henry and Victoria Alonso. "The Private Lands Conservation Initiative in Chile." In *Conservation Capital in the* Americas, edited by James Levitt. Cambridge, Massachusetts: Lincoln Institute of Land Policy, 2010. 57.
116. Clark, Story. *A Field Guide to Conservation Finance*. 81.
117. Conner, Mile. Interviewed by Brian McFarland. April 2017.
118. Speth, James Gustave and Peter M. Haas. *Global Environmental Governance*. 30
119. Food and Agriculture Organization of the United Nations. "Global Forest Resources Assessment 2015: Desk Reference." Accessed November 23, 2016. http://www.fao.org/3/a-i4808e.pdf. 235–240.

120. Marquez, Javier and Oscar Nuñez. Interviewed by Brian McFarland. April 2017.
121. Conner, Mile. Interviewed by Brian McFarland. April 2017.
122. Marquez, Javier and Oscar Nuñez. Interviewed by Brian McFarland. April 2017.
123. Ibid.
124. Database of State Incentives for Renewables & Efficiency. "Home." Accessed April 14, 2017. http://www.dsireusa.org/
125. Graham, Peter. Interviewed by Brian McFarland. January 2017.
126. Ibid.
127. Ibid.
128. Clark, Story. *A Field Guide to Conservation Finance*. 228.

9

Government International Budgetary Allocations

"Tropical rainforests are global treasures. Their protection and restoration are indispensable to reach both the Sustainable Development Goals and the Paris Agreement climate change goals. Protecting and restoring these forests, however, comes with a significant financial and political cost, and to which it is unreasonable to expect tropical forest governments to carry out on their own. Therefore, international partnerships between dedicated tropical forest country governments and more affluent partner country governments are important to success. These should be genuine partnerships, led by tropical forest countries and based on mutual commitments and obligations, rather than as donor-recipient aid relationships. Designed well, such partnerships can catalyze both unprecedented environmental benefits, but also catalyze large scale private sector investments in deforestation free agriculture and sustainable forestry, yielding stronger, more equitable, more sustainable rural economic development."[1]

—Per Fredrik Ilsaas Pharo

B.J. McFarland, *Conservation of Tropical Rainforests*, Palgrave Studies in Environmental Policy and Regulation, https://doi.org/10.1007/978-3-319-63236-0_9

Introduction

While governments can utilize the budgetary process to finance domestic conservation of tropical rainforests or to reduce one's tax liability, governments can also utilize a part of their budget to finance tropical rainforests in other countries or to finance domestic protected areas that link to international peace parks.

International peace parks, also known as transfrontier conservation areas (TFCAs) or transboundary protected areas (TBPAs), are protected areas that span more than one country. Such international peace parks can help establish positive relationships between countries, while expanding protected areas to increase wildlife habitat and preserve the territories of Indigenous Peoples and local communities.

Historical Overview

It is difficult to discern when the first government international budgetary expenditures for conservation were allocated. However, it is likely this mechanism was first established in the late nineteenth or early twentieth century, when the world's first international conservation organizations were established. This includes the Society for the Preservation of the Wild Fauna of the Empire (now known as Fauna and Flora International or FFI), which was founded in 1903,[2] and the International Council for Bird Preservation (ICBP; the ICBP would rebrand as BirdLife International in 1993) was established in 1922.[3]

Other notable dates for government international budgetary allocations and international peace parks include:

- 1932: Waterton Glacier International Peace Park was established between the United States and Canada and marked the world's first international peace park.[4]
- 1982: Parque Internacional La Amistad (La Amistad International Park) is established between Costa Rica and Panama.[5]

- 1991: The "Bhutan Trust Fund for Environmental Conservation (BTF)—the first environmental fund in the developing world—established a solid foundation for biodiversity conservation through enduring legal, institutional, and technical frameworks."[6]
- 1997: The Peace Parks Foundation is established to facilitate the creation of peace parks in Southern Africa.[7]
- 2000: The Tri-National de la Sangha, an international peace park between Cameroon, Central African Republic, and the Republic of the Congo, is established through a cooperation agreement signed in 2000.[8]
- 2002: International treaty signed to establish Great Limpopo Transfrontier Park.[9]
- 2002: The Amazon Region Protected Areas (ARPA) Program is established and is now the largest collection of protected areas in the world.[10]
- 2006: Memorandum of Understanding (MOU) is signed, which established the "cornerstone" for the creation of the Kavango-Zambezi Transfrontier Conservation Area.[11]
- 2007: The Tri-National de la Sangha Trust Fund is established.[12]
- 2007: Prime Minister Jens Stoltenberg of Norway launches the Norwegian International Climate and Forest Initiative (NICFI) with an annual budget goal of approximately USD$500 million.[13]
- 2008: Amazon Fund is established via Federal Decree in Brazil.[14]

Mechanisms of Instrument

The source of funds for a government's international budgetary allocations is likely similar to, or the same as, the source of funds for a government's domestic budgetary allocations. These sources may include taxes (e.g., property taxes or income taxes), environmental fines, or lottery sales. However, structuring government international budgetary allocations comes in a variety of distinct forms. Such financing mechanisms can be done via bilateral or multilateral channels. Funds can also go toward the establishment of a local, state, or federal protected area in a different country, as well as toward the establishment of a joint, international peace park. Often, a designated entity such as a trust fund is established to receive and manage the funds.

Another aspect of government assistance to a foreign country is via sister parks, which might involve financial and/or non-financial, technical assistance. For instance, protected areas may enter into a partnership to exchange ideas through in-person trainings, share experiences with new technology (e.g., provide drawings and list of manufacturers for composting toilets or conservation drones), and pay for food and lodging for visiting rangers. One such sister park agreement exists between Yosemite National Park in the United States and Chile's Parque Nacional Torres del Paine.

Size of Instrument

As of 2007, the World Database on Protected Areas identified "227 TBPA complexes incorporating 3,043 individual protected areas or internationally designated sites."[15]

With respect to bilateral assistance for tropical rainforest conservation, the Organisation for Economic Co-operation and Development (OECD) reports that the:

> Biodiversity-related ODA by members of the OECD DAC {Development Assistance Committee} reached USD 8.7 billion per year in 2014–15, representing 6% of total bilateral ODA commitments. Nearly half of these activities (46%, USD 4.0 billion) targeted biodiversity as a primary or 'principal' objective, implying that they would not have been funded but for their biodiversity-related goals. This represents what can be considered a 'lower bound' of biodiversity-related ODA.
>
> The rest (54%, USD 4.7 billion), targeted biodiversity as a secondary or 'significant' objective, indicating that biodiversity is being mainstreamed into development co-operation activities with other primary objectives.[16]

With respect to multilateral assistance for tropical rainforest conservation, there are several initiatives established. This includes:

- The Central African Forest Initiative (CAFI) which had a total source of funds in 2016 of USD$49.24 million[17];

- The UN-REDD Programme had a total source of funds in 2016 of USD$282.85 million[18];
- The Global Environmental Facility has USD$4.43 billion in committed contributions for 2014–2018[19];
- The World Bank's Forest Carbon Partnership Facility (specifically the Readiness Fund and the Carbon Fund) had more than USD$1.1 billion committed at the end of Fiscal Year 2016[20];
- The Forest Investment Program of the Climate Investment Funds is a USD$775 million fund[21];
- The BioCarbon Fund initially committed USD$90 million to approximately 25 projects.[22] The new BioCarbon Fund Initiative for Sustainable Forest Landscapes, with support from Germany, Norway, the United Kingdom, and the United States, has approximately USD$360 million in fund capital.[23]

More specifically, the BioCarbon Fund has structured Emission Reduction Purchase Agreements (ERPAs) for emission reductions (i.e., carbon offsets) from projects such as:

- The Carbon Emissions Reduction Project in the Corridor Ankeniheny-Zahamena Protected Area in Madagascar[24];
- The Coopeagri Agroforestry Project in Costa Rica[25]; and
- The Precious Woods in Nicaragua.[26]

The Green Climate Fund has the potential to dwarf all these multilateral sources of financing as:

> {…} advanced economies have formally agreed to jointly mobilize USD 100 billion per year by 2020, from a variety of sources, to address the pressing mitigation and adaptation needs of developing countries. Governments also agreed that a share of new multilateral funding should be channeled through the newly established Green Climate Fund.[27]

While USD$10 billion has been raised, it is important to note that not all of this USD$100 billion may be raised and not all of the funding will be specifically committed to conserving tropical rainforests.

Nevertheless, the Green Climate Fund offered USD$500 million to private sector climate change projects via a Request for Proposal (RFP) issued in May 2017,[28] and although the details were not disclosed, the Green Climate Fund transferred its first set of funds for REDD+ in July 2017.[29]

Introduction to Case Studies

The following case studies look at the Amazon Region Protected Areas (ARPA) Program of Brazil, Norway's International Climate and Forest Initiative (NICFI), and the Tri-National de la Sangha of the Congo Basin.

The ARPA Program, which was established in 2002, is now the largest collection of protected areas in the world, and due to international support, ARPA has a 25-year fund. After these 25 years, the Brazilian Government will be able to fully finance the protected areas by themselves.

The NICFI, since its launch in 2007, has committed nearly USD$3 billion to international tropical rainforest conservation initiatives. Since 2009, the NICFI has provided funding to over 125 civil society projects. Furthermore, the NICFI has provided significant financing for Brazil's Amazon Fund (i.e., which in turn supported the ARPA Program), helped establish the continuing Peatland Moratorium in Indonesia, and played a critical role in getting Reducing Emissions from Deforestation and Degradation (REDD+) into the Paris Agreement.[30]

The Tri-National de la Sangha (TNS) is an international peace park established between Cameroon, the Central African Republic, and the Republic of the Congo. The TNS is the first landscape, multi-country protected area in the Congo Basin, and in 2007, the Sangha Tri-National Trust Fund was established. The Trust Fund's endowment, which was provided for via governments' international budgetary allocations, was approximately €23.5 million as of 2014.

Case Study #1: Amazon Region Protected Areas Program of Brazil

Introduction

The Amazon Region Protected Areas (ARPA) Program, established in 2002, is now the largest collection of protected areas in the world. The ARPA Program consists of state and national parks, along with extractive reserves, located throughout the Brazilian Amazon. The ARPA Program does not include Indigenous Peoples' areas nor does the Program include every protected area in Brazil.[31] Nonetheless, the protected areas in the ARPA Program include Parque Nacional de Tumucumaque (Tumucumaque National Park) and the Ecological Station of Grão-Pará. Tumucumaque National Park is currently, as of 2016, the largest tropical rainforest park in the world. The Ecological Station of Grão-Pará, which is approximately 4.2 million hectares (10.4 million acres), is the "the world's largest strict nature reserve in a tropical forest."[32] This Ecological Station, located in the Amazon Biome, is a part of the Calha Norte Mosaic of Pará which encompasses in total "{…} an area of almost thirty million hectares. This mosaic connects to protected areas of the Guiana Shield countries and is likely to constitute the largest protected pristine tropical rainforest area of the world."[33]

Furthermore, the ARPA Program also includes the Jaú National Park, Anavilhanas National Park, and the Amanã Sustainable Development Reserve. These 3 protected areas, in addition to the Mamirauá Sustainable Development Reserve, make up the Central Amazon Conservation Complex. This Complex "makes up the largest protected area in the Brazilian Amazon Basin (over 6 million hectares) and is one of the planet's richest regions in terms of biodiversity. It also includes an important sample of varzea ecosystems, igapó forests, lakes and channels which take the form of a constantly evolving aquatic mosaic that is home to the largest array of electric fish in the world. The site protects key threatened species, including giant arapaima fish, the Amazonian manatee, the black caiman and two species of river dolphin."[34]

The following map outlines the ARPA-supported protected areas including both state lands (e.g., state-managed strict protected areas and

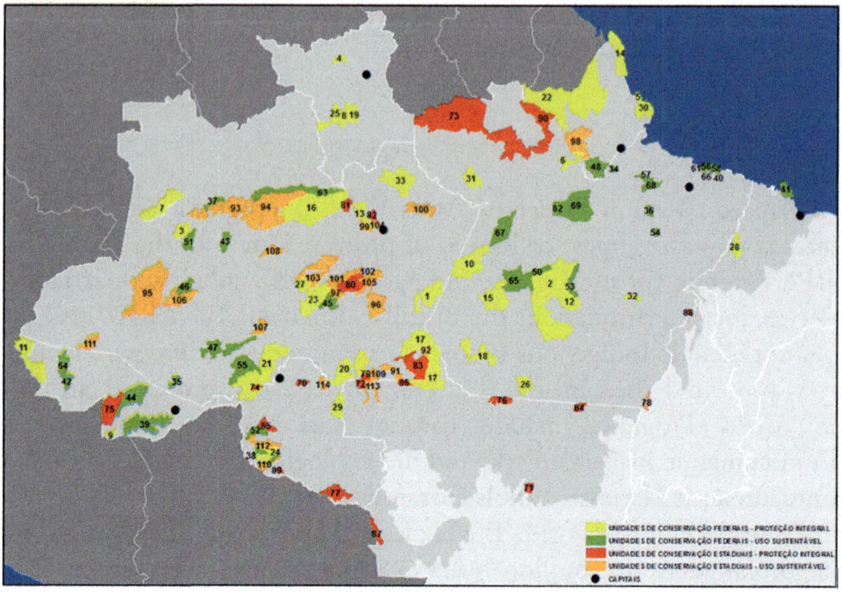

Fig. 9.1 Map of ARPA Program (Credit: FUNBIO). Yellow: areas of strict protection (federal areas). Green: sustainable use (federal areas). Red: strict protection (state-managed areas). Orange: sustainable use (state areas)

state-managed sustainable use areas) and federal lands (i.e., federal areas of strict protection and federal areas allowing for sustainable use) (Fig. 9.1).

When initially planned, the ARPA Program covered 12 million hectares (29.6 million acres) and only included areas of strict protection. Over time, sustainable use areas were included, and now, the ARPA Program covers 59 million hectares (145.7 million acres) of protection which is twice the size of Germany. The final target of the ARPA Program is to cover 60 million hectares (148.2 million acres).

The primary contributors to the ARPA Program were the World Bank through the Global Environment Facility (GEF) and the German development bank KfW, along with WWF International together with WWF Brazil (with assistance from the Ford Foundation).[35] The Amazon Fund, which has received substantial funding from Norway through the Brazilian Development Bank, supported the second phase of the ARPA Program, beginning in 2010 (i.e., the Amazon Fund did not exist when the ARPA Program was first established). The ARPA Program is:

> {…} divided into three phases, the first of which began in 2003 and ended in 2010, disbursing a total of USD$55.2 million, helping to create 46 PAs {protected areas} on over 24 million hectares of Amazonian land. The second phase got underway in 2010 and is expected to reach completion in 2017, handling a further USD$54.2 million. For Phase three, which will last twenty-five years, a Transition Fund was set up using a seed donation of USD$123 million with the aim of leveraging new resources and gradually phasing-in total government funding for ARPA and its PAs.[36]

The ARPA Trust Fund is being managed by FUNBIO (Fundo Brasileiro para a Biodiversidade or the Brazilian Biodiversity Fund in English).

FUNBIO is a financial mechanism which manages several funds. FUNBIO initially started the Program with an endowment from these international donors, and the funds were invested with professional investors who have a committee of investors and asset managers. The interest earned from the investments is to be used for conservation activities at the aforementioned protected areas.

The ARPA Trust Fund is now a sinking 25-year fund, and after these 25 years, the Brazilian Government will be able to fully finance the protected areas by themselves. Herein lies one of the most fascinating aspects of the ARPA Program and that is the Program is structured for long-term, permanent financing.[37] It is important to note that the protected areas within the ARPA Program are still receiving funds from the Brazilian Government (e.g., FUNBIO does not pay for any ARPA Program staff, including government staff). However, there are significant additional funds required for training, mapping, monitoring, and establishing park infrastructure which are financed via the ARPA Trust Fund.

Identify the Problem

The main problem that the ARPA Program sought to address was the lack of financial resources, the lack of protected areas, and deforestation—particularly in the Arc of Deforestation along the Amazon's southern border.

Whether it is Brazil, Colombia, Mexico, or Peru, governments very often do not have the information to define a common structure for each protected area and to determine how much it will realistically cost to manage the protected areas over the long-term. Often, it is difficult to define a budget for 5 to 10 years, let alone for the next 20 to 25 years.[38]

Also, despite being the eighth largest economy in the world, Brazil has undergone several crises. In the 1970s, there were the shocks due to oil embargoes, then in the 1980s there was the foreign debt crisis, in the 1990s there were continuing economic and political crises, and as recently as 2017, Brazil is facing new political and economic crises.

Why the Problem Is Important

The problem of tropical rainforest deforestation and degradation, in conjunction with a lack of financial resources and lack of protected areas, was important to stakeholders because the Amazon Rainforest is the largest tropical rainforest on Earth. Tropical rainforest deforestation and degradation in Brazil during the 1990s and early 2000s was among the highest rates in the world.

In addition, there was a target established at the Convention on Biological Diversity's (CBD) Seventh Conference of the Parties (COP7) in 2004 to protect 10% of each biome.[39] WWF, along with other stakeholders, took the goal to the President of Brazil, who agreed with the goal, and then there needed to be a process and funding to uphold the goal and protect 15% (i.e., more than the 10% target) of the Amazon Basin.

How Problem Was Identified

The problems associated with the lack of financial resources, the lack of protected areas, and deforestation—particularly in the Arc of Deforestation along the Amazon's southern border—were identified by a wide range of local conservation organizations such as FUNBIO and Imazon, international conservation organizations such as WWF, and both the Brazilian Government and multilateral agencies (i.e., the World Bank).

Effectiveness of Process for Identifying Problem

The process for identifying these problems has been relatively effective. This is primarily due to establishing breakeven points between conservation outcomes and financial commitments as well as the structure that allows time for the Brazilian states and federal government to allocate more money over time to their budgets for the protected areas (i.e., to fund the ongoing monitoring and consolidation of protected areas).[40]

Steps Taken to Address the Problem

There were several steps taken to address the aforementioned problems. First, political will was garnered when the then President of Brazil supported the ARPA Program. Next, the President of the World Bank was willing to commit funds, and shortly thereafter, Germany's KfW and WWF's private donors committed additional funds. With the political will and funding in place, there was then lots of technical staff needed to map the areas, identify the needs, and refine the overall costs.[41]

Results

The primary result is the ARPA Program is the largest collection of protected areas in the world, covering 59 million hectares. The ARPA Program, in conjunction with other non-ARPA protected areas in the Amazon, totalled more than 10% of the Amazon Biome.

Nowadays, the protected areas within the ARPA Program have little-to-no deforestation, with annual deforestation rates less than 1%. However, there is still a lot of work to be done to further consolidate the protected areas.[42]

Challenges and How They Were Met

At that time—20 years ago—the ARPA Program was a very novel idea with a very ambitious target. This said, it took 5 years to get the design approved through all the levels of government. Similarly, it is hard to change minds and took a lot of perseverance.

There were concerns that only a 10% section of the Amazon would be protected and then the remaining 90% of the Amazon would be deforested. However, the ARPA Program has helped to protect more than 10% of the Amazon Biome. In addition, from 2005 to 2013, the country of Brazil made remarkable progress on slowing deforestation by an estimated 70% from its 10-year average from 1996 to 2005.[43] Yet it is important to note that Brazil's deforestation—not necessarily within ARPA-funded protected areas—has recently increased in 2015 and 2016.

Another challenge was that the ARPA Program had a board that consisted of both government and civil society members, but the board members were not always in agreement. There were challenges of political influence, managing different viewpoints, and challenges of establishing a public-private program. Such programs rarely survived more than 15 years in Brazil.

In addition, another challenge was the sheer size of the resources, including staff, but mainly the large sums of money required. A private financing mechanism was a prerequisite and donors said they would support a large program, but the resources donated would need to be privately managed.

Lastly, there is ongoing difficulty of demonstrating some indicators such as the total number of flora and fauna species identified within the various protected areas of the ARPA Program.[44]

Beyond Results

The ARPA Program should be considered a sustainable model that could be replicated elsewhere. This is in part due to the long-term nature of the program which will be 15 years old in 2017 and there is still approximately 20 to 25 years of funding left. The institutional arrangement is very unique and the Program is less expensive than comparable models—especially considering the full benefits, including carbon sequestration and carbon storage benefits.

If you plot all of the protected areas in Brazil (e.g., municipal, state, and federal lands, along with Indigenous Peoples' lands), including the protected areas that are a part of the ARPA Program, there is significant connectivity between all of them. Specifically, there are three large corridors going from East to West in Brazil.[45] This shall help with the ecological sustainability of the protected areas.

It is important to note that going forward, there is still a need to further consolidate some of the protected areas into a comprehensive management plan with detailed plans and personnel.

Lessons Learned

The greatest lesson learned is that financial resources, especially of this magnitude, need to be managed by a private, nonprofit organization. Credibility and continuity are absolutely important, especially for donors, who require transparency and need to know where each dollar is spent.

In addition, a well-designed, long-term financial plan needs to be developed. At a minimum, such a plan needs to specify how much is needed and who is going to fund the plan in the long-term.

Additional lessons learned are that it is vitally important to build the capacity of park managers. FUNBIO believes, from the perspective of a program and fund manager, it is also important to continuously improve, to develop better processes, and to recognize one's limitations. FUNBIO is constantly looking to improve upon by their own internal efficiencies by defining specific targets and distributing responsibilities for each target to staff members.[46]

Other Resources on ARPA

Federal Government of Brazil's Programa ARPA

- http://programaarpa.gov.br/en/

FUNBIO's Programa ARPA Page

- http://www.funbio.org.br/en/programaarpa/

KfW's State of the 3rd Phase of ARPA

- https://www.kfw-entwicklungsbank.de/International-financing/KfW-Development-Bank/News/News-Details_296960.html

The Global Environment Facility's ARPA Page

- http://thegef.org/project/amazon-region-protected-areas-program-arpa

World Bank's Amazon Region Protected Areas (GEF) Page

- http://www.worldbank.org/projects/P058503/amazon-region-protected-areas-gef?lang=en

World Wildlife Fund's ARPA Page

- http://wwf.panda.org/what_we_do/where_we_work/amazon/vision_amazon/models/amazon_protected_areas/financing/arpa/
- https://www.worldwildlife.org/magazine/issues/summer-2014/articles/the-amazon-region-protected-areas-program-is-the-single-largest-tropical-forest-conservation-program-in-history

Case Study #2: Norway's International Climate and Forest Initiative

Introduction

In 2007, Prime Minister Jens Stoltenberg of Norway launched the Norwegian International Climate and Forest Initiative (NICFI) with an annual budget goal of approximately USD$500 million at the UNFCCC COP in Bali, Indonesia.[47] NICFI is overseen by the Norwegian Ministry of Climate and Environment (e.g., budget and policy responsibilities), and some of the Ministry's responsibilities are delegated to the Norwegian Agency for Development Cooperation (NORAD) as well as to several Norwegian embassies.[48] Since 2007, Norway has made significant financial and technical commitments to numerous tropical rainforest countries including Brazil, Indonesia, Guyana, and Tanzania.

The two overarching goals for NICFI are to (1) achieve the global climate change target of limiting temperature rise by 1.5–2 C° by reducing emissions and (2) sustainable development. More specifically, the key objectives of NICFI, as decided by the Norwegian Parliament, are:

1. The international climate regime is an effective instrument for reducing emissions;
2. Achieve reduced emissions from forest and land use in partner countries; and
3. Help partner countries safeguard natural forests.

The following map depicts the countries and subnational jurisdictions that received support from NICFI in 2015 (Fig. 9.2).

NICFI has also contributed to the development of numerous multilateral initiatives under the World Bank as well as the UN-REDD Programme, and runs a large-scale civil society support program, including support for Indigenous Peoples as well as collaboration with the private sector.[49]

Identify the Problem

The main problem is that while forests are a public good, there is a market failure that inadequately values standing forests and there are insufficient mechanisms to channel financing toward tropical rainforest conservation. Thus, the related problems which NICFI seeks to help address are:

- The mitigation of global climate change;
- The high rates of tropical rainforest deforestation in many of the recipient countries; and

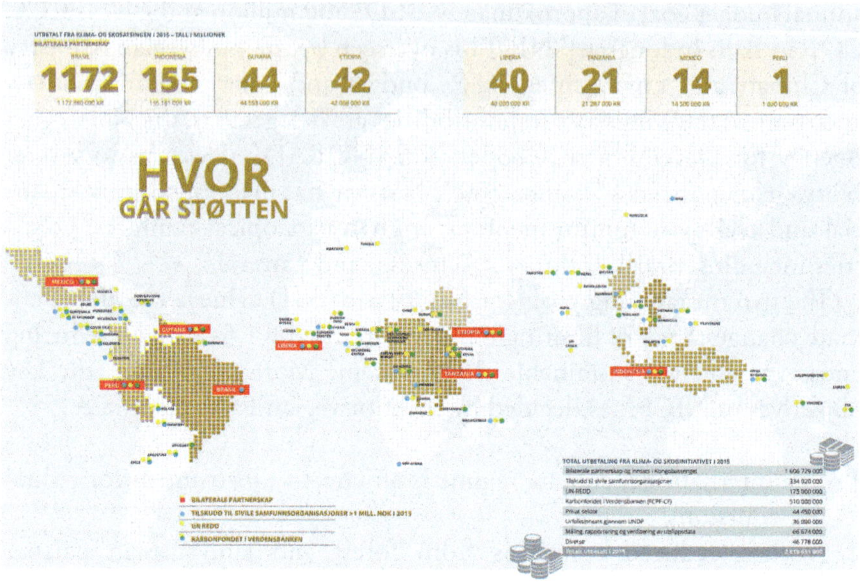

Fig. 9.2 Map of NICFI's support in 2015 (Credit: Itera AS)

- The lack of technical and financial capabilities of the recipient countries, particularly the lack of predictable and long-term REDD+ (Reducing Emissions from Deforestation and Degradation) financing.[50]

Why the Problem Is Important

The aforementioned problems were important to Norway because of Norway's desire to fulfill its Official Development Assistance (ODA) commitments and to help the world meet the 1.5–2° C global climate change goal. This said, Norway is dependent on its fossil fuel economy, and it is more cost-effective to support initiatives to reduce tropical rainforest deforestation and its associated emissions.

However, please note that while reducing deforestation is a cost-effective way to achieve emission reductions, from the Norwegian government's perspective, NICFI is not related to the Norwegian fuel economy. This said, the emission reductions that NICFI buys via its payment for performance structure do not fulfill any Norwegian obligations under the

United Nations Framework Convention on Climate Change (UNFCCC); rather these voluntary purchases are only for development aid.[51]

How Problem Was Identified

The problems were identified by Norway through its participation in the UNFCCC,[52] via its participation in the UN-REDD Programme,[53] and via its participation in the World Bank's Forest Carbon Partnership Facility (FCPF).[54] In addition, Norway has world-class research facilities such as the Center for International Climate and Environmental Research (CICERO).

Effectiveness of Process for Identifying Problem

The process of identifying global climate change and tropical rainforest deforestation, along with the lack of technical and financial capabilities in recipient countries, has been relatively effective. While negotiations within multilateral forums can be slow, Norway has been a key architect of both multilateral and bilateral partnerships.[55]

Steps Taken to Address the Problem

The 2 main steps to address the aforementioned problems were the demonstration of ambitious, political leadership and the successful process of budgeting for NICFI.[56]

Several elements came together for Norway's ambitious, political leadership on tackling tropical rainforest deforestation and degradation. First, Norway's civil society, particularly the strong national lobby of the NGO Rainforest Foundation Norway, encouraged ambitious, political leadership. Second, Norway is a rich country with a long history of traditional ODA and international engagement. One of Norway's largest economic sectors is exporting petroleum, and there was an emerging recognition that Norway had generated significant revenue from something that was considered to be the heart of the climate change problem.[57] In fact, in 2016, approximately 350 billion Norwegian Kroner

(NOK; approximately USD$40 billion)—equivalent to 47% of Norway's total export value—was derived from the export of crude oil and natural gas.[58] Third, there was a realization—as pointed out in the seminal Stern Report—that tropical forests were a big part of the global climate change problem and an even bigger part of the solution.[59]

During the 2007 UNFCCC Conference of the Parties (COP) in Bali, Indonesia, there was emerging international consensus to include tropical forests, and subsequently, there was an Agreement on the Climate Report (i.e., the Climate Consensus) between all the political parties in Norway in 2008.[60] As part of this Agreement, there was a consensus that the long-term goal was to budget NOK3 billion to allocate to REDD+. In 2008, the program began with approximately NOK400 million, then grew to approximately NOK1.5 billion in 2009, and in 2015, Norway reached its annual goal of around NOK3 billion (approximately USD$350 million). Parliament has allocated funds each year since 2008 and since 2015, the annual funds are around NOK3 billion.[61] This NOK3 billion for REDD+ financing is part of Norway's commitment to 1% of its Gross Domestic Product (GDP) allocated to ODA.

Results

Since the launch of NICFI in 2007, there have been numerous results. As of April 2017, the greatest results to date include:

- The creation of the Amazon Fund and all of the projects funded by the Amazon Fund;
- The establishment and continuation of the Peatland Moratorium in Indonesia; and
- Getting REDD+ incorporated into the Paris Agreement.[62]

Since 2007, NICFI has committed nearly USD$3 billion to international REDD+ initiatives and this financing includes primarily grants.[63] Since 2009, the NICFI has provided funding to over 125 civil society projects including 46 projects from 2009 to 2012, 42 projects from 2013 to 2015, and 39 projects will be supported from 2016 to 2020.[64]

From 2013 to 2015, some of the larger funding initiatives for NICFI's civil society grant scheme (i.e., which is approximately 10% of the overall NICFI grant total) included NOK60 million to CIFOR's global comparative analysis of REDD+, phase II of IPAM's Building Bridges Between Commodity Roundtables, and WWF's Forest and Climate Initiative.[65]

The NICFI has also established bilateral partnerships with Brazil (committed up to USD$1 billion), Colombia, Ethiopia, Guyana, Indonesia (up to USD$1 billion pledged), Liberia (up to USD$150 million pledged), Mexico, Peru (up to USD$300 million pledge), Tanzania, and Vietnam.[66] It is important to note that in June 2017, Norway announced that it will reduce its 2017 payments to Brazil due to the increase in Brazil's deforestation.[67]

Additional results of the NICFI include:

- In 2013, "the Indonesian Constitutional Court recognised the customary rights of Indigenous Peoples to forests which the state had claimed unilaterally as state forests."[68]
- In the Mai-Ndombe Province of the Democratic Republic of the Congo, "NICFI grantees jointly ensured the participation of previously marginalised groups. This offers an opportunity for local populations to provide input to programmes that may affect them. As a result of major progress on a provincial emission reduction programme in Mai Ndombe, it will be the first province in Africa to sign an agreement with the World Bank's Carbon Fund. The Agreement will ensure green development and the authorities will receive payments from the Carbon Fund when reaching agreed-upon milestones (pay for performance). The 10-year finance plan amounts to USD 176 million."[69]
- Funding of the Forest Compass Project, which was "instrumental in building monitoring capacity and piloting new monitoring systems among forest communities in Guyana and Brazil."[70]
- Funding Project Leaf, which helped INTERPOL track down a cargo vessel with illegal timber. It is estimated that "the load when leaving Peru consisted of 1.2 million cubic metres illegal timber, which is an equivalent of more than 480 Olympic swimming pools. The value of the load is estimated to almost half a million US dollars."[71]
- The "Indonesian Ministry of Environment and Forestry has released public forest information related to logging permits after organisations advocated for access to such information."[72]

- The Observatório do Código Florestal (OCF; Forest Code Observatory) was established with assistance from the NICFI.[73] Thus, if the Brazilian government does not enforce compliance with the Brazilian Forest Code, the system of remote sensing technology can be utilized by the NGO community to identify the ongoing problems of non-conformance. These NGOs include domestic organizations such as Imazon and IPAM, along with international organizations such as WWF, TNC, and Conservation International.[74]

Challenges and How They Were Met

The major challenge facing NICFI, and the global community, is the overall level of financing and the expectation of long-term financing for tropical rainforest conservation. This said, the circumstances since 2007 have changed. Back then, there were great expectations for the UNFCCC COP15 in Copenhagen, Denmark, and in the United States, the American Clean Energy and Security Act of 2009 (commonly referred to as the Waxman-Markey Bill) had passed the House of Representatives. Thus, there were general expectations of an international climate change compliance regime (i.e., not necessarily a market-based approach) with an important role for a payments for ecosystem services (PES) model, and there were expectations of large-scale climate finance commitments beyond traditional ODA.

A related problem is that:

> Since its inception, NICFI has pledged funds in line with its political commitment of NOK three billion annually, however its disbursements have fallen short of this, totalling around 50%. This is partly due to the fact that the majority of funds allocated have been in the form of results-based payments, where payments are made ex-post for demonstrated results, which conflicts with the financial regulations of the Norwegian government, based on a system of ex-ante payments for activities to be carried out.[75]

To help address this challenge, Norway, which has been a global leader with respect to REDD+ financing through NICFI, acheived a consensus of annually spending NOK3 billion. However, this creates a cap on the amount of annual funding from Norway, and one cannot expect this figure to increase. To overcome this particular challenge, Norway works closely with other donor countries, such as the United Kingdom and Germany, to increase the overall level of financing. For instance, the Central African Forest Initiative (CAFI) was established in 2015 and helped to bring in other pledged financing from the European Union, Germany, France, the United Kingdom, the Netherlands, and South Korea.[76] CAFI and the DRC's Minister of Finance have signed a letter of intent for USD$200 million.[77]

A related challenge to NICFI is that budgeting is done on an annual basis and there is uncertainty over the Program's future. The Norwegian Parliament decided to maintain the same budget level for NICFI until 2020, and the government has announced its wish to prolong NICFI until 2030. However, there are elections in Fall 2017 and the Program's post-2030 future is undecided.[78]

There are also numerous challenges that are general international development challenges. For instance, the sheer size of these countries is enormous and presents a challenge when working at the national level or at a large, subnational jurisdictional level, especially when the process is not always democratic. This said, the Brazilian Legal Amazon covers approximately 5 million square kilometers,[79] and Indonesia's archipelago covers approximately 17,000 islands.[80] In addition, there are various implementation challenges such as bureaucratic politics, competing priorities, and the challenge of working through bilateral agreements with rainforest countries. Many of these issues are public policy issues and the policies are governed by the countries themselves. NICFI, due to national sovereignty, can be nothing more than play a supporting role. Similarly, NICFI cannot create or control the momentum, but NICFI can support the momentum. It takes a lot of patience to work through these issues while keeping the ultimate goal in mind and that is to identify the solutions, effectively channel the finance, and facilitate the large shift for rural economies.[81]

Beyond Results

Although it is difficult to say how long things will stay in place—particularly with annually renewed government funding—it appears as though NICFI's results are indeed sustainable. This said, NICFI has focused on a structural approach and works with a variety of stakeholders. Norway announced during the UNFCCC COP21 event in Paris, France, that the NICFI was extended until 2030.[82]

While the Peatland Moratoriums in Indonesia could be temporary, NICFI works through several channels. For example, NICFI works with civil society to address appropriate reform agendas, with governments to enforce laws, and with the private sector to support deforestation-free supply chains and to demonstrate it is possible for countries to generate revenue without deforestation. Similarly, by working to include REDD+ in the Paris Agreement, it is a good way to make REDD+ financing sustainable and make obligations for the future.

Lessons Learned

According to Mr. Per F.I. Pharo, Director of Norway's International Climate and Forest Initiative, some lessons learned include:

- It is important to identify what a project, program, or initiative did right. With respect to NICFI, it was right to focus on the national level and to pursue a political reform agenda. Otherwise, if you do not have political reform, you can waste a tremendous amount of time and money on-the-ground.
- NICFI has increased its private sector engagement and this engagement is now a core part of the NICFI agenda. However, it took some time to engage the private sector, and there are some thoughts that maybe this engagement should have been done earlier. Then again, the relationships and engagements have to mature overtime and this role is more of a job for NGOs; plus, there is no assurance that earlier engagement would have helped more.
- After hundreds of conversations with financiers, who generally love the basic idea of a transition to high productive, no-deforestation agriculture, their concerns tend to be that the big transition and investments will be into pipelines of relatively risky, unproven business models. This is where the public good comes in and where initiatives, such as NICFI, can play a leading role. Similarly, practitioners must think deeply about pipeline development and specifically about what type of investments

are needed to drive this transformation (i.e., it is not just about forest protection and enforcement), along with the types of investments needed to drive a transition in the private sector to a no-deforestation, high-productivity, high-yield model. This concept works for high-finance practitioners (e.g., finance ministers, private equity groups, and multinational corporations), but many people working on REDD+ are policy experts. Similarly, "if you see REDD+ debate, turned into a supply chain debate—then you are talking to different types of people."[83]

- A payments for ecosystem services (PES) model is essentially a good idea and acts as a risk management model. Thus, tying payments to political progress is essential, and otherwise, Norway would have potentially squandered up to NOK1 billion in Indonesia.[84]

Other Resources on NICFI

Amazon Fund

- http://www.amazonfund.gov.br/FundoAmazonia/fam/site_en

CGD Policy Paper 56, February 2015, The Indonesia-Norway REDD+ Agreement: A Glass Half-Full

- http://www.cgdev.org/sites/default/files/CGD-Climate-Forest-Paper-Series-20-Seymour-Savedoff-Birdsall-Indonesia-Norway_1.pdf

Climate Funds Update's Overview of the Norwegian International Climate and Forest Initiative

- http://www.climatefundsupdate.org/listing/norway-s-international-climate-and-forest-initiative#TOC-Summary

Guyana REDD+ Investment Fund

- http://www.guyanareddfund.org/

NORAD's Real-Time Evaluation of Norway's International Climate and Forest Initiative

- https://www.norad.no/en/toolspublications/publications/2016/real-time-evaluation-of-nicfi-literature-review-and-programme-theory/

Norway's International Climate and Forest Initiative (NICFI)

- https://www.regjeringen.no/en/topics/climate-and-environment/climate/climate-and-forest-initiative/id2000712/

Norway and Indonesia

- http://www.norway.or.id/Norway_in_Indonesia/Environment/

The Norwegian International Climate and Forest Funding to Civil Society

- https://www.norad.no/en/front/funding/climate-and-forest-initiative-support-scheme/grants-2013-2015/projects/
- https://www.norad.no/contentassets/05152b063fb0432ea65a729f643e1127/the-norwegian-climate-and-forest-funding-to-civil-society---key-results-2013--2015.pdf

Case Study #3: Tri-National de la Sangha of the Congo Basin

Introduction[85]

Three national parks—Dzanga-Ndoki National Park in the Central African Republic (established in 1990),[86] Nouabalé-Ndoki National Park in the Republic of the Congo (established in 1993),[87] and Lobéké National Park in Cameroon (established in 2001)[88]—collectively make up the Tri-National de la Sangha (TNS) international peace park.

The TNS, which was established through a cooperation agreement signed in 2000,[89] covers 746,309 hectares (approximately 1.8 million acres). In addition, a buffer zone of 1,787,950 hectares (approximately 4.4 million acres) has been established. This buffer zone, which is sometimes referred to as the TNS Landscape, includes the Dzanga-Sangha Forest Reserve in the Central African Republic (CAR), which provides connectivity between the two parcels of Dzanga-Ndoki National Park.[90] Importantly, the TNC is the first landscape, multi-country protected area in the Congo Basin (Fig. 9.3).

In March 2007, the Sangha Tri-National Trust Fund was established.[91] Most of this funding comes directly from the German Government via KfW. The Trust Fund was originally set up with a goal of €35 million. The total amount for the Tri-National de la Sangha was supposed to be covered by 75% from the Trust Fund, and the remaining 25% was supposed to come from a combination of tourism (e.g., some gorillas in the region have been habituated), the 3 host governments, and donors. However, the previous calculations are now out of date, the current costs have increased, and the overall budget for the Trust Fund should be higher. The current endowment—as of 2014—was at approximately of €23.5 million (€20.5 million from KfW and €3 million from the French Development Agency).[92] The interest earned from the principal of the Trust Fund is being used for the protected areas and to set up the necessary institutions. For instance, stakeholders are currently working on a management plan and finalizing the new budget for the Dzanga-Ndoki National Park in the CAR.[93]

In addition to the Trust Fund, the Tri-National de la Sangha Foundation (TNS Foundation) also undertakes additional projects that require extra funding and assistance from organizations such as FUNBIO and the Global Environment Facility (GEF).[94] Furthermore, the TNS Foundation "has benefited from buffer funds or projects through grants from the KfW (€8 million), the European Union via the Centre for the UNESCO World Heritage (€400,000) and the Congo Basin Forest Fund housed within the African Development Bank and funded by the Governments of the United Kingdom and Norway (€661,000)."[95]

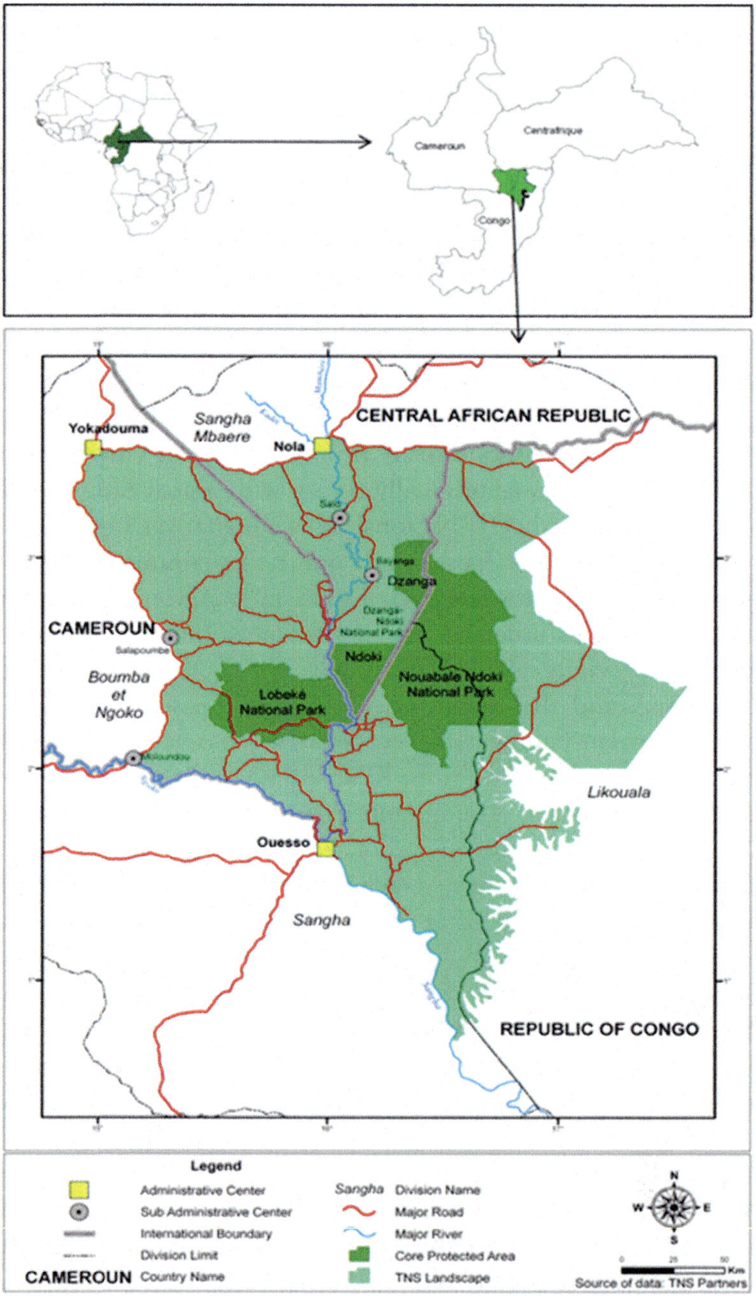

Fig. 9.3 Map of Tri-National de la Sangha (Credit: Sangha Tri-National Partners)

Identify the Problem

Deforestation is not a major issue within TNS, as just the high value trees are selectively logged (e.g., high grading) and this leads to minor degradation. Degradation, particularly defaunation of large mammals, is a major issue. The parts of TNS in the Republic of the Congo and the CAR, although not Cameroon, have relatively stable elephant populations. There are still major problems associated with elephant poaching and bushmeat trade; however, it is important to recognize that elephants are disappearing at much faster rates outside of protected areas.[96] The main problem with poachers is that these are not small-scale, operations. Rather, organized poachers show up with 20+, heavily armed men in trucks who are battle-hardened and willing to use their weapons. Even the Janjaweed from Sudan have crossed hundreds, if not thousands, of miles to poach wildlife from the TNS.[97] In addition, due to the civil war in the Central African Republic, which dates back to 2012, there are a lot of weapons readily available in the region, which can be relatively easily acquired by poachers.[98]

Another problem is that the CAR is a developing nation with little infrastructure. For instance, the CAR ranks #185 in the world for ease of doing business according to the World Bank.[99] This includes #182 for enforcing contracts, #145 for protecting minority investors, #138 for trading across borders, and #167 for registering property. According to the global rankings by Transparency International of the Corruption Perceptions Index 2016,[100] the CAR is tied for #159 in the world. The global rankings by the United Nations Development Programme of the Human Development Index (HDI),[101] which combines life expectancy, education, and per capita income indicators, rank the CAR second to last at #187. These factors have contributed to the fact that tourism has not reached its potential as a funding source for TNS.

The World Heritage Convention (WHC) notes the additional, following problems at the TNS:

> Whereas in Lobéké National Park (Cameroon) there are use zones within the park, in the Central African Republic and the Republic of Congo, local resource use, including indigenous hunting and gathering, is not permitted in the protected areas thereby affecting local livelihoods and creating the

> potential for conflict. This illustrates the crucial importance of finding an overall balance between nature conservation and local resource use in the broader landscape. {…} Maintaining the ecological values of the property will not only depend on law enforcement but eventually both on the standards of commercial resource extraction in the buffer zone and the acceptance and support of parks by the local and indigenous communities in the surrounding landscape.[102]

In addition to civil unrest in the region, other problems identified by the WHC include ground transport infrastructure, illegal activities, major linear utilities, and mining.[103]

Lastly, there is a need to set up additional, institutional arrangements such as a management foundation in the CAR to transition the protected area's management responsibility from the government. This said, the World Wildlife Fund (WWF) has been working in the region since the late 1980s and has a co-management arrangement, but there is no official mandate.[104]

Why the Problem Is Important

The TNS is a special place that still boasts a tremendous amount of biodiversity. As highlighted by the WHC:

> The property is characterised by its large size, further supported by the very large buffer zone, minimal disturbance over long periods and intactness thereby enabling the continuation of ecological and evolutionary processes at a huge scale. This includes the continuous presence of viable populations and natural densities of wildlife, including top predators and large mammals which are often affected by hunting and poaching elsewhere. There is a fully connected mosaic of very diverse habitats, including numerous types of ecologically remarkable forest clearings attracting major wildlife aggregations and countless plant species otherwise not found in the forest landscape. Unlike many other forest protected areas, the property is not a remaining fragment but continues to be part of a much larger intact and landscape with good conservation prospects. This is increasingly rare and significant at a global scale.[105]

Such wildlife includes African forest elephants (*Loxodonta cyclotis*), bongos (*Tragelaphus eurycerus*), common chimpanzees (*Pan troglodytes*), and sitatunga antelopes (*Tragelaphus spekii*). Similar to the Salonga National Park in the Democratic Republic of the Congo, the African forest elephants of TNS are key seed dispersers, and if lost, the composition of the local forests will be permanently altered. This is happening across the Congo Basin where large mammals are being lost, the forests are dying, and significant ecological changes are underway to the extent that very different forests will result in the next 100 to 200 years.[106]

How Problem Was Identified

The problems at TNS were identified by local and international NGOs (such as the TNS Foundation and WWF), as well as by governments. Likewise, the TNS Foundation works with the 3 host governments to not only identify the ongoing problems but to work toward finding and implementing solutions.[107]

Effectiveness of Process for Identifying Problem

Although one cannot say that all the difficulties have been overcome, the process for identifying the problems has been effective. For instance, the TNS Foundation works everyday with local communities and is in frequent communication with all three host countries.[108]

Steps Taken to Address the Problem

There are numerous steps that have been taken to address the problems at TNS. Such steps include but are not limited to:

- The Dzanga-Ndoki National Park in the CAR was established in 1990[109];
- The Nouabalé-Ndoki National Park in the Republic of the Congo was established in 1993[110];

- The Lobéké National Park in Cameroon was established in 2001[111];
- These 3 parks formed the TNS via a cooperation agreement signed in 2000[112];
- In 2007, the TNS Trust Fund was established.[113]

In addition, the TNS Foundation has engaged local communities and the 3 host countries, along with poachers and logging companies in the region, to attempt to address the core problems. For instance, one of the host governments signed a permit that would have allowed an international company to log the forests within its portion of the TNS. However, the TNS Foundation successfully discussed the situation with the government and the government canceled the permit. The TNS Foundation has also successfully engaged logging companies in the region. Thus, instead of seeing these companies as the enemy, the TNS Foundation has partnered with some of these companies to help protect the forest. This said, while it is illegal to commercially log in the TNS, logging is permitted in the buffer zone and some companies will illegally log without a permit. In addition, there are issues with corruption, where one government will say no to a permit, but then the company will simply ask the next government.

Another important step taken is that the TNS Foundation has helped raise funds from all around the world and developed international partnerships.[114] Furthermore to counteract poaching, the TNS has at least 50 armed rangers, and reinforcement from the United Nations is available when the situation gets really bad.[115]

Results

In addition to establishing the national parks and the collective TNS international peace park, one of the greatest results thus far is the establishment of the TNS Trust Fund. The Trust Fund took approximately 14 years to be established, but is now incorporated in the United Kingdom and has an international board with representatives from the 3 host countries, and a majority of board members are from nongovernmental organizations. In addition, the Trust Fund has a director and a small staff providing day-to-day functions.[116]

Similarly, a joint committee has been established:

> Management, law enforcement, research, monitoring and tourism all require coordination across the national boundaries. There is a Trinational Monitoring and Action Committee (Comité Trinational de Suivi et d'Action), bringing together the three countries at the ministerial level. A Trinational Monitoring Committee unites the three countries at the level of regional administrations. These mechanisms are effective in providing a joint protection and management approach to the property, and will need to be maintained and built upon.[117]

Another result of the TNS is that with an official agreement in place between the 3 countries, there is joint infrastructure on the border of the 3 countries with a rotating commander and rangers. This said, the rangers are allowed to cross the national borders. As a result, there have been some successes due to this collaboration between the countries, including prisoners who have been sent across borders.[118]

Furthermore, the stabilization of the TNS after the civil war in the CAR was a critical result, which should not be understated.[119]

Challenges and How They Were Met

One major challenge has been to work with the 3 host country governments and to convince the governments to do something they do not want to do. This said, the governments need economic development, and the conversion of forests—whether by granting timber concessions and diamond mining permits in CAR or palm oil concessions in Cameroon[120]—can generate a lot of money, including foreign exchange.

In addition, to work with the governments and help them overcome poaching, and to teach communities more productive forms of agriculture (e.g., use of technology and fertilizers), is challenging. Similarly, bringing the governments together is another challenge. For example, there was a protocol adopted back in 2013–2014 that would have allowed tourists to visit the TNS with a single visa. Without this protocol, the movement of tourists was restricted unless they obtained 3 separate visas, which is both expensive and time-consuming due to the bureaucracy.

However, despite the fact that the protocol was adopted 3 years ago, the protocol still needs to be officially signed by the 3 governments.[121]

Working in places, such as the CAR, can be extremely challenging. For instance, there are language barriers and the in-country capacity is almost nonexistent. This said, another challenge is that WWF has taken over many traditional government roles such as building and maintaining both schools and healthcare facilities.[122]

Militias also present a major challenge. The Séléka Militia took over Bangui, the capital of the CAR, in 2013. The militia installed a president for less than a year, but despite the short-term, this resulted in major poaching events because there was an official declaration made to allow poaching. During this time, WWF basically could not operate, since the militia essentially became the government. This challenge was met as local communities prevented major looting when the militia took over. The local communities stood up against poachers and even returned stolen laptops which demonstrates the communities' commitment to conservation.[123]

Beyond Results

The Africa Parks Network is an organization that sets up local NGOs to manage protected areas on behalf of host governments. The Africa Parks Network now manages approximately 20 protected areas across Africa. To help the future sustainability of Tri-National de la Sangha, WWF would like to adopt a similar model as the Africa Parks Network, without actually using the Network.[124]

As another testament to the potential sustainability of TNS, local communities originally showed lots of hostilities as WWF was seen as keeping people away from bushmeat and elephants. Yet, as the logging companies in the area went bankrupt and WWF starting hiring local communities, the vast majority of people now support WWF's conservation work. In fact, WWF employs approximately 250 local people and is the largest employer in the region, if not the country (CAR). In addition to employment opportunities, approximately 40% of tourism revenue goes to the communities.[125]

Furthermore, the establishment of the Trust Fund and the long-term financing generated via interest on the Trust Fund's principal has greatly contributed to the sustainability of the TNS.[126]

Lessons Learned

Two important lessons learned from TNS are:

- First, organizations dedicated to conservation need a long-term commitment, a vision of 20+ years, and a plan to maintain that vision. Despite civil wars, organizations such as WWF who desire to see the protected area to be sustainable in every aspect, kept their eyes on the goal, never ever left, and stuck to the commitment they made.
- Second, we need leadership from both expats and nationals. Such leaders are people with the right mindset and organizations need to invest in them.[127]

Another important lesson learned is that it was a very good idea to create the TNS and additional transfrontier protected areas should be established.[128]

Other Resources on Tri-National de la Sangha

CIFOR Report: Current vulnerability in the Tri-National de la Sangha landscape, Cameroon

- http://www.cifor.org/library/4084/current-vulnerability-in-the-tri-national-de-la-sangha-landscape-cameroon/

Dzanga Sangha National Park

- http://www.dzanga-sangha.org/node/309

Tri-National de la Sangha Foundation

- http://www.fondationtns.org/dev/index.php/en/

WWF's Congo Basin Page

- http://www.wwf-congobasin.org/

Return

The returns generated from a government's international budgetary allocations can come in a few different forms. For instance, this transfer of payment could generate a financial return or a non-financial return, such as a change in policy. If the funding from one government to another is in the form of a loan, then the lender may earn interest. In addition, the host government could earn a financial return if the protected area incorporates revenue streams (e.g., visitor fees for tourists or researchers) as outlined in the domestic budgetary allocations chapter.

Risk

Business Risk

The specific business risk of international budgetary allocations depends on the underlying program or project that is being invested in—such as large-scale ecotourism operations, setting up national REDD+ programs, and/or establishing certification standards for products with zero-deforestation supply chain commitments. Additional business risks are that funds may not be renewed in future years and the underlying on-the-ground activities that are being funded may not be designed to become a commercially viable business in the future.

Strategic Risk

Strategic risks for international budgetary allocations include:

- Choosing which country or countries to support and choosing the specific partners among local nonprofit organizations, government agencies, and international organizations; and
- Choosing what approach to take (e.g., setting up in-country trust fund, establishing international peace park, or financing watershed conservation).

There is also "donor drift" when donor countries may shift their funding to other countries, other conservation financing instruments, or to different biomes.

Reputation Risk

Reputation risks may arise for international budgetary allocations if:

- Funds are squandered due to mismanagement, corruption, or improper budgeting;
- The underlying conservation activities or the conservation financing instrument (i.e., monitoring shows that deforestation has increased) does not succeed; or
- Local stakeholders (e.g., Indigenous Peoples and/or forest-dependent communities) are not properly consulted.

Liquidity Risk

Similar to a government's domestic budgetary allocations, a government's international budgetary allocations are often illiquid. Thus, a government's contribution to a multilateral agency such as the World Bank's Forest Carbon Partnership Facility, or the creation of transfrontier conservation area, or funding an international trust fund, is likely to have few options for liquidation.

Also from the recipients' perspective, the disbursements from a government's international budgetary allocation may come at a later time than when their expense (i.e., paycheck for forest patrols and park administration) is due. In turn, this may create another liquidity risk, for instance, if a park ranger gets paid yearly, but the park ranger needs to pay his or her child's school fees on a monthly basis.

Operational Risk

One operational risk for international budgetary allocations is that there might be delays in disbursements (i.e., similar to the previous liquidity risk) and the lack of a sufficient project pipeline. For instance:

> The GCF aims to distribute $2.5 billion of its $10 billion pot over the course of 2016, meaning it will need to approve between $850 million and $930 million during each of three meetings this year. But there are a mere 29 projects currently in the GCF pipeline, and only one accredited partner has the ability to handle projects valued at more than $250 million.[129]

Additional operational risks can be the challenges associated with getting funding to the ground (i.e., lack of internet or few passable roads to the protected area that in turn creates a rural, informal banking sector) and the lack of audited financial statements for the protected area.

Legal and Regulatory Risk

Multilateral agencies often have near universal national government support and, thus, are less likely to involve legal and regulatory risks.

However, significant legal and regulatory risks can occur with transfrontier conservation areas when one party attempts to exit the partnership. Similarly, the protected area may entirely dissolve or there could be disputes between which level of government (i.e., provincial versus federal) has authority over the protected area.

There can also be legal and regulatory risk associated with bilateral policy and funding priority changes with the donor country.

Furthermore, there can be overlapping legal title claims, such as between a mining concession and a national park or between local communities and a timber concession.

Credit Risk

Credit risk includes default risk, bankruptcy risk, downgrade risk, and settlement risk. One example of settlement risk is that the entity's bank or credit union (i.e., remote outpost for an international peace park) may not be able to receive an international transfer of funds. In contrast, multilateral agencies often have near universal national government support and, thus, are less likely to have credit risks.

Market Risk

Market risk includes interest rate risk, equity price risk, foreign exchange risk, and commodity price risk. A specific market risk for a government's international budgetary allocation is that, due to foreign exchange fluctuations, the amount committed to another country may rise or fall overtime. A similar situation could occur due to commodity price risk, where a donor country's level of support is tied to the revenue they earn from a globally traded commodity.

Risk, Return, Time (Horizon), Taxes, Liquidity, Legal and Unique (RRTTLLU)

Risk and Return

Please see above for the risk and return associated with government international budgetary allocations. It is important to note that returns should be viewed not just as financial returns but also returns from net positive climate, community, and biodiversity outcomes.

Time Horizon

The time horizon for government international budgetary expenditures will depend on the nature of the commitment. For instance, many countries have contributed to the World Bank and the United Nations for several decades. In contrast, other government international budgetary allocations can be more short-term in nature such as the United States Agency for International Development's multiyear commitment to the Forest Carbon, Markets, and Communities (FCMC) program. Furthermore, the ARPA Trust Fund is a sinking 25-year fund, while the Trust Fund for the Tri-National de la Sangha is set up for perpetual financing.

Taxes

There are likely limited, if any, taxes due to the donor government or host government if the governments are the sole providers and recipients of the investments. However, there could be local taxes such as sales tax levied on concessions, employment taxes, and/or some form of tourism tax for visitors (e.g., if visitors stay at onsite facilities). Depending on what exactly is being financed (i.e., a strict protected area versus an ecotourism operation), there may be lost tax revenue such as lost excise taxes, royalties, or income taxes—especially when compared to a timber, mining, or palm oil concession.

Liquidity

Similar to government domestic budgetary allocations, liquidity might exist with concessions or other onsite facilities (e.g., tourist attraction or lodging) that could be transferred to another owner. However, the actual underlying protected area is likely to be illiquid unless the federal government can transfer the protected area to a state or province (i.e., or vice versa).

If a protected area has multiple cash flows via different conservation financing instruments (e.g., ecotourism, international budgetary allocations, and forest carbon offsets), there could be a liquidity risk if one of the more significant sources of financing slows down.

Legal

Legal considerations associated with government international budgetary expenditures will vary from country to country. When considering the establishment of a trust fund, it is important to consider the jurisdiction of the legal entity that will manage the trust fund.

Unique

Unique aspects include the establishment of multi-country, joint protected areas. Most of the other conservation financing instruments—such as debt-for-nature swaps, payments for ecosystem services, or ecotourism—focus on one country.

Policy Analysis

Defining the Problem

Governments, whether its domestic budgetary expenditures or international budgetary expenditures, are facing competing interests such as healthcare, energy access, and education. Tropical rainforest deforestation and degradation continues today, and many host countries do not have adequate funds, hence, the need for international support. Furthermore, many tropical rainforests stretch across national borders, and there is a need for countries to establish cross border, international peace parks.

Establishing Goals

The goal of developed countries earmarking a certain percentage of their revenue to Official Development Assistance (ODA) dates back to 1958, when the World Council of Churches proposed a target of at least 0.7%. In 1969,

> {..} the Pearson Commission—in its report Partners in Development—proposed a target of 0.7% of donor GNP. {…} With the revised System of National Accounts in 1993, gross national product was replaced by gross national income (GNI), an equivalent concept. DAC members' performance against the 0.7% target is therefore now shown in terms of ODA/GNI ratios.[130]

Thus, a subset goal is to have a significant portion of this ODA be allocated to tropical rainforest conservation and toward effective policies and projects to reduce the pressure on these forests.

Selecting a Policy

On the international level, the Convention on International Trade in Endangered Species of Wild Fauna and Flora (CITES) was very relevant to the establishment of the Tri-National de la Sangha.[131]

With respect to Official Development Assistance (ODA), the largest providers of ODA in 2015,[132] and the largest recipients of ODA in 2014,[133] in terms of a percentage of Gross National Income (GNI), are given in Table 9.1.

In contrast, the largest providers of ODA in 2015,[134] and the largest recipients of ODA in 2014,[135] in US dollars are given in Table 9.2.

With respect to biodiversity conservation, Brazil, India, and Turkey were the largest recipients, on average from 2011 to 2015, of bilateral biodiversity-related ODA. Germany, the United States, and Japan, on average from 2011 to 2015, were the largest providers of biodiversity-related ODA. In addition, "Norway, Iceland and Belgium dedicated the highest shares of their ODA portfolios to biodiversity activities."[136]

In part because of Norway's Climate Consensus and the NICFI, Norway is widely viewed as the leading nation for international budgetary allocations to reduce tropical rainforest deforestation and degradation. In addition Germany, in part due to the commission of the REDD Early Movers (REM) Programme, has also become a large financial supporter of Reducing Emissions from Deforestation and Degradation (REDD+) activities in tropical rainforest countries.

Table 9.1 Largest ODA providers and recipients as a percentage

	Country	Net ODA in 2015 (% of GNI)		Country	Net ODA received in 2014 (% of GNI)
1	Sweden	1.4	1	Tuvalu	69.2
2	United Arab Emirates	1.2	2	Liberia	44.3
Tied-3	Norway	1.0	3	Central African Republic	35.7
Tied-3	Luxembourg	1.0	4	Micronesia	33.9
5	Denmark	0.8	5	Marshall Islands	24.6
Tied-6	Netherlands	0.7	6	Afghanistan	23.8
Tied-6	United Kingdom	0.7	7	Kiribati	23.0
8	Finland	0.6	8	Somalia	21.3
Tied-9	Germany	0.5	9	Sierra Leone	18/6
Tied-9	Switzerland	0.5	10	Solomon Islands	18.1
Tied-9	Turkey	0.5			

Source: OECD and World Bank

Table 9.2 Largest ODA providers and recipients in US dollars

	Country	Net ODA in 2015 (billions of USD$)		Country	Net ODA received in 2014 (current USD$, billions of USD$)
1	United States	USD$30.68	1	Afghanistan	USD$4.82
2	Germany	USD$21.04	2	Vietnam	USD$4.22
3	United Kingdom	USD$19.75	3	Syria	USD$4.20
4	France	USD$10.70	4	Pakistan	USD$3.61
5	Japan	USD$10.29	5	Ethiopia	USD$3.59
6	Sweden	USD$8.52	6	Egypt	USD$3.53
7	Saudi Arabia	USD$7.53	7	Turkey	USD$3.44
8	Netherlands	USD$6.83	8	India	USD$2.98
9	Norway	USD$5.53	9	Jordan	USD$2.70
10	Canada	USD$4.95	10	Kenya	USD$2.67

Source: OCED and World Bank

Implementing a Policy

The Norwegian Ministry of Climate and Environment is implementing Norway's International Climate and Forest Initiative (NICFI). Germany's REM Programme was "commissioned by the German Federal Ministry for Economic Cooperation and Development (BMZ) and implemented by the KfW Development Bank and the Gesellschaft für Internationale Zusammenarbeit (GIZ)."[137]

The CITES convention has been adopted, among many countries, by Cameroon, the Central African Republic (CAR), and the Republic of the Congo, which are the 3 host countries of the Tri-National de la Sangha (TNS).

Evaluating the Policy

Norway is tied for third with Luxembourg for ODA as a percentage of their GNI, ranked ninth for ODA in US dollars, and alongside Iceland and Belgium, Norway "dedicated the highest shares of their ODA portfolios to biodiversity activities."[138] A significant portion of Norway's ODA has been allocated to NICFI (see case study in Chap. 9), which has accomplished a wide range of results including the Peatland Moratorium in Indonesia and financing of the Amazon Fund in Brazil and facilitated the inclusion of REDD+ in the Paris Agreement.

Similarly, Germany is tied for ninth with Switzerland and Turkey for ODA as a percentage of their GNI, ranked second for ODA in US dollars, and was the largest provider of biodiversity-related ODA from 2011 to 2015. Germany's policy has, among many results, enabled payments for performance to the subnational jurisdiction of Acre, Brazil, for its reduction in deforestation.[139] Germany has also collaborated with Norway through REM.[140] Thus, the policies in both Germany and Norway seem to be very effective.

The success of CITES is mixed. While the countries of Cameroon, CAR, and the Republic of the Congo adopted CITES, its adoption was a driving force for the creation of the TNS, and this enabled the creation of the TNS Trust Fund, there are still major problems with poaching at TNS and elsewhere of endangered species.

Future Outlook for Instrument

There are elements that are within the control of a government and there are elements that are out of their control. Similarly, there are many lessons learned and other lessons that are still to be learned and incorporated. As previously mentioned, governments in developing countries often own the majority of their country's land, as opposed to the private sector. This creates a potential opportunity for the expansion of existing national parks and the creation of additional protected areas, if adequate and long-term financing is available to offset the opportunity costs.

It is important to note that many governments are facing budget constraints, and if structured properly, international budgetary allocations to protected areas will provide opportunities for the private sector. With this in mind, there is a shift from the old aid structure to a results-based financing approach. This transition continues to be a struggle, both for the donor and the recipient. The least developed countries still do not have the capacity or upfront funding. From the donors' side, there are leaders such as Norway and Germany, for whom it has been difficult to mainstream the concept. One difficulty in this context of limited public funds has been the attempt to move the funding out of the aid budgets and over to the Ministries of Finance because this funding is more budgetary support than traditional aid.

With respect to international peace parks, it was a very good idea to create the TNS, and additional transfrontier protected areas should also be established. Such future transfrontier protected areas, for instance, could include collaborations between Cameroon, Gabon, and/or Equatorial Guinea. One reason, among many reasons, that establishing transfrontier protected areas is a good idea, and for why they should continue into the future, is because animals do not recognize national borders and such protected areas allow for animals to migrate.[141]

Other Resources on Government International Budgetary Allocations

Central Africa Regional Program for the Environment (CARPE)

- http://carpe.umd.edu/index.php

Global Transboundary Conservation Network

- http://www.tbpa.net/page.php?ndx=12

IUCN's Report entitled, "Transboundary Conservation: A systematic and integrated approach"

- http://www.tbpa.net/newsletters/39_TB_BPG_-_FINAL_WEB_COMPLETE3.pdf

OECD Statistics on External Development Finance Targeting Environmental Objectives Including the Rio Conventions

- http://www.oecd.org/dac/stats/rioconventions.htm?utm_source=social-media&utm_medium=twfbgppin&utm_campaign=devtworg/

Notes

1. Pharo, Per Fredrik Ilsaas. Email Message to author. April 2, 2017.
2. Fauna & Flora International. "110 Years of Fauna & Flora International." Accessed November 23, 2016. http://www.fauna-flora.org/timeline/
3. BirdLife International. "Our History." Accessed November 23, 2016. http://www.birdlife.org/worldwide/partnership/our-history
4. UNESCO. "Waterton Glacier International Peace Park." Accessed December 13, 2016. http://whc.unesco.org/en/list/354
5. INBIO. "Resumen." Accessed December 13, 2016. http://www.inbio.ac.cr/pila/parque.htm
6. Gonzales, José and Hermilio Rosas. "The Caral Archeological Site and Supe Valley Restoration." In *Conservation Capital in the Americas*, edited by James N. Levitt. Massachusetts: Lincoln Institute of Land Policy, 2010. 91.
7. Peace Parks Foundation. "Origins of Peace Parks Foundation." Accessed December 13, 2016. http://www.peaceparks.org/story.php?pid=1&mid=2
8. The Sangha Tri-National Trust Fund. "Historical Background of the Sangha Tri-National (TNS)." Accessed February 15, 2017. http://www.

fondationtns.org/dev/index.php/en/background-historical-background-of-the-sangha-tri-national-tns/

9. Great Limpopo Transfrontier Park. "About GLTP." Accessed December 13, 2016. http://www.greatlimpopo.org/about-gltp/
10. ARPA. "Full List: PAs Supported by ARPA." Accessed December 8, 2016. http://programaarpa.gov.br/en/full-list-pas-supported-by-arpa/
11. Kavango-Zambezi Transfrontier Conservation Area. "About Us." Accessed December 13, 2016. http://www.kavangozambezi.org/about-us
12. The Sangha Tri-National Trust Fund. "History of the Trust Fund." Accessed December 13, 2016. http://www.fondationtns.org/dev/index.php/en/history-of-the-trust-fund/
13. Elliott, Chris. "REDD+ Is a Little More Complex than 'Not Cutting Down a Tree'." In *On the Edge*, written by Claude Martin. Berkeley: Greystone Books, 2015. 239.
14. Amazon Fund. "Amazon Fund/Purposes and Management." Accessed November 23, 2016. http://www.amazonfund.gov.br/FundoAmazonia/fam/site_en/Esquerdo/Fundo/
15. Global Transboundary Conservation Network. "UNEP-WCMC Transboundary Protected Areas Inventory-2007." Accessed December 13, 2016. http://www.tbpa.net/page.php?ndx=78
16. Organisation for Economic Co-operation and Development. "Biodiversity-Related Official Development Assistance 2015." Accessed April 4, 2017. http://www.oecd.org/dac/environment-development/Biodiversity-related-ODA.pdf. 4.
17. Multi-Partner Trust Fund Office. "Central African Forest Init." Accessed February 20, 2017. http://mptf.undp.org/factsheet/fund/AFI00?fund_status_month_to=&fund_status_year_to=2016
18. Multi-Partner Trust Fund Office. "UN REDD Programme Fund." Accessed February 20, 2017. http://mptf.undp.org/factsheet/fund/CCF00?fund_status_month_to=&fund_status_year_to=2016
19. Global Environmental Facility. "Funding." Accessed April 2, 2017. https://www.thegef.org/about/funding
20. Forest Carbon Partnership Facility. "2016 Annual Report." Accessed April 2, 2017. https://www.forestcarbonpartnership.org/sites/fcp/files/2016/Sep/FCFP%20Annual%20Report%20FY16.pdf . 59.
21. Climate Investment Funds. "Funds & Programs: Forest Investment Program." Accessed April 10, 2017. http://www-cif.climateinvestmentfunds.org/fund/forest-investment-program

22. The World Bank's Carbon Finance Unit. "About the BioCarbon Fund (BIOCF)." Accessed April 11, 2017. https://wbcarbonfinance.org/Router.cfm?Page=BioCF&FID=9708&ItemID=9708&ft=About
23. BioCarbon Fund Initiative for Sustainable Forest Landscapes. "Funding." Accessed April 11, 2017. http://www.biocarbonfund-isfl.org/funding
24. The World Bank's Carbon Finance Unit. "Madagascar: Madagascar: Ankeniheny–Zahamena—Mantadia Biodiversity Conservation Corridor (REDD) Project." Accessed April 11, 2017. https://wbcarbonfinance.org/Router.cfm?Page=Projport&ProjID=9638
25. The World Bank's Carbon Finance Unit. "Costa Rica: Coopeagri Agroforestry." Accessed April 11, 2017. https://wbcarbonfinance.org/Router.cfm?Page=Projport&ProjID=9632
26. The World Bank's Carbon Finance Unit. "Nicaragua: Precious Woods." Accessed April 11, 2017. https://wbcarbonfinance.org/Router.cfm?Page=Projport&ProjID=9640
27. Green Climate Fund. "Global Context: Response Description." Accessed April 4, 2017. http://www.greenclimate.fund/about-gcf/global-context#the-big-picture
28. Cooper, Graham. "Green Climate Fund offers 500m for private sector climate projects." May 25, 2017. *Environmental Finance*. https://www.environmental-finance.com/content/news/green-climate-fund-offers-$500m-for-private-sector-climate-projects.html
29. Carbon Pulse. "GCF transfers first REDD cash ahead of talks on increased forest funding." July 5, 2017. www.carbon-pulse.com/36837/
30. Graver, Ane Broch. Interviewed by Brian McFarland. March 2017.
31. ARPA. "Full List: PAs supported by ARPA." Accessed December 8, 2016. http://programaarpa.gov.br/en/full-list-pas-supported-by-arpa/
32. Conservation International. "Protecting the Heart of the Amazon in Calha Norte, Brazil." Accessed December 8, 2016. http://www.conservation.org/projects/Pages/Protecting-Heart-of-the-Amazon-in-Calha-Norte-Brazil.aspx
33. Maretti, Cláudio. "The Amazon: There is Hope!—If We All Do the Right Thing." In *On the Edge*, written by Claude Martin. Berkeley: Greystone Books, 2015. 209–210.
34. UNESCO World Heritage Centre. "Central Amazon Conservation Complex." Accessed December 8, 2016. http://whc.unesco.org/en/list/998

35. World Wildlife Fund. "Amazon Region Protected Areas Programme." Accessed December 8, 2016. http://wwf.panda.org/what_we_do/where_we_work/amazon/vision_amazon/models/amazon_protected_areas/financing/arpa/
36. FUNBIO. "Amazon Region Protected Areas Program (Arpa)." Last modified June 26, 2015. Accessed April 10, 2017. http://www.funbio.org.br/en/arpa/
37. FUNBIO. Interviewed by Brian McFarland. October 2016.
38. Ibid.
39. Convention on Biological Diversity. "Decision Adopted by the Conference of the Parties to the Convention on Biological Diversity at Its Seventh Meeting." Accessed December 8, 2016. https://www.cbd.int/doc/decisions/cop-07/cop-07-dec-30-en.pdf. 9.
40. FUNBIO. Interviewed by Brian McFarland. October 2016.
41. Ibid.
42. Ibid.
43. Earth Innovation Institute. "Brazilian Agreement on Amazon Deforestation." Accessed December 12, 2016. http://earthinnovation.org/our-work/case-studies/amazon-deforestation/
44. FUNBIO. Interviewed by Brian McFarland. October 2016.
45. Ibid.
46. Ibid.
47. Elliott, Chris. "REDD+ Is a Little More Complex than 'Not Cutting Down a Tree'." In *On the Edge*, written by Claude Martin. Berkeley: Greystone Books, 2015. 239.
48. Forest 500 Project. "The Government of Norway's International Climate and Forest Initiative." Accessed April 1, 2017. http://forest500.org/rankings/other-powerbrokers/government-norways-international-climate-and-forest-initiative
49. Norwegian Government. "Norwegian International Climate and Forest Initiative Extended Until 2030." Last modified December 11, 2015. https://www.regjeringen.no/en/aktuelt/norwegian-international-climate-and-forest-initiative-extended-until-2030/id2467031/
50. Graver, Ane Broch. Interviewed by Brian McFarland. March 2017.
51. Ibid.
52. United Nations Framework Convention on Climate Change. "Norway." Accessed April 1, 2017. http://unfccc.int/tools_xml/country_NO.html
53. UN-REDD Programme. "Our Donors." Accessed April 1, 2017. http://www.un-redd.org/donors

54. Forest Carbon Partnership Facility. "Donor Participants." Accessed April 1, 2017. https://www.forestcarbonpartnership.org/donor-participants#overlay-context=donor-participants-0
55. Graver, Ane Broch. Interviewed by Brian McFarland. March 2017.
56. Ibid.
57. Pharo, Per Fredrik Ilsaas. Interviewed by Brian McFarland. March 2017.
58. Norwegian Petroleum. "Exports of Oil and Gas." Accessed April 1, 2017. http://www.norskpetroleum.no/en/production-and-exports/exports-of-oil-and-gas/
59. Pharo, Per Fredrik Ilsaas. Interviewed by Brian McFarland. March 2017.
60. Government of Norway. "The Agreement on Climate Policy." Last modified December 12, 2014. https://www.regjeringen.no/en/topics/climate-and-environment/climate/innsiktsartikler-klima/agreement-on-climate-policy/id2076645/
61. Pharo, Per Fredrik Ilsaas. Interviewed by Brian McFarland. March 2017.
62. Graver, Ane Broch. Interviewed by Brian McFarland. March 2017.
63. Ibid.
64. The Norwegian Agency for Development Cooperation. "The Norwegian Climate and Forest Funding to Civil Society." Accessed December 29, 2016. https://www.norad.no/en/front/funding/climate-and-forest-initiative-support-scheme/
65. NORAD. "List of Grant Recipients Norad/NICFI Grant Scheme for Civil Society 2013–2015." Accessed April 1, 2017. https://www.norad.no/globalassets/import-2162015-80434-am/www.norad.no-ny/filarkiv/3.-sivsa-2012---/cfi/list-of-grant-recipients-2013-2015.pdf
66. Government of Norway. "Our Partnerships." Accessed April 1, 2017. https://www.regjeringen.no/en/topics/climate-and-environment/climate/climate-and-forest-initiative/kos-innsikt/samarbeidspartnere/id2345203/
67. Thomson Reuters. "Norway to halve forest protection payments to Brazil to about $50 mln." June 23, 2017. http://news.trust.org/item/20170623102123-g79q1/
68. NORAD. "Historical Constitutional Court in Indonesia Recognises Indigenous Territories." Accessed April 1, 2017. https://www.norad.no/en/front/funding/climate-and-forest-initiative-support-scheme/grants-2013-2015/achievements/achievements-2016/

historical-constitutional-court-in-indonesia-recognises-indigenous-territories/

69. NORAD. "A More Inclusive Process is Ensured in Africa's First Emissions Reductions Programme." Accessed April 1, 2017. https://www.norad.no/en/front/funding/climate-and-forest-initiative-support-scheme/grants-2013-2015/achievements/achievements-2016/a-more-inclusive-process-is-ensured-in-africas-first-emissions-reductions-programme/
70. NORAD. "Community Based Forest Monitoring Improves Decision-Making." Accessed April 1, 2017. https://www.norad.no/en/front/funding/climate-and-forest-initiative-support-scheme/grants-2013-2015/achievements/achievements-2016/community-based-forest-monitoring-improves-decision-making/
71. NORAD. "International Law Enforcement Cooperation Leads to Detention of Illegally Exported Timer from Peru." Accessed April 1, 2017. https://www.norad.no/en/front/funding/climate-and-forest-initiative-support-scheme/grants-2013-2015/achievements/achievements-2016/international-law-enforcement-cooperation-leads-to-detention-of-illegally-exporterted-timer-from-peru/
72. NORAD. "Improved Access to Forest Information in Indonesia." Accessed April 1, 2017. https://www.norad.no/en/front/funding/climate-and-forest-initiative-support-scheme/grants-2013-2015/achievements/achievements-2016/improved-access-to-forest-information-in-indonesia/
73. NORAD. "Forest Code Observatory: Carrying Out Dormant Forest Legislation." Accessed April 1, 2017. https://www.norad.no/en/front/funding/climate-and-forest-initiative-support-scheme/grants-2013-2015/projects/forest-code-observatory-carrying-out-dormant-forest-legislation/
74. Moura Costa, Pedro of BVRio. Interviewed by Brian McFarland. March 2017.
75. Forest 500 Project. "The Government of Norway's International Climate and Forest Initiative." Accessed April 1, 2017. http://forest500.org/rankings/other-powerbrokers/government-norways-international-climate-and-forest-initiative
76. Central African Forest Initiative. "The CAFI Declaration." Last modified September 29, 2015. http://www.cafi.org/content/cafi/en/home/our-work/the-cafi-declaration.html

77. Ferguson-Mitchell, Jennifer. "US$ 200 Million Agreement Signed Between CAFI and the DRC Set to Prevent Tree Loss and Ensure Sustainable Development." April 25, 2016. *UN-REDD Programme*. Accessed April 10, 2017. http://www.unredd.net/index.php?option=com_content&view=article&id=2406:us-200-million-agreement-signed-between-cafi-and-the-drc-set-to-prevent-tree-loss-and-ensure-sustainable-development&catid=98&Itemid=749
78. Graver, Ane Broch. Interviewed by Brian McFarland. March 2017.
79. The World Bank Group. "Brazil Protects the Amazon." Last modified October 9, 2013. http://www.worldbank.org/en/results/2013/10/09/Brazil-protects-Amazon-increasing-size-protected-areas
80. National Geographic. "Indonesia Facts." Accessed March 29, 2017. http://travel.nationalgeographic.com/travel/countries/indonesia-facts/
81. Pharo, Per Fredrik Ilsaas. Interviewed by Brian McFarland. March 2017.
82. Norwegian Government. "Norwegian International Climate and Forest Initiative Extended Until 2030." Last modified December 11, 2015. https://www.regjeringen.no/en/aktuelt/norwegian-international-climate-and-forest-initiative-extended-until-2030/id2467031/
83. Kusumaatmadja, Rezal. Interviewed by Brian McFarland. August 2016.
84. Pharo, Per Fredrik Ilsaas. Interviewed by Brian McFarland. March 2017.
85. The Sangha Tri-National Trust Fund. "Historical Background of the Sangha Tri-National (TNS)." Accessed February 12, 2017. http://www.fondationtns.org/dev/index.php/en/background-historical-background-of-the-sangha-tri-national-tns/
86. The Sangha Tri-National Trust Fund. "Dzanga Sangha—Central African Republic (CAR)." Accessed February 15, 2017. http://www.fondationtns.org/dev/index.php/en/dzanga-sangha-central-african-republic-car/
87. The Sangha Tri-National Trust Fund. "Nouabale Ndoki—Congo." Accessed February 15, 2017. http://www.fondationtns.org/dev/index.php/fr/nouabale-ndoki-congo/
88. The Sangha Tri-National Trust Fund. "Lobéké—Cameroon." Accessed February 15, 2017. http://www.fondationtns.org/dev/index.php/en/lobeke-cameroon/

89. The Sangha Tri-National Trust Fund. "Historical Background of the Sangha Tri-National (TNS)." Accessed February 15, 2017. http://www.fondationtns.org/dev/index.php/en/background-historical-background-of-the-sangha-tri-national-tns/
90. World Heritage Convention. "Sangha Trinational." Accessed February 15, 2017. http://whc.unesco.org/en/list/1380
91. The Sangha Tri-National Trust Fund. "History of the Trust Fund." Accessed December 13, 2016. http://www.fondationtns.org/dev/index.php/en/history-of-the-trust-fund/
92. TNS Foundation. "Mobilization of Financing Capitalization." Accessed April 7, 2017. http://www.fondationtns.org/dev/index.php/en/mobilization-of-financing-capitalization/
93. Blom, Dr. Allard. Interviewed by Brian McFarland. January 2017.
94. Zognou, Théophile. Interviewed by Brian McFarland. April 2017.
95. TNS Foundation. "Mobilization of Financing Capitalization." Accessed April 7, 2017. http://www.fondationtns.org/dev/index.php/en/mobilization-of-financing-capitalization/
96. Blom, Dr. Allard. Interviewed by Brian McFarland. January 2017.
97. Ibid.
98. Zognou, Théophile. Interviewed by Brian McFarland. April 2017.
99. World Bank Group. "Doing Business: Economy Rankings." Accessed January 12, 2017. http://www.doingbusiness.org/rankings
100. Transparency International. "Corruption Perceptions Index 2016." Accessed February 8, 2017. http://www.transparency.org/news/feature/corruption_perceptions_index_2016
101. United Nations Development Programme. "Human Development Report 2015: Work for Human Development." Last modified 2016. http://hdr.undp.org/sites/default/files/2015_human_development_report.pdf. 211.
102. World Heritage Convention. "Sangha Trinational." Accessed February 15, 2017. http://whc.unesco.org/en/list/1380
103. World Heritage Convention. "State of Conservation: Sangha Trinational." Accessed February 15, 2017. http://whc.unesco.org/en/soc/3228
104. Blom, Dr. Allard. Interviewed by Brian McFarland. January 2017.
105. World Heritage Convention. "Sangha Trinational." Accessed February 15, 2017. http://whc.unesco.org/en/list/1380
106. Blom, Dr. Allard. Interviewed by Brian McFarland. January 2017.

107. Zognou, Théophile. Interviewed by Brian McFarland. April 2017.
108. Ibid.
109. The Sangha Tri-National Trust Fund. "Dzanga Sangha—Central African Republic (CAR)." Accessed February 15, 2017. http://www.fondationtns.org/dev/index.php/en/dzanga-sangha-central-african-republic-car/
110. The Sangha Tri-National Trust Fund. "Nouabale Ndoki—Congo." Accessed February 15, 2017. http://www.fondationtns.org/dev/index.php/fr/nouabale-ndoki-congo/
111. The Sangha Tri-National Trust Fund. "Lobéké—Cameroon." Accessed February 15, 2017. http://www.fondationtns.org/dev/index.php/en/lobeke-cameroon/
112. The Sangha Tri-National Trust Fund. "Historical Background of the Sangha Tri-National (TNS)." Accessed February 15, 2017. http://www.fondationtns.org/dev/index.php/en/background-historical-background-of-the-sangha-tri-national-tns/
113. The Sangha Tri-National Trust Fund. "History of the Trust Fund." Accessed December 13, 2016. http://www.fondationtns.org/dev/index.php/en/history-of-the-trust-fund/
114. Zognou, Théophile. Interviewed by Brian McFarland. April 2017.
115. Blom, Dr. Allard. Interviewed by Brian McFarland. January 2017.
116. Ibid.
117. World Heritage Convention. "Sangha Trinational." Accessed February 15, 2017. http://whc.unesco.org/en/list/1380
118. Blom, Dr. Allard. Interviewed by Brian McFarland. January 2017.
119. Zognou, Théophile. Interviewed by Brian McFarland. April 2017.
120. Cannon, John C. "'This is Not Empty Forest': Africa's Palm Oil Surge Builds in Cameroon." March 30, 2016. *Mongabay*. Accessed April 10, 2017. https://news.mongabay.com/2016/03/this-is-not-empty-forest-africas-palm-oil-surge-builds-in-cameroon/
121. Zognou, Théophile. Interviewed by Brian McFarland. April 2017.
122. Blom, Dr. Allard. Interviewed by Brian McFarland. January 2017.
123. Ibid.
124. Ibid.
125. Ibid.
126. Zognou, Théophile. Interviewed by Brian McFarland. April 2017.
127. Blom, Dr. Allard. Interviewed by Brian McFarland. January 2017.
128. Zognou, Théophile. Interviewed by Brian McFarland. April 2017.

129. Ecosystem Marketplace. "The Carbon Chronicle: Carbon Finance: Crisis line bling." February 10, 2016. Accessed April 10, 2017. http://www.ecosystemmarketplace.com/wp-content/themes/ecosystem_marketplace/newsletter/carbon_chronical.php?newsletterID=23
130. OECD. "The 0.7% ODA/GNI Target—A History." Accessed February 20, 2017. http://www.oecd.org/dac/stats/the07odagnitarget-ahistory.htm
131. Zognou, Théophile. Interviewed by Brian McFarland. April 2017.
132. OECD. "Net ODATotal, % of Gross National Income, 2015." Accessed February 20, 2017. https://data.oecd.org/oda/net-oda.htm
133. World Bank. "Net ODA Received (% of GNI)." Accessed February 20, 2017. http://data.worldbank.org/indicator/DT.ODA.ODAT.GN.ZS?year_high_desc=true
134. OECD. "Net ODATotal, Million US Dollars, 2000–2015." Accessed March 23, 2017. https://data.oecd.org/oda/net-oda.htm
135. World Bank. "Net Official Development Assistance Received (Current US$)." Accessed March 23, 2017. http://data.worldbank.org/indicator/DT.ODA.ODAT.GN.ZS?year_high_desc=true
136. OECD's Development Assistance Committee. "Biodiversity-Related Official Development Assistance 2015." Accessed April 8, 2017. http://www.oecd.org/dac/environment-development/Biodiversity-related-ODA.pdf. 7–8.
137. The REDD Desk. "Germany's REDD+ Early Movers Programme." Last modified March 8, 2016. Accessed April 8, 2017. http://theredddesk.org/markets-standards/germanys-redd-early-movers-programme
138. OECD's Development Assistance Committee. "Biodiversity-Related Official Development Assistance 2015." Accessed April 8, 2017. http://www.oecd.org/dac/environment-development/Biodiversity-related-ODA.pdf. 7–8.
139. The REDD Desk. "Germany's REDD+ Early Movers Programme." Last modified March 8, 2016. Accessed April 8, 2017. http://theredddesk.org/markets-standards/germanys-redd-early-movers-programme
140. GIZ. "REDD Early Movers—Tools and Instruments." Accessed April 8, 2017. https://www.giz.de/en/worldwide/33356.html
141. Zognou, Théophile. Interviewed by Brian McFarland. April 2017.

10

Impact Investing

"Sustainable capital is an area of public and private financing that utilizes access to favorable equity, credit, and microfinance as an incentive to promote environmental sustainability and responsible business practices. {…} Examples of sustainable capital {…} include environmental investment or biodiversity enterprise funds; forest securitization, or more broadly, eco securitization; favorable credit or loan terms tied to performance incentives or standards required by public or private lenders; and emerging microfinance programs to promote community conservation and social development."[1]
—World Wildlife Fund

Introduction

As defined by the Forum for Sustainable and Responsible Investment, which is a member of the Global Sustainable Investment Alliance, "sustainable, responsible and impact investing (SRI) is an investment discipline that considers environmental, social and corporate governance (ESG) criteria to generate long-term competitive financial returns and positive societal impact."[2] Impact investing, particularly related to tropical rainforest conservation and natural capital, is a relatively new class of

B.J. McFarland, *Conservation of Tropical Rainforests*, Palgrave Studies in Environmental Policy and Regulation, https://doi.org/10.1007/978-3-319-63236-0_10

emerging conservation financing instruments. Such instruments include impact investments structured by nonprofit and for-profit organizations, along with through dedicated funds such as Althelia Ecosphere's Althelia Climate Fund. Furthermore, there are emerging techniques to develop listing standards for publicly traded equities and an emerging divestment movement.

Historical Overview

The modern socially responsible investment movement, which is related to impact investing, has its origins in the "1970s as part of grassroots campaigns to boycott firms in South Africa as well as the nuclear weapons and tobacco industries."[3]

Other notable dates associated with impact investing, as it relates to the conservation of tropical rainforests, include:

- 2006: Terra Global Capital is founded by Leslie Durschinger.[4] Terra Global Capital would later launch the Terra Bella Global Fund.
- 2009: BioCarbon Group is founded (initial capital raised in 2011).[5]
- 2010: The World Bank's Wealth Accounting and the Valuation of Ecosystem Services (WAVES) partnership is launched.[6]
- 2010: TNC developed an impact investment strategy and formed a dedicated team with support from the Robertson Foundation. TNC would later form NatureVest in 2014 with assistance from JPMorgan Chase & Co.[7]
- 2011: Althelia Ecosphere is launched.[8]
- 2011: Bolsa de Valores Ambientais do Rio de Janeiro (BVRio, Rio de Janeiro Environmental Exchange) and the BVRio Institute are launched by Pedro Moura Costa and Mauricio Moura Costa.[9]
- 2012: Natural Capital Declaration is launched.[10]
- 2012: The Economics of Ecosystems and Biodiversity (TEEB) for Business Coalition is launched. The Coalition would later become the Natural Capital Coalition in 2014.[11]
- 2012: TNC's Conservation Notes are launched.[12]
- 2014: Credit Suisse's Nature Conservation Notes are issued.[13]

Mechanisms of Instrument

On a general level, impact investing involves an equity investment or interest bearing loan to a particular entity, project, or program that is generating a return while also having a positive impact on the environment. However, there are seemingly endless possibilities for how impact investments can be structured. Several examples follow.

The Cambridge, Massachusetts-based organization Root Capital offers "{...} trade credit, which is available for up to one year and oriented around a production cycle such as a harvest."[14] Oftentimes for farmers there is a gap between when upfront costs are due, such as to purchase seeds and hire help to plant crops, and when the farmers are paid (i.e., when crops are mature and eligible for sale). Groups, such as Root Capital, can help bridge this gap. This said, Root Capital recently secured a USD$50 million lending facility from the Overseas Private Investment Corporation (OPIC) to expand its lending to smallholder farmers in Latin America and Africa.[15]

There are also "investment practitioners {who} are buying conservation properties from bankrupt companies, restructuring corporate loans conditioned on conservation outcomes, and partnering with environmentally responsible investors to purchase vast acres of farms and forests that otherwise would be developed and in the process lost to the working landscape."[16]

Another example of impact investing is the Caral-Supe Trust Fund, which is structured in such a way that "investors will receive an average market return for participating in the investment fund, and all residual capital gains over the average return will be donated to the trust fund."[17] While this is not specifically related to the conservation of a tropical rainforest, the Caral-Supe Trust Fund funds the "5,000-year-old, 626-hectare archaeological site of The Sacred City of Caral-Supe {which} is situated on a dry desert terrace overlooking the green valley of the Supe river. It dates back to the Late Archaic Period of the Central Andes and is the oldest centre of civilization in the Americas."[18]

The Environmental Quality Incentives Program (EQIP), which is a voluntary program of the US Department of Agriculture, "can fund a broad range of conservation practices and projects: funding storage systems to better contain manure, fencing animals out of streams, and developing alternative watering sources are among the initiatives to protect water and air quality. Other practices frequently funded include conservation tillage, pest management, grassed waterways, wetland habitat restoration, brush management, proscribed burning, forest-stand improvement, and stream habitat improvement."[19]

Yet another example that could be considered an impact investment is when the Overseas Private Investment Corporation (OPIC) provided political risk insurance to Terra Global Capital's Oddar Meanchey REDD+ Project in Cambodia. This transaction between OPIC and Terra Global Capital won the 2012 Sustainable Forestry Transaction of the Year award from *Environmental Finance*.[20] OPIC also extended an anchor investment to Terra Global Capital's Terra Bella Fund for up to USD$40 million.[21]

Furthermore, there are numerous sustainable forestry funds. This includes:

- The Sydney, Australia-based firm New Forests which recently sold a 22.5% stake to Tokyo, Japan-based Mitsui & Co.[22] In 2013, New Forests Tropical Asia Forest Fund (TAFF) closed with USD$170 million raised and was the "first private investment vehicle dedicated to sustainable plantation timber production in Southeast Asia using long-term, institutional capital[23];
- SLM Partners with its €50 million (approximately USD$55.20 million) SLM Silva Fund "will convert 'clear-fell' forests—which are chopped down all at once and then replanted—into 'continuous cover' forests, which are chopped down gradually, leaving a large number of trees standing at all times"[24]; and
- Finance in Motion which is seeking to close a USD$200 million forestry fund.[25]

Size of Instrument

There are several different ways to assess the size of the impact investing sector and its ability to assist with the financing of tropical rainforest conservation.

According to the Global Impact Investing Network (GIIN) 2016 Annual Impact Investor Survey, there was an estimated USD$15.2 billion committed by 157 respondents to 7,551 distinct impact investments throughout 2015.[26]

However as of February 2017, a coalition of 87 investors with USD$5.3 trillion of assets under management "support the UN Guiding Principles Reporting Framework. {…} to incentivize improved disclosure and see it as an essential tool that enables investors to review companies' understanding and management of human rights risks."[27] This said, if avoiding tropical rainforest deforestation in conjunction with providing sustainable economic opportunities for local communities, or at least not deforesting their communal lands, is viewed as a human right, then the size of interested investors will increase by magnitudes from the current size of the impact investing sector.

Furthermore, approximately 1,700 signatories from more than 50 countries, representing US$62 trillion, agreed to the following 6 voluntary principles of the United Nations' Principles for Responsible Investment:

Principle 1: We will incorporate ESG issues into investment analysis and decision-making processes.

Principle 2: We will be active owners and incorporate ESG issues into our ownership policies and practices.

Principle 3: We will seek appropriate disclosure on ESG issues by the entities in which we invest.

Principle 4: We will promote acceptance and implementation of the Principles within the investment industry.

Principle 5: We will work together to enhance our effectiveness in implementing the Principles.

Principle 6: We will each report on our activities and progress toward implementing the Principles.[28]

This said, ensuring investments are not leading to tropical deforestation at the expense of local communities and biodiversity applies to each of the six principles.

However, conservation activities, in order to attract impact investments and especially funding from institutional investors, will need to demonstrate their activities, among many things, are scalable, have a proven track record, and can absorb large sums of funding.

Introduction to Case Studies

The following case studies will examine Althelia Ecosphere's Althelia Climate Fund and Credit Suisse's Nature Conservation Notes.

Althelia Ecosphere's Althelia Climate Fund successfully closed its first round of financing with €60 million raised from a wide variety of investors including the Church of Sweden, the European Investment Bank, and the Finnish Fund for Industrial Cooperation. In addition, the Althelia Climate Fund also received investments via Credit Suisse's Nature Conservation Notes. Taken together, the Althelia Climate Fund has financed on-the-ground projects including a USD$10 million, 8-year investment to preserve 200,000 hectares of forest and savannah grasslands in Kenya and a USD$12 million, 7-year investment to finance the conservation of 570,000 hectares in Madre de Dios, Peru.[29]

Credit Suisse raised €15 million in funds by issuing the Nature Conservation Notes. The Notes, which were a first-of-their kind for both Credit Suisse and the banking industry, were originally placed into a portfolio of 10–20 green bonds, and then once the on-the-ground projects were lined up, the bonds were liquidated and the funds were reinvested into the Althelia Climate Fund.[30]

Case Study #1: Althelia Ecosphere's Althelia Climate Fund

Introduction

The Althelia Ecosphere was launched by Sylvain Goupille and Christian del Valle[31] in 2011 and closed its first round of financing in June 2013 with €60 million (USD$80 million at the prevailing exchange rate) for the Althelia Climate Fund. The Althelia Climate Fund "is managed by its Luxembourg-based General Partner, Althelia Climate Fund GP s.á.r.l. and advised by its London-based subsidiaries Ecosphere Capital Partners LLP and Ecosphere Capital Limited."[32] Some of the initial investors into the Althelia Climate Fund were:

- Church of Sweden;
- The European Investment Bank (EIB);
- The Dutch Development Bank (FMO, *Financierings-Maatschappij voor Ontwikkelingslanden*); and
- The Finnish Fund for Industrial Cooperation (Finnfund).[33]

In June 2017, it was reported that "Mirova, a unit of French asset manager Natixis Global, had entered into exclusive talks to buy a majority stake"[34] in Althelia Ecosphere.

More specifically, the Althelia Climate Fund "invests in projects that reduce deforestation, mitigate climate change, protect biodiversity and provide a fair and sustainable living to rural communities, whilst offering investors a fair return on capital."[35]

Identify the Problem

The essential problems are global climate change, the role of tropical deforestation, and the lack of investable opportunities for institutional investors. This said, the vision of Althelia Ecosphere is that there is a lot of money available from the private sector, and specifically from institutional investors, that want to invest in the conservation finance asset class. Such investments would lead to tropical rainforest conservation and help mitigate global climate change. However, this is a challenging asset class and investment proposition for a variety of reasons including the fact that many of the local projects are too small in size and many of their business models are unproven when viewed from a traditional commercial standpoint.[36]

Why the Problem Is Important

The problems of global climate change, the role of tropical deforestation, and the lack of investable opportunities for institutional investors are explicitly linked, and these issues are personal to the founders of Althelia Ecosphere, Sylvain Goupille and Christian del Valle.

The Althelia Ecosphere, as intended by the founders, is:

> mission driven to design and implement a new model that will reverse this trend- one that works for the benefit of present and future generations of life on Earth. This model aims to address the drivers of deforestation and unsustainable land-use. Through a focus on blended value investments that deliver the highest calibre social, environmental and economic performance, we aim to demonstrate that financial performance can be fully aligned with sound environmental stewardship and social development.[37]

Sylvain Goupille has been involved in carbon finance for 20 years.[38] Christian del Valle, in addition to an advanced degree in conservation biology and community-level voluntary work in Africa, has been working with environmental investment products since at least 2005.[39]

How Problem Was Identified

The problems of global climate change, the role of tropical deforestation, and the lack of investable opportunities for institutional investors were identified by Althelia Ecosphere's visionary founders, a team of world-class experts, and their extensive network.

Effectiveness of Process for Identifying Problem

With visionary founders, a team of world-class experts, and their extensive network, the Althelia Ecosphere has been very effective at identifying the problems of global climate change, the role of tropical deforestation, and the lack of investable opportunities for institutional investors.[40]

Steps Taken to Address the Problem

The first step taken was to establish Althelia Ecosphere. The Althelia Ecosphere plays sort of an incubator role to address these problems and was designed to bridge the gap and channel funds to on-the-ground activities by scaling up investments and by trying out different business models.[41]

Althelia Ecosphere then launched the Althelia Climate Fund. To tackle the problem, the Althelia Climate Fund is seeking to transform unsustainable land use in tropical countries—particularly areas with high carbon value and high biodiversity. To achieve this, the Althelia Climate Fund:

> {…} generates revenues through the purchase and distribution of both environmental assets (including carbon and payment for ecosystem services) and environmentally-certified, sustainably-sourced commodities. The Fund has built its pipeline of investment opportunities over the past 3 years through its network of partners (including global and local NGOs, public sector entities, project developers, etc.) and secures distribution channels for produced assets through long-term partnerships with public, private and civil society sectors.[42]

Thus, the Althelia Climate Fund is always looking for a conservation component that is paired with sustainable commodity production to diversify revenue sources. By using both sets of commodities (i.e., sustainable agriculture and carbon), the Althelia Climate Fund is addressing the drivers of deforestation by creating value for standing forests and by creating economic alternatives for local communities. For instance, sustainable cacao can be a good alternative for other commodities that traditionally result in deforestation such as cattle, palm oil, or soy.[43]

Furthermore, Althelia Ecosphere took steps to identify strategic partners, identify investors, and assess projects alongside the development of their Environmental, Social and Governance (ESG) Policy[44] and an Impacts Monitoring Framework.[45]

Results

The Althelia Ecosphere was launched by Sylvain Goupille and Christian del Valle in 2011 and closed its first round of financing in June 2013 with

€60 million (USD$80 million at the prevailing exchange rate) for the Althelia Climate Fund. The Althelia Climate Fund reportedly raised approximately USD$117 million by the end of 2014 with a goal to reach USD$150 million.[46]

Some of the recent projects financed by the Althelia Climate Fund include:

- Amazon Sustainable Beef;
- The Cordillera Azul National Park REDD+ Project;
- The Guatemalan Caribbean Forest Corridor;
- The Taita Hills Conservation and Sustainable Land Use Project; and
- The Tambopata-Bahuaja REDD+ and Agroforestry Project.[47]

Althelia Climate Fund's first investment was for USD$10 million over 8 years to preserve 200,000 hectares of forest and savannah grasslands at Wildlife Works' Taita Hills Conservation and Sustainable Land Use Project (i.e., known as the Chyulu Hills REDD+ Project) in Kenya.[48] The Fund's second investment was a USD$12 million investment over 7 years to finance the conservation of 570,000 hectares in Madre de Dios, Peru, at the Tambopata-Bahuaja REDD+ and Agroforestry Project (i.e., also known as the Reduction of deforestation and degradation in Tambopata National Reserve and Bahuaja-Sonene National Park within the area of Madre de Dios region, Peru).[49,50]

In addition to the Althelia Climate Fund, Althelia Ecosphere also established several other initiatives including:

- The Madagascar Climate and Conservation Fund[51];
- The Sustainable Ocean Fund; and[52]
- The Nature Conservation Notes (see following case study).[53]

These projects and funds will achieve the industry's highest level of assurance. As stated by Althelia Ecosphere:

> As an expression of its commitment to define excellence in environmental, social and governance standards and safeguards and portfolio performance, Althelia has worked with investors and NGO partners to design a proprietary Environmental, Social and Governance (ESG) policy and management system, incorporating the International Finance Corporation's

> Performance Standards on Environmental and Social Sustainability (2012) and the EIB Statement of Environmental and Social Principles and Standards (2009). Forest-based emissions reductions financed by the Fund will also be validated and verified to the Climate, Community, and Biodiversity (CCB) Standards, using the Gold Level of the Standards for projects delivering smallholder/community-led equitable benefits and exceptional biodiversity benefits, as appropriate.[54]

Another major result of Althelia Ecosphere was the establishment of a risk-sharing loan guaranteed from the US Agency for International Development's (USAID) Development Credit Authority for up to USD$133.8 million.[55]

Challenges and How They Were Met

One challenge has been the limited types of projects and investments available. For instance, there are some local NGOs that understand how cash flows come in and how local community development projects are done, but these local NGOs are unfamiliar with how to generate revenue from their projects. In contrast, there are commercial entities that have detailed plans on commodity supply chains and agricultural intensification, but these entities have very limited knowledge about tropical rainforest conservation. This creates a challenge to deploy capital, and thus, there is a need to connect the dots between impact investment funds, commercial entities, and NGOs. Similarly, there are lots of opportunities that are at too early of a stage and are not investable.[56] To overcome this challenge, the Coalition for Private Investment in Conservation (CPIC) was launched in September 2016 by its founding members Cornell University, Credit Suisse, the International Union for Conservation of Nature (IUCN), and The Nature Conservancy (TNC).[57]

There is also a challenge with the forest carbon offset market, particularly for REDD+ offsets. For instance, soon after the Althelia Climate Fund was established, the carbon markets entered a period of uncertainty due largely to the after effects of the global financial crisis, as well as continuing regulatory uncertainty around international climate change mitigation commitments. This challenge was partially addressed by supporting projects that integrated other revenue models, including sustainable commodities. In addition, Althelia Ecosphere established Ecosphere+

which will act as a downstream marketing group to help grow the market for environmental assets, including the market for carbon offsets.[58]

Lastly, there is a challenge with a lack of money flowing into forest conservation. There is an increasingly accepted model of "Produce & Protect," but there is still a challenge for money to flow to the protection side of the equation. This said, there is a false belief that projects receiving investments from the Althelia Climate Fund are financially all set; however, these projects still need to monetize the carbon asset from the protection of standing forests. This monetization of carbon assets could come from a market-based payments for ecosystem services (PES) scheme, from a premium on sustainably sourced commodities, from voluntary corporate contributions, or via results-based financing (RBF) from governments. To address this challenge, Althelia Ecosphere is working with partners to ensure that once the Paris Agreement is operationalized, the private sector has a role and its early actions are recognized.[59]

Beyond Results

The results appear to be thus far sustainable, especially due to the long-term financial commitment to projects. The Althelia Climate Fund is an 8-year, closed-ended fund (CEF). While 8 years might not be long for conventional commercial agriculture and timber concessions, 8 years does create a sustainable cash flow for conservation projects linked to sustainable agricultural commodities. A sustainable cash flow allows for local projects to weather the lack of a market and/or market fluctuations and to achieve the land use transformations sought by the Althelia Climate Fund. The goal is for the theory of change to take place over the Althelia Climate Fund's period of intervention, and then after the Althelia Climate Fund exits, the projects should be well-established with community development, sustainable commodity operations, and be financially sustainable without carbon revenues.[60]

Althelia Ecosphere, beyond its Althelia Climate Fund, is also attracting investments for its other funds. For example, in October 2016, the:

> Green Climate Fund made its first equity investment in a landscape conservation strategy, committing $35 million to the Althelia Madagascar Sustainable Landscapes Fund, managed by Althelia Ecosphere. Althelia

Ecosphere expects to deploy $50 million across 15 to 25 investments in conservation, renewable energy and eco-friendly agriculture, with target returns of eight percent. The profits will be reinvested in additional conservation and climate initiatives.[61]

Lessons Learned

Some lessons learned by Edit Kiss, the Director of Althelia Ecosphere's Business Development and Operations, are:

- It is very important to provide technical assistance to projects in the very beginning via a Technical Assistance Facility (TAF). This technical assistance would not necessarily be an investment opportunity for the private sector but rather could come as a grant opportunity from NGOs and/or governments.
- You can do a lot of work with communities and agricultural intensification in buffer zones surrounding forests, but this may not end up saving the forest. The question is how do you do these activities and, simultaneously, monetize the value of forest protection?
- A key success factor for Althelia Ecosphere and its Althelia Climate Fund is that both were set up to be flexible. The Althelia Climate Fund has an open mandate on how to finance conservation and, thus, is adaptable and can seek new opportunities as conditions (e.g., government policies) change.
- While an 8-year fund is good, it could be a good idea to have a longer time horizon of 10–15 years. For instance, the Impact Investment Fund for Land Degradation Neutrality (LDN Fund) has a fund life of 15 years.[62]
- More needs to be done with blended financing models and layered financing. For instance, the Althelia's blended financing model involved launching with Conservation International in the beginning, and then over time, additional investors included the European Investment Bank and BNP Paribas, along with the risk-sharing loan guarantee by USAID. Thus, the blended financing included the private sector, the public sector, and nonprofit organizations. With respect

to layered financing, public capital is often more patient than private capital. By layering, the public capital can be structured as a junior tranche, and this enables you to de-risk your portfolio and attract more private sector investors. While reducing risk is important when examining the risk-return profile, there is still a need to push the market for better returns—especially by valuing the carbon in standing forests. If more progress is made on this note and an economic system is created that recognizes the value of forests, then funds such as the Althelia Climate Fund can channel financing and scale up investments into forest conservation and into sustainable businesses operating within the buffer zones.[63] Although not necessarily related to the Althelia Climate Fund, the eco.business fund provides another example of a public-private partnership fund "offering different capital tranches adapted to each investor profile."[64]

Other Resources on Althelia Climate Fund

Althelia Ecosphere's News & Press

- https://althelia.com/news-press/

Bloomberg

- http://www.bloomberg.com/research/stocks/private/snapshot.asp?privcapId=264550904

Chyulu Hills REDD+ Project

- http://www.vcsprojectdatabase.org/#/project_details/1408

European Investment Bank's Althelia Climate Fund Page

- http://www.eib.org/projects/pipelines/pipeline/20100720

Reduction of deforestation and degradation in Tambopata National Reserve and Bahuaja-Sonene National Park within the area of Madre de Dios region –Peru

- http://www.vcsprojectdatabase.org/#/project_details/1067

USAID

- https://www.usaid.gov/news-information/press-releases/may-28-2014-us-government-althelia-climate-fund-mobilize-1338-million-forest-conservation

Case Study #2: Credit Suisse's Nature Conservation Notes

Introduction

The Nature Conservation Notes, which resulted from a collaboration between the Althelia Ecosphere and Credit Suisse, along with several other partners, "provide exposure to (1) Althelia Climate Fund, a unique fund focused on sustainable agro-forestry and ecosystem conservation and (2) a portfolio of selected Green Bonds with high credit quality. The 6-years notes are listed and offer secondary trading."[65]

These Nature Conservation Notes are not to be confused with The Nature Conservancy's (TNC) Conservation Notes, which were previously designed by TNC. TNC's Conservation Notes were first issued in 2012, and the Notes explicitly offered a below market return of approximately 0–2% depending on factors such as the maturity chosen.[66] The market for these Conservation Notes is probably individuals and organizations that—for whatever reason—do not want to donate their capital to TNC. Instead, this group of individuals and organizations would like to support TNC and want to put their money to work for TNC, but do not require a market return. As explained by TNC:

> The $25 million offering was the first investment-grade retail product focused on conservation and was backed by the full faith and credit of the Conservancy, which carries an Aa2 Moody's bond rating. With a minimum investment of $25,000 and terms of 1, 3 and 5 years, the small offering has consistently sold out to investors looking to generate a financial return as well as deep conservation impact.[67]

Furthermore, the Conservation Notes "make loans to state, country and regional programs of The Nature Conservancy that:

- Conserve important natural areas and essential resources;
- Help protect wildlife;
- Restore drinking water supplies; and
- Improve people's lives."[68]

In contrast, Credit Suisse's Nature Conservation Notes were issued in December 2014 and did not offer a below market rate.[69] The €15 million funds raised by the Nature Conservation Notes were originally placed into a portfolio of 10–20 green bonds from the Barclays MSCI Green Bond Index, and then once the profit-making conservation projects were lined up, the bonds were liquidated and the funds were reinvested into the Althelia Climate Fund.[70]

These profit-making conservation projects that will be:

> {…} financed by the notes have been selected by the Althelia Climate Fund, based in Luxembourg, which calls itself a "sustainable land-use impact fund." Althelia finances projects that create carbon offset credits through the protection of forests, and also generate revenues from sustainably certified products such as cocoa, coffee and wood. Althelia announced the sale of the notes as part of the close of its overall fundraising of 105 million euros, or about $122 million.[71]

Identify the Problem

One major problem is that the world needs an estimated USD$200–USD$300 billion per year in additional capital to conserve the world's biodiversity "and private investment capital may be the main source of additional capital. Attracting such a level of private capital will require attractive risk-adjusted rates of return, in addition to clear and measurable conservation impacts."[72]

As reported in the *State of Private Investment in Conservation 2016*, approximately USD$8.2 billion was invested by the private sector between 2004 and 2015 in the following "three groups of conservation-oriented investments:

- Sustainable food and fiber production (e.g., forestry, agriculture, and fisheries)
- Habitat conservation (e.g., mitigation banking and forest carbon trading)
- Water quality and quantity protection (e.g., watershed protection and water rights trading)."[73]

Thus, there is a significant shortfall of private sector investments into the conservation of tropical rainforests.

Another major problem is that many impact investments are only available to high-net-worth and ultra-high-net-worth individuals. For instance, the minimum investment in the Althelia Climate Fund was believed to be approximately €1–5 million.

Why the Problem Is Important

The lack of private sector investments was an important problem for Credit Suisse for a variety of reasons. First, there was money to be made by issuing these Nature Conservation Notes. Second, and related, is that Credit Suisse was interested in getting on the ground floor of what could be a very large market in the coming 5–10 years. Third, Credit Suisse was able to meet their clients' needs and demands, as some of their clients were asking about impact investing and specifically about conservation finance. Fourth, there were reputation dividends to be had by being a part of a unique transaction that invested into what a lot of people see as a good cause.[74]

How Problem Was Identified

The problem of few private sector conservation finance opportunities was identified by the investment professionals of Credit Suisse and particularly the Bank's global sustainability group.

Effectiveness of Process for Identifying Problem

The investment professionals appear to have a very effective process in place to structure innovative products, such as the Nature Conservation Notes, along with identifying investor demand and strategic partners.[75]

Steps Taken to Address the Problem

The general steps taken to address the problem of few private sector conservation finance opportunities and to create the Nature Conservation Notes were:

- Credit Suisse partnered with the World Wildlife Fund (WWF) and McKinsey & Company to publish a report in 2014 entitled, *Conservation Finance: Moving beyond donor funding toward an investor-driven approach*.
- John Tobin and Fabian Huwyler of Credit Suisse's Sustainability function devoted a lot of time studying and thinking about the next big thing. The next big thing, the Holy Grail, was going to be if Credit Suisse could issue its own product.
- The team explored possible partnerships and organizations they could work with and started a short list of products they might pursue.
- Finally, the team worked with the structured notes/structured finance professionals from Credit Suisse on how to put together the Nature Conservation Notes to sell to clients.[76]

Results

The Nature Conservation Notes raised financing in 10 business days, generated an innovation premium, and received positive impressions from bank colleagues and from media coverage such as the *Financial Times*.[77] The Nature Conservation Notes raised €15 million (approximately USD$17.5 million at the time of issuance). These funds will be invested in the Althelia Climate Fund, and subsequently, "Althelia's first two investments aim to improve livelihoods for local residents and generate cash flows from premium cacao in Peru and sustainable charcoal in Kenya to augment sales of carbon credits."[78]

Furthermore, the Nature Conservation Notes were Credit Suisse's "{…} first environmental impact investing product and the first conservation finance product of any kind launched by a big bank. The Notes have received the *Sustainable Forestry Deal of the Year 2015 Award* by *Environmental Finance*."[79]

Challenges and How They Were Met

A few challenges faced by the Nature Conservation Notes were the fact that the Notes were a first-of-its kind and the Notes were not for sale in the United States.

The Notes were the first such issuance for Credit Suisse and the first such issuance for the banking industry. Thus, the first time you do anything, it is difficult. Essential questions, such as how do you do it and why should we do it, had to be answered.[80]

The particular regulatory requirements and the litigation risk in the United States are so high, that oftentimes, if you think you can sell your entire issuance—whether it is selling debt, equity, or IPOs—without tapping into the US market that is what you do. To overcome this challenge, investor interest was explored, and it became apparent that there was already a lot of interest for the Nature Conservation Notes outside of the United States. Furthermore, a Special Purpose Vehicle (SPV) was set up by Credit Suisse. What normally happens in structured deals is the issuer is a SPV entity and the SPV is created for a very clear reason. Thus, Credit Suisse was not going to become a limited partner in the Althelia Climate Fund. Rather, the SPV entity, in turn, became a limited partner in the Althelia Climate Fund.[81]

Another challenge was that the €$15 million Notes were small with limited impacts and limited liquidity and both the legal and structuring works were high costs from the bank's perspective. Similarly, the returns and the size of the investment would not exactly make Credit Suisse's investment bankers wealthy. However, hopefully, these Notes will open up future issuances from other banks and additional issuances from Credit Suisse that will be available to an even larger group of people.

Beyond Results

Overall, the issuance of the Nature Conservation Notes went very well. The Nature Conservation Notes, at €15 million, was a relatively small issuance, and ideally, the Notes would have been larger. However, the Notes represented more than just a pilot and are a trailblazer in the industry.[82] Thus, "the demand for the notes should send a signal to other banks that individual investors, in addition to institutions, have an appetite for competitive, market-rate instruments that also have demonstrable and measurable environmental and social benefits."[83] Also when combined with the larger capital raising campaign of the Althelia Climate Fund, the on-the-ground projects should be sustainable beyond the lifetime of the Nature Conservation Notes.

Lessons Learned

The essential lesson learned is that not only did it help to have a cross section of the bank on board—this would never have happened without a cross section of the bank on board. Likewise, the Sustainability function did not have capacity, the expertise, the relationships, or the place in the bank to set up and issue the Nature Conservation Notes by themselves.[84]

Other Resources on Nature Conservation Notes

Credit Suisse, Our Role in Conservation Finance

- https://www.credit-suisse.com/ch/en/about-us/responsibility/banking/sustainable-products-services.html

Environmental Finance

- https://www.environmental-finance.com/content/deals-of-the-year/sustainable-forestry-credit-suisse-althelia-ecosphere.html

The David & Lucile Packard Foundation's Conservation Notes Page

- https://www.packard.org/what-we-fund/stories-of-progress/the-nature-conservancy-conservation-notes/

The Nature Conservancy's The Conservation Note

- Investment Opportunities: http://www.naturevesttnc.org/get-involved/investment-opportunity/
- 2017 Prospectus: https://www.nature.org/about-us/conservation-note-prospectus-2017-final.pdf
- 2016 Impact Report: https://www.nature.org/about-us/conservation-note-report-2016.pdf

Return

According to the 157 respondents to GIIN's 2016 Annual Impact Investor Survey, "nearly 60% of respondents primarily target 'risk-adjusted, market rate returns,' while a quarter of respondents primarily target 'below market rate returns: closer to market rate' and 16% target 'below market rate returns: closer to capital preservation'."[85] Thus, it is important to know your investor audience when it comes to proposing the impact investment and its expected returns—both financial returns and non-financial returns.

Risk

There are a wide variety of risks associated with impact investing depending on the exact nature of the investment, including the specific underlying assets, the actual structure of the investment (e.g., debt financing or equity investment), and the overall time horizon of the investment.

Business Risk

Business risks are based around supply and demand for a given product. This said, depending on the underlying investment, does the demand exist—now and into the future—for example, for an ecotourism lodge, visiting a given national park, or a sustainably sourced Brazilian nut from a forest conservation project?

Manufacturing costs also present a risk. Is the cost to produce a sustainably sourced Brazilian nut cost competitive, relative to its premium price, to compete against a more conventionally sourced and produced Brazilian nut? Likewise, is there even a premium price that can be charged?

Furthermore, there could be material risk of companies having to switch suppliers due to consumer pressure focused on their supplier's deforestation. Switching suppliers can involve costs (e.g., unwinding contracts and search costs for new suppliers), higher costs of goods, and competitive pressure as other firms secure exclusive rights with deforestation-free suppliers.

Strategic Risk

Strategic risks for an impact investor in forest conservation can be mitigated through a diversified portfolio consisting of a range of different projects. Likewise, the Althelia Climate Fund did not invest into just a single project, in one country. There are also strategic risks associated with impact investors changing their investment philosophy or investment direction. In addition, strategic risks may arise when choosing the host country to invest in, the actual project and its partners, or when lining up financing. Furthermore, there can be strategic risks associated with asking for too much, or too little, funding.

Reputation Risk

Reputation risks can arise from investing in fraudulent and/or inexperienced organizations that are undertaking the conservation activities on the ground. This includes organizations that are unable to deliver on the financial and environmental expectations, along with the social concerns of fair labor practices, gender equity, and fair wages.

Liquidity Risk

Investments in large, underlying assets—such as advanced manufacturing facilities for Brazilian nuts or sustainable cattle operations—can pose more of a liquidity risk, than purchasing shares in a publicly traded company. In addition, impact investments into new industries (e.g., REDD+ offsets or Nature Conservation Notes) can also pose a liquidity risk. However, as these projects and companies become more mainstream, their liquidity will likely increase.

Operational Risk

Operational risks for impact investors include inadequate technology (e.g., inadequate in-country computer systems, intermittent energy, or internet access), incompetent management, and possible fraud (e.g., illegal sourced timber or falsely claimed, sustainability certifications).

Legal and Regulatory Risk

Legal and regulatory risk is potentially significant in host countries where the ease of doing business is often more challenging and corruption is more significant. This can make getting fair court hearings, recovered investments from bankrupt companies, or export licenses approved more difficult. In essence, there can be a lack of legal recourse.

Credit Risk

Credit risk includes default risk, bankruptcy risk, downgrade risk, and settlement risk. From the impact investors' perspective, there are credit risks, and specifically default or bankruptcy risks, associated with the company or project receiving the impact investment.

Market Risk

Market risk includes interest rate risk, equity price risk, foreign exchange risk, and commodity price risk. Interest rate and equity price risks pose unique risks depending on the nature of the investments. For example, if debt financing is structured over phases with a variable rate, the borrower may experience hardship and its operations could be jeopardized.

For impact investments in an export-oriented product (e.g., reduction in tropical rainforest deforestation through developing more sustainable, rotational cattle pastures or sustainably sourced Brazilian nuts), foreign exchange risk can be significant depending on the host country's currency relative to the importing country's currency. For instance, a commodity and foreign exchange risk could occur if buyers are willing to switch to cheaper oil from Brazilian soybeans as a result of the change in foreign exchange, instead of Indonesian palm oil. In turn, the palm oil plantations in Indonesia may face bankruptcy risk, while deforestation may increase in Brazil for the expansion of additional soybean plantations specifically because of the foreign currency exchange (i.e., depreciation of Brazilian Reals versus an appreciation of the Indonesian Rupiah).

Risk, Return, Time (Horizon), Taxes, Liquidity, Legal and Unique (RRTTLLU)

Risk and Return

Please see above for the risk and return associated with impact investing. The type of return required by the impact investor may change overtime as the composition of their portfolio changes over time. This said, it is very important to understand both the financial and non-financial returns expected by the impact investors.

Time Horizon

The time horizon for impact investing is likely to be a short-term to medium-term investment. Likewise, TNC's Conservation Notes had 1-, 3-, and 5-year terms, while Credit Suisse's Nature Conservation Notes were 6-year notes, and the Althelia Climate Fund was an 8-year, closed-ended fund.

Taxes

There are likely taxes, such as taxes on capital gains. Depending on the structure of the investment, there could be sales taxes or value-added taxes levied on the underlying products and employment taxes, along with export and/or import taxes and duties.

Liquidity

Depending on the nature of the investment, the investment might be more or less liquid. For instance, the Nature Conservation Notes were fairly liquid, while investments in sustainable forestry operations in an unstable region are likely to be less liquid.

Legal

Legal considerations associated with impact investing will vary from country to country. This said, it is important to consider the jurisdiction of the impact investment fund.

Unique

Unique aspects include the divestment movement, which recently focused on fossil fuels and is starting to focus on deforestation. For instance, "Dimensional Fund Advisors, a U.S. investment firm with USD$445 billion of assets under management, has divested two of its portfolios of all palm oil plantation companies."[86]

Another unique aspect is that the environment is only one of the ESG (Environmental, Social and Governance) components and there are lots of other competing investments. Furthermore, forest conservation is only one sector of the environment which also includes water, transportation, fisheries, and clean energy.

In addition, impact investors need to assess the risk-adjusted return after deducting management fees and taxes.

Policy Analysis

Impact investing, similar to green bonds and ecotourism, depends on a wide variety of public policies to be in place.

Defining the Problem

Choosing a tropical rainforest country to make an impact investment depends on factors such as the overall ease of doing business, the level of corruption, and the rule of law. While impact investors are willing to make more risky investments, such investors are not willing to throw away their money. In addition, some investors—particularly institutional investors—have a legal and fiduciary obligation to only invest in countries and/or companies that have an investment grade rating. This presents a particular challenge for some of the more unstable rainforest countries.

Establishing Goals

The goals of impact investing are to get one's investment back, plus a financial return, while also making a positive environmental and social impact. With this in mind, impact investments should ideally also contribute to the conservation of tropical forests and the preservation of biological diversity and promote economic development for local communities.

Selecting a Policy

When choosing which country or countries to make an impact investment, one should assess the countries in their totality. If the investment grade rating is the primary concern, then as of February 28, 2017, the rainforest countries of Mexico, Peru, Colombia, and India were rated as investment grade. Angola, Bolivia, Zambia, Brazil, and the Democratic Republic of the Congo (DRC) were rated as non-investment grade. Indonesia was rated as investment grade by Moody's, but was rated as non-investment grade by S&P (Standard and Poor's) and Fitch.

The following rainforest countries were ranked in order of the Trading Economics Credit Rating (TERating), which "scores the credit worthiness of a country between 100 (riskless) and 0 (likely to default) (Table 10.1)."[87]

If one's firm can make an impact investment in countries without an investment grade rating, then there are additional policies to consider such as policies related to the ease of doing business, corruption, the Human Development Index, and the rule of law.

Table 10.1 Credit ratings of countries with largest rainforests

	Country	S&P		Moody's		Fitch		TERating
1	Mexico	BBB+	Negative	A3	Negative	BBB+	Stable	60
2	Peru	BBB+	Positive	A3	Stable	BBB+	Stable	60
3	Colombia	BBB	Negative	Baa2	Stable	BBB	Stable	52
4	India	BBB-	Stable	Baa3	Positive	BBB-	Stable	48
5	Indonesia	BB+	Positive	Baa3	Positive	BBB-	Positive	46
6	Angola	B	Negative	B1	Negative	B	Negative	41
7	Bolivia	BB	Stable	Ba3	Negative	BB-	Stable	38
8	Zambia	B	Negative	B3	Negative	B	Negative	34
9	Brazil	BB	Negative	Ba2	Negative	BB	Negative	34
10	DRC	B-	Negative	B3	Stable	–	–	25

Credit: Trading Economics

For instance, the global rankings for ease of doing business, according to the World Bank,[88] for the top 10 countries with the largest tropical rainforests are as follows:

- Tied for #47: Mexico
  - #40 for enforcing contracts
  - #53 for protecting minority investors
  - #61 for trading across borders
  - #101 for registering property
- #53: Colombia
  - #13 for protecting minority investors
  - #53 for registering property
  - #121 for trading across borders
  - #174 for enforcing contracts
- #54: Peru
  - #37 for registering property
  - #53 for protecting minority investors
  - #63 for enforcing contracts
  - #86 for trading across borders
- #91: Indonesia
  - #70 for protecting minority investors

  - #108 for trading across borders
  - #118 for registering property
  - #166 for enforcing contracts
- #98: Zambia
  - #87 for protecting minority investors
  - #135 for enforcing contracts
  - #145 for registering property
  - #161 for trading across borders
- #123: Brazil
  - #32 for protecting minority investors
  - #37 for enforcing contracts
  - #128 for registering property
  - #149 for trading across borders
- #130: India
  - #13 for protecting minority investors
  - #138 for registering property
  - #143 for trading across borders
  - #172 for enforcing contracts
- #149: Bolivia
  - #98 for trading across borders
  - #128 for enforcing contracts
  - #137 for protecting minority investors
  - #139 for registering property
- #182: Angola
  - #81 for protecting minority investors
  - #170 for registering property
  - #183 for trading across borders
  - #186 for enforcing contracts
- #184: DRC
  - #156 for registering property

  - #162 for protecting minority investors
  - #171 for enforcing contracts
  - #188 for trading across borders

The global rankings by Transparency International of the Corruption Perceptions Index 2016[89] for the top 10 countries with the largest tropical rainforests were as follows:

- Tied for #79: Brazil
- Tied for #79: India
- Tied for #87: Zambia
- Tied for #90: Colombia
- Tied for #90: Indonesia
- Tied for #101: Peru
- Tied for #113: Bolivia
- Tied for #123: Mexico
- Tied for #156: DRC
- Tied for #164: Angola

The global rankings by the United Nations Development Programme of the Human Development Index (HDI),[90] which combines life expectancy, education, and per capita income indicators, for the top 10 countries with the largest tropical rainforests were as follows:

- #74: Mexico
- #75: Brazil
- #84: Peru
- Tied for #97: Colombia
- Tied for #110: Indonesia
- #119: Bolivia
- #130: India
- #139: Zambia
- #149: Angola
- #176: DRC

The World Justice Project has designed a rule of law ranking, which primarily utilizes eight factors (e.g., Constraints on Government Powers,

Absence of Corruption, Open Government, Fundamental Rights, Order and Security, Regulatory Enforcement, Civil Justice, and Criminal Justices).[91] Of the 113 countries assessed, the following global rankings are for the top 10 countries with the largest tropical rainforests:

- #52: Brazil
- #61: Indonesia
- #65: Peru
- #66: India
- #71: Colombia
- #81: Zambia
- #88: Mexico
- #104: Bolivia
- Unranked: Angola
- Unranked: DRC

Implementing a Policy

With these problems in mind, host countries need to implement a suite of public policies to better position themselves for impact investments. The reason impact investments are more prominent in the European Union and the United States is because of the strong rule of law, transparency, relative price discovery, the range of financial intermediaries, scalability of projects, and both managerial and technical expertise, along with other local capacities.

Evaluating the Policy

Despite mixed signals based off their rankings on the ease of doing business, corruption, the HDI, and the rule of law, the countries of Mexico, Peru, Colombia, and India have investment grade ratings and seem to be better choices for impact investments.

Brazil and Indonesia—despite at least one non-investment grade rating—still rank relatively high among the countries with tropical rainforests in the rule of law, HDI, and transparency. It is also important to note that Brazil and Indonesia have the largest and third largest rainforests in the world.

In contrast, Angola, Bolivia, Zambia, and the DRC are rated as non-investment grade and rank very low on some of the other indicators, which suggests that impact investments in these countries would be especially risky.

Working with Ministries of Finance and structuring underwriting policies can help mitigate risk and allow impact investment funds to attract investors. For instance, the Overseas Private Investment Corporation (OPIC) provided political risk insurance for Terra Global Capital's Oddar Meanchey REDD+ Project in Cambodia,[92] and the Althelia Climate Fund received a risk-sharing loan guaranteed from USAID's Development Credit Authority for up to USD$133.8 million.[93]

Future Outlook for Instrument

Demand for impact investments including "demand for interim financing will continue to grow as land becomes more expensive, the scale of projects increases, market competition intensifies, interest rates rise, organizational sophistication grows, and greater numbers of conservation projects require immediate capital."[94]

The essential question is how do impact investments scale to the size required to mitigate tropical rainforest deforestation? There are several nice ideas, such as TNC's Conservation Notes, but they are not transformative. In general, the common understanding and wisdom is that there is a lot of demand for these types of investments, but not a good supply. The way you can increase the supply is if you can get more certainty of selling PES offsets and/or sustainable products. However, there needs to be more regulatory schemes to give certainty to these markets. In the absence of regulatory action, what is it that is needed to scale? Similarly, how do you reach the mainstream investors and pension funds and not "just" small, boutique ESG investment funds?[95]

Other Resources on Impact Investing

BlueOrchard

- http://www.blueorchard.com

Case Foundation's Impact Investing Network Map

- https://casefoundation.org/networkmap/

Conservation International's Verde Ventures

- http://www.conservation.org/projects/Pages/Verde-Ventures.aspx

EcoEnterprises Fund

- https://www.ecoenterprisesfund.com/

Five Ways to Advance Conservation Entrepreneurship

- https://ssir.org/articles/entry/five_ways_to_advance_conservation_entrepreneurship

Hancock Natural Resources Group

- http://hancocknaturalresourcegroup.com/

Impact Finance Center

- http://www.impactfinancecenter.org/

iPAR

- https://iparimpact.com/

IRIS is an initiative of the Global Impact Investing Network (GIIN)

- https://iris.thegiin.org/

Mission Investors Exchange

- https://www.missioninvestors.org/

Motif Impact Portfolios

- https://www.motifinvesting.com/products/impact-portfolio

New Forests

- https://www.newforests.com.au/

OpenInvest

- https://www.openinvest.co/how-it-works/

Private Capital for Nature Conservation: Could Impact Investments be a Solution?

- www.globalnature.org/bausteine.net/f/8525/Privatecapitalfornature conservation_WEB.pdf?fd=3

Rockefeller Foundation's Bellagio Center

- https://www.rockefellerfoundation.org/our-work/bellagio-center/

Toniic SDG Impact Theme Framework

- http://www.toniic.com/sdg-framework/

Notes

1. World Wildlife Fund. *Guide to Conservation Finance*. 36.
2. The Forum for Sustainable and Responsible Investment. "SRI Basics." Accessed April 14, 2017. http://www.ussif.org/sribasics
3. Daily, Gretchen C. and Katherine Ellison. *The New Economy of Nature*. 116.
4. Terra Global Capital. "Team and Advisors." Accessed November 23, 2016. http://www.terraglobalcapital.com/team-and-advisors
5. Ebeling, Johannes of BioCarbon Group. Interviewed by Brian McFarland. October 2016.
6. WAVES. "About Us." Accessed December 13, 2016. https://www.waves-partnership.org/en/about-us
7. NatureVest. "About Us." Accessed December 13, 2016. http://www.naturevesttnc.org/about-us/
8. Ecosystem Marketplace. "Althelia Raises $80 Million For REDD and Ecosystem Services." Accessed December 13, 2016. http://www.ecosystemmarketplace.com/articles/althelia-raises-80-million-for-br-redd-and-ecosystem-services/
9. Moura Costa, Pedro of BVRio. Interviewed by Brian McFarland. March 2017.
10. Natural Capital Finance Alliance. "The NCD and NCFA Story." Accessed December 13, 2016. http://www.naturalcapitalfinancealliance.org/about-the-natural-capital-declaration/
11. Natural Capital Coalition. "History, Vision & Mission." Accessed December 13, 2016. http://naturalcapitalcoalition.org/who/history-vision-mission/

12. NatureVest. "Frequently Asked Questions." Accessed December 13, 2016. http://www.naturevesttnc.org/about-us/faq/
13. Credit Suisse. "Sustainable Products & Services: Conservation Finance: Our Role in Conservation Finance." Accessed March 29, 2017. https://www.credit-suisse.com/us/en/about-us/responsibility/banking/sustainable-products-services.html
14. Milder, Brian. "Conservation Finance in the Galápagos Islands: Capitalizing the 'Missing Middle'." In *Conservation Capital in the Americas*, edited by James N. Levitt. Massachusetts: Lincoln Institute of Land Policy, 2010. 120.
15. Price, Dennis. "Root Capital Secures $50 Million from OPIC to Lend to Smallholder Farmers." April 28, 2016. *Impact Alpha*. Accessed April 10,2017. http://impactalpha.com/root-capital-secures-50-million-from-opic-to-lend-to-smallholder-farmers/
16. Ginn. *Investing in Nature*. 7.
17. Gonzales, José and Hermilio Rosas. "The Caral Archeological Site and Supe Valley Restoration." In *Conservation Capital in the Americas*, edited by James N. Levitt. Massachusetts: Lincoln Institute of Land Policy, 2010. 92.
18. World Heritage Centre. "Sacred City of Caral-Supe." Accessed April 14, 2017. http://whc.unesco.org/en/list/1269
19. Ginn. *Investing in Nature*. 140.
20. OPIC. "OPIC Receives Environmental Finance Award for Sustainable Forestry." Last Modified July 25, 2012. https://www.opic.gov/blog/opic-in-action/opic-receives-environmental-finance-award-for-sustainable-forestry-2
21. Terra Global Capital. "Terra Global Secures Investment Capital For REDD and Land-Use Carbon Fund." Last modified November 3, 2011. http://www.terraglobalcapital.com/news/terra-global-secures-investment-capital-redd-and-land-use-carbon-fund
22. Cripps, Peter. "Sustainable Forestry Firm Attracts Japanese Investor." February 29, 2016. *Environmental Finance*. Accessed April 10, 2017. https://www.environmental-finance.com/content/news/sustainable-forestry-firm-attracts-japanese-investor.html?utm_source=229na&utm_medium=email&utm_campaign=alert
23. New Forests. "New Forests Announces Final Close of the Tropical Asia Forest Fund with $170 Million of Capital Commitments and Completes

Fund's First Transaction." July 3, 2013. Accessed April 12, 2017. https://www.newforests.com.au/wp-content/uploads/2013/03/20130703-New-Forests-Final-Close-of-Tropical-Asia-Forest-Fund.pdf

24. Ali, Hamza and Peter Cripps. "Irish Sustainable Forestry Fund Eyes First Close." March 29, 2016. *Environmental Finance.* Accessed April 10, 2017. https://www.environmental-finance.com/content/news/irish-sustainable-forestry-fund-eyes-first-close.html?utm_source=329na&utm_medium=email&utm_campaign=alert
25. Cooper, Graham. "Impact Investment Specialist Prepares $200 m Forestry Fund." May 11, 2016. Environmental Finance. Accessed April 10,2017.https://www.environmental-finance.com/content/news/impact-investment-specialist-prepares-200m-forestry-fund.html?utm_source=511na&utm_medium=email&utm_campaign=alert
26. Global Impact Investing Network. "Annual Impact Investor Survey, 2016, The Sixth Edition." Last modified May 2016. https://thegiin.org/assets/2016%20GIIN%20Annual%20Impact%20Investor%20Survey_Web.pdf. II.
27. UN Guiding Principles. "UN Guiding Principles Reporting Framework Investor Statement." Last modified February 2017. http://www.ungpreporting.org/early-adopters/investor-statement/
28. United Nations' Principles for Responsible Investment. "About the PRI." Accessed April 4, 2017. https://www.unpri.org/about
29. Bank, David. "Beyond Carbon: Althelia Climate Fund Attracts Conservation Investors." October 24, 2014. *Impact Alpha.* Accessed April 11, 2017. http://impactalpha.com/beyond-carbon-althelia-climate-fund-attracts-conservation-investors/
30. Environmental Finance. "Sustainable Forestry: Credit Suisse/Althelia Ecosphere's Nature Conservation Notes." Last modified April 1, 2015. https://www.environmental-finance.com/content/deals-of-the-year/sustainable-forestry-credit-suisse-althelia-ecosphere.html
31. Althelia Ecosphere. "Our Team." Accessed February 20, 2017. https://althelia.com/about-us/our-team/#all
32. Althelia Ecosphere. "Our Vision." Accessed March 31, 2017. https://althelia.com/about-us/our-vision/
33. Althelia Ecosphere. "Althelia Ecosphere Completes First Closing for Althelia Climate Fund with over EUR60 Million (US$80 m) Raised." Last modified June 11, 2013. https://althelia.com//wp-content/uploads/2014/07/ACFclose_PR_FINAL.pdf

34. Carbon Pulse. "Natixis unit to buy natural capital firm Althelia Ecosphere." June 30, 2017. Accessed August 23, 2017. http://carbonpulse.com/36647/?utm_source=CP+Daily&utm_campaign=b6354bffc7-CPdaily30062017&utm_medium=email&utm_term=0_a9d8834f72-b6354bffc7-36342993
35. Althelia Ecosphere. "Althelia Climate Fund." Accessed February 20, 2017. https://althelia.com/althelia-climate-fund/
36. Kiss, Edit from Althelia Ecosphere. Interviewed by Brian McFarland. March 2017.
37. Althelia Ecosphere. "About Us." Accessed March 31, 2017. https://althelia.com/about-us/
38. Althelia Ecosphere. "Sylvain Goupille." Accessed March 31, 2017. https://althelia.com/team/sylvain-goupille/
39. Althelia Ecosphere. "Christian del Valle." Accessed March 31, 2017. https://althelia.com/team/christian-del-valle/
40. Kiss, Edit from Althelia Ecosphere. Interviewed by Brian McFarland. March 2017.
41. Ibid.
42. Althelia Ecosphere. "Climate Fund." Accessed March 31, 2017. https://althelia.com/initiatives/climate-fund/
43. Kiss, Edit from Althelia Ecosphere. Interviewed by Brian McFarland. March 2017.
44. Althelia Ecosphere. "ESG Approach." Accessed March 31, 2017. https://althelia.com/esg-approach/
45. Althelia Ecosphere. "Our Impacts Monitoring." Accessed March 31, 2017. https://althelia.com/our-approach/our-impacts-monitoring/
46. Bank, David. "Beyond Carbon: Althelia Climate Fund Attracts Conservation Investors." *ImpactAlpha*. Last modified October 24, 2014. http://impactalpha.com/beyond-carbon-althelia-climate-fund-attracts-conservation-investors/
47. Althelia Ecosphere. "Althelia Climate Fund." Accessed March 31, 2017. https://althelia.com/althelia-climate-fund/
48. VCS Project Database. "Chyulu Hills REDD+ Project, Kenya." Accessed April 14, 2017. http://www.vcsprojectdatabase.org/#/project_details/1408
49. Bank, David. "Beyond Carbon: Althelia Climate Fund Attracts Conservation Investors." October 24, 2014. *Impact Alpha*. Accessed April 11, 2017. http://impactalpha.com/beyond-carbon-althelia-climate-fund-attracts-conservation-investors/

50. VCS Project Database. "Reduction of Deforestation and Degradation in Tambopata National Reserve and Bahuaja-Sonene National Park within the Area of Madre de Dios Region, Peru." Accessed April 14, 2017. http://www.vcsprojectdatabase.org/#/project_details/1067
51. Althelia Ecosphere. "Madagascar Fund." Accessed February 20, 2017. https://althelia.com/initiatives/madagascar-fund/
52. Althelia Ecosphere. "Oceans Fund." Accessed February 20, 2017. https://althelia.com/initiatives/oceans-fund/
53. Althelia Ecosphere. "Conservation Notes." Accessed February 20, 2017. https://althelia.com/initiatives/conservation-notes/
54. Althelia Ecosphere. "About Us." Accessed March 31, 2017. https://althelia.com/about-us/
55. US Agency for International Development. "U.S. Government, Althelia Climate Fund Mobilize $133.8 Million for Forest Conservation and Alternative Livelihoods." Last modified May 28, 2014. https://www.usaid.gov/news-information/press-releases/may-28-2014-us-government-althelia-climate-fund-mobilize-1338-million-forest-conservation
56. Kiss, Edit from Althelia Ecosphere. Interviewed by Brian McFarland. March 2017.
57. International Union for Conservation of Nature. "New Coalition Launches to Scale Private Conservation Investment at IUCN World Conservation Congress." Last modified September 3, 2016. https://www.iucn.org/news/secretariat/201609/new-coalition-launches-scale-private-conservation-investment-iucn-world-conservation-congress
58. Kiss, Edit from Althelia Ecosphere. Interviewed by Brian McFarland. March 2017.
59. Ibid.
60. Ibid.
61. Pothering, Jessica. "#Dealflow 11.1.16: Green Climate Fund Backs Althelia, IFC is a Startup Catalyst and a Reach Fund for Social Enterprises." *ImpactAlpha*. Last modified November 1, 2016. http://impactalpha.com/dealflow-10-5-16-green-climate-fund-backs-althelia-ifc-is-a-startup-catalyst-and-a-reach-fund-for-social-enterprises/
62. United Nations Convention to Combat Desertification. "Land Degradation Neutrality Fund: An Innovative Fund Project Dedicated to Sustainable Land Use." Last modified February 2017. http://www2.unccd.int/sites/default/files/relevant-links/2017-02/LDN%20Fund%20brochure%20Feb2017_0.pdf. 6.

63. Kiss, Edit from Althelia Ecosphere. Interviewed by Brian McFarland. March 2017.
64. eco.business fund. "About the Fund." Accessed April 14, 2017. https://www.ecobusiness.fund/about-the-fund/
65. Althelia Ecosphere. "Conservation Notes." Accessed February 20, 2017. https://althelia.com/initiatives/conservation-notes/
66. The Nature Conservancy. "Conservation Note Fact Sheet." Accessed March 29, 2017. https://www.nature.org/about-us/conservation-note-fact-sheet-2016.pdf
67. NatureVest. "Frequently Asked Questions." Accessed December 13, 2016. http://www.naturevesttnc.org/about-us/faq/
68. The Nature Conservancy. "The Conservation Note." Accessed March 29, 2017. https://www.nature.org/about-us/conservation-note.xml?redirect=https-301
69. Credit Suisse. "Sustainable Products & Services: Conservation Finance: Our Role in Conservation Finance." Accessed March 29, 2017. https://www.credit-suisse.com/us/en/about-us/responsibility/banking/sustainable-products-services.html
70. Environmental Finance. "Sustainable Forestry: Credit Suisse/Althelia Ecosphere's Nature Conservation Notes." Last modified April 1, 2015. https://www.environmental-finance.com/content/deals-of-the-year/sustainable-forestry-credit-suisse-althelia-ecosphere.html
71. Bank, David. "'Nature Conservation Notes' Bring Individual Investors into Climate Fund." *Huffington Post*. Last modified March 24, 2015. http://www.huffingtonpost.com/david-bank/nature-conservation-notes_b_6516576.html
72. Huwyler, Fabian, Jürg Käppeli and John Tobin. "Conservation Finance: From Niche to Mainstream: The Building of an Institutional Asset Class." Last modified January 2016. https://www.credit-suisse.com/media/assets/corporate/docs/about-us/responsibility/banking/conservation-finance-en.pdf. 3.
73. Hamrick, Kelley. "State of Private Investment in Conservation 2016." *Ecosystem Marketplace*. December 2016. http://www.forest-trends.org/documents/files/doc_5474.pdf#. vii.
74. Tobin-de la Puente, Dr. John. Interviewed by Brian McFarland. April 2016.
75. Ibid.
76. Ibid..
77. Huwyler, Fabian. Interviewed by Brian McFarland. March 2016.

78. Bank, David. "'Nature Conservation Notes' Bring Individual Investors into Climate Fund." *Huffington Post.* Last modified March 24, 2015. http://www.huffingtonpost.com/david-bank/nature-conservation-notes_b_6516576.html
79. World Forum on Natural Capital. "Edinburgh 2015: Fabian Huwyler." Accessed March 27, 2017. http://naturalcapitalforum.com/news/article/fabian-huwyler/
80. Tobin-de la Puente, Dr. John. Interviewed by Brian McFarland. April 2016.
81. Ibid.
82. Ibid.
83. Bank, David. "'Nature Conservation Notes' Bring Individual Investors into Climate Fund." *Huffington Post.* Last modified March 24, 2015. http://www.huffingtonpost.com/david-bank/nature-conservation-notes_b_6516576.html
84. Tobin-de la Puente, Dr. John. Interviewed by Brian McFarland. April 2016.
85. Global Impact Investing Network. "Annual Impact Investor Survey, 2016, The Sixth Edition." Last modified May 2016. https://thegiin.org/assets/2016%20GIIN%20Annual%20Impact%20Investor%20Survey_Web.pdf. 3.
86. Thoumi, Gabriel. "The Chain: Dimensional Fund Advisors Divests From Some Palm Oil Positions to Mitigate Risks." December 22, 2016. *Chain Reaction Research.* Accessed April 11, 2017. https://chainreaction-research.com/2016/12/22/the-chain-dimensional-fund-advisors-divests-from-some-palm-oil-positions-to-mitigate-risks/
87. Trading Economics. "Credit Rating." Accessed February 28, 2017. http://www.tradingeconomics.com/country-list/rating
88. World Bank Group. "Doing Business: Economy Rankings." Accessed January 12, 2017. http://www.doingbusiness.org/rankings
89. Transparency International. "Corruption Perceptions Index 2016." Accessed February 8, 2017. http://www.transparency.org/news/feature/corruption_perceptions_index_2016
90. United Nations Development Programme. "Human Development Report 2015: Work for Human Development." Last modified 2016. http://hdr.undp.org/sites/default/files/2015_human_development_report.pdf. 210.
91. World Justice Project. "Rule of Law Index 2016." Accessed January 12, 2017. http://data.worldjusticeproject.org/#table
92. OPIC. "OPIC Receives Environmental Finance Award for Sustainable Forestry." Last Modified July 25, 2012. https://www.opic.gov/blog/opic-

in-action/opic-receives-environmental-finance-award-for-sustainable-forestry-2
93. US Agency for International Development. "U.S. Government, Althelia Climate Fund Mobilize $133.8 Million for Forest Conservation and Alternative Livelihoods." Last modified May 28, 2014. https://www.usaid.gov/news-information/press-releases/may-28-2014-us-government-althelia-climate-fund-mobilize-1338-million-forest-conservation
94. McBryde, Mary, Peter R. Stein and Story Clark. "External Revolving Loan Funds: Expanding Interim Financing for Land Conservation." In *From Walden to Wall Street*, edited by James Levitt. Massachusetts: Lincoln Institute of Land Policy, 2005. 86.
95. Settelmyer, Scott. Interviewed by Brian McFarland. February 2017.

11

Payments for Ecosystem Services

"Compensation payments are an effective way to hold companies accountable for the impact they have on ecosystems and biodiversity. They finance conservation by collecting revenue from fines for pollution, royalty fees for natural resource use, compensation for environmental impacts, or even voluntary contributions. Although compensation payments don't necessarily reflect the actual environmental impact or provide one-for-one compensation, they pay for the extraction or use of one natural resource by investing in the conservation of another."[1]

—World Wildlife Fund

Introduction

Payments for ecosystem services come in several different conservation financing instruments such as biodiversity offsets, species banking, watershed protection payments, mitigation banking, and forest carbon markets, particularly Reducing Emissions from Deforestation and forest Degradation (REDD+). The "+" in REDD+ includes REDD along with forest conservation, sustainable forest management, and the enhancement

B.J. McFarland, *Conservation of Tropical Rainforests*, Palgrave Studies in Environmental Policy and Regulation, https://doi.org/10.1007/978-3-319-63236-0_11

of carbon stocks. There are also several other types of banks including "grassbanks, forest banks, and water banks."[2]

Historical Overview

In the late 1970s, the concept of payments for ecosystem services (PES, historically referred to as payments for environmental services) first appears.[3] For instance, Costa Rica's "first forest-incentive program, established in 1979, gave an income tax deduction to landowners who engaged in reforestation projects."[4]

Other notable dates for the various payments for ecosystem services include:

- 1988: AES Corporation supported reforestation in "denuded highlands of Guatemala in October 1988" with $2 million. "The new trees wouldn't do that work along, however: AES's carbon calculations included the labor done by trees already standing, which were to be protected from land-hungry farmers and loggers." {…} The mitigation project's "total cost, about $14 million, was shared by AES, the Guatemala forestry service, the U.S. Agency for International Development, CARE, and the Peace Corps."[5]
- 1995: Dennis O'Regan, "in charge of GPU's environmental auditing, said he met with Janzen in the fall of 1995 to review his proposal, which to his surprise turned out to be a multimillion-dollar project."[6] This project, which refers to a forest carbon project in Costa Rica, may again precede the Noel Kempff Mercado Climate Action Project.
- 1996: Costa Rica "earned an international reputation for reforestation when it introduced a unique piece of legislation, the 1996 Forestry Law, that promotes conservation through payments for environmental services, reforestation, and conservation programs."[7]
- 1996: TNC and "Fundación Amigos de la Naturaleza (FAN) created the Noel Kempff Mercado Climate Action Project to reduce carbon emissions by protecting 1.5 million acres of tropical forest that were threatened by timber harvesting and deforestation."[8] This is arguably the world's first REDD+ project.

- 1998: The "New South Wales state legislature had taken an aggressive step to forward this vision in November 1998, adopting the world's first law establishing 'carbon rights'."[9]
- 1999: In 1999, "State Forests was able to announce the biggest climate change investment thus far. The Tokyo Electric Power Company, the world's largest power firm, had signed up for the carbon rights from new pine and eucalyptus forests expected to absorb 200,000 to 800,000 metric tons of CO2 per year, offsetting the company's emissions in Japan."[10]
- 2000 and 2002: "In 2000 and 2002, PFT {Pacific Forest Trust} also brokered a sale to Future Forests {now Natural Capital Partners} of United States carbon emissions reductions produced by forest conservation and management."[11] This is perhaps the first transaction of forest conservation offsets.
- 2003: Climate, Community, & Biodiversity Alliance (CCBA) is founded.[12]
- 2003: Chicago Climate Exchange (CCX) is established.[13]
- 2004: First Clean Development Mechanism (CDM) Project (Brazil NovaGerar Landfill Gas to Energy Project) is registered.[14]
- 2005: Verified Carbon Standard Association (VCSA), which manages the Verified Carbon Standard (then known as the Voluntary Carbon Standard) and which is now the world's largest voluntary carbon market certification standard, is founded.[15] The VCSA started with more quality assurance/quality control (QA/QC) than its predecessors, had more structured risk management, and had more approved methodologies (i.e., including acceptance of CDM Methodologies). Large-scale carbon projects started coming on board shortly thereafter, including REDD+ projects.
- 2006: Boulder, Colorado, passes the first carbon tax in the United States.[16]
- 2006: First CDM Forestry Project (Facilitating Reforestation for Guangxi Watershed Management in Pearl River Basin) is registered.[17]
- 2007: First Tropical Climate, Community & Biodiversity Alliance (CCBA) Project (Small-scale Reforestation for Landscape Reforestation Project) is validated against the Climate, Community & Biodiversity Standard (CCBS).[18]

- 2007: Second Tropical CCBA Project (Panama Native Species Reforestation Project) is validated against the CCBS.[19]
- 2007: In October 2007, "the first such measure—the use of the CCAR {California Climate Action Registry} Forest Project Protocols—was adopted. The protocols are the first rigorous governmental accounting standards in the United States for climate projects that embrace forest management and avoid deforestation."[20]
- 2007: In December 2007, "the Van Eck Forest became the first project from any emissions sector to register with the CCAR {California Climate Action Registry} as a carbon emissions reduction project."[21]
- 2007: The "concept of REDD was taken up by environmental groups and some tropical countries, the latter informally organized into a Coalition for Rainforest Nations, and in 2007 at the United Nations Climate Change Conference in Bali, Indonesia, the Bali Road Map was adopted as a two-year plan to finalize a binding international agreement in December 2009 in Copenhagen."[22]
- 2008: First tropical forest conservation CCBA Project (Reducing Carbon Emissions from Deforestation in the Ulu Masen Ecosystem) is validated against the CCBS.[23]
- 2008: Norway and Brazil sign "a Letter of intent, also called MoU, for the climate and forest project. Norway pledged to contribute up to 1 billion US$ to the Amazon Fund in Brazil until 2015, if Brazil could show that deforestation in the Amazon went down."[24]
- 2009: Norway's "contract with the Brazilian Development Bank BNDES {Banco Nacional de Desenvolvimento Econômico e Social}, which administers the Amazon Fund, was signed."[25]
- 2009: The first CDM Tropical Forestry Project (Carbon Sequestration through Reforestation in the Bolivian Tropics by Smallholders of "The Federación de Comunidades Agropecuarias de Rurrenabaque" Project) (FECAR, The Federation of Agricultural Communities of Rurrenabaque in English) is registered.[26]
- 2009: The American Clean Energy and Security Act of 2009 (i.e., commonly referred to as the Waxman-Markey Bill) passed the US House of Representatives and included language referencing REDD+. The bill would not be signed in law.[27]

- 2010: Norway and Indonesia enter into a partnership whereby Norway pledged up to USD$1 billion to support Indonesia's national REDD+ work.[28]
- 2010: The Intercontinental Exchange (ICE) acquires Climate Exchange PLC., the parent company of the Chicago Climate Exchange (CCX). The CCX would later be closed down.
- 2011: Wildlife Works Carbon's Kasigau Corridor REDD Project—Phase I Rukinga Sanctuary becomes the first VCS REDD+ project to complete verification and issuance.[29]
- 2011: Face the Future registers the world's first tropical improved forestry management project,[30] known as the INFAPRO Rehabilitation of logged-over dipterocarp forest in Sabah, Malaysia, Project,[31] under the VCS.
- 2012: The Suruí Forest Carbon Project becomes the first Indigenous-led REDD+ project to achieve dual validation to the VCS and CCBS. The Project would later become the first Indigenous-led REDD+ project to achieve verification to the VCS and CCBS.[32]
- 2013: The Green Climate Fund "opened its doors" in South Korea.[33]
- 2013: The Purus Project becomes the first private REDD+ project to be validated to the VCS and CCBS in Acre, Brazil; Acre is largely seen as the most advanced subnational jurisdiction in the world working on REDD+-related initiatives.
- 2013: Québec begins its cap-and-trade program in 2013 and links with California's in 2014.
- 2015: The Paris Agreement of the UNFCCC is signed by almost every country on Earth.[34]
- 2016: The "Central African Forest Initiative (CAFI) and the Minister of Finance of the DR Congo signed a letter of intent (LOI) for 200 million US dollars to address deforestation and forest degradation in the country and to promote sustainable development. This LOI is the first signed between CAFI and a country of the Central Africa region, and the largest one ever concluded on REDD+ in Africa."[35]
- 2016 (September): The world's first global policy on biodiversity offsets is adopted.[36]

Mechanisms of Instrument

As Dr. Naeem and his co-authors wrote in *Science*:

> Payments for Ecosystem Services (PES) mechanisms leverage economic and social incentives to shape how people influence natural processes and achieve conservation and sustainability goals. Beneficiaries of nature's goods and services pay owners or stewards of ecosystems that produce those services, with payments contingent on service provision. Integrating scientific knowledge and methods into PES is critical. Yet many projects are based on weak scientific foundations, and effectiveness is rarely evaluated with the rigor necessary for scaling up and understanding the importance of these approaches as policy instruments and conservation tools. Part of the problem is the lack of simple, yet rigorous, scientific principles and guidelines to accommodate PES design and guide research and analyses that foster evaluations of effectiveness.[37]

All of the PES instruments work differently, but essentially PES instruments compensate the owner and/or steward of natural resources for its services. These ecosystem services include carbon sequestration and storage, water storage and purification, wildlife habitat, and nutrient recycling.

One could categorize PES in two categories: those ecological assets which already have value (e.g., water) and those ecological assets without a market value (e.g., the reduction of greenhouse gas emissions from tropical forests). Water needs some regulatory support to establish a price that convinces a hydroelectric company, a large bottling company, or a local government to support the preservation of an upstream watershed. In contrast, greenhouse gases—particularly carbon dioxide stored and being sequestered in tropical forests—are often undervalued or considered to have no market value. This situation creates unclear and/or very unstable pricing, and thus, regulation creates the potential for monetization.

There is also the possibility of bundling or stacking ecosystem services where a project developer could generate revenue for the protection of a tropical forest by selling forest carbon offsets, biodiversity offsets, and watershed protection payments from the same tract of forest. As an example, although not focused on the conservation of tropical rainforests, the Ohio River Basin Trading Project in the United States is exploring the possibility of including carbon offsets alongside its preexisting water quality credits initiative.[38]

Biodiversity Offsets

Biodiversity offsets, known also as species banking or conservation banking, "represents a free market mechanism for mitigating development impacts on endangered species' habitats."[39] The mitigation hierarchy starts with a firm avoiding impact to an endangered species' habitat, then minimizing any impacts if avoidance is impossible, followed by rehabilitating or restoring degraded habitats, and concluding with the use of biodiversity offsets as a last mechanism.[40]

A comparable mechanism is known as mitigation banking, where:

> Once all parties involved approve the plans and the bank meets its ecological success criteria (defined by the government, not the landowner), the mitigation banker earns credits that regulatory agencies will recognize. The landowner can then sell these mitigation credits to permittees and others who must compensate for adversely affected wetlands or other natural areas. The sale of wetland credits legally transfers the liability for wetland mitigation from the permitee to the wetland banker.[41]

Watershed Payments

According to the *State of Watershed Investment 2016* Report, a watershed investment is defined as "any transaction between a buyer and a seller where financial value is exchanged for activities or outcomes associated with the maintenance, restoration, or enhancement of watershed services or natural areas considered important for watershed services."[42]

Watershed payments are often structured between downstream users of water (e.g., hydroelectric facilities, manufacturing facilities that utilize significant quantity of water such as bottlers, and agricultural interests) and upstream stewards of watersheds (e.g., farmers, forest landowners, and/or local communities). Similarly, "much of the funding for the 'green' solutions to flood control and stormwater quality comes from existing stormwater management monies, in addition to a stormwater utility fee within water bills that is based on the amount of impervious surface on a property."[43]

The Mechanisms of Developing Forest Carbon Projects

The author is most familiar with the conservation financing instrument known as forest carbon offsets and particularly REDD+ projects.

REDD stands for Reducing Emissions from Deforestation and forest Degradation. The "+" includes REDD along with forest conservation, sustainable forest management, and the enhancement of carbon stocks (e.g., afforestation or reforestation activities).

With respect to REDD+, there is government-to-government REDD+ financing which is often for REDD+ Readiness activities. REDD+ Readiness is the funding for host governments to set up the infrastructure for subnational or national REDD+ initiatives, and this can take the form of legal or regulatory reform (i.e., assistance with establishing enabling legislation or regulation and the creation of a REDD agency or department), national forest monitoring and accounting systems, and financial reform.

There are also a few ways to categorize project-level REDD+ projects. First, there are avoided unplanned deforestation and avoided planned deforestation REDD+ projects. An avoided planned deforestation REDD+ project is essentially when the landowner has met all the requirements (e.g., legal permissibility, intent, and ability) to demonstrate they would have deforested their land if it was not for a forest carbon project. In contrast, an avoided unplanned deforestation REDD+ project is when someone other than the landowner—such as local communities or illegal loggers—is responsible for the deforestation.

Another way to categorize REDD+ projects are those that are connecting REDD+ with other sustainable products and REDD+ projects that are located in more remote areas. The former are REDD+ projects that are located in marginal forest, and these projects need to be near non-forest for sustainable, commercial agriculture (i.e., there is already deforestation there and the project needs to be close to markets). The latter are projects in very pristine areas, often with high conservation values such as protecting Indigenous Peoples and conserving wildlife, which are too remote for sustainable, commercial agriculture.[44]

While payments for performance have been made to a few select countries and subnational jurisdictions (e.g., Acre, Brazil), payments for REDD+ performance are often for voluntary forest carbon projects. These voluntary forest carbon projects must follow the general steps from initial project idea through project implementation and ending with the issuance of verified emission reductions (also known as certified carbon offset credits). This process is similar whether the project is being designed and implemented against a compliance carbon standard such as the Climate Action Reserve (CAR) or against a voluntary carbon standard such as the Verified Carbon Standard (VCS) or the American Carbon Registry (ACR). Similarly, this process is relatively the same whether you are developing a forest carbon project (i.e., reforestation, improved forestry management, or forest conservation project) or another eligible carbon offset project such as a renewable energy project.

1. *Due Diligence and Assessment of Project Potential*

Similar to any other type of project, it is very important to undertake an initial assessment of the project potential and conduct sufficient due diligence. Some aspects of the due diligence and assessment for forest carbon projects include:

- Review of stakeholders and evaluation of potential partners;
- Hold initial meetings with stakeholders and potential partners including, but not limited to, local communities and community leaders; local, state, and/or federal government officials; and strategic partners such as:
  - Universities (e.g., professors, graduate students, and/or academic institutes);

  - Nonprofit organizations (e.g., land trusts and/or conservation organizations); and
  - Subject matter experts (e.g., biologists, community specialists, forestry firms, and/or mapping firms);

- Conduct financial calculations (essentially how much potential revenue and costs are involved). More complex calculations such as Net Present Value (NPV), Internal Rate of Return (IRR), payback periods, and sensitivity analyses can also be performed;
- Risk assessment, legal analysis, and potentially drafting of contracts;
- Analysis of landownership titles and carbon ownership;
- Review of safety issues. Forestry projects, especially remote tropical forestry projects, present a wide range of safety concerns such as getting lost, poisonous snakes, mosquito-borne illnesses, and availability of clean drinking water.

Two of the most important aspects of tropical forestry carbon projects involve the analysis of landownership titles to ensure no overlapping title claims and to ensure clear carbon ownership is, or can be, granted to the Project Proponents.

2. *Selection of Certification Standard and Methodology*

All compliance forest carbon projects and more than 99% of voluntary forest carbon projects are designed and audited against third-party certification standards.

The leading forest carbon certification standard in the compliance market is the Climate Action Reserve (CAR), followed by the American Carbon Registry (ACR). The European Union Emissions Trading System (EU ETS), which is the world's largest carbon market, currently does not allow REDD+ projects, and its rules are very restrictive to afforestation and reforestation projects under the Clean Development Mechanism (CDM).

The leading forest carbon certification standard in the voluntary carbon market is the Verified Carbon Standard (VCS) which is often coupled with the Climate, Community, & Biodiversity Standard (CCBS) to demonstrate net positive community and biodiversity benefits. Additional voluntary carbon certification standards that accept forest carbon projects are the Gold Standard (due to its acquisition of the CarbonFix Standard), ACR, and Plan Vivo. There is a future possibility the projects designed to, and verified to, a voluntary carbon certification standard (i.e., the VCS) may be eligible for future compliance markets.

Here are the most frequently used carbon certification standards:

- ACR: http://americancarbonregistry.org/
- CAR: http://www.climateactionreserve.org/
- CCBS: http://www.climate-standards.org/
- CDM: https://cdm.unfccc.int/about/index.html
- Gold Standard: http://www.goldstandard.org/
- Plan Vivo: http://www.planvivo.org/
- VCS: http://www.v-c-s.org/

These certifications, such as the VCS, are designed in conjunction with the International Organization for Standardization (ISO).

A methodology, also known as a protocol, is essentially the blueprint for how various parameters of a project are calculated and/or determined. This includes guidance on how to design the forest carbon inventory, how to determine the project's start date and longevity, and how to develop a deforestation baseline. Each carbon certification standard—whether it is a compliance or voluntary standard—will have a wide variety of approved methodologies for use under the given standard. This includes methodologies such as for livestock methane projects, tropical forest conservation projects, and improved forestry management projects.

Here are links for some of the approved methodologies and protocols under the CAR, the CDM, and the VCS:

- CAR Protocols: http://www.climateactionreserve.org/how/protocols/
- CDM Methodologies: http://cdm.unfccc.int/methodologies/index.html
- VCS Methodologies: http://database.v-c-s.org/methodologies/find

Some of the technical forestry firms capable of working through these complex methodologies or protocols include:

- Carbon Decisions International: http://carbondecisions.com/
- Climate Focus: http://www.climatefocus.com/
- Finite Carbon: http://www.finitecarbon.com/
- Silvestrum: http://www.silvestrum.com/
- TerraCarbon: http://terracarbon.com/
- Terra Global Capital: http://www.terraglobalcapital.com/
- Wildlife Works Carbon: http://www.wildlifeworks.com/
- Winrock International: https://www.winrock.org/

3. *Selection of Independent Auditors*

The next step is to choose independent auditors, which are also known as validation/verification bodies (VVBs). The auditors are subject matter experts. For example, forest carbon projects are often audited by professional foresters, agriculturalists, or community development experts. In contrast, renewable energy or energy efficiency carbon projects are often audited by trained engineers.

Auditors are often accredited to the American National Standards Institute (ANSI), which is the US representative of the International Organization for Standardization (ISO). This accreditation helps to ensure auditors have proper management systems in place including a conflict of interest policy and proper document version control.

Auditors, such as Environmental Services, Inc. (ESI) or Rainforest Alliance, then must be formally accepted by the carbon certification standard. This acceptance is based off an application process and possibly includes a more rigorous training specific to the given certification standard. In addition, many auditors are accepted across several carbon certification standards. ESI, for instance, is an accepted auditor for ACR, CAR, CCBS, Plan Vivo, and VCS.

Approved VVBs for the CAR and the VCS can be found here:

- CAR: http://www.climateactionreserve.org/how/verification/connect-with-a-verification-body/
- VCS: http://database.v-c-s.org/verification-validation/find-vvb

4. *Implementation of Project Activities*

After a forest carbon project has passed due diligence, chosen a certification standard and a methodology or protocol, and lined up an independent auditor, it is then time to start implementing the on-the-ground activities.

For an afforestation or reforestation carbon project, these activities include:

- Conducting a carbon sequestration/forest growth curve analysis;
- Choosing the composition of tree species to be planted;
- Identifying a source for seeds or saplings; alternatively, a tree nursery could be established; and
- Planting the seeds or saplings at the project site.

For REDD+ forest conservation projects, the on-the-ground activities include:

- Designing and implementing a forest carbon inventory to determine a statistically significant quantification of the project area's biomass and carbon stock;
- Developing a computer model to forecast future deforestation and degradation and to establish a deforestation and degradation baseline; and
- Starting to implement the specific activities which will help reduce and stop deforestation and degradation at the project site. This might include the landowner foregoing conversion of the forest to another land use, offering agricultural extension training courses, and developing alternative revenue and/or income activities for local communities.

5. *Write Initial Project Documents*

While the project activities are being implemented, it is a wise idea to start writing the initial set of project documents. These project documents, particularly for the VCS and CCBS, are comprehensive and require an elaboration on various aspects of the project including:

- Background information on the project and Project Proponents;
- Incorporation of the technical measurements derived from the chosen methodology;

- Maps of the project location and project activities;
- Discussion of all relevant laws and regulations;
- Demonstration of Free, Prior, and Informed Consent and engagement with stakeholders; and
- Risk assessment of the project.

Many of the carbon certification standards offer templates for developing these project documents. See here for some of these templates:

- CAR (see section *Forestry Projects*): http://www.climateactionreserve.org/how/program/documents/
- VCS (see section *Project Templates & Forms*): http://database.v-c-s.org/program-documents
- Gold Standard (see section *Templates*): http://www.goldstandard.org/resources/afforestation-reforestation-requirements

6. *Register/List Project; Undertake Validation Audit*

The Project, upon completion of the draft project documents and after a sufficient amount of project activities are implemented on the ground, will register or list with an approved registry. A registry for carbon projects is somewhat comparable to an online bank account in the sense that you have a username and password, and upon logging in, you can view your current holdings and then either receive a transfer or initiate a transfer of VERs from your account to another account holder.

Some of the carbon certification standards use one specific registry. For example, the American Carbon Registry (as a voluntary carbon certification standard) uses its own registry. In contrast, some of the carbon certification standards, such as the VCS, have approved a few different registry providers.

Here are a few of the registry providers:

- ACR: http://americancarbonregistry.org/how-it-works/registry-reports
- APX: http://www.apx.com/

- IHS Markit's Environmental Registry: http://www.markit.com/Product/Registry

The validation audit is often conducted by the previously secured auditor. The validation audit involves the auditor reading through the initial project documents, checking all calculations, interviewing the Project Proponents and other relevant stakeholders, and visiting the project site.

Upon successful completion of the validation audit, the auditor will issue a validation report and validation statement. These documents essentially confirm that, in the auditor's professional opinion, the project has been designed and thus far implemented in accordance with the chosen carbon certification standard.

7. *Conduct Periodic Monitoring, Reporting, and Verification with Issuance of Carbon Offset Credits*

The project, after achieving validation, will continue to implement the on-the-ground project activities and periodically conduct monitoring, reporting, and verification (MRV).

Monitoring involves the monitoring or measurement of certain metrics associated with the project. For a REDD+ project, this mostly involves monitoring deforestation and degradation relative to the established baseline, along with monitoring the project's impact on local communities and biodiversity within the Project Zone (i.e., the project area and its buffer zone).

Reporting involves the preparation of another set of project documents. Whereas the initial set of project documents are more of a forward-looking management plan, this next set of project documents are prepared for a specific, historical time period—usually the previous calendar year.

Verification involves another audit, known as the verification audit. Depending on the rules of the carbon certification standard, this audit can be performed by the same auditor or a different auditor and involves a similar process of the auditor reading through the new project documents, checking all new calculations derived from the monitoring, (re)interviewing the Project Proponents and other relevant stakeholders, and (re)visiting the project site.

Upon successful completion of the verification audit, the auditor will issue a verification report and verification statement. These documents confirm that, in the auditor's professional opinion, the project has mitigated or sequestered a specific quantity of carbon dioxide equivalent (CO_2e) emissions, in addition to community and biodiversity benefits. The Project Proponents would then deliver the project documents and auditor reports to the registry, and the registry will issue the specific quantity of offsets—each with a distinct serial number—into the Project Proponent's registry account.

8. *Sales*

Sales should not be left to the very end, but a common structure in the forest carbon market is for buyers to only purchase offsets after the Project Proponents have successfully completed the verification audit and are ready to transfer the issued VERs. As of April 2017, there are no compliance markets that accept REDD+ projects from tropical rainforest countries. In contrast, the largest buyers of voluntary REDD+ offsets include Disney and Microsoft. For instance, "the government of Cambodia, through a partnership with the Wildlife Conservation Society (WCS), has sold several million voluntary carbon credits from a REDD+ project to entertainment giant Disney."[45]

Size of Instrument

The following outlines the approximate sizes of the biodiversity offsets market, watershed payments, and the forest carbon sector.

Biodiversity Offsets

According to the last *State of Biodiversity Markets* Report dated 2010, there are

> {…} 39 existing compensatory mitigation programs around the world, ranging from programs with active mitigation banking of biodiversity

> credits to programs channeling development impact fees to policies that drive one-off offsets. There are another 25 programs in various stages of development or investigation. Within each active offset program, there are numerous individual offset sites, including over 600 mitigation banks worldwide.
>
> The global annual market size is $1.8–$2.9 billion at minimum, and likely much more, as 80% of existing programs are not transparent enough to estimate their market size. And the conservation impact of this market includes at least 86,000 hectares {212,000 acres} of land under some sort of conservation management or permanent legal protection per year.[46]

It is important to note that not all of these commitments went to mitigation programs and biodiversity offsets in tropical rainforests and some of this funding may also be included as overlapping figures reported in the section on domestic government expenditures for protected areas.

Watershed Payments

According to the *State of Watershed Investment* 2016 Report, approximately USD$25 billion was spent on green infrastructure water projects across 62 countries and 419 distinct programs in 2015. Of this USD$25 billion, the vast majority (USD$23.7 billion) came in the form of direct subsidy payments from governments "to landowners to protect and restore water-critical landscapes and promote a green economy."[47] These landowners received an estimated USD$16 billion to reward their stewardship of watersheds. Furthermore, these projects "protected, rehabilitated, or created new habitat on more than 486 million hectares of land around the world, an area nearly 1.5 times the size of India."[48] Again, it is important to note that not all of these commitments went to watersheds consisting of tropical rainforests and some of this funding may also be included as overlapping figures reported in the section on domestic government expenditures for protected areas.

Forest Carbon

Forest Trends, through its Ecosystem Marketplace, produces annual reports on the state of the voluntary and forest carbon markets. Although not all targeted to tropical rainforests, in 2015, Ecosystem Marketplace reported:

> $888 million (M) in new results-based finance for forests—the largest influx ever tracked in this report series. Of this committed money, $88 M flowed through voluntary markets; $73 M flowed through compliance markets including California's and New Zealand's; $588 M flowed through Australia's Emissions Reduction Fund {60.7 million tonnes transacted}; and $126 M flowed through non-market agreements. Voluntary markets channeled money into emissions reductions in at least 44 countries, while compliance markets—or those mandated by law—channeled money to domestic initiatives in the United States (through the California cap-and-trade program), New Zealand (through its government Emissions Trading System), and Canada (through the British Columbia government's Carbon Neutral Commitment). On the non-markets side, Norway, Germany, and the Brazilian oil company Petrobras paid Brazil $126 M for reducing deforestation in the new and growing form of results-based payments. Pledges to 20 additional countries are 'on the table'.[49]

Furthermore, Ecosystem Marketplace identified:

> More than 800 forest and land-use projects are currently operational or under development around the world, with the vast majority in two countries that have signaled compliance-driven markets—namely, Australia (428) and the United States (217).
>
> {...} Outside of compliance markets, Peru, Brazil, Kenya, and Uganda were the most common project locations. Beyond emissions reductions, voluntary forest carbon projects that reported on co-benefits in 2015 employed almost 8,000 people, protected habitat for 376 endangered species, provided benefits to vulnerable groups, contributed to water security, built resilience to climate change impacts, and helped to clarify land tenure.
>
> {...} Forest carbon finance may be increasingly channeled through compliance markets, as 15 current and future compliance markets include an offsetting mechanism, and 11 of those have developed protocols for land-use and forestry offsets. The International Civil Aviation Organization (ICAO) is also in the process of designing a global market-based measure that may include forests. The only non-market payments for forest-based

emissions reductions made to-date have flowed from the Amazon Fund (Norway, Germany, and the Brazilian oil company Petrobras) to Brazil; from the REDD+ Early Movers program (Germany, Norway, and the United Kingdom) to Colombia and the Brazilian state of Acre; and from the Norwegian government to Guyana. The growing, results-based finance pipeline is likely to flow through the World Bank's Carbon Fund and through continued bilateral agreements. The Green Climate Fund also could serve as a major catalyst for forest carbon emissions reductions, if it chooses to.[50]

Forest carbon projects, according to data from the *State of Forest Carbon Finance 2016* Report, are protecting 28 million hectares (69 million acres).[51]

REDD*X*, an initiative of Forest Trends, identified approximately USD$6 billion in "total REDD+ finance commitments between 2009 and 2016, spanning ten different countries throughout Latin America, Africa, and Asia Pacific. Of this amount, $2.9 billion was allocated toward national–level REDD–readiness initiatives, while $3 billion was earmarked for subnational initiatives that target specific regions or provinces in individual countries."[52] This said, a lot of the funding was for REDD+ Readiness activities (e.g., setting up national forest inventories, harmonizing domestic legislation, and developing remote sensing capabilities) and not necessarily financing on-the-ground, direct conservation activities. Furthermore, some of this funding may overlap with the figures reported in the government international expenditures chapter.

Introduction to Case Studies

The following 4 case studies will examine the Purus Project, the Rimba Raya Biodiversity Reserve REDD+ Project, the Kasigau Wildlife Corridor REDD Project—Phase I Rukinga Sanctuary, and the Rio Tinto QMM Biodiversity Offset Project.

Please note, the author works for both Carbonfund.org Foundation and CarbonCo, which is one of the Project Proponents of the Purus Project. Nevertheless, the Purus Project is the first private REDD+

project in Acre, Brazil—which is widely seen as the leading subnational jurisdiction working on REDD+-related initiatives—to achieve dual validation and verification to the VCS and CCBS with Gold Distinction for exceptional biodiversity benefits.

The Rimba Raya Biodiversity Reserve REDD+ Project is located on the island of Borneo within the Central Kalimantan province of Indonesia. The Project was the first REDD+ project in the world to be validated to the Verified Carbon Standard (VCS) and also the first REDD+ project in the world to achieve Triple Gold validation under the Climate, Community and Biodiversity Standard (CCBS) for exceptional climate change adaptation benefits, along with exceptional biodiversity and community benefits.[53] InfiniteEARTH, the Project Proponent, is also the author of the first REDD+ forest carbon accounting methodology. Furthermore, this was the first REDD+ project to be approved in Indonesia and addresses all 17 of the United Nations' Sustainable Development Goals (SDGs).[54]

The Kasigau Wildlife Corridor REDD Project—Phase I Rukinga Sanctuary is the first VCS REDD+ project to complete verification and issuance.[55] The Project is located in the county of Taita Taveta, Kenya, between Tsavo West and Tsavo East National Parks. Among the Project's many industry-leading accomplishments, the Project is in the process of reducing 7.5 million mtCO_2e over the first 30 years and helped spur the adjacent Kasigau Corridor REDD Project—Phase II, The Community Ranches, the relatively close-by Chyulu Hills REDD+ Project, and the Mai Ndombe REDD+ Project in the Democratic Republic of the Congo. Collectively, these four projects are protecting up to 859,400 hectares (approximately 2.1 million acres).

The Rio Tinto QMM Biodiversity Offset Project is located in Madagascar. Rio Tinto, to reduce the impact at the QMM Project, has followed the mitigation hierarchy to avoid impacts on endangered species and threatened habitats, then minimized any impacts when avoidance was impossible, followed by rehabilitating or restoring degraded habitats, and then used biodiversity offsets when necessary.[56] With this in mind, Madagascar is a widely recognized biological hotspot with a tremendous number of endemic species and highly threatened ecosystems. However, much of the original forest cover has been lost and, for instance, only 10% of the original littoral forest remains, much of it highly degraded.

Case Study #1: Purus Project: A Tropical Forest Conservation Project in Acre, Brazil

Introduction

The Purus Project is a payments for ecosystem services forest conservation project, otherwise known as a Reduced Emissions from Deforestation and Degradation (REDD+) project, on approximately 34,702 hectares (approximately 85,714 acres) of privately owned land in Acre, Brazil. The Project is within the Purus River basin which stretches from Peru through the State of Acre, and the River eventually becomes a tributary to the Amazon River. The Project Area is considered lowland tropical rainforest (Fig. 11.1).

The Purus Project was successfully validated by Scientific Certification Systems, Inc. (now doing business as SCS Global Services) in January 2013 to the Verified Carbon Standard (VCS) and to the Climate, Community, & Biodiversity Standard (CCBS) with Gold Distinction for exceptional biodiversity benefits. The Project was subsequently verified by Environmental Services, Inc. (ESI) to the VCS and CCBS with Gold Distinction in December 2013, October 2014, November 2015, and November 2016.

The three main Project Proponents are CarbonCo, LLC ("CarbonCo"); Freitas International Group, LLC (doing business as "Carbon Securities"); and Moura e Rosa Empreendimentos Imobiliários LTDA ("Moura & Rosa"). CarbonCo, the wholly owned subsidiary of Carbonfund.org Foundation, is responsible for getting the Project certified and for early-stage Project finance. Carbon Securities acts as a liaison between CarbonCo and Moura & Rosa, along with acting as a translator and assisting with logistics for site visits. Moura & Rosa is an Acre, Brazil-based organization created by the landowners and is primarily responsible for day-to-day management of the Project and the implementation of activities to mitigate tropical rainforest deforestation and degradation.

The ultimate project activities of the Project are to undertake a forest carbon inventory, model regional deforestation and land use patterns,

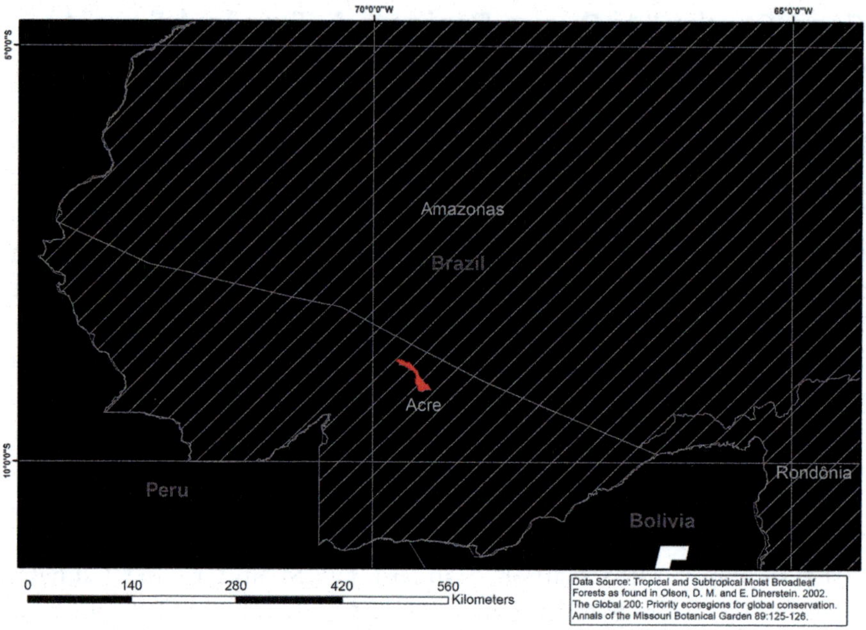

Fig. 11.1 Map of Purus Project (Credit: James M. Eaton)

and mitigate deforestation and degradation pressures by utilizing payments for the Project's ecosystem services, along with ongoing monitoring of the climate, community, and biodiversity impacts. Social projects and activities to mitigate deforestation and degradation range from engaging local professionals for agricultural extension training, to patrols of potential deforestation sites, to building an onsite health clinic and granting land title to local communities to help strengthen their land tenure.

While the Project is a privately owned, voluntary REDD+ project, the State of Acre is working on jurisdictional REDD+ initiatives and the Project is seeking registration with the State. This process of nesting the Project within the subnational jurisdiction of Acre could enable the Purus Project to become one of the first REDD+ projects in the world to offer compliance grade carbon offsets in the event that Acre links to a compliance market, such as in the State of California.

Identify the Problem

The essential problem is the Project faces deforestation pressures from the recent construction of the nearby highway BR-364, along with the relatively unsustainable slash-and-burn agriculture and cattle ranching by local communities. In addition, the local communities are relatively poor and lack many basic services, and the habitat for biodiversity, which includes several endangered species, is shrinking.

A few other problems that are, or were in the past, facing the Project include:

- There are unclear landownership rights. More specifically, many communities have lived on the land for more than 20 years. This entitles the communities to some rights such as compensation for their assets (e.g., house, pasture fencing, and/or cattle), but the communities do not have full land tenure rights and, thus, could be removed by the landowners. The landowners have no intention of doing this, but the risk could arise in the future if the land was ever sold.
- At the time of project initiation, there were no other REDD+ projects in the State of Acre, Brazil. This created some first-mover disadvantages.
- Diversifying project revenue with additional revenue from agroforestry or ecotourism has proved difficult. This is because the Project is a bit too remote—the nearest town is Manoel Urbano, which is a several-hour boat ride from the Project; the capital city of Rio Branco is another several hours away via car; and Brazil's major cities including Brasilia, São Paulo, and Rio de Janeiro are several days away for ground transportation.

Why the Problem Is Important

The tropical rainforest deforestation and degradation of the Purus Project is important because:

- The Project provides pristine habitat for rare, threatened, and endangered species. There were at least two endangered flora species identified at the Purus Project as classified on the International Union for Conservation of Nature (IUCN) Red List. These endangered flora species are Car-cara (*Aniba rosaeodora*) and Baboonwood (*Virola surinamensis*). In addition, numerous IUCN vulnerable and near threatened species, including short-eared dogs (*Atelocynus microtis*), jaguars (*Panthera onca*), giant anteaters (*Myrmecophaga tridactyla*), and lowland tapirs (*Tapirus terrestris*), have been photographed at the Project. In fact, the second known photograph in Acre of a short-eared dog was taken at the Project.

- The Project protects a significant tract of primary forest which is within the watershed of the Purus Basin, one of the most important rivers in Acre and an important tributary to the Amazon River. The Project also acts as a barrier against deforestation and provides connectivity to areas south of the Project.
- The forests within the Project provide a vitally important resource to local communities for food, fuel, and fodder for their animals.
- The forests within the Project store a significant quantity of carbon. It is estimated the Project—through its conservation and community development activities—will mitigate the release of nearly 1 million $mtCO_2e$ emissions over the Project's first 10 years.

How Problem Was Identified

The problem of tropical rainforest deforestation at the Project has been identified via computer modeling of future deforestation; review of satellite imagery, including the use of Landsat; and visual ground-truthing. In addition, numerous stakeholder meetings have taken place since 2011 with local and state government officials, local and international nongovernmental organizations, and with local community members living in, and around, the Project.

In addition, there were numerous community surveys conducted including Basic Necessities Surveys (BNSs), Participatory Rural Assessments or Appraisals (PRAs), and an Agricultural Survey.

The initial PRA was conducted by the Project Proponents in March 2012. The Project Proponents attempted to sample each community living within the Purus Project Area, along with all adjacent communities living along the Purus River and within the Project Zone. A total of 16 communities were interviewed as part of this initial PRA. The PRA helped to show that all community members practice agriculture and nearly two-thirds participate in cattle ranching. In addition, the PRA revealed that although no communities sell timber or charcoal outside of the community, a significant majority of the communities sell either crops or cattle and a significant majority also make charcoal. The PRA helped to establish a baseline of local economic activities and land use practices, along with a mechanism to assess leakage (i.e., offsite impacts attributable to the Project). In April 2013, the Project Proponents administered a follow-up PRA that focused on fuelwood consumption, charcoal consumption, and leakage to better understand if degradation was occurring throughout the Project Area. In June 2015, the Project Proponents administered another follow-up PRA. Similar to the April 2013 survey, the June 2015 PRA survey found that there is little, if any, leakage occurring as a result of the Project and that the majority of communities continue to produce charcoal, while fewer communities collect fuelwood.

The Project Proponents also conducted a Basic Necessities Survey (BNS) in March 2012 among the aforementioned 16 communities. One way to define poverty is those without access to the basic necessities. For the BNS, a focus group was created among the Project Proponents and the community to identify the top 25 assets or services which were believed to be basic necessities, or things that no one in the community should have to live without. In addition, a community price was established if a particular good was to be locally purchased. The Project Proponents then individually surveyed each of the 16 communities, and only those assets or services which at least 50% of the communities deemed a basic necessity were included in the final calculations of a poverty index and poverty score. The assets or services which had a higher percentage of communities considering them a basic necessity, compared to the number of communities actually possessing those assets or services, were considered higher priority social projects or programs. For example, this included the access to a telephone (i.e., installed at the Project's headquarters in May 2014), access to school (i.e., a school bus boat was secured and supplies donated in 2014), and access to a health clinic (Moura & Rosa facilitated visits by doctors in 2014 and a health clinic was built in 2016). A follow-up BNS was conducted in 2014 and 2016. These follow-up BNSs revealed the communities' local needs remained essentially the same as in 2012, yet access (i.e., such as to the newly built health clinic) was improving.

The Agricultural Survey involved the same communities in and around the Purus Project being surveyed in March 2012 to better understand which agricultural extension training courses would be of the most interest. A total of 32 courses, ranging from rotational pasture management to organic coconuts, were offered. In July 2013, courses on the production/cultivation of bananas, rural property management, and the sustainable use of legumes were taught by a local consultant. In August 2013, courses on the sustainable management of cattle pastures and alternative systems for establishing chicken coops were taught by another local consultant.

The course on bananas focused on:

- The origin of bananas;
- The different species of bananas;
- Proper planting, spacing, and tillage systems;
- Identification and control of diseases;
- Harvesting and transportation techniques; and
- Market access.

The rural property management course taught how to manage one's property and the need to plan each phase including site preparation, implementation, procurement, sales, and receipts.

The course on sustainable use of legumes taught how to fertilize using locally sourced manure, humus, and compost and how legumes are nitrogen-fixation species.

The sustainable management of cattle pastures course focused on how to use the pasture without degrading the pasture, how to implement a sustainable pasture rotation system, and proper division of grazing paddocks.

The course of chicken coops taught:

- How to raise chicks;
- Slaughtering techniques;

- How to maintain a semi-confined and confined chicken coop;
- Types of feed; and
- Treatment of diseases.

Furthermore, the communities that participated in the chicken course were built a chicken coop and received free chicks and free feed.

In 2016 and 2017, Moura & Rosa began to structure a partnership with the Federal University of Acre (UFAC) in Rio Branco to collaborate with the departments of agronomy, forestry, and medicine. This collaboration will include facilitating additional agricultural extension work.

Effectiveness of Process for Identifying Problem

The process was very successful at identifying the underlying problems at the Purus Project. The satellite imagery, coupled with ground-truthing, accurately identified deforestation. The computer modeling of deforestation provided an essential guide to where deforestation was most likely to occur in the future. In addition, the community surveys helped to identify the challenges the local communities were facing, along with their concerns about the Project and their desired outcomes.

Steps Taken to Address the Problem

There were numerous steps taken to address the problem of deforestation, while simultaneously improving the quality of life for local communities and preserving the area's rich biodiversity. These steps included:

- Initial community consultations and ongoing community consultations;
- Signing of a Tri-Party Agreement between three Project Proponents;
- Administering the aforementioned Basic Necessities Survey (BNS), Participatory Rural Assessment or Appraisal (PRA), and Agricultural Survey;
- Conducting a forest carbon inventory and modeling of deforestation pressures, which are required components of a validated and verified REDD+ project;

- Installation of wildlife cameras to monitoring the Project's impact on biodiversity;
- Implementation of several projects and programs for communities; and
- The purchase of a drone to regularly monitoring deforestation and acquisition of satellite imagery on an annual basis.

Results

The Purus Project became the first project in the State of Acre, Brazil, to achieve validation and verification to the VCS and CCBS with Gold Distinction. This Gold Level Distinction to the CCBS was earned for the Project's exceptional biodiversity benefits.

The Project has secured long-term financing via the purchase of verified emission reductions (VERs, also known as carbon offset credits) by a large aggregator.

This carbon financing has enabled the Project to implement numerous on-the-ground activities including:

- Hiring local project managers;
- Offering free agricultural extension courses and building chick coops with hundreds of free chicks given away;
- Paying for the visits of doctors, nurses, and dentists to access the Project site;
- Building the community's first, local onsite health clinic;
- Granting the first land titles to local families to secure their land tenure;
- Achieving a reduction for 6 straight years in deforestation relative to the deforestation baseline;
- Establishing a Project headquarters with a phone tower for community use, acquisition of a boat to provide free transport for the communities' agricultural goods, and a potable water system;
- Installation of motion-sensitive wildlife cameras which photographed numerous IUCN vulnerable and near threatened species, including short-eared dogs, jaguars, giant anteaters, and lowland tapirs.

Furthermore, the Project Proponents have built broad relationships with the State of Acre and support the Chico Mendes Foundation with a portion of the Project's revenue, and comradery has developed across diverse stakeholders.

Challenges and How They Were Met

There are several ongoing challenges with the Purus Project.

First, deforestation and degradation—although at a level below the established baseline—continues by a few families, particularly the wealthier families who do not need the Project benefits. This problem is being met by ongoing community consultations, engagement of stakeholders, and seeking legal assistance, along with continuing to implement local projects and programs for the other community members.

The second challenge is that the voluntary carbon market, particularly the subsector of REDD+ projects, is oversupplied with VERs with steady or declining demand. This supply-demand imbalance has resulted in declining prices for VERs. While grateful that the Project has received long-term financing, an active compliance market with natural buyers would facilitate more liquidity and higher prices.

Additional challenges include understanding the complex mix of subnational, national, and international policies alongside the technical, methodological requirements.

Beyond Results

There are a variety of measures, both in place and planned, to ensure the Project's climate, community, and biodiversity benefits are maintained and enhanced beyond the Project Lifetime. This includes:

- The Tri-Party Agreement's longevity;
- Creation of Moura e Rosa Empreendimentos Imobiliários LTDA;
- Social projects; and
- Ongoing education and outreach.

The Tri-Party Agreement between Carbonfund.org, Carbon Securities, and Moura & Rosa was signed on March 17, 2011, and stipulates a minimum 60-year Project Lifetime, followed by two renewable terms of 25 years each. The Project Proponents are committed to maintaining forest cover within the Project beyond the initial Project Lifetime. Furthermore in 2013, the Project and its project documents (both VCS and CCBS) were submitted to the State of Acre's Climate Change Institute and were officially preregistered with the IMC in 2014 and again in 2017.

The landowners Normando Sales and Wanderley Rosa created the legal entity Moura e Rosa Empreendimentos Imobiliários LTDA on February 27, 2009, to specifically ensure the Purus Project is managed beyond their lifetime by their children.

The social projects are designed to provide long-lasting climate, community, and biodiversity benefits beyond the Project Lifetime. Such social projects include offering agricultural extension training courses and granting official land title to local communities. This said, even if the Project were to fail in the future, the community's title—aside from if they sell their title to someone else—will be retained by the community member and their family in perpetuity.

There are also a variety of education and outreach activities which will both maintain and enhance the climate, community, and biodiversity benefits beyond the Project Lifetime. In addition, it is the Project Proponents' hope that such benefits will not only extend temporally (i.e., beyond the Project Lifetime) but also in a spatial manner (i.e., beyond Project Zone, across State of Acre, across the country of Brazil, and internationally). For example, CarbonCo and Carbon Securities have successfully completed the validation and verification of 3 more private REDD+ projects in Acre known as the Russas, Valparaiso, and Envira Amazonia Projects. Collectively, these 3 Projects and the Purus Project are conserving approximately 300,000 hectares (nearly 750,000 acres).

An additional educational activity involves the agricultural extension courses that will provide benefits to local communities beyond the Project Lifetime. Agricultural courses were offered in 2013, and now Moura & Rosa is developing a partnership with UFAC through which knowledge, new techniques of work, and research will be brought to the Project via small pilot projects between university staff and local communities. This collaboration will ideally lead to improvements in the communities by teaching them cultivation techniques of fruits and vegetables, implementation of septic tanks, construction of semi-artesian wells for drinking water, and better forms of personal hygiene (i.e., teaching children how to wash their hands).

Furthermore, Carbonfund.org and its subsidiary CarbonCo are both climate change organizations and regularly educate the public about climate change and the vital role of forest conservation projects through blog postings, newsletters, presentations, and during outreach discussions with partners. For example, the Project Proponents presented the Purus Project at the United Nations Conference on Sustainable Development, Rio+20 Conference, in Rio de Janeiro, Brazil, in June 2012. The author has also presented at numerous industry events, which included highlighting the Purus Project.

Lessons Learned

The Purus Project has provided many valuable lessons since 2011. These lessons include:

- Community consultations and engagement are essential. If communities are not a part of the process and directly benefit from the project, then the project will not succeed in the long-term.
- Hiring local assistance (e.g., local biologists, foresters, and professors) is very advantageous. Local labor costs are often less, while local assistance is often more familiar with the local context and this provides an opportunity for technical, cultural, and knowledge transfer.
- When designing private REDD+ projects, it is vitally important to consult government bodies to outline the parameters that will help align the project(s) for possible nesting.
- Hiring world-class, technical expert(s) is also essential. In the REDD+ sector, such firms include Carbon Decisions International, Climate Focus, Silvestrum, TerraCarbon, Terra Global Capital, Wildlife Works Carbon, and Winrock International. CarbonCo's hiring of TerraCarbon was one of the best decisions it ever made.
- The Project Proponents must identify the root cause of deforestation and degradation. Otherwise, a lot of money and effort may be allocated to addressing a secondary cause without actually addressing the primary, underlying cause.

- Choosing an auditor is an important decision. Ideally, the auditor should be familiar with the project area, along with the chosen certification standard and methodology.
- Finding a secure, ideally long-term, buyer of carbon offsets is very important, and this activity should be sought after from the beginning stages of the project.

Other Resources on Purus Project

Carbonfund.org Foundation

- www.carbonfund.org

Climate Change Institute of Acre, Brazil

- http://imc.ac.gov.br/

Markit Environmental Registry

- https://mer.markit.com/br-reg/public/project.jsp?project_id=103000000005599

Verified Carbon Standard's Project Database

- http://www.vcsprojectdatabase.org/#/project_details/963

Case Study #2: The Rimba Raya Biodiversity Reserve REDD Project in Indonesia

Introduction

The Rimba Raya (meaning "Infinite Forest") Biodiversity Reserve REDD+ Project is located on the island of Borneo within the Central

Kalimantan province of Indonesia. The Project was developed by, and is being managed by, InfiniteEARTH and its local affiliate PT Rimba Raya Conservation, in addition to a variety of partners including the Orangutan Foundation International (OFI). Furthermore, InfiniteEARTH's founder Mr. Todd Lemons is also the CEO of the EnVision Corporation which is structuring innovative derivatives related to the Project including EarthVEST ETF, InfiniteRETURNS, and BitNatura.[57]

The Project was the first REDD+ project in the world to be validated to the Verified Carbon Standard (VCS) and also the first REDD+ project in the world to achieve Triple Gold validation under the Climate, Community, & Biodiversity Standard (CCBS) for exceptional climate change adaptation benefits, along with exceptional biodiversity and community benefits.[58] InfiniteEARTH is also the author of the first REDD+ forest carbon accounting methodology. Furthermore, this was the first REDD+ project to be approved in Indonesia and addresses all 17 of the United Nations' Sustainable Development Goals (SDGs) (Fig. 11.2).[59]

The total Project Area, and thus the total area at risk of deforestation, is approximately 64,500 hectares (159,315 acres), consisting mainly of lowland, tropical peat swamp forests. Within this Project Area, the Project is mitigating the release of 130 million tonnes of carbon dioxide equivalent (mtCO_2e) emissions and working with 16 local villages.[60] The release of 130 million mtCO_2e, which is stored in the carbon-rich peat soils, is equivalent to:

- GHG emissions from 311.6 billion miles driven by an average passenger vehicle;
- GHG emissions from 27.5 million passenger vehicles driven for 1 year;
- CO_2 emissions from 14.6 billion gallons of gasoline consumed;
- CO_2 emissions from 138.7 billion pounds of coal consumed; or
- The amount of carbon sequestered by 123.1 million acres of US forests in 1 year.[61]

The Project is also providing refuge to more than 600 species of flora and fauna. This includes at least 122 species of mammals, 300 species of birds, 180 species of plants and trees, and more than 100 IUCN Red Listed Endangered and Threatened Species such as the Borneo orangutan

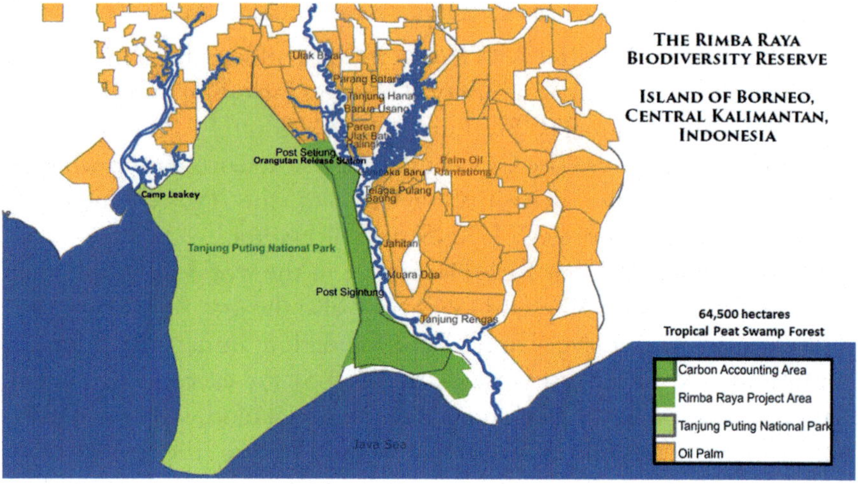

Fig. 11.2 Map of the Rimba Raya Biodiversity Reserve REDD Project (Credit: InfiniteEARTH)

(*Pongo pygmaeus*).[62] In fact, "Rimba Raya is also the world's largest privately-funded orangutan sanctuary."[63]

Identify the Problem

The island of Borneo, particularly within Central Kalimantan, is experiencing widespread tropical rainforest deforestation and degradation due to the expansion of palm oil plantations. Specific to the Rimba Raya Biosphere REDD Project, the Project Zone was facing planned deforestation as a result of government-approved palm oil plantations.

A related problem was the shrinking Project Area, which went from approximately 101,000 hectares (249,470 acres) down to 64,500 hectares (159,315 acres), even after getting the concession approved, after conducting local community engagements, and after receiving both local and provincial licenses. Shortly after this shrinking Project Area issue was "resolved" and the final concession boundary lines were drawn, there were illegal incursions into the Project Area which resulted in 80 hectares (198 acres) of forest being illegally burnt.[64]

In the 1980s, there was illegal logging in the adjacent Tanjung Puting National Park. Local communities were funded by palm oil companies (e.g., companies financed the chainsaws, gasoline, and labor) to illegally log the area, and these communities built 4 kilometer paths with rails through the forest to extract timber. The palm oil companies would then secretly purchase the logs back from the communities. In Indonesia, there are laws that define a forest as an area with approximately 100 cubic meters or more of "commercial timber" (narrowly defined as a dozen or so high value species) per hectare. This process of illegal logging led to sufficient degradation by the legal definition and resulted in the palm oil companies being able to lobby for the areas to be reclassified as agricultural lands, rather than as forest, and thus to then be converted into palm oil plantations. While poaching of timber was historically a major issue, the poaching of wildlife is not a major problem in and around the Project.[65]

Another major problem in the region is mining. Mining started happening upstream from the Project, and soon thereafter, the main river was silted up. This siltation and ensuing water pollution killed fish in the main river. The communities learned that where there had been fires, artificial lakes were created, and this presented a place for the communities to fish and so the communities created more. However, the communities were basically depleting the fish nurseries and the local fish populations began to plummet.[66]

Finally, fires—as a result of land clearing—were a major problem and lasted for upward of 5 years during the 1990s.

Why the Problem Is Important

These problems are important because there were lots of degradation by fire, most of the commercially viable timber species were gone, and communities were becoming desperate as their local fisheries were beginning to collapse. The collapse of the fisheries reduced an important food source for the local communities, and soon many bird species also disappeared due to their dependence on the fisheries.[67]

In addition, tropical rainforest deforestation and degradation, particularly of carbon-rich peat swamp forests, releases a significant amount of greenhouse gas emissions.

Such deforestation and degradation also contributes to forest fragmentation and habitat loss, which is a leading factor to the endangered and threatened status of the more than 100 IUCN Red Listed species at the Project. In addition to the Borneo orangutan, which is one of the closest relatives to humankind, the Project provides habitat for Asian sun bears (*Helarctos malayanus*), Sunda clouded leopards (*Neofelis diardi*), Bornean agile gibbons (*Hylobates albibarbis*), Sunda pangolins (*Manis javanica*), and proboscis monkeys (*Nasalis larvatus*), among other endangered and keystone species. The Project also provides a critical buffer for the Tanjung Puting National Park which borders the Project to the West.

Similarly, much of the accessible, high-grade timber is gone. For instance, meranti (*Shorea* spp.) and Bornean ironwood (*Eusideroxylon zwageri*) are both locally extinct.[68]

How Problem Was Identified

Many of the problems were identified by InfiniteEARTH and the Orangutan Foundation International (OFI). OFI, led by the world famous Dr. Biruté Mary Galdikas, is responsible for Tanjung Puting National Park being designated as a national park. Originally, the area was classified as production forest, and over the years, OFI—through sheer perseverance—was able to get the area reclassified as a national park.[69]

Dr. Biruté Mary Galdikas had, and still does to this day have, a very strong presence in the Park, and this sphere of influence extended into the Park's buffer zone, which now includes the Rimba Raya Biodiversity Reserve REDD Project. More specifically, OFI's patrols were being conducted on a regular basis, and the team identified the illegal deforestation and was also working on classifying species throughout the region.[70]

Furthermore, InfiniteEARTH and its partners installed thousands of test plot measurements throughout the Reserve and engaged the local communities. Through community engagement, InfiniteEARTH was able to identify opportunities for community enterprises by learning from the communities what their skill sets were and then taking those skills and seeing how to build an enterprise from there.

Effectiveness of Process for Identifying Problem

InfiniteEARTH, in conjunction with OFI, was very effective at identifying the problems. InfiniteEARTH initially focused on identifying the root causes and drivers of deforestation. The team did not make any assumptions as to what the drivers were, based on either micro or globally preconceived notions. As a result of this process, InfiniteEARTH soon learned that there were a unique set of circumstances for the communities living in the North, Central, and South regions of the Project. For instance, the communities in the North region were more in support of palm oil plantations because of the direct benefits (i.e., employment) the communities received from these plantations. In contrast, the communities in the South region were far more remote, had fewer overall resources, and were in a more desperate situation. The communities in the Central region were in a mixed situation, where they received some benefits from the palm oil plantations (i.e., although not as much as the communities in the North region), but were also in a fairly desperate situation (i.e., although not as much as the communities in the South region). The communities in the Central region were/are much more open to InfiniteEARTH because they were not as dependent on palm oil, preferred their traditional way of life, and were not so desperate that they had to resort to extreme measures to survive.[71]

It is important to note that at the end of the day, most of the ideas for local, sustainable enterprises came directly from the communities.

Steps Taken to Address the Problem

There were numerous steps taken by Todd Lemons and InfiniteEARTH. This includes Todd Lemons founded InfiniteEARTH as a Hong Kong-based company and subsequently founded PT Rimba Raya Conservation as a local Indonesian company. PT Rimba Raya Conservation relatively soon thereafter obtained a 60-year concession to operate the Rimba Raya Biodiversity Reserve REDD Project.[72]

Todd Lemons formed a strategic partnership with Dr. Biruté Mary Galdikas and the Orangutan Foundation International (OFI). With

OFI's assistance, InfiniteEARTH undertook extensive community engagements, and due to the unique circumstances of the 3 main communities in the Project, InfiniteEARTH hired a unit manager for each of the 3 sections.[73]

To mitigate the problems associated with illegal expansion into the Project Area, InfiniteEARTH installed cement markers every 500 meters along the Northern border of the Project, which the neighboring communities regularly patrol to help keep palm oil incursions at bay.[74]

In conjunction with the aforementioned activities, InfiniteEARTH developed the first REDD+ accounting methodology and simultaneously designed, financed, and developed a VCS and CCBS validated and verified REDD+ project. Furthermore, InfiniteEARTH was able to conclude a few sales with large companies including Gazprom Marketing & Trading[75] and Allianz.[76]

Results

The Rimba Raya Biodiversity Reserve REDD Project has thus far achieved numerous results. Such results include, but are not limited to:

- Continuing to protect upward of 64,500 hectares (159,315 acres) from deforestation and degradation.
- Approximately 10.26 million Verified Carbon Units (VCUs; i.e., carbon offset credits) have been issued as of March 21, 2017.[77]
- The Project was validated to the VCS in August 2011 and to the CCBS in October 2011.[78]
- The Project then achieved its first VCS verification in May 2013 and its first CCBS verification in January 2014.[79]

Another major result and the single goal for the Project, according to Todd Lemons, was to change the communities' relationship to the forest—which historically was a purely extractive relationship—to become a more productive relationship. To this end, numerous community projects have been developed to date. A local women's cooperative was established to raise egg-laying hens. This community now has approximately

200 hens and the women either trade or sell the eggs to neighboring communities. Another women's cooperative was established by the communities in the South region which now produces shrimp paste. InfiniteEARTH helped the cooperative get in business, to acquire the necessary equipment and proper certifications, and to design sanitary packaging, and now, the cooperative is selling the fish paste in major Indonesian cities such as Bali and Jakarta. In addition, there are also community projects that weave and carve handicrafts, along with local community fish farms.[80]

While, for instance, the egg farm was not directly related to conservation, the communities received compensation for not cutting down the forest. The Project did not take anything away from these communities and did not penalize the communities. Rather, through a results-based or performance-based financing structure, the money to expand community enterprises comes from conservation. Word of this concept traveled up and down the river and now, in aggregate, InfiniteEARTH is starting to see changes in the behavior in all the communities.

Another result of the project has been the planting of approximately 400,000 native, cash crop trees including the reintroduction of the gembor tree (*Alseodaphne* spp.). The gembor tree produces an essential oil that can be used as an insect repellent. The communities were previously cutting down the tree. However, most of the oil is in the tree's bark, and InfiniteEARTH hired a local expert to teach the communities how to sustainably extract the oils from the tree's bark without cutting down the entire tree.[81]

Challenges and How They Were Met

The two main challenges that Todd Lemons faced were the palm oil plantation "PT Best, which would use its economic and political muscle to try to block the project any way it could. The other challenge came from a source he hadn't anticipated: old-school environmentalists, that opposed the market based approach to conservation."[82]

Additional challenges associated with the Project involved launching community enterprises. For instance, the first attempt at raising chickens resulted in lots of the chickens dying because the chickens were not inoculated. Fortunately, the second batch of inoculated chickens worked out well. Similarly, the fish farms were codesigned by a local university that originally selected a rather delicate species that was very difficult to raise. The second fish species chosen was an exotic species. Finally, the third attempt resulted in the identification of a suitable, native fish species to farm raise, but then there were issues associated with obtaining fish food. This lack of adequate fish food then led to the funding of another community project that involved raising natural fish food in the form of local insects.[83]

Beyond Results

These results thus far appear to be very sustainable. This said, InfiniteEARTH never wanted the communities to be solely dependent on the sale of carbon offset credits, and this is why the Project's focus has always been on establishing locally designed and locally operated enterprises.[84]

However at the very moment, none of these local enterprises will sustain the communities by themselves. This said, the communities still rely on the neighboring palm oil plantations for employment. The communities like having access to the forest and the communities' choice is to keep their traditional way of life. Their work for the palm oil plantations is out of necessity. Yet, it appears as though the communities would rather work for their microenterprises, and for this reason, the communities will likely continue to grow their microenterprises.[85]

Lessons Learned

There are several important lessons learned by Todd Lemons. The first is that, ideally, project developers would secure a corporate sponsor at the very beginning. This would allow project developers to focus on the implementation of project activities and for the corporate sponsor to leverage a higher return on their investment.[86]

The second lesson learned is that there are enormous opportunities for companies to leverage REDD+ more than they currently do.

InfiniteEARTH has sent high-quality photographs and videos, and regularly proposes new and interesting campaigns for companies to reach their customers and employees. Yet, still to this today, and even with their own customers and employees, these companies often do not leverage their support of REDD+. In contrast, these companies should, and could, build brand awareness around their support rather than seeing the support as a line item cost and a footnote in their annual report. To achieve this more intimate relationship, project developers need to connect with the companies' human resources, marketing departments, supply chain managers, and the C-suite.[87]

Third, palm oil is not going away because there is no good substitute. Rather, palm oil—and other commodities—needs to reflect proper pricing by incorporating the price of externalities into products.[88]

Other Resources on Rimba Raya Project

Code REDD's Page on Rimba Raya, Indonesia

- http://www.coderedd.org/redd-project/infiniteearth-rimba-raya-indonesia/#.WKuIrH9WfeQ

InfiniteEARTH's Rimba Raya Biodiversity Reserve Page

- http://infinite-earth.com/rimba-raya-biodiversity-reserve/

Markit Environmental Registry

- https://mer.markit.com/br-reg/public/project.jsp?project_id=100000000000762

Natural Capital Alliance

- http://www.naturalcapitalalliance.com/

Orangutan Foundation International

- https://orangutan.org

Rimba Raya Biodiversity Reserve's Official Page

- http://rimba-raya.com/

The Road to Hell is Paved with Abundance

- https://www.linkedin.com/pulse/road-hell-paved-abundance-todd-lemons

Verified Carbon Standard's Project Database

- http://www.vcsprojectdatabase.org/#/project_details/674

Case Study #3: The Kasigau Wildlife Corridor REDD Project in Kenya

Introduction

On February 3, 2011, the Kasigau Wildlife Corridor REDD Project—Phase I Rukinga Sanctuary became the first Verified Carbon Standard (VCS, then known as the Voluntary Carbon Standard) REDD+ project to complete verification and issuance.[89]

Although the Project is protecting a dryland Acacia-Commiphora forest, and thus not a tropical rainforest, the Project still serves as a valuable case study. The Project is located in the county of Taita Taveta, Kenya, between Tsavo West and Tsavo East National Parks and is being managed by Wildlife Works Carbon, LLC (Fig. 11.3).

The initial Project Area is approximately 30,169 hectares (approximately 74,517 acres) and aims to reduce 7.5 million metric tonnes of

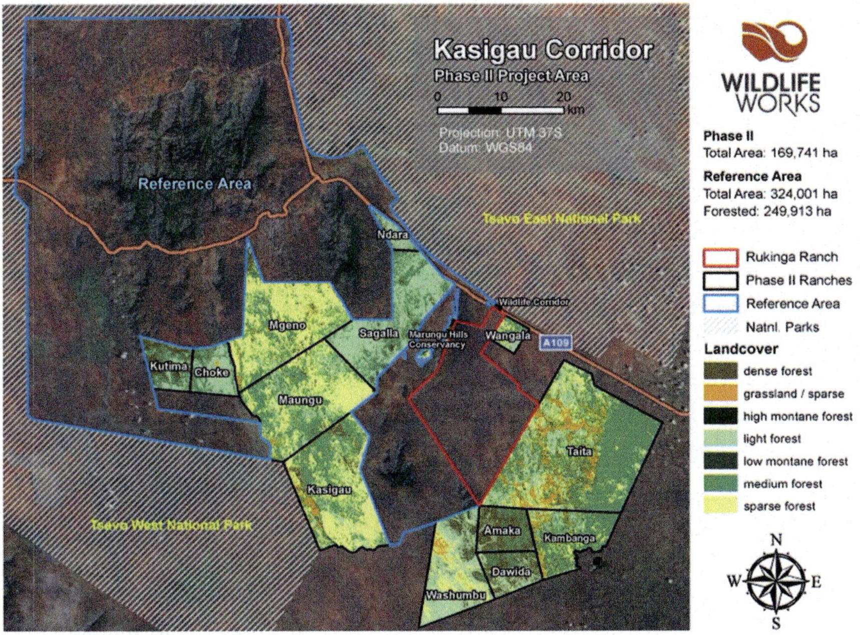

Fig. 11.3 Map of the Kasigau Wildlife Corridor REDD Project (Credit: Wildlife Works)

carbon dioxide equivalent ($mtCO_2e$) emissions over the first 30 years of the Project Lifetime.[90]

As a verified REDD+ project, the Project generates 95% of its revenue from the sale of carbon offsets. The other activities at the Project—including ecotourism and an onsite organic cotton textile facility—generate approximately 5% of the overall revenue. Yet, it is important to note these alternative activities are far more important than 5%, because these activities generate good jobs, are good for the psyche of the communities, and provide educational opportunities for outsiders.

Identify the Problem

Prior to the initiation of the Project, there were few problems. This lack of problems was due to the fact that the Project Area was not very inhabited. The Taita People previously inhabited the area, but were evicted by the British during World War I. The Taita People began to run out of room on the Taita Hills and moved back into the area during the 1960s and 1970s. Upon coming back into the lowlands, including within the now Project Area, the Taita People practiced slash-and-burn agriculture. There was also subsistence poaching for bushmeat.

The main problem prior to the initiation of the Project was deforestation by slash-and-burn agriculture. Due to the lack of water, past economic activities—including cattle ranching and ecotourism—have failed. Large-scale afforestation activities and/or commercial agriculture also do not appear to be viable due to the general lack of water.[91] The longer-term threat would have been cattle ranches. Cattle ranches would have been fine in the short-term, but there are long-term problems associated with cattle because cattle eat all the grass and saplings while also compacting the soil which prevents seeding. Over time, the trees are not replaced and the forests are converted to grasslands.[92]

Today, slash-and-burn agriculture, poaching, and charcoal production are the problems throughout the region. This said, both food security and water security are everyone's first, and second, thoughts.

The short-term threat to wildlife is subsistence poaching. Growing crops, such as maize, is a risky proposition due to the sporadic rainfall. There have been years when 6 out of 22 harvests fail, and when harvests fail, the communities use snares to catch and eat bushmeat. There is also a problem with commercial poaching, particularly for elephants and their tusks. Unfortunately, the poaching of elephants has come back and in 2010 is when the poaching started to get bad again.[93] As recently as March 2017, a large 50-year-old, so-called "giant tusker" elephant—1 of less than 30 such "giant tuskers" left in the world—was reportedly poached in neighboring Tsavo National Park.[94]

Such poaching is driven, in large part, due to the high demand and subsequently high prices paid for ivory. It is estimated that the ivory from elephant tusks can fetch USD$1,500 per pound and "horns from endangered rhinoceros bring up to USD$25,000 per pound in China."[95] While these prices are not paid directly to the poachers (i.e., there are several layers of middlemen), there is still enough money paid to poachers relative to their other economic opportunities.

Charcoal production in East Africa, including Kenya, is another huge problem. Charcoal is a very efficient way to produce a transportable fuel. Essentially when producing charcoal, you can burn away three-fourths of the wood's weight and then you only need to transport one-fourth of the weight, and the remaining charcoal still retains 80% of its fuel value. The cities throughout East Africa have run out of trees, and there is basically a concentric circle around the cities of disappearing forests as people are going further and further out of the cities to access forests for charcoal.[96] This said, the Project is approximately 120 kilometers from Mombasa, which has a high charcoal demand due to its population of more than 1 million people. The local communities near the Project Area do not really use a lot of charcoal, because if you can use the trees—especially if you do not have to carry the tree too far—then you can utilize 100% of the fuel value.[97]

Social and political unrest in Kenya, along with incursions from neighboring countries such as Somalia and the Al-Qaeda affiliated Al-Shabaab terrorist organization, propose yet more problems in the region. Furthermore, additional problems throughout East Africa include droughts, false promises by politicians during election years, and corruption.

Why the Problem Is Important

Deforestation as a result of slash-and-burn agriculture is a problem because of the Project's tremendous biodiversity and the poverty situation facing the local communities. The Project provides habitat to over 20 species of bats and over 50 species of large mammals including many IUCN Red Listed species such as Grevy's zebra (*Equus grevyi*), cheetah (*Acinonyx jubatus*), African lion (*Panthera leo*), and African elephants (*Loxodonta africana*). Furthermore, more than 300 species of birds have been identified.[98]

In addition, the Project establishes an important ecological corridor between Tsavo West and Tsavo East National Parks. The Tsavo ecosystem has Kenya's largest elephant populations, and as a result of other countries doing a worse job, Kenya is starting to have a relatively higher population of elephants.[99]

Charcoal is a major fuel source for local communities, and without charcoal, the communities would have to seek alternative fuel sources. Similarly, the problem of deforestation and degradation as a result of charcoal production is important because it can take hundreds of years for trees to grow in Kenya due to the arid environment.[100]

Poaching of elephants, and the related violence, impacts tourism. Kenya generated USD$785 million in 1995 from international tourism, and 20 years later, in 2014, Kenya generated USD$1.8 billion.[101] The poaching of elephants is a problem, in part, because the land is very open and very flat which makes for easy access by poachers.

How Problem Was Identified

Initially the problems facing the Kasigau Wildlife Corridor REDD Project—Phase I Rukinga Sanctuary—particularly human-wildlife conflicts—were identified by Mr. Mike Korchinsky, the Founder of Wildlife Works Carbon, during a vacation to Kenya.[102] Ongoing community engagement and a local presence have allowed Wildlife Works Carbon to further identify existing problems such as poaching and charcoal production. In addition, land use change is observed through satellite imagery.[103]

Effectiveness of Process for Identifying Problem

The process for identifying problems has been very effective. As explained by Mr. Korchinsky, the farther away you get from the land, the less obvious the root problems are to observers and practitioners. In contrast, if you live at the project site, the problems and threats are right in front of your face. This is precisely why Wildlife Works Carbon went straight to the communities and established an office in the bush, rather than far away in Nairobi or Mombasa.[104]

Steps Taken to Address the Problem

There are numerous steps taken over the last 20 years to address the underlying problems. Such steps include:

- Mike Korchinsky created Wildlife Works in 1997.
- Wildlife Works Carbon engaged local communities, government officials in Kenya, and international investors.
- An organic cotton textile facility was built.
- A new Verified Carbon Standard methodology, VM0009: Methodology for Avoided Ecosystem Conversion, was created.[105]
- Wildlife Works designed, financed, and now implements a VCS and CCBS validated and verified REDD+ project.
- Ongoing review of satellite imagery is conducted to assess land use change.

To address the problem of poaching, particularly elephant poaching, Wildlife Works Carbon works closely with the Kenya Wildlife Service (KWS) and occasionally provides resources (e.g., fuel) for KWS patrols. KWS has national responsibility for wildlife and is permanently assigned to the corridor between Tsavo East and Tsavo West National Parks. Oftentimes, KWS guards are colocated with non-armed community guards from the Project. Wildlife Works Carbon also collaborates with international NGOs such as Save the Elephants, which has done a great job of receiving donations to address the recent spike in elephant poaching.[106] Using bloodhounds, or another tracking dog such as Malinois, to find poachers is difficult because such dogs struggle to survive in the African bush due to the heat, parasites, poisonous snakes, and leopards. In addition, the dogs cannot run very far before overheating in the midday sun. Furthermore, the dogs—especially in Kenya—can cost more than USD$20,000 each, and this does not include ongoing costs such as food, housing, and training. Yet, the Kenya Wildlife Service (KWS) does have a few tracking dogs due to the escalation of poaching in Kenya which proves the dogs may work, despite the high demand and costs for the dogs.[107] Furthermore, KWS conducts aerial surveillance with gyrocopters.

Charcoal is still an issue in the area, but less so in the Project. Because the canopy is not closed nor is the canopy very high, it is easy to put a ranger on a hill to keep their eyes open in the morning for smoke rising above the canopy from charcoal production. These days if people are caught producing charcoal, they are not arrested; rather their charcoal gets destroyed. If you let people destroy the forest for charcoal, there will be no more trees and no more charcoal, and then the price of charcoal will go up, which will spur more charcoal production elsewhere. Wildlife Works Carbon is addressing this problem through a sustainable charcoal program. The Program involves harvesting the excess branches of trees to make charcoal briquettes. The trees, which only need 25% of their canopy to survive, are left intact.[108]

The local communities within the Project Area agreed to stop slash-and-burn agriculture. Nowadays, the communities are an integral part of the REDD+ project and help to police the area for illegal deforestation and poaching. In return, the communities receive training and funding for health clinics, school buildings, and scholarships.[109]

Results

The Project has accomplished many industry-leading results. This includes, the Project:

- Became the first verified VCS REDD+ project in February 2011.
- Is in the process of reducing 7.5 million $mtCO_2e$ over the first 30 years of the Project. As of March 3, 2017, approximately 1.57 million $mtCO_2e$ have been verified and issued.[110]
- Helped spur the adjacent Kasigau Corridor REDD Project—Phase II, The Community Ranches, the relatively close-by Chyulu Hills REDD+ Project (which received financing from the Althelia Climate Fund, see case study in Chap. 10), and the Mai Ndombe REDD+ Project in the Democratic Republic of the Congo. Collectively, these four projects are protecting up to 859,400 hectares (approximately 2.1 million acres).
- Built an organic cotton textile facility.[111]
- Became the first forest carbon project to sign an agreement to supply offsets to the International Finance Corporation's Forests Bond with REDD+ Coupon Payments (see case study in Chap. 16).[112]
- Contributed to the production of Ivory Wars, a multi-part television series broadcast on Animal Planet and Discovery Channel.[113]
- In April 2017, Wildlife Works founded Everland, a specialized marketing company focused on selling forest carbon offsets from global REDD+ projects[114]

Challenges and How They Were Met

One of the initial challenges for Wildlife Works Carbon, during the pre-REDD+ days, was to try and imagine a new economic model that could replace the economic system (e.g., slash-and-burn agriculture and charcoal production) that was at the time being followed by communities. The new economic model had to be appealing to the community, had to create good local jobs, and had to promote wildlife conservation.

Two challenges facing Wildlife Works Carbon today are leakage and opportunity costs. The Project Area is the only patch of woodland forest remaining between Mombasa and Nairobi. By protecting this forest, the collection of charcoal leaks, or shifts, to other areas. Wildlife Works Carbon has established a sustainable charcoal briquette program, regularly monitors for charcoal production, and also tracks the price of charcoal to help mitigate this challenge.

With respect to the other challenge of opportunity costs, as economies grow and local communities improve, landowners are constantly reevaluating their opportunities and other land use practices. For instance, Kenya's Gross Domestic Product (GDP) in the year 1995 was USD$9.0 billion, and then, 20 years later, Kenya's GDP in 2015 was USD$63.4 billion.[115] To address this challenge, Wildlife Works works very closely with communities, has established mutual trust and loyalty, and tries to give as much value as possible. This scenario is a challenge around the world. In Indonesia, for example, if someone were to say that you can earn USD$4,000 a year per hectare for palm oil or USD$40 per hectare for forest carbon conservation, the decision is most often palm oil.[116]

Beyond Results

The Project appears to be very sustainable as it has consistently demonstrated—via independent assessments against leading carbon certification standards—net positive climate, community, and biodiversity benefits.

The initial Project has expanded by incentivizing the development of the Kasigau Corridor REDD Project—Phase II, The Community Ranches. This larger project covers 169,741 hectares (i.e., 419,260 acres) and aims to reduce a gross total of 48.8 million mtCO_2e over the project's 30-year lifetime.[117] A third project, known as the Chyulu Hills REDD+ Project, is being undertaken by several project partners, including Wildlife Works Carbon, in relatively close proximity to Kasigau Corridor REDD Project—Phases I and II. The Chyulu Hills REDD+ Project covers 410,534 hectares (i.e., approximately 1.01 million acres) and aims to reduce 37.8 million mtCO2e over the Project's initial 30 years.[118]

Wildlife Works Carbon subsequently entered into a joint agreement with ERA Ecosystem Restoration Associates Inc. to implement the Mai Ndombe REDD+ Project in Northwest Democratic Republic of the Congo. This Project will protect up to 248,956 hectares (nearly 615,000 acres).[119] Furthermore, in June 2016, the provincial REDD+ program design of Mai-Ndombe was conditionally approved by the World Bank's Carbon Fund, making Mai-Ndombe and Costa Rica among the first two jurisdictions that may receive payments for performance via the World Bank's Carbon Fund.[120]

Along with partners Wildlife Alliance and Cambodia's Ministry of the Environment, Wildlife Works is assisting with the development of a REDD+ project in Cambodia's Southern Cardamoms. The Project, "which supports the protection of 465,000 ha {1.15 million acres} of forest and critical Asian elephant {*Elephas maximus*} habitat, is Wildlife Works' first Asia REDD+ project."[121]

Wildlife Works Carbon also created Code REDD to help increase the awareness and support of REDD+ projects. This includes the Stand for Trees campaign, which aims to facilitate smaller donations from a wide range of actors.[122] In addition, Wildlife Works advised the Althelia Ecosphere and spurred the creation of Everland.[123]

Specific to the Kasigau Corridor REDD Project—Phase I, it appears as though long-term financing is in place, as the VCS Project Description states that there are "existing offtake contracts for the life of the project with BNP Paribas and Nedbank, two major international banks."[124]

Furthermore, the Kasigau Corridor REDD Project—Phase I has focused a lot on education, good governance, and economics which helps with the Project's overall sustainability. As Mr. Korchinsky explains, conservation is all about thinking about the future. Hence, education is vitally important to incorporate children into the Project. Good governance helps to ensure good representation by minorities and by those who traditionally have less influence such as women. Increased representation helps empower people to make collective decisions. The economic focus is mainly a focus on jobs. However, you need the whole community to support the Project and not just the people that are employed by the Project; otherwise, those who are not employed by the Project would go deforest. Similarly, jobs are important because if there is a drought, you have money to buy food.[125]

In addition to working with KWS, Wildlife Works Carbon is also expanding their work by helping the Kenya Forest Service (KFS) assess long-term financing opportunities for their national forests.

Lessons Learned

There are several lessons learned from Mike Korchinsky, Founder of Wildlife Works[126]:

- Non-timber, non-forest products such as ecotourism is a nice idea; it is important and can play a vital role, but it is absolutely insufficient at scale.
- There is no one act, and early on, you need a basket of opportunities (i.e., a portfolio of options) in order to solve the underlying problem(s) and to scale the activities to protect 200,000 hectares and to alter a process that people have been economically and culturally connected to over many years.
- You can tie job creation and job earning with a conservation outcome and then try to introduce ways for the communities to be more successful themselves (e.g., finance and help build a nursery that provides fruit and cash crops, or teach agricultural intensification techniques). However, this may only work if you have the right to somehow enforce forest conservation. If there is no active forest conservation with the communities, this idea can backfire as the communities become more productive and then clear more land which results in deforestation.
- Every site is different and you cannot use a cookie cutter approach from one site to the next, whether it is between projects within a country or projects across countries, and certainly not for projects across continents. However, a model needs to have underlying principles and then every

site will share some common characteristics. For instance, the projects of Wildlife Works Carbon in Kenya and the DRC are very different. The project in the DRC is protecting a tropical wet forest, while Kenya is a dry forest. The DRC, unlike Kenya, is facing commercial logging as opposed to slash-and-burn agriculture. The DRC is Francophone and Kenya is Anglophone, while the communities in the DRC are forest dwellers and the communities in Kenya live in the grasslands. However, both project sites—with different designs and approaches—ultimately need food security, incomes, housing, healthcare, and education.

- It is important to understand the emergence of a pay-for-performance model. Wildlife Works Carbon, as a for-profit, was originally not seen as a competitor to environmental nonprofits as Wildlife Works Carbon was not seeking money from the same pot. However, as Wildlife Works Carbon's model of pay-for-performance has evolved, now the model is seen as a threat to the traditional donor model. Pay-for-performance is what private sector REDD+ projects are all about, and this can be frightening to the donor community. Traditionally, the donor model is a "get and give" model. A government will get money from taxes and give to an aid agency. The aid agency will get money from the government and give to NGOs. The NGOs get money from aid agencies and they give money to communities. Everyone takes their piece (e.g., administrative fees and operating costs) along the way, and for years, the model worked fine. There is now a hurdle for governments to give out funds without performance-based targets and accountability. If aid agencies do not give out all their funds, they will have smaller budgets the following year. NGOs, because they are not used to doing it, traditionally did not have firm proof of what they performed. With a pay-for-performance approach, resources are allocated efficiently and practitioners need to understand that some projects will succeed and some will fail, but one can only give funding to those that perform. This can lead to an under-allocation of funding or actors may ignore performance; yet, results really matter and results especially matter for our climate and tropical forests. Ultimately, the world needs a new system, and there needs to be a complete shift among those who have given most of the money for conservation over the years. The marketplace understands this concept of pay-for-performance, yet there is less money in the private sector for conservation as of today. It is also important to note that similar models—known as performance-based financing (PBF) and results-based financing (RBF)—are being adapted to other sectors including the international development health sector.

Other Resources on Kasigau Corridor REDD Project—Phase I

Code REDD's Page on Kasigau Corridor, Kenya

- http://www.coderedd.org/redd-project/wildlife-works-carbon-kasigau-corridor/#.WKuIvn9WfeQ

Elephant Protection Trust

- https://www.elephantprotectiontrust.org/

Everland

- http://www.everlandmarketing.com/

Markit Environmental Registry

- https://mer.markit.com/br-reg/public/project.jsp?project_id=100000000000585

Verified Carbon Standard's Project Database: Chyulu Hills REDD+ Project

- http://www.vcsprojectdatabase.org/#/project_details/1408

Verified Carbon Standard's Project Database: The Kasigau Corridor REDD Project—Phase I Rukinga Sanctuary

- http://www.vcsprojectdatabase.org/#/project_details/562

Verified Carbon Standard's Project Database: The Kasigau Corridor REDD Project—Phase II: The Community Ranches

- http://www.vcsprojectdatabase.org/#/project_details/612

Verified Carbon Standard's Project Database: The Mai Ndombe REDD+ Project

- http://www.vcsprojectdatabase.org/#/project_details/934

Wildlife Works' Press Room

- http://www.wildlifeworks.com/company/pressroom.php

Case Study #4: Rio Tinto QMM Biodiversity Offset Project in Madagascar

Introduction

Approximately 10 years ago, the global mining company Rio Tinto purchased the QMM (QIT Madagascar Minerals) Project from the previous owner, a Canadian mining company formerly known as Quebec Iron & Titanium (QIT). The QMM Project is majority owned by Rio Tinto (80%), and a minority stake (20%) is owned by the Government of Madagascar. The project will impact approximately 6,000 hectares (approximately 14,820 acres).[127] Rio Tinto, to reduce the impact at the QMM Project, has followed the mitigation hierarchy to avoid impacts on endangered species and threatened habitats, then minimize any impacts if avoidance is impossible, followed by rehabilitating or restoring degraded habitats, and then using biodiversity offsets if necessary.[128]

As part of minimizing impacts, QMM excluded significant mineral deposits from the area to be mined, which are now legally recognized, protected areas. With respect to biodiversity offsetting, the gold standard is to use a like-for-like approach. For instance, if 10 hectares of littoral forests will be impacted by mining operations, then at least 10 hectares of littoral forest will be preserved. This preserved area should be of the same quality, but if that is not possible, a discount rate can be applied by incorporating, among other things, habitat quality and a risk factor attributable to the viability of the needed conservation activities, including the ability of the implementing organizations to successfully carry out these activities. Because only a limited area of littoral forest remains in Southeastern Madagascar, the QMM Biodiversity Offset Project—as part of the overall QMM Biodiversity Action Plan—included a significant area with different ecosystems located in the region.[129]

Identify the Problem

The problem in Southeastern Madagascar is deforestation, in part due to mining activities and infrastructure development and due to poverty. As explained by the International Union for Conservation of Nature (IUCN) and Rio Tinto:

> The Rio Tinto QMM project will mine ilmenite ore as a mineral sand, to provide titanium dioxide to the world market. Titanium is a major ingredient in steels and other alloys; titanium dioxide is the white pigment found in most paints and plastics. The Rio Tinto QMM project consists of three sites to be mined sequentially (Mandena, Ste Luce and Petriky) over a period of c.40–50 years. A new deepwater port has been constructed at Fort Dauphin. Ancillary infrastructure includes a dedicated port industrial zone, road networks, housing areas and a stone quarry. Ilmenite is mined using a dredge situated on artificial ponds that moves across the ore body as mining progresses; the mining process entails the removal of all vegetation cover along with the soil. Approximately 100 ha of the deposit will be mined each year; the mine itself occupies about 50 ha of land as it progresses slowly through the deposit area. Rehabilitation of the mined area will be carried out once the dredge has moved on to the next part of the deposit.[130]

With this in mind, Madagascar is a widely recognized biological hotspot with a tremendous number of endemic species and highly threatened ecosystems. However, much of the original forest cover has been lost, and, for instance, only 10% of the original littoral forest remains, much of it highly degraded. Littoral forests are a unique ecosystem comprising humid tropical forests on white sand and are restricted to a narrow belt along the island's east coast (Fig. 11.4).

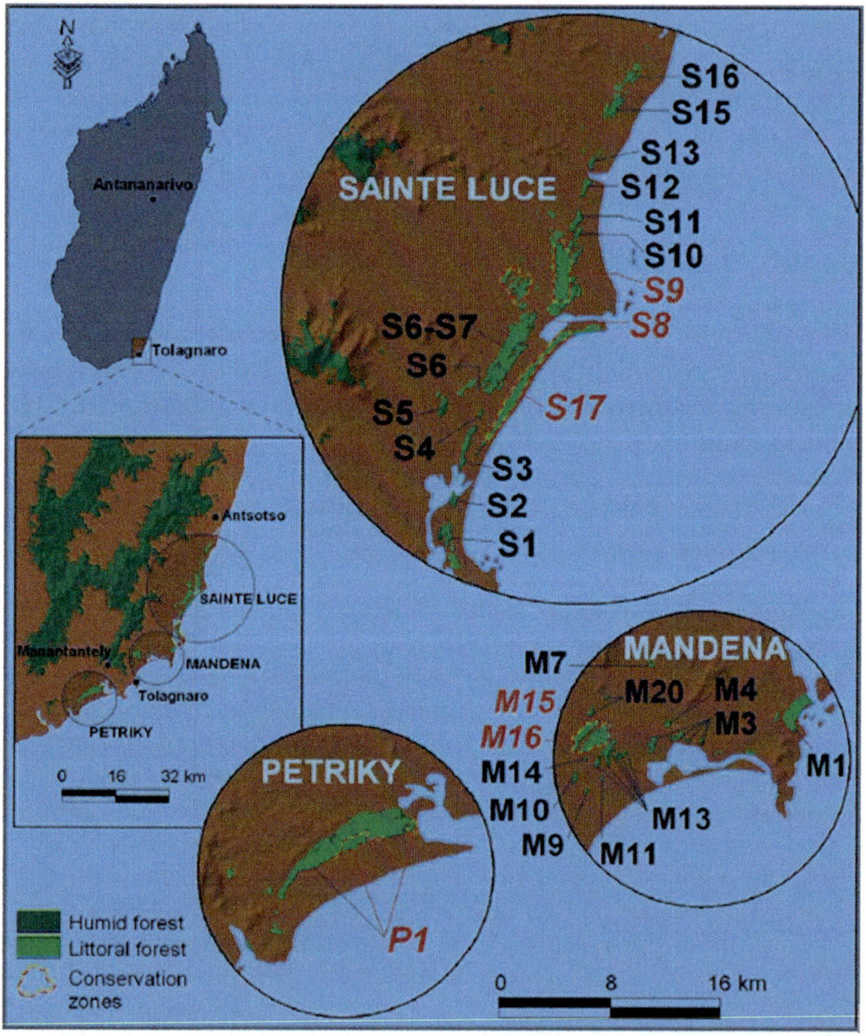

Fig. 11.4 Map of remaining littoral forests in Madagascar (Credit: Dr. Pete Lowry)

The small area of remaining littoral forests is fragmented and threatened by deforestation. Likewise, as of a presentation by Matthew Bateson of Rio Tinto in November 2013,[131] it is estimated that 3.5% of this remaining littoral forest will be lost due to mining activities.[132]

Why the Problem Is Important

The problem of deforestation in Southeastern Madagascar is important because infrastructure projects—including mining operations—can have a significant, detrimental impact on biodiversity. Soon after Madagascar started politically opening back up to Western nations in the mid-1980s, QIT entered into an agreement with the Government of Madagascar to explore potential mining operations. Simultaneously, several major conservation organizations began operating in Madagascar and awareness grew about the highly threatened ecosystems and endemic species on the island. The Government of Madagascar was also becoming aware of the issue and of the need for conservation.[133] In fact, Norman Myers et al. have recognized Madagascar as 1 of 8 "hottest hotspots" with respect to biodiversity.[134]

How Problem Was Identified

The problems of mining and deforestation, particularly by the QMM Project, were identified by the conservation community in Madagascar, which was quite vocal. In addition, the operator of the Project at the time, QIT, was under pressure from shareholders and others and thus concluded that they had to do more than just pay lip service to biodiversity conservation and minimizing the impacts of the proposed mine. The Missouri Botanical Garden (MBG) had conducted flora vegetation baseline studies in the late 1980s and early 1990s for QIT and found the issues at stake were far more significant than previously thought.[135] MBG realized that littoral forests were distinct ecosystem and that they contained dozens of rare plant species, including "40 taxa recorded within the QMM mining area (Sainte Luce, Mandena and Petriky) that have at one time or another been regarded as Priority 1 or Priority 2 taxa" for plant conservation.[136]

Effectiveness of Process for Identifying Problem

For the most part, the process of identifying the problems of mining's impact on Madagascar's forests and biodiversity was effective. For instance, it took longer for QIT, and then Rio Tinto, to understand the dynamics of the local communities living nearby with respect to deforestation and biodiversity offsetting. In Madagascar, one cannot effectively manage forests and conserve biodiversity without involving communities. While these communities are responsible for most of Madagascar's deforestation as they try to grow food to feed their families, they also are a key part of the solution. The rural communities near the QMM Project, like everywhere in Madagascar, are poor and are in need of alternative, sustainable economic development in order to be able to change the way they use the land.[137]

In addition, the buy-in process for the QMM Biodiversity Action Plan, including the QMM Biodiversity Offsets Project, involved dozens of stakeholders at all levels including the regional and national governments of Madagascar, conservation organizations, business elites, and the local communities. At any given time, some of these stakeholders were actively participating in the process, others were skeptical but engaged, while others still remained hostile and opposed to the Project.[138]

One of the advantages working with a mining company, and what can make for an effective process, is that it will be in operation at a given site for a long time with a significant upfront investment. Thus, you can have constructive conversations with the company about medium-term and long-term impacts, and it makes it possible to envision using the mitigation hierarchy and biodiversity offsets. Similarly, if, for example, the local communities are up in arms about the company's operations or the government says the company violated its environmental permits, the problems will have to be addressed. If the company then attempts to ignore the problem, or get out of it by using political influence or bribery, they face the possibility of losing their license to operate, which in turn could lead to the subsequent loss of their investment. In contrast, a forestry company operating a short-term, one-rotation concession is much less likely to be concerned with long-term impacts.[139]

Steps Taken to Address the Problem

As summarized by Matthew Bateson, the then Global Practice Leader of Environment and Climate Change at Rio Tinto, the following general steps, which comprise the mitigation hierarchy, were taken to address the problem:

1. Establishment of avoidance zones;
2. Minimizing unavoidable impacts;
3. Rehabilitation and restoration; and
4. Biodiversity offsets.[140]

Results

There are several notable results from the QMM Biodiversity Action Plan and the QMM Biodiversity Offsets Project. First, 3 protected areas were established, all of which are now officially protected areas recognized by the Government of Madagascar.[141]

QMM invested a lot of time considering what a no-net-loss or net-positive-impact policy would look like in practice. By thinking through the process of what this might mean, the company came to the realization that there would be an unavoidable, residual impact from its operations, even after avoidance, minimization, and restoration. The next conversation was how to make good decisions, particularly with respect to what type of biodiversity offsetting the company should pursue and how to quantify the gains relative to the losses resulting from the residual impacts of the mine. This was a positive process, one that had not historically taken place with the mining sector and certainly not with mining operations in tropical countries. Likewise, the QMM site was one of only a few locations around the world where this type of conversation was taking place.[142]

Furthermore, the identification of the Mahabo site (i.e., the only substantial remaining area of littoral forest within several hundred kilometers of the QMM Project) as a suitable like-for-like offset (indeed, the only one) led to a commitment to provide financing for the Missouri Botanical Garden's community-based conservation work there.[143]

Challenges and How They Were Met

As summarized by Matthew Bateson, the following challenges were faced at the QMM Biodiversity Action Plan and the Biodiversity Offsets Project:

- "Implementation
  - Implementation challenges included additionality, community engagement, measurement and metrics, long-term management, and financial security.
- Costs
  - Cost challenges included the multiple overlay of legislation, the ratios and multipliers, the perverse applications and use of Key Biodiversity Areas (KBAs), and the transaction costs.
- Integration
  - Integration challenges included combining financial lender requirements with national frameworks.
- Severance of liability
  - The severance of liability challenge mainly related to placing a cap on liability under appropriate legal frameworks."[144]

Beyond Results

There are significant questions about what parts of the QMM Biodiversity Action Plan—and the Biodiversity Offsets Project component—are sustainable and for how long.

For instance, while QMM remains committed to a no-net-loss policy, funding has not been promised beyond 5 years. Thus, there is no guarantee that the QMM Biodiversity Offsets Project will be sustainable beyond that time unless the case can be made that it remains necessary to maintain the company's license to operate.[145]

There is also a risk to the QMM Project's sustainability due to the low global price of minerals, including the price of ilmenite ore. With a low market price for ilmenite, it makes for a difficult sell to continue the QMM Biodiversity Offsets Project since offsetting is not directly related to the company's main activities (i.e., getting the ore out of the ground). However, the QMM president is committed and proactive, and if stakeholder pressure continues to hold the company responsible, the QMM Biodiversity Offsets Project could be sustainable.[146]

The activities of the MBG at the Mahabo site will likely enable the littoral forest there to regenerate, and thus, there is a reasonable chance of success, in part due to reduced pressure on the forest. MBG has been active at the location for at least 12 years and is committed to long-term engagement, as opposed to the traditional 3–4-year funding cycle of most conservation and development projects. In contrast, the 2 littoral forest offset sites located adjacent to the mine are a different story. The mining operators have been there for approximately 30 years (e.g., 15 years of mineral exploration, 7–8 years of construction, and 7–8 years of operations), and there is a lot of pressure on these forests from local communities, in part because there are more settlements in the area, there are roads, and most of the forest that occurred in the area has been cleared for charcoal production.[147]

Furthermore, after closure of the QMM Project, the company will no longer be extracting ore. There is a legal obligation for project shutdown and a requirement for site cleanup (e.g., the site must be left in a certain way with restored and protected areas), but no commitment has been made about how to finance the offset sites post closure. Such financing could take the form of a trust fund, but the capitalization of a trust fund takes a lot of money upfront. This said, if the price of ore goes up and the QMM Project becomes profitable, then it might make sense for the company to establish a trust fund to reduce the managerial oversight from the company, except for management of the trust fund.[148]

Lessons Learned

Some of the lessons learned by Dr. Pete Lowry, Senior Curator and Director of the Africa & Madagascar Program at the Missouri Botanical Garden (MBG), are:

A lot can be gained for science, conservation, and local communities by engaging with a mining company and working with them to explain the context of where they are operating, the stakes involved, and how to adopt a responsible stance. Similarly, there is a lot to lose if this is not done and no action is taken.

Some might say the MBG is sleeping with the devil. The reality is that people all around the world use titanium in white paint, cell phones, computers, and many other things. Unfortunately, resource extraction will continue on our planet as long as we buy these products. This resource extraction can have negative consequences and be done with total disregard for the local communities and social and environmental impacts. Alternatively, this resource extraction can be done in a way that acknowledges and minimizes the impacts. There is good to be done with engaging those companies that chose to act in a responsible way, but there is a fairly narrow sweet spot with only a few players seeing the strategic value of limiting their impact and supporting the mitigation hierarchy, as opposed to just doing what is strictly, legally required.

MBG is a science and conservation organization—not a consulting company—with a mandate to conduct and promote science and conservation. There are some situations, such as the QMM Project, in which MBG involvement clearly enables them to generate scientific knowledge, contribute to learning about the flora of Madagascar, and help achieve conservation. Funding science is becoming more difficult and complicated to obtain, but scientists can play a useful role in other ways. For instance, the development of a specific palm oil project involves clearing grasslands in Gabon. Under most circumstances, MBG would probably not partner with such a company. However, Gabon required the palm oil concession to have large set-asides where no clearing will take place. The company's initial approach was that forests are important for conservation and that savannas are wastelands. However, this is a huge oversimplification. Logged forests might have comparatively low biodiversity value, while some savannas have very rich biodiversity. MBG helped the palm oil company understand that the situation was more nuanced, with heterogeneous (not homogenous) forests and savannas of varying conservation importance, and that proper management of the area should aim to conserve high biodiversity forests and savannas. This evolution toward better practice is being made largely because MBG stepped up and made a clear and compelling case. If MBG did not engage the company, they would have continued clearing savannas and leaving forests, with no regard to their conservation value of each individual area.[149]

Other Resources on Rio Tinto's QMM Biodiversity Offsets Project

Madagascar Conservation & Development Journal

- http://journalmcd.com/index.php/mcd/index

Missouri Botanical Garden's Madagascar Page

- http://www.missouribotanicalgarden.org/plant-science/plant-science/africa/madagascar.aspx

Rio Tinto's Biodiversity Page

- http://www.riotinto.com/energyandminerals/biodiversity-15520.aspx

Return

Compensation payments—whether it is supporting forest carbon offsets, watershed protection, or biodiversity offsets—may be voluntary or mandated by law.

Although not related to a tropical rainforest project,

> Other studies have estimated the economic value of recreational opportunities and drinking water supply as forest ecosystem services in Chile, using methods such as travel cost and production function, respectively. These studies estimated a value of \$1.60 to \$6.30 per hectare per year for recreation opportunities, when all the area of a national park is considered. When only the intensively used areas of the national park are considered, the economic value is in the range of \$35 to \$178 per hectare per year, which compares favorably to operating costs of \$1.50 to \$9.30 required to maintain relevant ecosystem services in the area. The value of drinking water supply as an ecosystem service was estimated as \$61 to \$162 per hectare per year for a watershed that provides water to Valdivia City.[150]

With respect to forest carbon projects, the average price for transactions in 2015 of voluntary forest carbon offsets was approximately

USD\$3.30 per offset with a range from under USD\$1.00 per offset to more than USD\$8.00 per offset. This wide price range depends on the perceived quality of the project; the quantity of offsets purchased; whether the transaction was to an end user or to an intermediary; and based on other transaction-specific details (e.g., vintage of offsets, project size, and project location).[151] In 2015, the average price of a REDD+ offset was slightly higher at USD\$3.40.[152] In contrast:

> In 2015, Ecosystem Marketplace tracked 6.5 MtCO_2e in California compliance demand for forestry tonnes at a total value of \$63.2 M. This represents a 6% increase in volume over 2014 demand and a 16% increase in value as the average price for forestry offsets on the California-Québec cap-and-trade program rose from \$8.9/tonne in 2014 to \$9.7/tonne last year. In purchasing offsets, entities can choose from different offset types; however, forestry-based offsets have dominated California issuances. The tracked demand, however, represents only a fraction of the potential demand for offsets in California. Compliance entities may use offsets to cover up to 8% of their cap. Last year's cap topped 394.5 MtCO_2e, setting potential demand for offsets at more than 30 MtCO_2e.

Thus, compliance grade offsets, particularly compliance grade forest carbon offsets, fetch a significantly higher price than their voluntary forest carbon offset counterpart.

Risk

There are a variety of risks facing payments for ecosystem services, most notably policy risk, overlapping land title claims, and clarity over the ownership of the underlying asset (e.g., biodiversity offsets and/or forest carbon offsets).

Business Risk

The core business risk of forest carbon projects is whether "the price of carbon is lower than the opportunity costs, and {how to overcome the fact that} funding comes later than typical farming alternative uses."[153]

Because REDD+ offsets are only accepted in the voluntary carbon markets, the transaction costs of identifying a potential buyer and closing a financial transaction are significantly higher than listing REDD+ offsets on an exchange.

From the supplier or project developer's perspective and the buyer's perspective, there is business risk associated with how to price the asset. In addition to the intrinsic value of nature, it is difficult to put an extrinsic value on nature. Likewise, what is the proper price of a biodiversity offset or the fair price of a forest carbon offset bundled with a water quality credit?

Strategic Risk

Some of the many strategic risks associated with PES projects include choosing the appropriate methodology and certification standard; choosing the appropriate partners, especially those who are able to meet all the technical requirements and deliver verified offsets; and strategically choosing the right, successful project to support.

A few strategic risks associated with biodiversity offsets include identifying a suitable area for the biodiversity offset project and negotiating with the sponsoring company to commit to the terms of the project and being the provider of funds (i.e., which could include community components). It is also important that there are a lot of reasons why a company may later back out of its biodiversity commitment (e.g., company bankruptcy, change in internal policies, or lack of profits) and create an orphan site, so project developers need to think through strategic options (i.e., diversifying revenue sources or engaging shareholders and host governments to pressure the company to maintain its commitment).

Reputation Risk

Because there is an oversupply of forest carbon offsets, particularly REDD+ offsets, relative to stagnant or declining demand, there is downward pressure on the price of REDD+ offsets. These low prices have resulted in some firms selling inexpensive offsets to generate cash flow, while other firms refuse to sell at such a low price (i.e., whether because it is a loss-making transaction or out of principle). Selling at such low prices creates a reputation risk for the industry, as it undermines the viability of many projects.

One optic risk, though not necessarily fact-based, is the related issue of permanence. For instance, forest carbon projects are set up based off current and projected prices. However, as REDD+ projects are scaled up, a tremendous volume of VERs are issued, the market subsequently gets flooded with VERs, and this depresses the market price.

Managing expectations, particularly the expectations of local communities, is another reputational risk. Communities share a lot of interest in the beginning stages of the projects as local benefits (e.g., building infrastructure such as health clinics and local employment opportunities) are discussed. Then 5 years, let alone 10 or 20 years, later, the communities have a new standard of living and desire even more benefits, yet this depends on a rising cost of VERs.

Another reputational risk is that "although mitigation banking clearly does enables development, the environmental ethics question hinges on whether the total production of ecosystem services was increased or decreased through the combined results of the development and the associated mitigation."[154]

With respect to biodiversity offsets, there is also a potential reputation risk for project developers to partner with large-scale mining, logging, and other extractive industries.

Liquidity Risk

There are relatively high upfront costs associated with developing a REDD+ project (e.g., preparation of project documentation, technical studies to develop a forest carbon inventory and deforestation baseline) and then ongoing costs—depending on the structure of the project and the specific activities to be implemented (e.g., community development projects, verification costs, and site visits), which can be lower, the same, or higher relative to the upfront costs. Nowadays, it is more difficult to secure upfront financing, so there is a liquidity risk of not being able to convert verified emission reductions into cash to recover sunk project costs.

Operational Risk

There are numerous operational risks. One of the biggest hurdles to overcome are the technical challenges with designing and implementing a REDD+ project. This includes navigating the complexities of a chosen methodology or protocol. Regarding the Clean Development Mechanism (CDM), the "1990 rule caused difficulties in site selection because many landowners couldn't provide proof that their land was deforested prior to 1990. Land use satellite data and multitemporal analysis typically were not available, costly to develop, and in some cases the sites were cloudy in satellite-generated data."[155]

The costs and time required to develop forest carbon projects pose another operational risk. For instance:

> project developers must consider specific unique aspects: (1) project design processes require time-consuming community consultations; (2) projects take 12 to 24 months to develop from concept through validation; (3) activity variation is high and results in dramatically different costs, which range from hundreds of thousands of dollars to several million dollars; and (4) land tenure and carbon ownership must be clarified across numerous land ownership situations.[156]

Furthermore:

> {…} forest restoration projects are relatively expensive due to the complexity of the activities involved. Second, forest restoration projects have high front-loaded costs because land clearing, nursery preparation, and planting costs occur early on (typically in years one through six). Third, because the ChoCO2 project was relatively small, certain fixed management costs comprised a higher percentage of the overall expenses than is normally desirable. Lastly, the transaction costs associated with defining, monitoring, and verifying the carbon credits made up a relatively high percentage of the project costs as well.[157]

To overcome these risks, Carbonfund.org Foundation has the marketing arm, and its lead contractor, TerraCarbon, has the backend technical expertise to present a complete picture. However, if a firm is too academic-heavy and lacks the necessary marketing, or vice versa, this will present significant operational risks.

Legal and Regulatory Risk

Legal and regulatory risks are significant for PES initiatives.

In Indonesia, for instance, acquiring an ecosystem restoration license is a significant hurdle that can take a lot of time and money (i.e., due to research and legal assistance). Several projects have spent considerable sums of money and developed project documentation, yet the project never materialized due to an inability to get the required license for a successful verification.

In Brazil, there is a risk of unclear land tenure due to the fact that land titles may not be free and clear of overlapping title claims. This risk should be mitigated over time due to the new Forest Code in Brazil and its requirement of rural property owners to register with the CAR (Cadastro Ambiental Rural or Rural Environmental Registry in English).

A unique legal and regulatory associated with carbon offsets used for the California cap-and-trade program is that the Air Resources Board (ARB), the regulatory body in California, can invalidate offsets for a certain amount of years after the carbon offsets are issued. To mitigate this risk, it was reported in May 2017 that "several forestry transactions related to carbon offset insurance took place in the California market. With ISU Environmental Insurance acting as retail broker and Parhelion Underwriting the facility underwriting manager, the globally recognized insurance market, Lloyd's of London, provided the security. The policy will insure owners of offsets against invalidation by regulators ARB, over the policy's three-year term."[158]

Establishing the rights and ownership of forest carbon is a concern in many developing countries, especially when it comes to community-based and Indigenous Peoples-based projects.

Around the world, many voluntary REDD+ projects have undertaken work to nest or link into higher levels of carbon accounting at the national or subnational level. Yet, how exactly projects will fit into the jurisdiction oftentimes remains a challenge. Furthermore, how will REDD+ projects fit into the Paris Agreement? Similarly, China recently froze its carbon offset pipeline, and while REDD+ projects may not have been initially eligible, this demonstrates the type of risks associated with a market premised off a public policy.[159] This said, working with state and/or multilateral agencies can help mitigate political risk such as Wildlife Works and Wildlife Alliance's partnership with Cambodia's Ministry of the Environment.

Taxation on forest carbon offsets is often unclear and the specific regulations on the sale of such forest carbon offsets can also present bureaucratic barriers to trade. This said, are forest carbon offsets to be taxed as a commodity, a service (i.e., ecosystem service), or as a derivative (i.e., its value is derived based off the underlying forest)?

Credit Risk

Credit risk includes default risk, bankruptcy risk, downgrade risk, and settlement risk. Many project developers face default and bankruptcy risk due to the decline in carbon offset prices. These risks may be mitigated, in part, by diversifying project revenues from agroforestry, ecotourism, or by bundling other ecosystem service offsets.

Market Risk

Market risk includes interest rate risk, equity price risk, foreign exchange risk, and commodity price risk. If commodity prices increase, then risk of conversion of tropical forests to an alternative land use (e.g., palm oil, cattle ranch, or soybeans) can also increase. If a payment for ecosystem service project is implemented in conjunction with an agroforestry, sustainable forestry, or certified non-timber forest products project, and the underlying commodity price decreases, then the PES project may no longer be financially sustainable. This commodity price risk includes non-PES commodities (e.g., certified cacao), along with PES commodities such as verified emission reductions or quantified biodiversity offsets. Then again, as seen in the case of the Rio Tinto QMM Project, if the price of its ilmenite ore increases, then there may be an opportunity to establish a long-term trust fund for the biodiversity offset project.

Risk, Return, Time (Horizon), Taxes, Liquidity, Legal and Unique (RRTTLLU)

Risk and Return

Please see above for the risk and return associated with payments for ecosystem services.

Time Horizon

The time horizon for payments for ecosystem services is a long-term commitment. For instance, many forest carbon projects have a minimum Project Lifetime of 30 years, and many projects have Project Lifetimes up to 100 years. The minimum Project Lifetime of a carbon offset project is often spelled out by the certification standard. From a buyer's perspective, short time horizons often exist for the purchase of forest carbon offsets because there are no ongoing requirements (e.g., project management or monitoring, reporting, and verification of the project).

Taxes

There can be a variety of taxes associated with payments for ecosystem services. This includes possible sales tax on the underlying PES asset, along with property taxes and employment taxes.

Liquidity

There appears to be a low level of liquidity for transferring the entire project—whether it be a biodiversity offset, forest carbon offset, or watershed payments project. Furthermore, the liquidity of a given offset will depend on whether the offset is transacted in the voluntary market or the compliance market. For instance, there is a lower level of liquidity for voluntary forest carbon offsets when compared to the more liquid market for compliance forest carbon offsets in the California-Québec market.

Legal

Legal considerations associated with payments for ecosystem services will vary from country to country. In addition, these legal considerations may vary from state to state and may even vary depending on whether the Project Proponent is a private citizen or, for example, an Indigenous Peoples' association.

Unique

One unique aspect of payments for ecosystem service projects is the potential ability to stack or bundle several offset types. Bundling is when a single offset includes several ecosystem services (i.e., biodiversity and carbon offsets), while stacking is when the ecosystem services are differentiated and sold separately.

Another unique aspect is that arguably there are compliance markets and there are voluntary mechanisms. Likewise, it has been argued that voluntary PES instruments are more of a voluntary transfer payment as opposed to an actual market. This said, markets require—among many things—price discovery and liquidity, which is often lacking in PES instruments such REDD+ forest carbon and biodiversity offsets.

Policy Analysis

Defining the Problem

As previously mentioned, there are a variety of potential problems facing payments for ecosystem services, most notably policy risk, overlapping land title claims, and clarity over the ownership of the underlying asset (e.g., biodiversity offsets and/or forest carbon offsets).

Because the markets for PES projects are usually first established by a government policy, it is problematic if governments do not have PES laws in place, or if such laws are unclear, and/or short-term in nature.

Countries can have significant overlapping title claims which makes it difficult to discern who exactly owns the land. Such overlapping title claims can result due to:

- A mismatch between subnational and central government land title registries;
- Poor map-making and/or the lack of georeferenced property maps;
- Uninformed government decisions to establish or expand government lands;
- Corruption and fraudulent titling; and
- Disregard for communal and/or ancestral land rights such as for Indigenous Peoples.

Furthermore, some countries have not yet clarified who has the right—assuming all approvals from local authorities have been properly received—to claim ownership of the underlying PES assets (e.g., biodiversity and/or forest carbon offsets). For instance,

- Is a private project developer entitled to ownership of PES assets? Does it matter whether this private project developer is registered in the country or whether the private project developer is a foreign entity?
- Is a public project developer and/or financier (e.g., Germany's KfW, Norway's International Climate and Forest Initiative, or USAID) entitled to ownership of PES assets?
- Are nonprofit organizations—whether they be domestic or international (e.g., Conservation International, TNC, or WWF)—entitled to ownership of PES assets?
- Does a subnational jurisdiction retain ownership of PES assets vis-à-vis the central government? Or in contrast, does the central government retain ownership of PES assets unless otherwise specified?
- Are Indigenous Peoples and other forest-dependent communities owners of the PES assets within their territory? Under what conditions are such Indigenous Peoples and local communities able to sign contracts and sell, or transfer, their underlying PES assets?
- What, if any, government entity grants permission—such as a distinct license, permit, or certification—to undertake, develop, finance, and sell PES assets?

- What, if any, government charge (e.g., taxes, royalties, and/or registration costs) will be levied on landowners, project developers, financiers, and/or sellers of PES assets?
- Does it make a difference—and if so, how?—if the underlying PES asset is for biodiversity, water, and/or carbon?
- If a country has a water regulatory agency, a forest service, and a ministry of environment, who then approves a bundled offset consisting of water, forest carbon, and biodiversity?

Policies which address these three risks will help expand the market for PES projects, programs, and investments while simultaneously conserving tropical rainforests, preserving biodiversity, and providing local sustainable economic opportunities.

Establishing Goals

The goals for establishing desirable policies for PES are to:

- Establish robust, long-term, transparent markets with price discovery and liquidity.
- Clarify how exactly landownership can be demonstrated and the process for effectively and efficiently resolving overlapping land title claims.
- Clearly state who can own the underlying PES assets and the additional, if any, costs to be levied by the government.

Selecting a Policy

There have been a wide range of policies that have had varying impacts on the development of PES. For instance, Ecuador has been a leader in watershed PES schemes, while several countries including Madagascar and Panama have enabled biodiversity offset markets, and the State of Acre, Brazil, is a leader in its subnational work on REDD+.

On the state level, Acre's State Law 2.308 of October 22, 2010, entitled, The State System of Incentives for Environmental Services (SISA in Portuguese), is a pioneer.[160] California's Assembly Bill (AB) 32, entitled the California Global Warming Solutions Act of 2006, has enabled the creation of a GHG compliance market with a cap-and-trade program that accepts avoided conversion forest offsets. In addition, there is a future possibility that Acre and California will link their programs.

On a national level and with respect to forest carbon policies, Costa Rica has been a pioneer with its PES system. In contrast, Brazil's Federal Decree 8.576/2015 seems to challenge the states on their development of PES systems and may forbid the states from transferring emission reductions to other countries.[161] On the international level and with respect to policies designed to create a market for forest carbon projects is the UNFCCC's Paris Agreement.

With respect to policies designed to clarify landownership and overlapping title claims, Brazil's Cadastro Ambiental Rural (CAR, Rural Environmental Registry) has the potential to effectively resolve these issues. In addition, with respect to policies designed to clarify the underlying ownership of PES assets, Indonesia's legislation for REDD+ licensing has the potential to effectively resolve these issues. These licenses include a carbon sequestration and carbon storage (PAN/RAP) license and an alternative license known as the ecosystem restoration license (IUPHHK-RE).

Implementing a Policy

On the state level, the California Air Resources Board (ARB) and Acre's Climate Change Institute are responsible for implementing their domestic programs.

The federal government of Brazil, in conjunction with the states, is responsible for the CAR, and Indonesia's federal government is primarily responsible for granting Indonesia's licensing. More specifically, Indonesia's carbon sequestration and carbon storage (PAN/RAP) license is issued by the Minister of Forestry. Similarly, Costa Rica's federal government—particularly the Ministry of the Environment and Energy—is primarily responsible for implementation of its PES system.

The UNFCCC Paris Agreement, in contrast to the Kyoto Protocol, has taken a bottom-up approach, and the responsibility for implementing the Nationally Determined Contributions (NDCs) is the responsibility of each member country.

Evaluating the Policy

On the state level, Acre's state legislation has received praise including from groups such as Climate Focus,[162] there has been detailed analysis conducted from groups such as CIFOR,[163] and Acre has received technical support from groups such as the Earth Innovation Institute[164] and the Environmental Defense Fund.[165] Furthermore, the SISA law was passed, the Climate Change Institute was established, and since reducing deforestation, Acre has received payments for performance from Germany's REDD Early Movers Programme.[166]

Also on the state level, California's AB32 is the "first state-backed program anywhere in the world that recognized the value of conserving and managing standing forests for their climate benefits."[167] In the future California's cap-and-trade program may allow international sector-based offsets, such as REDD+ offsets after a link to the State of Acre, Brazil.[168] In addition, forest carbon offsets have been the most widely used offset in the California cap-and-trade program.

Costa Rica has continued to be a national leader with its PES system and currently has the goal of becoming carbon neutral by 2021.[169]

Brazil's Federal Decree 8.576/2015 is still relatively new and time will tell regarding its impact on tropical deforestation. Brazil's CAR, as discussed in the case study *Brazil's Forest Code and Legal Reserve Credits* (see Chap. 8), has the potential to clarify rural landownership and minimize overlapping title claims, but has been slow in its implementation.

Indonesia's licensing, although it took a considerable amount of time, has issued ecosystem restoration licenses such as for InfiniteEARTH's Rimba Raya Biodiversity Reserve REDD Project and the PT Rimba Makmur Utama's Katingan Peatland Restoration and Conservation Project. The policy would likely be better if licenses could be issued in a more efficient manner.

While it is still too early to evaluate the outcomes for tropical forests from the UNFCCC Paris Agreement, nearly every country on Earth submitted Intended Nationally Determined Contributions (INDCs) prior to the COP21 event in Paris, France. Almost 200 countries signed onto the Paris Agreement in December 2015, which now commits countries to carry out their NDCs. Imbedded in the UNFCCC Paris Agreement is an allowance for ITMOs (Internationally Transferred Mitigation Outcomes; i.e., possibly to be interpreted as code for international offsets) and REDD+:

> The use of internationally transferred mitigation outcomes to achieve nationally determined contributions under this Agreement shall be voluntary and authorized by participating Parties. {...}
>
> Recognizes the importance of adequate and predictable financial resources, including for results-based payments, as appropriate, for the implementation of policy approaches and positive incentives for reducing emissions from deforestation and forest degradation, and the role of conservation, sustainable management of forests and enhancement of forest carbon stocks; as well as alternative policy approaches, such as joint mitigation and adaptation approaches for the integral and sustainable management of forests; while reaffirming the importance of non-carbon benefits associated with such approaches; encouraging the coordination of support from, inter alia, public and private, bilateral and multilateral sources, such as the Green Climate Fund, and alternative sources in accordance with relevant decisions by the Conference of the Parties.[170]

Alongside COP21, 18 countries signed a declaration on international carbon markets. This group, which is being led by New Zealand, is also comprised of Australia, Canada, Chile, Colombia, Germany, Iceland, Indonesia, Italy, Japan, Mexico, Netherlands, Panama, Papua New Guinea, South Korea, Senegal, Ukraine, and the United States.[171]

Future Outlook for Instrument

PES initiatives, particularly the support for forest carbon projects via climate finance, are potentially very promising. While biodiversity offsets, nutrient markets, and watershed protection payments can deliver conservation outcomes, these are more niche markets and do not have the greatest potential to scale to conserve tropical forests like REDD+ and forest carbon offsets. This said, REDD+ still distinguishes itself from these other markets because it is a global market and the impacts (i.e., GHG emissions) are global impacts, rather than "just" local or regional impacts like nutrient markets.[172]

Emerging carbon compliance mechanisms that may include REDD+ offsets are the California cap-and-trade program, the Carbon Offsetting and Reduction Scheme for International Aviation (CORSIA) of the International Civil Aviation Organization (ICAO), and the Paris Agreement.

CORSIA may have its offsetting rules decided by 2018, with a pilot phase from 2021 to 2023, then a first compliance phase from 2024 to 2027, and a second compliance phase from 2027 to 2035. Thomson Reuters' Carbon Market Analyst estimates approximately 970 million offsets from 2021 to 2030 for the CORSIA program,[173] while the ICAO estimates that an optimistic scenario would require approximately 873 million offsets from 2020 to 2035.[174] In addition, another analysis in December 2016 estimated that "the cumulative number of tonnes of CO_2 emissions to be offset under CORSIA in the period 2021–2035 equals 2,711 Mt {2.711 billion tonnes of CO_2} of CO_2."[175] Numerous leading organizations such as Conservation International, Environmental Defense Fund, Forest Trends, the Global Canopy Programme, Sustainable Travel International, The Nature Conservancy, and the Wildlife Conservation Society have endorsed REDD+'s inclusion in CORSIA.

In addition, "South Africa's proposed carbon tax, China's emissions trading system, South Korea's emissions trading system, and other markets may post transaction volumes—if they make room for forestry offsets."[176]

Furthermore, while it is still far-off, there is a potential that the International Maritime Organization (IMO) may adopt a similar model as CORSIA for the international shipping industry. The IMO may have its preliminary suggestions by 2018 and then final suggestions by 2023.[177]

Other Resources on Payments for Ecosystem Services

Bundling and Stacking for Maximizing Social, Ecological, and Economic Benefits: A Framing Paper for Discussion at the "Bundling and Stacking Workshop"

- https://rmportal.net/library/content/translinks/translinks-2012/Wildlife%20Conservation%20Society/2012%20PES%20Bundling%20and%20Stacking%20Workshop%20(Washington,%20DC,%20USA)/Paper_PESBundlingandStacking.pdf

Business and Biodiversity Offsets Programme (BBOP)

- http://bbop.forest-trends.org/
- http://bbop.forest-trends.org/pages/our_work_standard

California Air Resources Board's Cap-and-Trade Program

- https://www.arb.ca.gov/cc/capandtrade/capandtrade.htm

Carbon Market Institute

- http://carbonmarketinstitute.org/

CIFOR's Forest Carbon Database

- http://carbonstock.cifor.org/user/HomeMap?utm_source=July+2013&utm_campaign=NEWS+UPDATE+English&utm_medium=email

Climate Finance Tracking at AfDB

- http://www.afdb.org/en/topics-and-sectors/sectors/climate-change/climate-finance-tracking-at-afdb/

ECOSTAR's Market Outlooks on European Voluntary Carbon, Biodiversity Offsets, and Watershed Investments

- http://www.ecostarhub.com/market-outlooks/

Fondo Nacional de Financiamiento Forestal or FONAFIFO (Costa Rica's National Forestry Financing Fund)

- http://www.fonafifo.go.cr/home/index.html

FONAG: Fondo para la Protección del Agua or FONAG (of Ecuador, The Water Protection Fund in English)

- http://www.fonag.org.ec/inicio/english-version.html

Forest Carbon Portal

- http://www.forestcarbonportal.com/

Identifying Priority Areas for REDD Projects with Co-benefits for Great Ape Populations

- http://apescarbon.eva.mpg.de/

International Database of REDD+ Projects

- http://www.reddprojectsdatabase.org/

Linking Flights and Forests: The Essential Role of Forests in Supporting Global Aviation's Response to Climate Change

- http://www.conservation.org/publications/Documents/CI_Linking-Flight-and-Forests-Briefing-Paper-Apr-2016.pdf

REDD+ Policy Assessment Center

- http://www.redd-pac.org/

REDDx Tracking Forest Finance

- http://www.reddx.forest-trends.org/

Sierra de las Minas Biosphere Reserve Water Fund of Guatemala

- www.watershedmarkets.org/casestudies/Guatemala_Sierra_Minas.html

Stacking Ecosystem Services Payments Risks and Solutions

- https://nicholasinstitute.duke.edu/sites/default/files/publications/stacking-ecosystem-services-payments-paper.pdf

Stimulating Interim Demand for REDD+ Emission Reductions: The Need for a Strategic Intervention from 2015 to 2020

- Executive Summary: http://d6.globalcanopy.org/materials/stimulating-interim-demand-redd-emission-reductions-need-strategic-intervention-2015-2020-
- Full Report: http://d6.globalcanopy.org/StimulatingInterimDemand-Report

The Implications of the Paris Climate Agreement for Private Sector Roles in REDD+

- http://www.forest-trends.org/documents/files/doc_5305.pdf

The Little REDD Book

- http://d6.globalcanopy.org/materials/little-redd-book

UN-REDD Programme's Resources

- http://www.un-redd.org/resources

U.S. Fish & Wildlife Service Guidance on Conservation Banking

- https://www.fws.gov/endangered/landowners/conservation-banking.html
- https://www.fws.gov/endangered/esa-library/pdf/Conservation_Banking_Guidance.pdf

Voluntary REDD+Database

- http://reddplusdatabase.org/

Notes

1. World Wildlife Fund. *Guide to Conservation Finance*. 20.
2. Ginn. *Investing in Nature*. 11.
3. Gómez-Baggethun, Erik, Rudolf de Groot, Pedro L. Lomas and Carlos Montes. "The History of Ecosystem Services in Economic Theory and Practice: From Early Notions to Markets and Payment Schemes." *Ecological Economics*. 69(6). April 2010. 1209–1218. doi:10.1016/j.ecolecon.2009.11.007; https://www.researchgate.net/publication/46490541_The_History_of_Ecosystem_Services_in_Economic_Theory_and_Practice_From_Early_Notions_to_Markets_and_Payment_Schemes
4. Meyer, Shannon. "Markets for Ecosystem Services and Reclaiming the Great Dismal Swamp." In *Conservation Capital in the Americas*, edited by James N. Levitt. Massachusetts: Lincoln Institute of Land Policy, 2010. 210.
5. Daily, Gretchen C. and Katherine Ellison. *The New Economy of Nature*. 40–41.
6. Daily, Gretchen C. and Katherine Ellison. *The New Economy of Nature*. 173–174.
7. Martin. *On the Edge*. 127.
8. TNC. "Climate Change: Noel Kempff: An Example of REDD Success." Accessed December 13, 2016. http://www.nature.org/ourinitiatives/urgentissues/global-warming-climate-change/places-we-protect/noel-kempff-mercado-national-park.xml

9. Daily, Gretchen C. and Katherine Ellison. *The New Economy of Nature*. 29.
10. Daily, Gretchen C. and Katherine Ellison. *The New Economy of Nature*. 43–44.
11. Wayburn, Laurie. "The Van Eck Forest: Carbon Markets and the New Economic Paradigm for Forest Sustainability." In *Conservation Capital in the Americas*, edited by James N. Levitt. Massachusetts: Lincoln Institute of Land Policy, 2010. 184.
12. Climate. Community & Biodiversity Alliance. "About the CCBA." Accessed November 23, 2016. http://www.climate-standards.org/about-ccba/
13. Intercontinental Exchange. "Chicago Climate Exchange." Accessed February 13, 2017. https://www.theice.com/ccx
14. Clean Development Mechanism. "Project Search." Accessed November 23, 2016. http://cdm.unfccc.int/Projects/projsearch.html
15. Verified Carbon Standard Association. "Who We Are." Accessed November 23, 2016. http://www.v-c-s.org/about-vcs/who-we-are/
16. World Wildlife Fund. *Guide to Conservation Finance*. 9.
17. Clean Development Mechanism. "Project Search." Accessed November 23, 2016. http://cdm.unfccc.int/Projects/projsearch.html
18. Climate. Community & Biodiversity Alliance. "Small-scale Reforestation for Landscape Reforestation." Accessed November 23, 2016. http://www.climate-standards.org/2007/01/01/small-scale-reforestation-for-landscape-reforestation/
19. Climate. Community & Biodiversity Alliance. "Panama Native Species Reforestation." Accessed November 23, 2016. http://www.climate-standards.org/2007/01/02/panama-native-species-reforestation/
20. Wayburn, Laurie. "The Van Eck Forest: Carbon Markets and the New Economic Paradigm for Forest Sustainability." In *Conservation Capital in the Americas*, edited by James N. Levitt. Massachusetts: Lincoln Institute of Land Policy, 2010. 188.
21. Wayburn, Laurie. "The Van Eck Forest: Carbon Markets and the New Economic Paradigm for Forest Sustainability." In *Conservation Capital in the Americas*, edited by James N. Levitt. Massachusetts: Lincoln Institute of Land Policy, 2010. 189.
22. Elliott, Chris. "REDD+ Is a Little More Complex than 'Not Cutting Down a Tree'." In *On the Edge*, written by Claude Martin. Berkeley: Greystone Books, 2015. 235.

23. Climate. Community & Biodiversity Alliance. "Reducing Carbon Emissions from Deforestation in the Ulu Masen Ecosystem." Accessed November 23, 2016. http://www.climate-standards.org/2007/11/02/reducing-carbon-emissions-from-deforestation-in-the-ulu-masen-ecosystem/
24. Norway Government. "Brazil." Accessed November 23, 2016. https://www.regjeringen.no/en/topics/climate-and-environment/climate/climate-and-forest-initiative/kos-innsikt/brazil-and-the-amazon-fund/id734166/
25. Ibid.
26. Clean Development Mechanism. "Project Search." Accessed November 23, 2016. http://cdm.unfccc.int/Projects/projsearch.html
27. Center for Climate and Energy Solutions. "The American Clean Energy and Security Act (Waxman-Markey Bill)." Accessed April 11, 2017. https://www.c2es.org/federal/congress/111/acesa
28. Norway Government. "Norway and Indonesia in Partnership to Reduce Emissions from Deforestation." Accessed November 23, 2016. https://www.regjeringen.no/en/aktuelt/norway-and-indonesia-in-partnership-to-r/id605709/
29. Warner, Maud and Molly Peters-Stanley. "Kenya Carbon Project Earns First-Ever VCS REDD Credits." *Ecosystem Marketplace*. Accessed November 29, 2016. http://www.ecosystemmarketplace.com/articles/kenyan-carbon-project-earns-first-ever-vcs-redd-credits/
30. Forest Carbon Asia. "Malaysia: First VCS Forest Project in Sabah, Borneo Registered." Last modified September 7, 2011. http://www.forestcarbonasia.org/in-the-media/first-forest-project-on-sabah-borneo-registered-under-the-vcs/
31. Verified Carbon Standard Association. "INFAPRO Rehabilitation of Logged-Over Dipterocarp Forest in Sabah, Malaysia." Accessed November 30, 2016. http://www.vcsprojectdatabase.org/#/project_details/672
32. Ecosystem Marketplace. "Indigenous Tribe Earns Validation for Landmark Forest Carbon Project." Last modified April 12, 2012. http://www.ecosystemmarketplace.com/articles/indigenous-tribe-earns-validation-for-landmark-forest-carbon-project/
33. The Green Climate Fund. "Behind the Fund." Accessed November 23, 2016. http://www.greenclimate.fund/the-fund/behind-the-fund
34. UNFCCC. "Paris Agreement." Accessed April 11, 2017. http://unfccc.int/paris_agreement/items/9485.php

35. Norway Government. "The Congo Basin." Accessed November 23, 2016. https://www.regjeringen.no/en/topics/climate-and-environment/climate/climate-and-forest-initiative/kos-innsikt/congo-basin-fund/id734156/
36. IUCN. "Biodiversity Offsets." Accessed April 14, 2017. https://www.iucn.org/theme/business-and-biodiversity/our-work/business-approaches-and-tools/biodiversity-offsets
37. Naeem, S., J. C. Ingram, A. Varga, T. Agardy, P. Barten, G. Bennett, E. Bloomgarden, L. L. Bremer, P. Burkill, M. Cattau, C. Ching, M. Colby†, D. C. Cook, R. Costanza, F. DeClerck, C. Freund, T. Gartner, R. Goldman-Benner, J. Gunderson, D. Jarrett, A. P. Kinzig, A. Kiss, A. Koontz, P. Kumar, J. R. Lasky, M. Masozera, D. Meyers, F. Milano, L. Naughton-Treves, E. Nichols, L. Olander, P. Olmsted, E. Perge, C. Perrings, S. Polasky, J. Potent, C. Prager, F. Quétier, K. Redford, K. Saterson†, G. Thoumi, M. T. Vargas, S. Vickerman, W. Weisser, D. Wilkie and S. Wunder. "Get the Science Right When Paying for Nature's Services." *Science*. 347(6227). 2015. 1206–1207. doi:10.1126/science.aaa1403; http://science.sciencemag.org/content/347/6227/1206
38. Barrett, Kelli. "Multi-State Water Program Hopes to Stand on Its Own Two Feet in 2016." *Ecosystem Marketplace*. Last modified January 20, 2016. http://www.ecosystemmarketplace.com/articles/multi-state-water-program-hopes-to-stand-on-its-own-two-feet-in-2016/
39. Meyer, Shannon. "Markets for Ecosystem Services and Reclaiming the Great Dismal Swamp." In *Conservation Capital in the Americas*, edited by James N. Levitt. Massachusetts: Lincoln Institute of Land Policy, 2010. 198.
40. BBOP. "Mitigation Hierarchy." Accessed December 13, 2016. http://bbop.forest-trends.org/pages/mitigation_hierarchy
41. Davis, Adam. "Mainstreaming Environmental Markets." In *From Walden to Wall Street*, edited by James Levitt. Massachusetts: Lincoln Institute of Land Policy, 2005. 158.
42. Bennett, Genevieve and Franziska Ruef. "Alliances for Green Infrastructure: State of Watershed Investment 2016." December 2016. *Ecosystem Marketplace*. Accessed April 4, 2017. http://forest-trends.org/releases/p/sowi2016. 1.
43. McQueen, Mike and Ed McMahon. *Land Conservation Financing*. 152.
44. Korchinsky, Mike. Interviewed by Brian McFarland. March 2017.
45. Carbon Pulse. "Disney in Multi-Million REDD+ VER deal Cambodia." Last modified July 25, 2016. http://carbon-pulse.com/22732/?utm_source=CP+Daily&utm_campaign=be99ebd6b0-CPdaily

22072016&utm_medium=email&utm_term=0_a9d8834f72-be99e-bd6b0-36342993

46. Madsen, Becca; Carroll, Nathaniel; Moore Brands, Kelly. 2010. "State of Biodiversity Markets Report: Offset and Compensation Programs Worldwide." http://www.ecosystemmarketplace.com/documents/acrobat/sbdmr.pdf. IV.
47. Bennett, Genevieve and Franziska Ruef. "Alliances for Green Infrastructure: State of Watershed Investment 2016." December 2016. *Ecosystem Marketplace*. Accessed April 4, 2017. http://forest-trends.org/releases/p/sowi2016. 2.
48. Ibid.
49. Ecosystem Marketplace. "State of Forest Carbon Finance 2016." Accessed November 23, 2016. http://www.forest-trends.org/documents/files/doc_5388.pdf. 2.
50. Ecosystem Marketplace. "State of Forest Carbon Finance 2016." Accessed November 23, 2016. http://www.forest-trends.org/documents/files/doc_5388.pdf. 3.
51. Ecosystem Marketplace. "State of Forest Carbon Finance 2016." Accessed November 23, 2016. http://www.forest-trends.org/documents/files/doc_5388.pdf. i.
52. Wolosin, Michael, Jessica Breitfeller, and Brian Schaap. "The Geography of REDD+ Finance: Deforestation, Emissions, and the Targeting of Forest Conservation Finance." *Forest Trends*. August 2016. Accessed April 4, 2017. http://www.forest-trends.org/documents/files/doc_5334.pdf. 3.
53. InfiniteEARTH. "Rimba Raya Biodiversity Reserve." Accessed December 29, 2016. http://infinite-earth.com/rimba-raya-biodiversity-reserve/
54. InfiniteEARTH. "Fact Sheet." Accessed April 2, 2017. http://rimba-raya.com/wp-content/uploads/Rimba-Raya-Initiatives-UN-SDGs-2016.pdf?utm_medium=EDM&utm_campaign=2016.11-Seeds-of-Change&utm_source=txt.DLRRFactSheet
55. Warner, Maud and Molly Peters-Stanley. "Kenya Carbon Project Earns First-Ever VCS REDD Credits." *Ecosystem Marketplace*. Accessed November 29, 2016. http://www.ecosystemmarketplace.com/articles/kenyan-carbon-project-earns-first-ever-vcs-redd-credits/
56. BBOP. "Mitigation Hierarchy." Accessed December 13, 2016. http://bbop.forest-trends.org/pages/mitigation_hierarchy
57. EnVision Corporation. "Environmental Finance Products." Accessed March 21, 2017. http://www.envisioncorporation.com/environmental-finance-products/

58. InfiniteEARTH. "Rimba Raya Biodiversity Reserve." Accessed December 29, 2016. http://infinite-earth.com/rimba-raya-biodiversity-reserve/
59. InfiniteEARTH. "Fact Sheet." Accessed April 2, 2017. http://rimba-raya.com/wp-content/uploads/Rimba-Raya-Initiatives-UN-SDGs-2016.pdf?utm_medium=EDM&utm_campaign=2016.11-Seeds-of-Change&utm_source=txt.DLRRFactSheet
60. InfiniteEARTH. "Rimba Raya Biodiversity Reserve." Accessed December 29, 2016. http://infinite-earth.com/rimba-raya-biodiversity-reserve/
61. US Environmental Protection Agency. "Greenhouse Gas Equivalencies Calculator." Accessed March 21, 2017. https://www.epa.gov/energy/greenhouse-gas-equivalencies-calculator
62. InfiniteEARTH. "Rimba Raya Biodiversity Reserve." Accessed December 29, 2016. http://infinite-earth.com/rimba-raya-biodiversity-reserve/
63. Rimba Raya Biodiversity Reserve. "What is Rimba Raya?" Accessed December 29, 2016. http://rimba-raya.com/
64. Lemons, Todd. Interviewed by Brian McFarland. March 2017.
65. Ibid.
66. Ibid.
67. Ibid.
68. Ibid.
69. Ibid.
70. Ibid.
71. Ibid.
72. Ibid.
73. Ibid.
74. Ibid.
75. PT Rimba Raya Conservation. "Rimba Raya Biodiversity Reserve Project: CCB Project Design Document." Accessed March 21, 2017. http://www.vcsprojectdatabase.org/#/project_details/674. vi.
76. Lemons, Todd. Interviewed by Brian McFarland. March 2017.
77. VCS Project Database. "VCU Search Results." Accessed March 21, 2017. http://www.vcsprojectdatabase.org/#/vcus/p_674
78. VCS Project Database. "Rimba Raya Biodiversity Reserve Project, Indonesia." Accessed February 20, 2017. http://www.vcsprojectdatabase.org/#/project_details/674
79. Ibid.
80. Lemons, Todd. Interviewed by Brian McFarland. March 2017.
81. Ibid.

82. Zwick, Steve. "Todd Lemons: Ecosystem Entrepreneur." *Ecosystem Marketplace*. Accessed March 21, 2017. http://www.ecosystemmarketplace.com/articles/todd-lemons-ecosystem-entrepreneur/
83. Lemons, Todd. Interviewed by Brian McFarland. March 2017.
84. Ibid.
85. Ibid.
86. Ibid.
87. Ibid.
88. Ibid.
89. Warner, Maud and Molly Peters-Stanley. "Kenya Carbon Project Earns First-Ever VCS REDD Credits." *Ecosystem Marketplace*. Accessed November 29, 2016. http://www.ecosystemmarketplace.com/articles/kenyan-carbon-project-earns-first-ever-vcs-redd-credits/
90. Wildlife Works Carbon, LLC. "VCS Project Description: The Kasigau Corridor REDD Project—Phase I Rukinga Sanctuary." Last modified January 25, 2011. http://www.vcsprojectdatabase.org/services/publicViewServices/downloadDocumentById/5885. 3.
91. Wildlife Works Carbon, LLC. "VCS Project Description: The Kasigau Corridor REDD Project—Phase I Rukinga Sanctuary." Last modified January 25, 2011. http://www.vcsprojectdatabase.org/services/publicViewServices/downloadDocumentById/5885. 4.
92. Korchinsky, Mike. Interviewed by Brian McFarland. March 2017.
93. Ibid.
94. MSN. "Satao, One of the Last 'Giant Tusker' Elephants, Killed in Kenya." Last modified March 7, 2017. http://www.msn.com/en-us/news/world/satao-one-of-the-last-giant-tusker-elephants-killed-in-kenya/ar-AAnVaaa?li=AAn4eAA&ocid=mailsignout
95. Fears, Darryl. "Antique Dealers say the New Federal Ivory Ban will Cost Owners up to $12 billion." *Washington Post*. June 2, 2016. https://www.washingtonpost.com/news/energy-environment/wp/2016/06/02/its-final-selling-just-about-any-item-containing-elephant-ivory-is-a-crime-in-the-u-s/?utm_term=.666779576ee6
96. Korchinsky, Mike. Interviewed by Brian McFarland. March 2017.
97. Ibid.
98. Wildlife Works Carbon, LLC. "VCS Project Description: The Kasigau Corridor REDD Project—Phase I Rukinga Sanctuary." Last modified January 25, 2011. http://www.vcsprojectdatabase.org/services/publicViewServices/downloadDocumentById/5885. 3.
99. Korchinsky, Mike. Interviewed by Brian McFarland. March 2017.

100. Ibid.
101. World Bank. "International Tourism, Receipts (Current US$)." Accessed March 1, 2017. http://data.worldbank.org/indicator/ST.INT.RCPT.CD?year_high_desc=true
102. Wildlife Works. "About the Founder." Accessed March 3, 2017. http://www.wildlifeworks.com/company/about-founder.php
103. Korchinsky, Mike. Interviewed by Brian McFarland. March 2017.
104. Ibid.
105. VCS. "VM0009: Methodology for Avoided Ecosystem Conversion, v3.0." Accessed March 1, 2017. http://database.v-c-s.org/methodologies/methodology-avoided-ecosystem-conversion-v30
106. Korchinsky, Mike. Interviewed by Brian McFarland. March 2017.
107. Korchinsky, Mike. Email Message to author. November 26, 2014.
108. Korchinsky, Mike. Interviewed by Brian McFarland. March 2017.
109. Ibid.
110. VCS Project Database. "The Kasigau Corridor REDD Project—Phase I Rukinga Sanctuary." Accessed March 3, 2017. http://www.vcsproject-database.org/#/project_details/562
111. Wildlife Works. "Mission." Accessed March 1, 2017. http://shop.wildlifeworks.com/pages/mission
112. International Finance Corporation. "IFC Issues Innovative $152 Million Bond to Protect Forests and Deepen Carbon-Credit Markets." Last modified October 31, 2016. http://ifcextapps.ifc.org/ifcext/pressroom/ifcpressroom.nsf/0/594A016A78A7B14E8525805D00461397
113. Wildlife Works. "Ivory Wars: Stand for Elephants." Accessed March 1, 2017. http://www.wildlifeworks.com/partners/ivorywars.php
114. Everland. "About Us." Accessed August 23, 2017. http://www.everlandmarketing.com/
115. World Bank. "Kenya: GDP (Current US$)." Accessed March 3, 2017. http://data.worldbank.org/country/kenya
116. Korchinsky, Mike. Interviewed by Brian McFarland. March 2017.
117. Wildlife Works Carbon, LLC. "VCS Project Description: The Kasigau Corridor REDD Project, Phase II." Last modified May 10, 2011. http://www.vcsprojectdatabase.org/services/publicViewServices/downloadDocumentById/6493. 3.
118. Wildlife Works Carbon, LLC. "VCS Project Description: The Chyulu Hills REDD+ Project." Last modified June 23, 2015. http://www.vcsprojectdatabase.org/services/publicViewServices/download DocumentById/20659

119. VCS Project Database. "The Mai Ndombe REDD+ Project." Accessed March 1, 2017. http://www.vcsprojectdatabase.org/#/project_details/934
120. The World Bank Group. "Taking Climate Action from Paris to the Rainforests," June 22, 2016. Accessed April 11, 2017. http://www.worldbank.org/en/news/feature/2016/06/22/taking-climate-action-from-paris-to-the-rainforests
121. Wildlife Works. "Wildlife Works Newsletter: January 2017." Accessed April 11, 2017. http://us9.campaign-archive1.com/?u=a613355d22bfd0d203935b7fe&id=e6ab731539&e=daa1a0d138
122. Code REDD. "About Code REDD." Accessed March 3, 2017. http://www.coderedd.org/about-code-redd/
123. PRWeb. "New $250 Million, Women-Led Climate Fund Joins Forces with Conservation Organizations to Save Vast Forests in Asia, Africa and Latin America." Accessed March 3, 2017. http://www.prweb.com/releases/2016/12/prweb13936152.htm
124. Wildlife Works Carbon, LLC. "VCS Project Description: The Kasigau Corridor REDD Project—Phase I Rukinga Sanctuary." Last modified January 25, 2011. http://www.vcsprojectdatabase.org/services/publicViewServices/downloadDocumentById/5885. 7.
125. Korchinsky, Mike. Interviewed by Brian McFarland. March 2017.
126. Ibid.
127. Rio Tinto. "About QIT Madagascar Minerals." Accessed April 4, 2017. http://www.riotinto.com/energyandminerals/qit-madagascar-minerals-4645.aspx
128. BBOP. "Mitigation Hierarchy." Accessed December 13, 2016. http://bbop.forest-trends.org/pages/mitigation_hierarchy
129. Lowry, Dr. Pete. Interviewed by Brian McFarland. April 2017.
130. Temple, H.J., S. Anstee, J. Ekstrom, J.D. Pilgrim, J. Rabenantoandro, J.-B. Ramanamanjato, F. Randriatafika and M. Vincelette. *Forecasting the Path towards a Net Positive Impact on Biodiversity for Rio Tinto QMM*. Gland, Switzerland: IUCN, 2012. x + 78 pp. 10. https://www.iucn.org/theme/business-and-biodiversity/our-work/business-partnerships/rio-tinto/natural-capital-project/valuing-tropical-forest-conservation-madagascar
131. Biodiversity Offsets Blog. "Biodiversity Offsets: Opportunities and Challenges for their Effective Design and Implementation—Resources from an OECD Workshop." Accessed April 5, 2017. http://www.biodiversityoffsets.net/biodiversity-offsets-opportunities-and-challenges-for-

their-effective-design-and-implementation-resources-from-an-oecd-workshop/

132. Bateson, Matthew. "Biodiversity Offsets: A Conservation Mechanism for the Private Sector." Accessed April 4, 2017. http://www.oecd.org/tad/sustainable-agriculture/Matthew%20Bateson%20Rio%20Tinto%20OECD%20Biodiversity%20Offset%20Presentation.pdf. 7.
133. Lowry, Dr. Pete. Interviewed by Brian McFarland. April 2017.
134. Myers, Norman, Russell A. Mittermeier, Cristina G. Mittermeier, Gustavo A. B. da Fonseca and Jennifer Kent. "Biodiversity Hotspots for Conservation Priorities." *Nature*. 403. 2000. 853–858. doi:10.1038/35002501. Received 22 September 1999; Accepted 22 December 1999. http://www.nature.com/nature/journal/v403/n6772/full/403853a0.html
135. Lowry, Dr. Pete. Interviewed by Brian McFarland. April 2017.
136. Lowry II, Porter P., Faly Randriatafika and Johny Rabenantoandro. "Conservation Status of Vascular Plant Species from the QMM/Rio Tinto Mining Area at Mandena, Tolagnaro (Fort Dauphin) Region, Southeast Madagascar." *Madagascar Conservation & Development*. 3(1). 2008. 59.
137. Lowry, Dr. Pete. Interviewed by Brian McFarland. April 2017.
138. Ibid.
139. Ibid.
140. Bateson, Matthew. "Biodiversity Offsets: A Conservation Mechanism for the Private Sector." Accessed April 4, 2017. http://www.oecd.org/tad/sustainable-agriculture/Matthew%20Bateson%20Rio%20Tinto%20OECD%20Biodiversity%20Offset%20Presentation.pdf. 7.
141. Lowry, Dr. Pete. Interviewed by Brian McFarland. April 2017.
142. Ibid.
143. Ibid.
144. Bateson, Matthew. "Biodiversity Offsets: A Conservation Mechanism for the Private Sector." Accessed April 4, 2017. http://www.oecd.org/tad/sustainable-agriculture/Matthew%20Bateson%20Rio%20Tinto%20OECD%20Biodiversity%20Offset%20Presentation.pdf. 7.
145. Lowry, Dr. Pete. Interviewed by Brian McFarland. April 2017.
146. Ibid.
147. Ibid.
148. Ibid.
149. Ibid.

150. Lara, Antonio and Rocío Urrutia. "The Growing Significance of Conservation: The Chilean Experiment." In *Conservation Capital in the Americas*, edited by James N. Levitt. Massachusetts: Lincoln Institute of Land Policy, 2010. 12.
151. Ecosystem Marketplace. "State of Forest Carbon Finance 2016." Accessed November 23, 2016. http://www.forest-trends.org/documents/files/doc_5388.pdf. 7.
152. Ecosystem Marketplace. "State of Forest Carbon Finance 2016." Accessed November 23, 2016. http://www.forest-trends.org/documents/files/doc_5388.pdf. 14.
153. Vitale, Ben, Tannya Lozada and Luis Suárez. "Ecuador ChoCO_2 Conservation Carbon Project: Conservation International." In *Conservation Capital in the Americas*, edited by James N. Levitt. Massachusetts: Lincoln Institute of Land Policy, 2010. 171.
154. Davis, Adam. "Mainstreaming Environmental Markets." In *From Walden to Wall Street*, edited by James Levitt. Massachusetts: Lincoln Institute of Land Policy, 2005. 161.
155. Vitale, Ben, Tannya Lozada and Luis Suárez. "Ecuador ChoCO_2 Conservation Carbon Project: Conservation International." In *Conservation Capital in the Americas*, edited by James N. Levitt. Massachusetts: Lincoln Institute of Land Policy, 2010. 169.
156. Vitale, Ben, Tannya Lozada and Luis Suárez. "Ecuador ChoCO_2 Conservation Carbon Project: Conservation International." In *Conservation Capital in the Americas*, edited by James N. Levitt. Massachusetts: Lincoln Institute of Land Policy, 2010. 170.
157. Vitale, Ben, Tannya Lozada and Luis Suárez. "Ecuador ChoCO_2 Conservation Carbon Project: Conservation International." In *Conservation Capital in the Americas*, edited by James N. Levitt. Massachusetts: Lincoln Institute of Land Policy, 2010. 166.
158. Bezbarua, Ronjoy. "De-risking Carbon Offsets: Forestry Credits Transact Under Innovative Insurance Policy." May 31, 2017. California Carbon Info. http://californiacarbon.info/de-risking-carbon-offsets-forestry-credits-transact-under-latest-insurance-policy/
159. Carbon Pulse. "ANALYSIS: Offset Freeze, Potential Exclusion Send Shivers through China's Carbon Market." Last modified March 31, 2017. Accessed April 11, 2017. http://carbon-pulse.com/32620/?utm_source=CP+Daily&utm_campaign=a19d9f4804-CPdaily31032017&utm_medium=email&utm_term=0_a9d8834f72-a19d9f4804-36342993

160. Governors Climate and Forest Task Force. "Unofficial Translation: State of Acre: Bill No 2.308 of October 22, 2010. Accessed April 11, 2017. http://www.gcftaskforce.org/documents/Unofficial%20English%20Translation%20of%20Acre%20State%20Law%20on%20Environmental%20Services.pdf
161. Federal Republic of Brazil. "Decreto Nº 8.576, de 26 de Novembro de 2015." Accessed April 8, 2017. http://pesquisa.in.gov.br/imprensa/jsp/visualiza/index.jsp?data=27/11/2015&jornal=1&pagina=1&totalArquivos=232
162. Climate Focus. "Acre, Brazil: Subnational Leader in REDD+." May 2013. Accessed April 11, 2017. http://www.climatefocus.com/sites/default/files/acre_brazil.pdf
163. Duchelle, Amy E., Maron Greenleaf, Denyse Mello, Maria Fernanda Gebara and Tadeu Melo. "Acre's State System of Incentives for Environmental Services (SISA), Brazil." *CIFOR*. Accessed April 11, 2017. http://www.cifor.org/redd-case-book/case-reports/brazil/acres-state-system-incentives-environmental-services-sisa-brazil/
164. Earth Innovation Institute. "Acre." Accessed April 11, 2017. http://earthinnovation.org/our-work/regional-initiatives/brazil/acre/
165. Environmental Defense Fund. "REDD+: Protecting Tropical Forests." Accessed April 11, 2017. https://www.edf.org/climate/redd
166. World Wildlife Fund. "Brazil's Acre State and German Development Bank Agree to Performance-Based Payments Linked to REDD+." January 11, 2013. Accessed April 11, 2017. http://wwf.panda.org/?207235/Brazils-Acre-state-and-German-development-bank-agree-to-performance-based-payments-linked-to-REDD
167. Wayburn, Laurie. "The Van Eck Forest: Carbon Markets and the New Economic Paradigm for Forest Sustainability." In *Conservation Capital in the Americas*, edited by James N. Levitt. Massachusetts: Lincoln Institute of Land Policy, 2010. 177.
168. California Air Resources Board. "International Sector-Based Offset Credits." Accessed April 11, 2017. https://www.arb.ca.gov/cc/capandtrade/sectorbasedoffsets/sectorbasedoffsets.htm
169. Salazar, Milagros. "Costa Rica Aims For Carbon Neutrality With Payments For Ecosystem Services." January 27, 2014. *Ecosystem Marketplace*. Accessed April 8, 2017. http://www.ecosystemmarketplace.com/articles/costa-rica-aims-for-carbon-neutrality-with-payments-for-ecosystem-services/
170. UNFCCC. "Paris Agreement." Accessed April 8, 2017. Available: https://unfccc.int/resource/docs/2015/cop21/eng/l09.pdf

171. Ali, Hamza. "18 Countries Sign Declaration on International Carbon Markets." December 14, 2015. *Environmental Finance*. Accessed April 8, 2017. Available: https://www.environmental-finance.com/content/news/18-countries-sign-declaration-on-international-carbon-markets.html
172. Settelmyer, Scott. Interviewed by Brian McFarland. February 2017.
173. Thomson Reuters. "Carbon Market Analyst: Will ICAO Save the CDM?." Accessed April 11, 2017. http://carbon-pulse.com/18995/
174. ICAO. "3. What would be the Impact of Joining CORSIA?" Accessed April 11, 2017. http://www.icao.int/environmental-protection/Pages/A39_CORSIA_FAQ3.aspx
175. Schep, Ellen, André van Velzen and Jasper Faber. "A Comparison between CORSIA and the EU ETS for Aviation." December 2016. *CE Delft*. Accessed April 11, 2017. https://www.transportenvironment.org/sites/te/files/publications/2016_12_CE_Delft_ETS_CORSIA_final.pdf. 6–7.
176. Ecosystem Marketplace. "State of Forest Carbon Finance 2016." Accessed November 23, 2016. http://www.forest-trends.org/documents/files/doc_5388.pdf. 5.
177. Kolos, Maria. Interviewed by Brian McFarland. March 2017.

12

Ecotourism and Private Reserves

> "In 1960, people around the world made 25 million trips outside their home countries. Last year, that number passed one *billion*. Tourism has become one of the most powerful, most influential and least-examined forces in the world. It produces $6.5 trillion of the global economy and employs one out of every 12 people on earth."[1]
>
> —Michael Caruso

Introduction

Costa Rica is often viewed as one of the most successful ecotourism places in the world. For instance, in 2010, Costa Rica generated more revenue from tourism than from cattle, coffee, and bananas combined.[2]

Historical Overview

Low-impact tourism likely began earlier, but in 1983, the term "ecotourism" was first coined by Mexican architect Héctor Ceballos-Lascuráin.[3]

B.J. McFarland, *Conservation of Tropical Rainforests*, Palgrave Studies in Environmental Policy and Regulation, https://doi.org/10.1007/978-3-319-63236-0_12

Other notable dates throughout the history of ecotourism include:

- 1923: In 1923, Barro Colorado Island (BCI), which is located within the Panama Canal's Gatun Lake, "became one of the first biological reserves in the New World, and is now the most intensively studied area in the tropics."[4]
- 1983 (November): Rara Avis S.A. is established. Rara Avis S.A. now manages the Rara Avis Rainforest Lodge, which is the oldest tropical rainforest lodge in Costa Rica.
- 1986: The Sian Ka'an Biosphere Reserve, a relatively popular ecotourism destination in Mexico, is established.
- 1990: The International Ecotourism Society (TIES) was established and was "launched by a team organized by Megan Epler Wood at a conference in Florida in 1989 as the world's first international nonprofit dedicated to ecotourism as a tool for conservation and sustainable development."[5]
- 1990: The Adventure Travel Trade Association (ATTA) is established.[6]
- 1993: "That reputation has brought definite payoffs, particularly in the tourism industry, which in 1993 surpassed bananas as the country's {i.e., Costa Rica} leading revenue source."[7]
- 2002: The year is "declared International Year of Ecotourism. Quebec (Canada) holds the World Ecotourism Summit, which adopts the Quebec Declaration on Ecotourism."[8]
- 2003: Center for Responsible Travel (CREST) is established.[9]
- 2007: Partnership for Global Sustainable Tourism Criteria was formed. In 2009, the Sustainable Tourism Stewardship Council was created. Both entities merged in August 2010 to form the Global Sustainable Tourism Council (GSTC).[10]

Mechanisms of Instrument

As explained by the World Wildlife Fund, "nature tourism provides a strong incentive for governments, communities and businesses to conserve species and their habitat since tourists demand a high-quality experience characterized by beautiful scenery and abundant wildlife."[11] Ecotourism operators, aside from private reserves or concession holders, often do not own the land. More often, ecotourism operators, rather than own land, have different arrangements. These different arrangements could be mutually exclusive deals between an ecotourism operator and the site owner (e.g., exclusive right to a lodge is fairly common in Botswana), partial exclusivity (i.e., ecotourism operator holds one of a few licenses to visit a park), or there could be some form of royalty payment to the site owner (e.g., commission to private landowners or a fee to the government). Furthermore, ecotourism operators may offer tours to a single location (i.e., Serengeti National Park), a particular region (i.e., the Amazon Rainforest), or offer tours around the world.

The destination—whether be a national park or a UNESCO World Heritage Site—is often public land. However, there are private reserves and concession holders which are specifically focused on the area they own or lease. Private reserves exist where there is a private landowner and there are also land grants or concession rights given to particular groups for a specified period of time. Private land conservation can work well, but it is often very limited in scope. Likewise, tourism can be fantastically high in revenue on a national level such as in Costa Rica, but small on a per unit scale (i.e., per acre scale). In Mexico, for example, some local communities own the land but these lands are not always well setup for tourism.

It is important to note that the visitors to such places are not necessarily tourists but may also be students and/or researchers. For instance, this includes undergraduate or graduate students participating in the School for Field Studies[12] and researchers visiting the Smithsonian Tropical Research Institute in Panama.[13]

Ideally, the destinations of ecotourism will have low impacts due to a restricted number of visitors that is both enforced and updated according to the sustainability of the location. Similarly, you often need a permit to visit the protected area and there are limits on the number of people who can visit.

Size of Instrument

The Center for Responsible Travel's *The Case for Responsible Travel: Trends & Statistics* 2016 Report estimated that:

> International tourist arrivals (overnight visitors) grew by 4.4% in 2015, reaching a total of 1184 million in 2015. Some 50 million more tourists traveled internationally in 2015 than in 2014, and 2015 marked the 6th consecutive year of above-average growth since the 2009 economic crisis. The UN World Tourism Organization (UNWTO) Confidence Index predicts a continuation of growth in international tourism in 2016.
>
> The travel industry contributed US$7.2 trillion or 9.8% to world GDP in 2015, and is forecast to grow by 4% per annum over the next ten years. Leisure spending represents 77% of travel & tourism GDP, with business spending contributing 23%. Travel and tourism also provided 284 million jobs (direct, indirect, and induced) in 2015, representing 9.5% of total employment or 1 in 11 jobs in the world.[14]

Nature-based tourism is estimated to account for approximately 20% of total international travel, which would equate to more than USD$1.4 trillion in 2015.[15] That is astounding and provides a potentially significant opportunity for conservation gains and financial investments. Although a bit outdated, the United Nations Environment Programme (UNEP) and Conservation International reported "that tourism in biodiversity hotspots more than doubled between 1990 and 2000."[16]

Introduction to Case Studies

The following case studies will examine the Rara Avis Rainforest Lodge & Reserve in Costa Rica, the Smithsonian Tropical Research Institute (STRI) in Panama, and the Sian Ka'an Biosphere Reserve in Mexico. The Rara Avis Rainforest Lodge is an ecotourism lodge, whereas the STRI in Panama raises a portion of its funding from visiting scientists, and the Sian Ka'an Biosphere Reserve raises a portion of its funding from ecotourism.

Rara Avis S.A. was established in November 1983 as a Costa Rican corporation which now manages the Rara Avis Rainforest Lodge. The Lodge is the oldest tropical rainforest lodge in Costa Rica, "and in April 1986, President Monge signed a decree expanding Braulio Carrillo National Park to the boundaries of Rara Avis and the La Selva Biological Station."[17] Collectively, the Braulio Carrillo National Park, the Zona Protectora La Selva (La Selva Protection Zone; includes the La Selva Biological Station), and Rara Avis Rainforest Lodge provide a contiguous protected area of more than 100,000 hectares (i.e., approximately 250,000 acres) and are a part of the larger Central Volcanic Range Conservation Area that covers more than 250,000 hectares (617,500 acres).[18]

In 1923, Barro Colorado Island (BCI), which is located within the Panama Canal's Gatun Lake, "became one of the first biological reserves in the New World, and is now the most intensively studied area in the tropics."[19] Barro Colorado Island would later become part of the Smithsonian Institution in 1946, and in 1966, "the organization changed its name to the Smithsonian Tropical Research Institute (STRI)."[20] Today, STRI is the longest standing and most modern research facility in the Neotropics—if not the world.

The Sian Ka'an Biosphere Reserve is located in the South of the Yucatan Peninsula in Mexico's State of Quintana Roo. The total area of the Biosphere Reserve was 528,148 hectares (1.3 million acres) with approximately 120,000 hectares (296,400 acres) of marine area, which includes a portion of the Mesoamerican Barrier Reef (i.e., the world's second largest barrier reef), and approximately 408,000 hectares (1 million acres) of tropical rainforests and palm savannas. However, based on studies from Amigos de Sian Ka'an, Sian Ka'an was expanded to the South in 1994 with the declaration of the Uaymil Flora and Fauna Protection Area, and to the East with the creation, in 1998, of the Sian Ka'an Reefs Biosphere Reserve. Now the 3 protected area complexes have 652,193 hectares (1.61 million acres). Several NGOs, including Amigos de Sian Ka'an, applied for and received approximately USD$1 million in funding from the Inter-American Development Bank (IADB) and the WWF-Fundación Carlos Slim Foundation to create a new destination in Quintana Roo: Maya Ka'an, one of the most ambitious ecotourism initiatives in the world.[21]

Case Study #1: Rara Avis Rainforest Lodge & Reserve in Costa Rica

Introduction

On the border of Braulio Carrillo National Park and the Zona Protectora La Selva (La Selva Protection Zone, which includes the La Selva Biological Station) is the oldest tropical rainforest lodge in Costa Rica known as the Rara Avis Rainforest Lodge. Collectively, these 3 areas provide a contiguous protected area of more than 100,000 hectares (i.e., approximately 250,000 acres) and are a part of the larger Central Volcanic Range Conservation Area that covers more than 250,000 hectares (617,500 acres) (Fig. 12.1).[22]

Rara Avis S.A. was established in November 1983 as a Costa Rican corporation which now manages the Rara Avis Rainforest Lodge, "and in April of 1986, President Monge signed a decree expanding Braulio

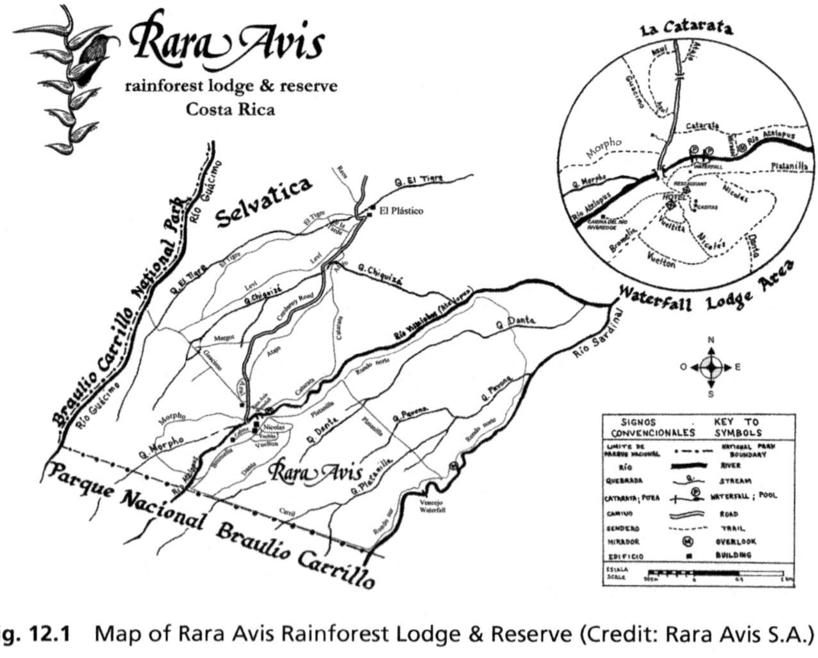

Fig. 12.1 Map of Rara Avis Rainforest Lodge & Reserve (Credit: Rara Avis S.A.)

Carrillo National Park to the boundaries of Rara Avis and the La Selva Biological Station."[23]

By 1986, Rara Avis S.A. had purchased the initial parcels and remodeled a building at El Plastico, which was formerly owned by Aserraderos Unidos (United Sawmills in English). Soon thereafter, in July 1987, United Sawmills offered to sell its remaining land in the area which totaled more than 700 hectares (approximately 1,700 acres) and buildings at El Plastico. By March 1988, Rara Avis S.A. was able to purchase the majority of the land.[24] Today, the Rara Avis Rainforest Lodge "directly conserves 485 hectares (1,200 acres) of tropical rainforest and has indirectly conserved an additional 1,000 hectares (2,500 acres)."[25]

Identify the Problem

Throughout the 1970s and 1980s, Costa Rica had some of the highest rates of deforestation throughout Central America and even throughout the world. As explained by the United Nations University in Costa Rica:

> In the 1940s, over 75 per cent of the country was covered in indigenous woodland, mostly tropical rainforest. In the subsequent decades, however, rampant and unchecked logging ensued as the nation's valuable forest resources were transformed into cash profits. By 1983 only 26 per cent of the country retained forest cover, and the deforestation rate had risen to 50,000 hectares per year.
>
> At this point, something amazing started to happen. By 1989 the annual deforestation rate had dropped to 22,000 hectares per year. The figure dropped even lower to 4,000 hectares per year by 1994 and in 1998 the deforestation rate had dropped to zero. Today forest cover has increased to 52 per cent (double 1983 levels), and the government has set the ambitious goal of further increasing this figure to 70 per cent and achieving carbon neutrality by 2021.[26]

The land now owned by Rara Avis S.A. was also facing deforestation pressures as the land was formerly owned by the sawmill consortium United Sawmills and other landowners.

Why the Problem Is Important

Tropical deforestation and degradation throughout Costa Rica was a problem because it was among the highest rates in the world. Furthermore, Costa Rica has some of the most biodiversity in the world which is dependent upon these very tropical forests.

How Problem Was Identified

The problem of deforestation was identified by satellite imagery and could be easily witnessed firsthand. In addition, the problems at the Rara Avis Rainforest Lodge were identified by its founder Mr. Amos Bien.

Effectiveness of Process for Identifying Problem

With more than 100,000 hectares of contiguous protected forest across Braulio Carrillo National Park, the La Selva Protection Zone, and the Rara Avis Rainforest Lodge, it appears as though the process was very effective at identifying the problems. Thus, this contiguous forest provides recreational opportunities for ecotourists at the Rara Avis Rainforest Lodge, a pristine study site for researchers at the La Selva Biological Station, and reduced pressure on the Braulio Carrillo National Park.

Steps Taken to Address the Problem

The steps Rara Avis S.A. took to help address the problem with deforestation included, but were not limited to:

- Established Rara Avis S.A. as a Costa Rican corporation;
- Engaged a wide range of stakeholders including government officials, local communities, and international NGOs;
- Acquired land and buildings from United Sawmills;
- Built the necessary infrastructure including access roads and the Lodge; and

- Worked with the Costa Rica Government to extend the park boundaries of Braulio Carrillo National Park.

Results

The greatest result, according to the founder Amos Bien, has been spreading the concept of using tourism—including ecotourism—as a tool for nature conservation and rural poverty alleviation. Rara Avis pioneered this idea, which was quite radical when it started, and has used it as a model to be copied all over Costa Rica. Costa Rica has, in turn, become a model for the rest of the world.[27]

Second, Rara Avis pioneered the use of non-timber forest products (NTFPs) as an income supplement for rural residents (e.g., native tree seedlings, butterfly farms, and crafts) and trained people from the community so that they could start their own small businesses. Rara Avis also trained the first generation of biologist and naturalist guides in Costa Rica, who in turn have trained many generations since.[28]

Additional noteworthy results include Rara Avis had the first commercial website in Costa Rica (1995), the first canopy tramway (1984), the first canopy rappelling (1986), and the first canopy treehouse (1990). Rara Avis was also the founder of the Costa Rican Network of Private Nature Reserves, which became an influential NGO in Costa Rica and helped start similar organizations throughout Latin America, which conserve hundreds of thousands of hectares.[29]

With respect to innovative financing results, Rara Avis "has been financed entirely by the purchase of stock by conservation-minded investors and by loans at market interest rates. Stock purchase offers time-sharing rights in the lodges and the possibility of building a cabin in the rainforest, as well as a proportional share of dividends." Ownership is in the form of 3,900 shares of common stock and 100 shares of preferred stock. Stockholders are entitled to "time-sharing in the lodge and the hotel if they hold three or more shares." In addition, holders of preferred stock "receive the same time-sharing privileges, as well as a tolerance to build a cabin or research station in the forest if they agree to comply with certain conditions."[30]

It is estimated that "time-sharing privileges for {common} stockholders in the Rara Avis tourist facilities provide a non-cash return on investment worth (at 1998 prices) 12.5–17.8% per year. This is of importance, as extremely difficult access, small number of rooms, and significant debt service make it unlikely that dividends will be declared without substantial new equity investment. These privileges are transferable and accumulate with certain restrictions."[31] To finance additional rainforest land purchase, "100 shares of common stock were converted into preferred stock selling at one thousand United States dollars per share. Ninety-five shares have been sold, while five shares are being held in reserve. The preferred stock carries the same time-sharing rights and earning as the common stock (0.025% of distributed net profit per share), as well as the right to construct a cabin on Rara Avis land, at stockholder's cost, with the purchase of five shares and agreement to abide by the *Bylaws for Preferred Stockholders*."[32]

Rara Avis also innovated with profit-sharing and employee stock ownership. For instance, "employees with two or more years of service receive a packet of three shares of common stock of the corporation, and the Board of Directors is authorized to reserve up to 5% of the distributed net profit as profit-sharing for those employees not on the Board."[33]

Furthermore, it is also important to note that Rara Avis has been able to help to conserve the tropical rainforest, preserve the area's rich biodiversity, partner with the surrounding communities, and provide an opportunity for ecotourists to see Costa Rica's biodiversity. Such biodiversity includes an estimated 367 bird species, along with jaguars (*Panthera onca*), ocelots (*Leopardus pardalis*), and Baird's tapirs (*Tapirus bairdii*).[34]

Challenges and How They Were Met

Three challenges that the Rara Avis Rainforest Lodge faced were start-up and learning how to run a business, underfinancing, and the "founder's effect."

It is important to note that the Rara Avis Rainforest Lodge was a first mover and started up an entirely new industry. Thus, there were no market studies, no competition studies, and no established bank financing. There were also challenges associated with learning how to run a business, such as understanding the business culture and legislation, along with learning how to be an hotelier.[35]

Underfinancing, since day one, was an issue due to the very difficult access, embezzlement by a former manager, near bankruptcy several times, and crises of one sort or another (i.e., as in any business).[36] For instance, before the access road was established and the Lodge was built, personnel and materials had to be carried in by either horse or by foot. In addition, cash flow challenges, specifically achieving profitability in the early years, were also present. The upfront costs of acquiring the land and establishing the infrastructure was expensive, while income was later generated via paying visitors and through the sale of seedlings.

As explained by Rara Avis S.A.:

> Until 1994, Rara Avis continued to grow at about 20 to 30% per year in income and occupancy. Starting in 1995, there was a substantial drop, reflecting the rapid expansion of the nature lodge market in Costa Rica, the removal of Donald Perry's experimental canopy tram from Rara Avis, the opening of the commercial Rainforest Aerial Tram by Perry in another location, and the decline of German tourism, which had become a mainstay of Rara Avis. As a result, the scholarship program and others were suspended.[37]

Lastly, there is a challenge of the "founder's effect." As described by Amos Bien, the founder of the Rara Avis Rainforest Lodge, "as the founder of a start-up 34 years ago, it is time for a new generation of management to step in as I step aside, in order to innovate. This is common to most start-ups."[38]

Beyond Results

The Rara Avis Rainforest Lodge should be considered an economically and ecologically sustainable enterprise as Rara Avis S.A. was established nearly 35 years ago and the Rainforest Lodge has been operating for more than 30 years. The forests within the Rara Avis Rainforest Lodge remain preserved.

Furthermore, Rara Avis S.A. has undertaken extensive educational initiatives, in addition to diversifying its revenues, which will likely contribute to the sustainability of its operations. For instance, a scholarship program was established, groups such as the School for Field Studies (i.e., including the author in 2002) have visited the area of Braulio Carrillo National Park, a butterfly chrysalises project was established in 1991 with funding from WWF that has been expended to surrounding local communities, and Rara Avis helped establish La Red Costarricense de Reservas Naturales (The Costa Rican Network of Natural Reserves).[39]

Lessons Learned

A few lessons learned by Amos Bien, founder of the Rara Avis Rainforest Lodge:

- For a business to be environmentally and socially sustainable, it must also be economically sustainable (i.e., profitable).
- Operating in sustainable manner also means working with and among one's neighbors. Without the help and support of the people of Horquetas, Rara Avis would never have gotten started, let alone continued to operate for 34 years—in good times and bad times.
- Finally, a business is a business, and clients need to be satisfied with their experience—just having great natural surroundings and good sustainability policies is not enough. Being "in a rainforest without a guide is like being in a library and not knowing how to read. Interpreting nature, not just naming trees and birds, but telling stories about them and their context, can satisfy clients almost as much as seeing monkeys or jaguars. The best guides combine the storytelling abilities of a campesino (i.e., a rural farmer) with the wide-ranging scientific knowledge of rainforest ecology and the social context."[40]

Other Resources on Rara Avis S.A.

Central Volcanic Range Conservation Area

- http://www.sinac.go.cr/EN-US/ac/accvc/Pages/default.aspx

La Red Costarricense de Reservas Naturales (The Costa Rican Network of Natural Reserves in English)

- http://reservasnaturales.org/

Organization of Tropical Studies' La Selva Biological Station

- http://ots.ac.cr/index.php?option=com_content&task=view&id=163&Itemid=348

Rara Avis S.A. Official Website

- http://rara-avis.com/

Case Study #2: The Smithsonian Tropical Research Institute (STRI) in Panama

Introduction

Founded in 1846 and based in Washington, DC, the Smithsonian Institution is a preeminent educational institution with museums—such as the Smithsonian National Museum of Natural History and the National Zoological Park—and international research facilities.[41] In 1923, Barro Colorado Island (BCI), which is located within the Panama Canal's Gatun Lake, "became one of the first biological reserves in the New World, and is now the most intensively studied area in the tropics."[42] Barro Colorado Island would later become part of the Smithsonian Institution in 1946, and in 1966, "the organization changed its name to the Smithsonian Tropical Research Institute (STRI)."[43] Today, STRI is the longest standing and most modern research facility in the Neotropics—if not the world.

BCI is 1,560 hectares (approximately 3,853 acres) and, together with five adjacent peninsulas, forms the greater Barro Colorado Nature Monument (BCNM), which totals 5,600 hectares (approximately 13,832 acres). These five adjacent peninsulas, which were added as part of the Panama Canal Treaty of 1977, are known as:

- Bohio Peninsula;
- Buena Vista Peninsula;
- Frijoles Peninsula;
- Gigante Peninsula; and
- Peña Blanca Peninsula.[44]

The BCNM abuts the Soberania National Park that protects an additional 22,000 hectares (approximately 54,340 acres) of lowland tropical rainforest. BCI and the larger Smithsonian set of facilities in Panama have some of the most modern tropical research facilities with numerous laboratories and computer rooms,[45] and consequently, BCI and surrounding areas are among the most studied tropical research sites in the world (Fig. 12.2).[46]

The annual budget of STRI is approximately USD$35 million with approximately two-thirds of this funding from direct US Congressional appropriations, and the remaining one-third is a mix of traditional philanthropy and major gifts, along with receipts from competitive funding (e.g., National Science Foundation funding).[47]

STRI also generates revenue via visitor fees including fees for visiting researchers and students as well as daily visitors. For a list of BCI's 2016 fees, please see here:

http://stri.si.edu/english/research/applications/fees/PDFs/BCI_Fees_FY2016.pdf

While the revenue from visitor fees is relatively minor, it presents a unique case study of the complimentary role visitor fees can play when

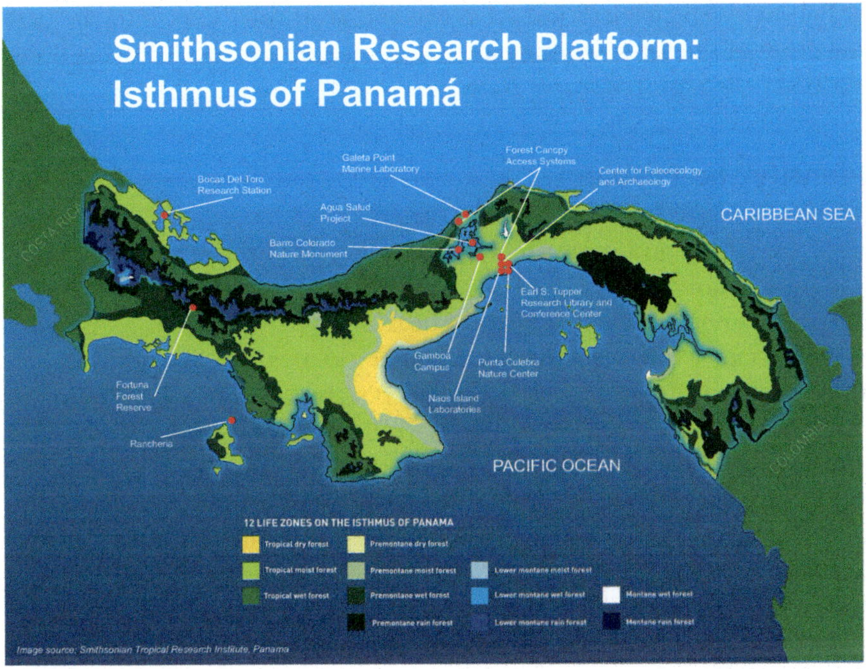

Fig. 12.2 Map of Smithsonian research platform: Isthmus of Panama (Credit: Smithsonian Institution)

operating a research facility and simultaneously preserving lowland tropical rainforest. STRI tries to keep these visitor fees as low as possible and the fees mainly cover the operating costs for the individual visitors. This said, BCI has dorm room space for 65 people and the visitor fees cover the basic costs. However, STRI's endowment, the annual Congressional appropriations, and donations help to cover all the remaining costs associated with transportation, park rangers, technology, and laboratory equipment. Also, while the adjacent Soberania National Park is separately funded and administered by the Panama Ministry of Environment, the research conducted by STRI in the BCNM and elsewhere in Panama is often used to inform land management and conservation decisions made by the government of Panama and by private landowners.[48]

For an excellent historical reference and an insider look into the personalities of scientists working at BCI, please see Elizabeth Royte's book entitled, *The Tapir's Morning Bath.*

Identify the Problem

Early on at BCI and in nearby forested areas, there was poaching of deer and other animals for food and for their pelts. Today, poaching and illegal logging are not problems at BCNM because of a 24-7 Smithsonian game warden staff[49]; however, there are still some areas of Panama that are threatened by illegal deforestation and poaching.[50] In 1952, "the canal watershed, which includes land beyond the Zone, was more than eighty percent forested; today {2002}, only forty-seven percent is forested."[51] Illegal logging and poaching have been long-standing problems throughout Central America, but have become less of an issue within Panama and specifically within the BCNM. This reality is in part because Panama has had a relatively stable economy for 100 years and its economy is more diverse than other Central American countries. For instance, the Panama Canal annually generates approximately USD$2 billion, with half of the revenue going to Panama's general fund and the other half of the revenue being used to maintain the Canal. In addition, the Panamanian airline company Copa Airlines is the largest airline in the region and employs nearly 9,000 people.[52] These economic activities underpin a generally robust and diverse economy. This diversified economy generates jobs which helps take pressure off Panama's lowland tropical rainforests.

Nevertheless, there are still some occurrences of illegal logging at Soberania National Park. One of the most sought after species is known as cocobolo (*Dalbergia retusa*) which is a vulnerable tropical hardwood species[53] that can be sold for as much as USD$10,000 per cubic meter—on par with the ivory trade.[54]

Furthermore, forest fragmentation is another problem in the region—particularly along the Mesoamerican Biological Corridor. Big cats, such as jaguars (*Panthera onca*) and pumas (*Puma concolor*), utilize ecological or biological corridors, but unfortunately, the Corridor is fragmented as even highways have been built across the Corridor.[55]

Why the Problem Is Important

Deforestation, forest degradation, and poaching within BCNM—if it were to occur—would be a problem because of the ongoing research that is dependent upon intact rainforests.

The Panama Canal is absolutely essential to the Panamanian economy. With respect to the greater BCNM area and the surrounding much larger Panama Canal Watershed (more than 3,000 km^2), if such areas were deforested, the Canal would experience increased sedimentation and flooding risk. Reduced forested watershed area leads to higher probability of flooding and damage to critical Canal infrastructure.[56]

How Problem Was Identified

The various problems facing BCNM and the watershed have been identified through research by both staff scientists and visiting scientists. For instance, an STRI Research Associate, Mr. Ricardo Moreno, works with jaguar conservation at the organization Yaguará Panamá Foundation.[57] Unfortunately, Mr. Moreno documented 40 jaguars that were shot and killed by ranchers and small landowners in 2016. The population of jaguars is already threatened and it is illegal to kill a jaguar in Panama. To identify these problems, Mr. Moreno meets with local communities; asks the communities for permission to install motion-sensitive, wildlife cameras; and then pays the landowners approximately USD$20 per jaguar sighting as an incentive to protect jaguars.[58]

Furthermore, land use change and its impact on the Panama Canal Watershed (PCW) are identified via satellite imagery, aerial photography, and on-the-ground research. This said, Dr. Robert Stallard and Dr. Jefferson Hall of STRI, along with fellow researchers, published a paper in 2013 entitled, *Effect of land cover and use on dry season river runoff, runoff efficiency, and peak storm runoff in the seasonal tropics of Central Panama*. The paper reported that between "1986 and 2003, the PCW lost 8% of its mature forest. {...) {and from} 2003 to 2008 in the PCW shows that the forested area has increased by approximately 4% as the result of forest conservation, reforestation, agroforestry, and the abandonment of pasture and crop lands."[59]

Effectiveness of Process for Identifying Problem

The process of identifying problems has been effective—particularly through the rigorous process of scientific research with peer reviewed journals.

Steps Taken to Address the Problem

When the Smithsonian Institution took over, a park ranger program—in conjunction with the national police force—was implemented. Nowadays, armed game wardens patrol the island on a 24-hour basis. These "Guardabosques are, literally, forest guards: rough-and-ready Panamanians who wear camouflage, carry sidearms, and valiantly patrol the island for poachers. (Looking for snares and listening for gunshots, they sometimes catch two or three poachers in a night, or none for an entire month)."[60] This ongoing monitoring and enforcement has helped to essentially eliminate occurrences of deforestation and forest degradation, along with most of the poaching, at least within the BCNM.[61]

Being a research institute, STRI's staff scientists and visitors help address the problems of the surrounding BCNM area through research. For instance, the Agua Salud Project is a multiyear study led by dozens of scientists who are researching how to best manage forests and water resources in the watershed to sustain the most environmental benefits and the most economic benefits without degrading the landscape.[62]

Results

There have been numerous results achieved at STRI and the greater BCNM in Panama, including, but not limited to:

- More than 1,500 people from the United States and from around the world visit STRI each year.[63] These 1,500 annual visitors include approximately 600 scientists and 900 students, representing 50 countries (i.e., more than 25% of all nations on Earth).

- A total of 225 ant species have been identified at BCI, along with 120 mammals (i.e., more than half of which are bats).[64]
- 12 terrestrial and marine research locations around Panama have now been established.[65]
- 400+ staff work for the Smithsonian Institution in Panama.[66]

Challenges and How They Were Met

The key challenge for STRI, despite its advantage over other research facilities because of its affiliation with the Smithsonian Institution, is funding.

To address this challenge, STRI has diversified its revenue sources to include US Congressional appropriations, traditional philanthropy, competitive funding, and visitor fees. STRI, due to it being a research institute that is part of the Smithsonian Institution, does not have legal authority to make policy; rather, STRI uses scientific research to inform policy. This neutral policy position has enabled STRI to maintain good relations with the people and government of Panama and to minimize the risk of Congressional budget cuts that could result from partisan affiliations.[67]

Beyond Results

The work of STRI is sustainable and significantly contributes to the growing body of scientific research, while training a new generation of researchers, and while working on issues important to tropical rainforests around the world.

This includes the former PRORENA initiative, which was led by "{…} the Smithsonian Tropical Research Institute and by the Yale Tropical Resources Institute at the Yale School of Forestry and Environmental Studies. It benefits from more than 80 years of STRI and Yale research on forest ecology and forest dynamics."[68] Nowadays, PRORENA has evolved into the Agua Salud Project along with the Environmental Leadership & Training Initiative (ELTI) of the Yale School of Forestry & Environmental

Studies.[69] In essence, the Environmental Leadership & Training Initiative entails Yale University leading workshops and trainings in Panama and neighboring countries for small landowners to optimize their forest resources. In addition, the initiative also trains local government officials such as park management staff in Costa Rica and government officials from Panama.[70] A comprehensive report has also been published based off recent watershed research and a 2-day conference in Panama in 2014 "under the auspices of ELTI and the PRORENA project (a collaboration between STRI and Yale FES with the goal of understanding the barriers to reforestation with native tree species) in collaboration with the Biodiversity and Ecosystem Services Program of the Inter-American Development Bank."[71]

The Center for Tropical Forest Science (CTFS), which is now known as the Center for Tropical Forest Science—Forest Global Earth Observatory (CTFS-ForestGEO), is a network of 64 forest plots in 25 countries around the world, led in partnership with more than 100 institutions, and is:

> {...} a program within the Smithsonian Tropical Research Institute that joins together—through formal memoranda—a voluntary association of natural and social scientists and institutions around the world. The mission of CTFS is to promote and coordinate long-term biological and socio-economic research within tropical forests and forest-dependent communities, and to translate this information into results relevant to forest management, conservation, and natural resource policies.[72]

In addition, the Biological Dynamics of Forest Fragments Project (BDFFP) "is the world's largest-scale and longest-running study of habitat fragmentation, operated cooperatively by the Smithsonian Tropical Research Institute and Brazil's National Institute for Amazonian Research."[73] This is located in Amazonas, rather than in Panama, but demonstrates the expanded reach of STRI.

STRI's research extends beyond tropical lowland rainforests and also focuses on marine conservation issues. Dr. Héctor Guzmán, for example, conducted research on humpback whales (*Megaptera novaeangliae*) by radio tagging the whales to understand how the whales interact with the high frequency of ships—a total of approximately 17,000 per year—that pass through the Panama Canal. Dr. Guzmán's research showed that if ships followed a narrower course while in shallower waters, the impact on humpback whales could be reduced by 95%.[74] Dr. Guzmán also identified offshore areas where the Ministry of Environment could establish marine protected areas (MPAs) and then helped the Ministry go through the legal steps (e.g., public forums), and now the Presidential Decree is in effect. This Decree was signed during a Panama governmental event held at STRI facilities, and Panama now has approximately 13.6% of its waters classified as MPAs.[75]

Lessons Learned

A key lesson described by Dr. Matthew C. Larsen, the fifth Director of STRI, is that it helps to start slow, build carefully, and be as strategic as you can. While not easy to apply and time-consuming, a lot of the strength and capacity of STRI come from the fact that the Institute has had a long, steady evolution. The idea started with a few scientists receiving permission to establish small wooden houses on a former hilltop that, with the damming of Lake Gatun as part of the Canal, formed BCI, where they could conduct research mainly during the winter months. Over time, the management of BCI was transitioned to the Smithsonian Institution (1946), and with time, each successive director had varying levels of success to increase funding, open new facilities, and make connections to new donors. In addition, another lesson described is the importance of being a part of a large entity such as the Smithsonian Institution.[76]

Other Resources on STRI

Center for Tropical Research Institute's Barro Colorado Island

- http://www.ctfs.si.edu/site/Barro+Colorado+Island

Chapman, Frank M. *My Tropical Air Castle: Nature Studies in Panama*. New York: D. Appleton and Co., 1929 (First Edition).

Leigh, Dr. Egbert Giles and A. Stanley Rand. Donald M. Windsor (Ed.). *The Ecology of a Tropical Forest: Seasonal Rhythms and Long-Term Changes*. Washington, DC: Smithsonian, 1986.

Leigh, Dr. Egbert Giles. *A Magic Web: The Tropical Forest of Barro Colorado Island*. United Kingdom: Oxford University Press Inc., 2003.

Leigh, Dr. Egbert Giles. *Tropical Forest Ecology: A View from Barro Colorado Island*. United Kingdom: Oxford University Press Inc., 1999.

STRI's Conservation Page

- http://www.stri.si.edu/english/conservation/index.php

STRI's Official Website

- https://www.stri.si.edu/

Royte, Elizabeth. *The Tapir's Morning Bath: Mysteries of the Tropical Rain Forest and the Scientists who are Trying to Solve Them*. Boston: Houghton Mifflin Company, 2001.

Case Study #3: Sian Ka'an Biosphere Reserve in Mexico

Introduction

The Sian Ka'an Biosphere Reserve is located in the South of the Yucatan Peninsula in Mexico's State of Quintana Roo. The total area of the Biosphere Reserve was 528,148 hectares (1.3 million acres) with approximately 120,000 hectares (296,400 acres) of marine area, which includes a portion of the Mesoamerican Barrier Reef (i.e., the world's second largest barrier reef), and approximately 408,000 hectares (1 million acres) of tropical rainforests and palm savannas.[77] However, based on studies from Amigos de Sian Ka'an, Sian Ka'an was expanded to the South in 1994 with the declaration of the Uaymil Flora and Fauna Protection Area, and to the East with the creation, in 1998, of the Sian Ka'an Reefs Biosphere Reserve. Now, taken together, the 3 protected area complexes have 652,193 hectares (1.61 million acres) (Fig. 12.3).

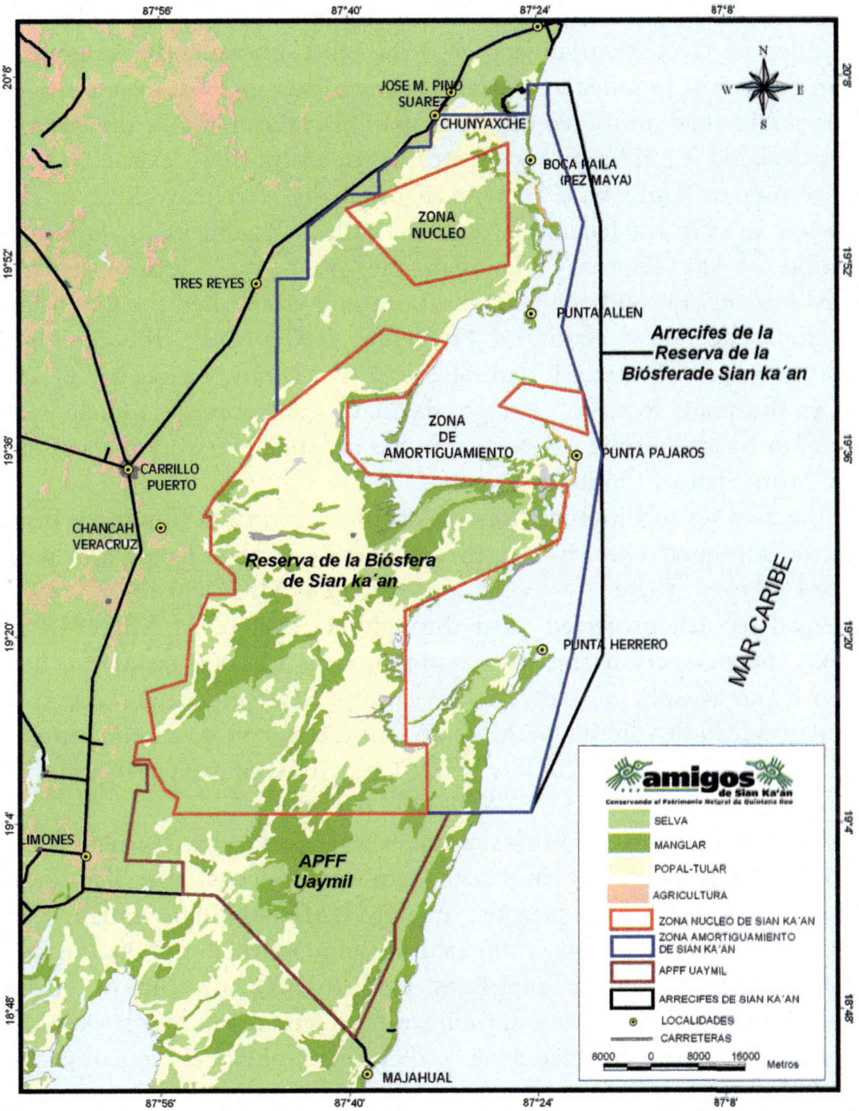

Fig. 12.3 Map of Sian Ka'an Biosphere Reserve (Credit: Amigos de Sian Ka'an A.C.)

The Sian Ka'an Biosphere Reserve was established via a Mexican Presidential Decree on January 20, 1986, and subsequently, Amigos de Sian Ka'an was founded as a private organization in 1986 by a group of citizens that had promoted the creation of Sian Ka'an. While the Reserve was declared a public protected area operated by the Mexican Federal Government, Amigos de Sian Ka'an originally started working in the Reserve to assist the federal government with management and administration of the Reserve.[78] Through the years, the Mexican Federal Government consolidated its institutions and established the Comisión Nacional de Áreas Naturales Protegidas (CONANP; The National Commission of Protected Natural Areas) which now manages the Sian Ka'an Biosphere Reserve.[79] Amigos de Sian Ka'an continues to help with the Sian Ka'an Biosphere Reserve, but has expanded its work throughout the entire State of Quintana Roo.

The Sian Ka'an Biosphere Reserve's primary source of funding is from Mexico's Federal Government to pay the Reserve's director and park guard salaries. There is a fee of approximately MXN$50 to MXN$60 charged to visit protected areas throughout Mexico. At MXN$50 to MXN$60 per person and with approximately 100,000 people visiting Sian Ka'an Biosphere Reserve each year, the Reserve generates approximately MXN$5 million to MXN$6 million per year. At the current exchange rate of MXN$20.28 = USD$1.00, this is equivalent to approximately USD$246,632 to USD$295,958 per year.

Approximately 80% of this revenue is supposed to be returned to the protected areas in the region that the fees were collected. The Sian Ka'an Biosphere Reserve receives approximately USD$400,000 annually, but this is not too much when compared to the more than 650,000 hectare size of the three Reserve complexes. Unfortunately, the current federal government is not honoring the obligation to send the money back to the region and to the protected areas; rather, this money now remains with the federal government.

However, there are additional sources of indirect and direct funds provided by NGOs such as WWF and Amigos de Sian Ka'an, Fondo Mexicano para la Conservación de la Naturaleza (Mexican Foundation for the Conservation of Nature), UNDP, the World Bank, and many others. Such NGOs receive donations to undertake projects both within, and around, the Reserve. For example, several NGOs including Amigos de Sian Ka'an applied for and received approximately USD$1 million in funding from the Inter-American Development Bank (IADB) and the WWF-Fundación Carlos Slim Foundation to create a new destination in Quintana Roo: Maya Ka'an, one of the most ambitious ecotourism initiatives in the world.[80] Such activities included the development of the brand Maya Ka'an and undertaking market studies to ultimately make central Quintana Roo a tourism destination.[81] These funds will not go directly to the Reserve, but will help the Reserve by increasing visitation to the Reserve and by reducing the impact of tourism in the buffer zones.

Identify the Problem

As explained by Amigos de Sian Ka'an, when the Sian Ka'an Biosphere Reserve was first established, there were not that many problems in the region. There were more potential, hypothetical problems, rather than actual problems, because Sian Ka'an was more of a visionary project that was responding to future problems, rather than reactionary and responding to current problems. In contrast to commercial agriculture or large-scale tourism, the traditional Mayan form of agriculture was common in the area. Under this system, the Mayans would stay in a given area for 3 to 4 years to grow corn, cabbage, and beans. When the soil was exhausted, the Mayans would move onto another area and let the forest recover. This traditional form of agriculture is viable with a low population density and has lower impact on the forests.[82]

Throughout the State of Quintana Roo, the main problems today are deforestation as a result of urban sprawl, water pollution, forest fires, and agriculture.[83] Such deforestation leads to a loss of freshwater, greenhouse gas emissions, and climate change, along with the threatened extinction of several species such as jaguars (*Panthera onca*) and ocelots (*Leopardus pardalis*) due to habitat loss and poaching.[84]

Similarly, the World Heritage Convention reports the following problems within the Sian Ka'an Biosphere Reserve:

> Sian Ka'an is susceptible to frequent and heavy tropical storms. The barrier reef provides natural protection for the coast, a telling example of

> conservation contributing to disaster preparedness. As for human impacts, the inaccessibility protects large tracts of the property. Besides the coastal fishing villages of Punta Allen and Punta Herrero, there are few permanent residents in the property. Hunting, fishing and collection of forest products, however, are widespread. Sport fishing and commercial fishing to supply nearby tourism centres has resulted in marked declines of some species, notably the Spiny Lobster. Management responses are needed. Agriculture north of the property bears pollution risks pollution and fires set to clear land have repeatedly spread into the property. Alien invasive species are reported, mostly along the dirt tracks on land but also in the ocean. The main economic sector directly and indirectly impacting on the property, however, is tourism. Fishing lodges and clubs, small hotels, cabins and trailer parks are the visible manifestations within the property. Tourism has reached proportions of mass tourism along parts of the Yucatan Coast and the property is in the vicinity of Tulum and Cancun, two of Yucatan's major tourist attractions. Associated coastal urbanisation with, for example, well-documented garbage and sewage problems, require monitoring and management responses. Attempts to encourage low impact forms of tourism in the property to promote public awareness and visitor education but also as a source of conservation funding deserves consolidation.[85]

One of the biggest problems within the Reserve is pollution, particularly water pollution[86] and the trash that is washed up on beaches.[87] With respect to trash on the beaches, a lot of people think the visitor fees go directly to the Reserve and then they complain about the trash. However, the Reserve receives trash from 2 primary sources. First, due to the currents, a lot of trash from Central and South America is washed up along the Reserve's 120 kilometers of shoreline. Second, the Reserve receives a lot of trash from passing boats, freightliners, and cruise liners due to the heavy traffic and throwing trash in the ocean. This trash destroys the natural beauty, hurts the local biodiversity, and presents an overall image problem. For instance, people complain to Amigos de Sian Ka'an about how they can have a Reserve with a lot of trash, yet Amigos de Sian Ka'an is not in charge of the Reserve but rather is working all throughout the State of Quintana Roo.[88]

Why the Problem Is Important

The aforementioned problems facing Sian Ka'an Biosphere Reserve are important because it provides a refuge for a tremendous amount of terrestrial and marine biodiversity. For instance, the roots of mangrove forests provide habitat for fish nurseries, and the mangrove forests also provide storm protection and have great capacity of carbon sequestration and storage.[89]

The Reserve also provides some important income to local communities.

In addition, Sian Ka'an Biosphere Reserve provides a lot of connectivity. For example, the coral reef within the Sian Ka'an Biosphere Reserve is connected to the entire Mesoamerican Reef, which stretches from Honduras, to Guatemala, to Belize, and to Mexico.[90] The hydrological system throughout the region, including the Reserve, connects the tropical forests, wetlands, and the coral reefs together via the aquifer. There is also a formal Mesoamerican Biological Corridor, and while not decreed by the Mexican Federal Government, there is still a dedicated institution in Mexico that receives funding to help support local communities within these corridors to manage their forest, to help do ecologically sound productive projects (e.g., the Maya Ka'an tourism initiative is within this corridor), and to help coordinate some of the actions between protected areas such as between the Sian Ka'an Biosphere Reserve, the Calakmul Biosphere Reserve (Southwest of Sian Ka'an), and the Bala'an K'aax Reserve (West of Sian Ka'an). It is also important to note that the Calakmul Biosphere Reserve is contiguous to Mexico's Montes Azules Biosphere Reserve and Guatemala's Maya Biosphere Reserve[91]; see case study *Naranjitos Conservation Easements in Guatemala* in Chap. 8 for more information on the Maya Biosphere Reserve in Guatemala.

How Problem Was Identified

On a regional level, satellite imagery has been used to identify land use change and deforestation. In addition, some aerial surveys for land use change are conducted via high-resolution photographs taken from airplanes. Such work allows for the identification of land use changes, authorities can be altered, and practitioners are able to target specific communities to work with to reduce deforestation.

As explained by Amigos de Sian Ka'an, throughout the years, the techniques change depending on the availability of funding. Currently, Amigos de Sian Ka'an is doing a lot of work on threat identification and technical, scientific work. The process has been undertaken with experts, such as TNC, along with local communities. As a result, Amigos de Sian Ka'an is now doing a lot of research on the region's aquifer, particularly to identify the threats to drinking water, threats to the coral reef, and threats directly to the Sian Ka'an Biosphere Reserve.[92]

Effectiveness of Process for Identifying Problem

In general, according to Amigos de Sian Ka'an, these strategies have been effective. The strategies are part of the institutional framework and allow for integrating new approaches and help identify threats to the region. In addition, mapping is a very useful tool as you can map just about everything.[93]

Steps Taken to Address the Problem

One step taken to help address the problems facing the Reserve, along with the adjacent areas, is to diversify funding. In addition to developing the Maya Ka'an concept, Amigos de Sian Ka'an is also exploring a payments for ecosystem services (PES) model for the entire State of Quintana Roo. Similar to the water fee that citizens already pay on their utility bill, the idea is to add a little percentage to create a sustainable financing mechanism to support the conservation of the tropical rainforest that is collecting the water that citizens drink in Quintana Roo.[94]

Amigos de Sian Ka'an has also scaled their work from working exclusively with the Sian Ka'an Biosphere Reserve to now working throughout the entire state of Quintana Roo, Mexico.

Results

There are numerous results achieved at the Sian Ka'an Biosphere Reserve. This includes the fact that the Reserve was designated a World Heritage Site in 1987[95] and 652,193 hectares (1.61 million acres) have been conserved between the Sian Ka'an Biosphere Reserve, the Uaymil Flora and Fauna Protection Area, and the Sian Ka'an Reefs Biosphere Reserve.[96]

Including the Sian Ka'an Biosphere Reserve, Amigos de Sian Ka'an has worked with the government and communities to establish 14 protected areas covering 1.2 million hectares (3 million acres) throughout Quintana Roo. On a side note, it takes 8 years on average to establish a protected area in Quintana Roo.

In addition, the first Permanent Beach Cleaning Program has been established at the Sian Ka'an Biosphere Reserve, and within the first 2 years, 29 tonnes (nearly 65,000 pounds) of waste were collected and removed from the Reserve.[97]

More recently, by studying the region's aquifer, Amigos de Sian Ka'an is helping to change paradigms. In addition to changing the hydrology models of the peninsula and changing how people see the aquifer, the results are also helping to promote development policies, planning tools, and policy actions.

Another result is Maya Ka'an, which is a brand that was created at the center of Quintana Roo as a tourist destination. This was a very important result for Amigos de Sian Ka'an because creating the Maya Ka'an site and positioning the Sian Ka'an Biosphere Reserve as a tourism destination will help spread the benefits beyond Sian Ka'an to the entire region. The initial goal is to include 15 rural communities and benefit an estimated 65,000 people.[98] Amigos de Sian Ka'an is planning to help give market viability to these communities and to link tourism with the conservation of the Sian Ka'an Biosphere Reserve, along with the other protected areas throughout Quintana Roo and with all the natural resources located outside of protected areas. Ultimately, this will help Quintana Roo diversify its tourism and the type of tourism offered. Such a model of tourism is undertaken in Costa Rica, and representatives of Amigos de Sian Ka'an traveled to Costa Rica to study their models for application to Maya Ka'an.[99]

Challenges and How They Were Met

Throughout Quintana Roo, with impacts on the Sian Ka'an Biosphere Reserve, is a general change in landholdings. In the past, more of the land tenure was in the form of ejidos. Ejidos was a legal structure created after the Mexican Revolution that gave farmers, especially poor farmers, the possession of the land as long as it was managed in a communal way and without the right to sell the land. This prevented the accumulation of large landholdings by a few wealthy landowners. In the 1990s, this began to change as the Mexico Constitution was reformed and the possessors of ejidos were then able to sell their land. Two situations have since arisen.

First, there are now big ejidos of 10,000–20,000 hectares with some even as big as 100,000 hectares. Land speculation is starting to happen now and some large companies are also buying land to expand their farms for biofuels production.

Second, the region is now faced with many thousands of people from all around the world who have become owners of small properties. These new landowners are diverse, not necessarily just wealthy people, because one can purchase a hectare of land for approximately MXN$100,000 which is the same price of a square meter in a nearby city. These 2 scenarios are leading to forest fragmentation and deforestation in areas that used to be under collective ownership, and this makes management of the tropical rainforest much more difficult. Amigos de Sian Ka'an, along with their partners, is now thinking about how to avoid the worst impacts of this development.

To address these changing landholdings and to counteract the currency movements of the peso which are making it more expensive to buy land for conservation, Amigos de Sian Ka'an is looking to diversify revenue. One such idea is to incorporate the PES charge for conservation. There are lots of social, legal, and political resistance to adding a little percentage for PES for the conservation of tropical rainforests. There are already some communities who are being paid to preserve their watersheds, which is an informal PES model covering 1,000 hectares (2,470 acres). By creating a more formal PES structure, an additional form of conservation financing could be added to the mix. It is important to note that in the North of Quintana Roo State, the water recharge service is a very important ecosystem service as the Riviera Maya development depends on water a lot. In contrast, in the South of Quintana Roo State, there is more potential for carbon sequestration ecosystem services. This said, there are not enough consumers in the South to pay for the water recharge services, but there is a greater potential for carbon sequestration.

Another big challenge is water pollution. Water pollution is a growing problem because many cities throughout the region are expanding without proper wastewater treatment facilities. To address this challenge, many organizations—including Amigos de Sian Ka'an—are investing a lot of time and funds into studying water and the aquifer.[100]

Beyond Results

The results at the Sian Ka'an Biosphere Reserve appear to be sustainable, particularly with respect to ecotourism. Amigos de Sian Ka'an, now with more than 30 years of experience, started out doing little tours in Sian Ka'an. The local communities saw the possibility that ecotourism could become a real opportunity. The communities, such as Muyil, asked permission to start their own tours. Amigos de Sian Ka'an changed their scale from operating the little tour to developing the business, doing training, and helping the local communities operate the tour. The tour was eventually turned entirely over to the local communities. Amigos de Sian Ka'an then scaled up again from working with a few cooperatives near the Reserve to the scale of the entire Quintana Roo State for all the people in the region interested in ecotourism to benefit.[101]

Lessons Learned

With 24 years of experience, including 11 years as the Executive Director, Mr. Gonzalo Merediz Alonso of Amigos de Sian Ka'an offers the following lessons:

- Having an NGO involved with all the projects, especially because NGOs can provide continuity, is invaluable. Establishing protected areas and all the associated conservation activities are long-term processes because it takes a lot of time to wait for the proper political moment, the right social mood, and sufficient financing opportunities—to have all of these three elements aligned is even more difficult. For instance, in Mexico, the municipal government changes every 3 years and the federal government changes every 6 years. NGOs provide continuity through (or despite) government change. If an NGO is unable to establish a protected area under a federal government, the NGO is able to adapt and change until the next administration when the possibility of establishing a protected area may arise. Some people complain about having to start all over with new administration, but NGOs can work behind the scenes on other opportunities.

- Another important lesson learned is that practitioners have to learn to read the social environment (i.e., understand the local culture) when implementing a project. For instance, in the Mayan region, there was a self-sustaining economy until the market economy came along. Unfortunately, many changes have taken place and poverty has come with the market economy.
- Similarly, many local communities want to get involved in projects, but they lack a business culture. Amigos de Sian Ka'an made the mistake early on, and the Mexican Federal Government keeps making the mistake, of putting too much pressure on local communities to establish cooperatives but the culture is more individual. Thus, there needs to be more patience and allow people to organize themselves.[102]

Other Resources on Sian Ka'an Biosphere Reserve

Amigos de Sian Ka'an

- http://www.amigosdesiankaan.org/en/

EcoColors' Tours to Sian Ka'an

- https://www.ecotravelmexico.com/

Mesoamerican Biological Corridor

- http://www.tbpa.net/page.php?ndx=65
- http://www.biodiversidad.gob.mx/v_ingles/corridor/mesoamerican-Cor.html

Maya Ka'an

- http://mayakaan.travel/

TNC's Valuing Nature in Mexico

- http://www.nature.org/ourinitiatives/regions/northamerica/mexico/explore/valuing-nature-english.pdf

World Heritage Convention's Site on Sian Ka'an

- http://whc.unesco.org/en/list/410

Return

A restricted supply of availability to a particular ecotourism destination can result in low population density, a high value for tourism, and reduced impact on the environment. Governments, particularly national governments, generate revenue from visitors through a variety of means. This includes visitor fees to protected areas, taxes on hotels, and airline departure charges. It is important:

> When establishing a program of protected area entry fees, planners should consider a number of feasibility issues, such as the annual number and origin of tourists; the potential economic value of the species, habitat, scenic beauty, or other natural attributes; and the accessibility of the protected

area. Foreign tourists are generally willing to pay substantially higher fees than many protected areas charge, but protected area authorities (particularly in developing countries) often choose not to increase the price due to an unsubstantiated fear of losing tourism (Krug et al. 2002). Many protected areas have implemented tiered systems in which foreign tourists, regional tourists, and national citizens are charged separate entry fees. By setting tiered fees according to visitors' ability to pay, rather than charging only foreign tourists, protected areas can increase the total amount of revenue collected.[103]

Private ecotourism operators primarily earn revenue from paying tourists. Some of this revenue may directly go to the protected areas, which are managed by private or public entities. Operators can also structure a revenue share with private reserves, while also selling advertisements on their website, selling or renting products (e.g., binoculars, field guides, t-shirts, travel mugs, water bottles, patches, and hats), or by offering additional services such as travel insurance.

Ecotourism operators, in conjunction with the destination, can help improve revenue through:

> {…} improved signage and trail maintenance; ranger exchanges and training opportunities on habitat conservation and management {invasive species identification, handling poachers, etc.); and English-language instruction programs designed specifically for {…} park rangers to improve their engagement with ever-growing number of international ecotourists.[104]

However it is important to note that in many places, such as a lot of parks throughout Africa, conservation is required to pay for itself (e.g., through ecotourism and/or visitor fees) and there are limited, or no, additional funds available through government domestic budgetary allocations.

Other unique mechanisms to generate revenue include species-related user fees, sport hunting fees, "green safaris," and voluntary contributions from tourists and/or tourism operators.[105]

With respect to species-related user fees, that are certain places—such as Rwanda—which charge an additional fee of approximately USD$750 per foreign tourist to walk into the protected area to see gorillas.[106]

One mechanism to solicit voluntary contributions from tourists and/or tourism operators involves offering certified carbon offsets to travelers and possibly incorporating a matching program between the ecotourism company and the traveler(s).

With respect to costs, one assessment:

> {…} of private-area costs from available data indicates that acquisition costs ranged between $294 and $409 per hectare for small areas of between 700 and 1,000 hectares. Per hectare costs for larger reserves were relatively lower (e.g., $126 for a 60,000-hectare reserve). Annual operating costs ranged between $9.30 and $18.90 per hectare for small reserves to between $1.50 and $24 per hectare for larger protected areas. The high value is probably explained by the large amount invested in research included in that figure.[107]

These costs will likely vary due to the nature of the business and its targeted clientele.

Risk

Ecotourism, and especially conventional forms of tourism, can have both positive and negative impacts on tropical rainforest conservation. For instance:

> While tourism can offer opportunities for wealth creation in low-income countries, in many cases growing tourist activity has exacerbated social inequality, cultural dislocations, and environmental degradation. The financial windfall generated by increasing numbers of visitors is often concentrated in the hands of the owners of a few large businesses, and only marginal economic benefits accrue to residents of the communities visited. A surge in visitors can also be a harbinger of related problems, such as human waste and noise pollution, deforestation, displacement of local people, and infringement on sensitive ecosystems.[108]

Unlike agricultural products, such as coffee and cocoa, for which certification systems like Rainforest Alliance and Fair Trade have gained traction in the market, sustainable tourism certifications are geographically fragmented. The panoply of more than 70 distinct programs worldwide that claim to certify sustainable tourism operators easily confuses both tourists and the travel agents who make bookings on their behalf. To address this concern, a new entity, the Sustainable Tourism Stewardship Council (STSC), was launched in 2009 by Rainforest Alliance, UN Foundation, UNEP, World Tourism Council, and private travel companies such as Expedia and Travelocity.[109]

A significant negative impact on the environment via unsustainable tourism can result in a variety of risks to local communities, ecotourism operations, and host governments, along with lost opportunities for tourists.

Business Risk

Many business risks associated with ecotourism and/or operating a private reserve are similar to other industries. Such business risks include:

- Theft of property or money from visitors or employees; and
- Staff morale, retention, recruitment, and training.

An additional business risk is that some ticket outlets, such as TripAdvisor and its service provider Viator, will no longer sell tickets to tourism experiences that allow "physical contact with captive, wild animals or endangered species, such as elephant rides, petting tigers, and swim-with-dolphin attractions."[110] This said, the business risk is that if you are operating a facility or protected area that involves the maltreatment of animals, there is a chance that your largest source of revenue—tourists that found your company online—may cease to exist.

Furthermore, a significant external event—such as a global recession or a terrorist attack—can result in a business risk due to a drop in tourism and travel.

Strategic Risk

Strategic risks for ecotourism companies and private reserve operators include:

- Ferocious competition from other ecotourism companies and private reserve operators;
- Inability to pay higher commissions and lost revenue share;
- Choosing the tour destinations or choosing the place to purchase a private reserve; and
- How to structure tours which require upward of 2 years of advance planning, such as trips to Antarctica.

Another strategic risk is for tour operators to choose the right type of tourism offering and be able to respond to market demands, such as whether to offer horseback riding, hot air balloon rides, and/or SCUBA diving. Otherwise, a tourist staying at an ecotourism lodge may book offsite activities with another business.

Reputation Risk

The level of corruption and the ease of doing business in many countries with tropical rainforest can be very challenging. For instance, there could be a tendency to pay bribes to the general manager of a hotel, to the owner of a private reserve, or to the park rangers in order to receive preferential treatment.

Another reputation risk—although more a risk to the protected area—is the fact that:

> In some countries, the fees collected do not always benefit conservation, as many park systems lack incentives for their staff to rigorously collect and account for entry fees. Additionally, in many countries entry fees are deposited into the general government treasury rather than allocated back to the park system. To ensure an effective financial stream, revenue from protected

area entry fees should be channeled directly back into the protected area system to cover operational needs such as staff salaries or investment needs such as infrastructure.[111]

This lost revenue can result in delayed maintenance and upkeep at a protected area, which in turn can lessen the visual appeal to tourists.

Furthermore, if visitors are harmed or lose their lives during the trip, which is possible in part due to a location's remoteness and its wildlife, this can present a reputation risk for both the tour operator and the protected area.

Greenwashing, in the sense of falsely stating or overstating one's sustainability practices, can also generate reputational risk. For instance, the cruise line industry—which is the opposite of sustainable ecotourism operators—has generated negative press due to dumping garbage at sea including a USD$40 million fine to Princess Cruise Lines Ltd.,[112] having numerous outbreaks of the illness norovirus,[113] and their generation of air pollution.[114]

Liquidity Risk

Oftentimes, there is a liquidity risk because ecotourism companies have few hard assets to sell. While lodges or websites may be transferred, it is unlikely that an ecotourism company could transfer their ecotourism operating license. This could be the case because the government may not allow the license, whether or not it is exclusive, to be transferred.

Liquidity risk is more pronounced for the owner of a private reserve, which could be more difficult to sell and recover the sunk costs—especially if the property is in a remote location and within a less stable country.

Operational Risk

In Mexico, a substantial amount of money is raised from visitors to protected areas and this money is collected by the federal government. The collected fees are then distributed to approximately 170 protected areas throughout Mexico. Thus, there is a disincentive for a particular protected area to increase visitation, when the funds are distributed throughout the country. Furthermore, oil prices have gone down and the Mexican Federal Government is cutting expenses, including expenses associated with the supervision and management of protected areas.

Legal and Regulatory Risk

There are numerous legal and regulatory risks associated with ecotourism and private reserves. This includes obtaining an operating license, obtaining the required permits to visit some protected areas, and following all the various rules, regulations, and policies.

Political risk can also be significant and can be detrimental to ecotourism operators and private reserves. For example, the Mara Conservancy in Kenya gets the great majority of its funds from entrance fees. This said, when political unrest occurs (i.e., whether due to terrorism, corruption, or government crackdown on opposition leaders) and if the number of visiting tourists drops, then subsequently local employment may go down and this increases demand for cheap bushmeat, while at the same time the Park's ability to patrol is constrained by the availability of funding.[115]

Credit Risk

Credit risk includes default risk, bankruptcy risk, downgrade risk, and settlement risk.

One of the biggest credit risks occurs when there is a long-term commitment to visit a particular location, without necessarily having a long-term commitment from clients. Some costs—such as lodge bookings might be refundable—whereas other costs (e.g., ship charters and/or airline tickets) might not be refundable. Imagine if a family of 7 plan on going to Africa for 3 weeks at a cost of USD$15,000 per person, and despite having reservations and activities booked, 2 people lose their jobs and are no longer able to afford the vacation. Alternatively, a family member could pass away and the remaining family members either cannot afford the trip, or do not feel right going on the trip without the family member who just passed away. Similarly, to charter a ship to Antarctica can cost upward of USD$1 million, and this fixed cost is levied whether the tour operator recruits 1 passenger or 50 passengers.[116]

Market Risk

Market risk includes interest rate risk, equity price risk, foreign exchange risk, and commodity price risk.

Market risks are particularly concerning for ecotourism and private reserves, since travel—especially travel to distant tropical rainforests—is a luxury item.

Currency risks, such as due to Brexit, can suddenly make it significantly more expensive for British tourists. Some currencies rise and fall with oil prices which can be a double whammy, such as for residents of Norway.[117]

Risk, Return, Time (Horizon), Taxes, Liquidity, Legal and Unique (RRTTLLU)

Risk and Return

Please see above for the risk and return associated with ecotourism and private reserves.

Time Horizon

The time horizon for ecotourism and private reserves is a medium-term to long-term commitment. Private reserves can involve capital intensive investments such as hotels, restaurants, and vehicles (e.g., safari tours, shuttle busses, and boats). Shorter time horizons may exist for smaller investments such as investments into the creation of a concession or for ecotourism companies that facilitate tours (i.e., as opposed to owning the underlying assets).

Taxes

There are likely a variety of taxes such as property taxes, business taxes, sales taxes, employment taxes, and/or some form of tourism tax for visitors (e.g., airline departure taxes or if visitors stay at onsite facilities).

Liquidity

Liquidity might exist with concessions or other onsite facilities (e.g., tourist attraction or lodging) that could be transferred to another owner.

Legal

Legal considerations associated with ecotourism and private reserves will vary from country to country. Similarly, different permits may be required to visit a national park versus a national forest, and many countries require visitors to obtain tourist visas.

Unique

One unique aspect is that the assets of ecotourism can be hard to sell, especially as opposed to sustainable, certified commodities. In addition, factors outside of the control of the ecotourism operators (such as changing consumer behaviors or a global recession) can present a unique risk.

Policy Analysis

Ecotourism, similar to green bonds and impact investing, depends on a wide variety of policies to be in place.

Defining the Problem

Vacation goers, including ecotourists, have numerous destinations to choose from, are likely traveling with friends and/or family, and will be spending a relatively large sum of disposable income. With these factors in mind, most vacation goers want to travel to a place that is safe, that is easy to get to, and that is relatively affordable. Unfortunately, many countries that host tropical rainforests are far from the European Union and/or the United States, are relatively difficult to travel to (e.g., due to distance, poor local infrastructure, and/or language barriers), and/or are relatively unsafe.

Establishing Goals

Some of the policy goals for ecotourism are to encourage travelers to visit your country while simultaneously preserving the natural landscape and providing local economic opportunities.

Selecting a Policy

One of the world's leading ecotourism destinations is Costa Rica, which is due to the country's rich biodiversity, safe environment (e.g., low crime rates, less violence, and less poverty), high literacy rates, and relatively modern facilities including an international airport, good highway systems, and quality healthcare. Furthermore, the global rankings of Central American countries against 190 countries as of June 2016 for ease of doing business, according to the World Bank, were as follows[118]:

- #62: Costa Rica
- #70: Panama
- #88: Guatemala
- #95: El Salvador
- #105: Honduras
- #112: Belize
- #127: Nicaragua

The global rankings by Transparency International of the Corruption Perceptions Index 2016 for Central American countries, out of 168 countries, were as follows[119]:

- Tied for #41: Costa Rica
- Tied for #87: Panama
- Tied for #95: El Salvador
- Tied for #123: Honduras
- Tied for #136: Guatemala
- Tied for #145: Nicaragua
- Unranked: Belize

The global rankings by the United Nations Development Programme of the Human Development Index (HDI), which combines life expectancy, education, and per capita income indicators, for Central American countries, were as follows[120]:

- Tied for #60: Panama
- Tied for #69: Costa Rica
- Tied for #101: Belize
- Tied for #116: El Salvador
- #125: Nicaragua
- #128: Guatemala
- #131: Honduras

Implementing a Policy

While several different ministries have a role in Costa Rica's economy, the primary entity responsible for tourism in Costa Rica is the Instituto Costarricense de Turismo (ICT; Costa Rican Tourism Board).[121] The ICT, among many responsibilities, also manages the Certificación para la Sostenibilidad Turística en Costa Rica (CST; the Certification of Sustainable Tourism in Costa Rica).[122]

Evaluating the Policy

The collection of policies that promotes ecotourism in Costa Rica has been very successful. This said, Costa Rica is a safe country, has successfully reversed its deforestation trend, and has generated a substantial amount of revenue from tourism. This said, in 2015, it was estimated that 1.4 million tourists visited Costa Rica,[123] and Costa Rica generated an estimated USD$3.67 billion in 2016 from tourism.[124]

In contrast, San Pedro Sula and Tegucigalpa are cities in Honduras with some of the highest homicide rates in the world,[125] and in El Salvador, there have been armed robberies in the country's national parks.[126]

Future Outlook for Instrument

As explained by Mr. Ted Cheeseman, anything that creates stability—be it political, social, or economic stability—helps tourism. Costa Rica receives far more tourists and generates significantly more funds, than other neighboring Central American countries, such as Honduras or Nicaragua, despite having similar ecosystems.[127] In addition to having higher levels of violence, the levels of corruption in Honduras and Nicaragua are both higher and are a major turnoff to visitors and tour operators.

Throughout the ecotourism industry, well-established associations are a major help to ecotourism operators. Although not directly related to tropical rainforest conservation, the International Association of Antarctica Tour Operators (IAATO) is a great example of an association that has made the prospect of getting to and from Antarctica, from either a tourist or operator point of view, very clear. In contrast, Alaska appears to have too many jurisdictions and overlapping entities that are inclined to troll operators with random bills for taxes and permits that questionably apply to operations. A key takeaway is that associations, when effective, on one hand can create much needed clarity for operations and, on the other hand, can also be a major resource through which conservation issues can be addressed.[128]

Other Resources on Ecotourism and Private Reserves

Center for Responsible Travel (CREST)

- http://www.responsibletravel.org/

Global Sustainable Tourism Council (GSTC)

- https://www.gstcouncil.org/en/

The Adventure Travel Trade Association

- http://www.adventuretravel.biz/

The International Ecotourism Society

- http://www.ecotourism.org/

Travel and Tourism Satellite Account Program

- http://travel.trade.gov/research/programs/satellite/

United Nations World Tourism Barometer

- http://www.unwto.org/facts/eng/barometer.htm

United Nations' World Tourism Organization (UNWTO)

- http://www2.unwto.org/en

World Travel & Tourism Council

- http://www.wttc.org/research/

WWF's Pay for Nature View

- http://assets.panda.org/downloads/paypernatureviewphotos.pdf

Notable Ecotourism Companies and Expeditions

Many of these ecotourism companies offer carbon neutral options to their clients, including options through Carbonfund.org Foundation.

Cascada Expediciones
Don Carlos 3227 C
Las Condes 755-0139
Santiago, Chile
Phone: +1 (888) 232-3813
Email: reservations@cascada.travel
Website: http://www.cascada.travel/index.php
Cheesemans' Ecology Safaris
555 North Santa Cruz Avenue
Los Gatos, California—95030-4336
United States of America
Phone: +1 (800) 527-5330
Email: info@cheesemans.com
Website: http://www.cheesemans.com/
Eagle-Eye Tours
4711 Galena Street
Windermere, British Columbia, V0B 2 L2, Canada
Telephone: +1 (800) 373-5678 or +1 (250) 342-8640
Email: travel@eagle-eye.com
Website: http://www.eagle-eye.com/home
Eagle Wing Tours

12 Erie Street—Fisherman's Wharf
Victoria, British Columbia, Canada
Phone: +1 (800) 708-9488
Email: info@eaglewingtours.com
Website: https://www.eaglewingtours.com/

Earthwatch Institute's Expeditions
114 Western Avenue
Boston, Massachusetts—02134
United States of America
Phone: +1 (978) 461-0081
Email: info@earthwatch.org
Website: http://earthwatch.org/expeditions/earthwatch-wildlife-and-ecosystems-expeditions

EcoColors in Mexico
Caucel Street, Lt 2, SM 307
Cancun, Quintana Ro, Mexico, CP 77560
Telephone: +52 (998) 884 9580 or +52 (998) 884 3667
Email: Info@ecotravelmexico.com
Website: https://www.ecotravelmexico.com/

Gondwana Ecotours
New Orleans, Louisiana
United States of America
Telephone: +1 (877) 587-8479
Email: info@gondwanaecotours.com
Website: http://www.GondwanaEcotours.com

Jackson Hole EcoTour Adventures
1585 Berger Lane
Jackson, Wyoming
United States of America
Phone: +1 (307) 690-9533
Email: taylor@jhecotouradventures.com
Website: http://www.jhecotouradventures.com

Maya Ka'an
Quintana Roo, Mexico
Email: info@mayakaan.travel
Website: http://www.mayakaan.travel/

Quasar Expeditions
Almagro Plaza
Ponce Carrasco E8-06 y Av. Diego de Almagro
Oficinas 1101-1106
Quito, Ecuador
Phone: (593-2) 382-5680 or +1 (866) 481-7790
Email: info@quasarex.com
Website: https://www.quasarex.com/
Rara Avis
P.O. Box 350-2070, Sabanilla
San José, Costa Rica—Cod. 11502
Phone: (+506) 2764 1111
Email: reservas@rara-avis.com
Website: http://rara-avis.com/
The Evolved Traveler
369 S. Doheny Drive, Suite 1232
Beverly Hills, California—90211
United States of America
Phone: +1 (323) 505-4414
Email: info@evolvedtraveler.com
Website: http://evolvedtraveler.com/
The School for Field Studies
100 Cummings Center, Suite 534-G
Beverly, Massachusetts—01915
United States of America
Phone: +1 (978) 741-3567
Emails, See: http://www.fieldstudies.org/about/contact
Website: http://www.fieldstudies.org/
Wild Planet Adventures
1001 Bridgeway, #455
Sausalito, California—94965
United States of America
Phone: +1 (800) 990-4376
Email: trips@wildplanetadventures.com
Website: http://wildplanetadventures.com/
Wow Borneo

Jalan Barito No 11, Palangkaraya 73112
Indonesia
Phone: +62 536-322-2099
Email: info@wowborneo.com
Website: http://www.wowborneo.com/

Notes

1. Caruso, Michael. "From the Editor—Apr 13." April 2013. *Smithsonian Magazine*. Accessed April 4, 2017. http://www.smithsonianmag.com/magazine/from-the-editor-apr-13-5482254/
2. Janzen, Daniel H. "How to Conserve a Large Tropical Wildland Through Biodiversity Development: Costa Rican Example." Presentation at Harvard University Center for the Environment, Cambridge, Massachusetts. September 30, 2010. Accessed March 17, 2017. https://vimeo.com/17675978
3. EcoClub. "International Ecotourism Monthly." Year 7. Issue 85. Accessed December 14, 2016. http://ecoclub.com/news/085.pdf. 2.
4. Smithsonian Tropical Research Institute. "History." Accessed November 23, 2016. https://www.stri.si.edu/english/about_stri/history.php
5. The International Ecotourism Society. "Our Story." Accessed December 13, 2016. https://www.ecotourism.org/our-story
6. Adventure Travel Trade Association. "Overview." Accessed December 14, 2016. http://www.adventuretravel.biz/about/
7. Daily, Gretchen C. and Katherine Ellison. *The New Economy of Nature*. 178.
8. World Tourism Organization. "History." Accessed December 14, 2016. http://www2.unwto.org/content/history-0
9. Center for Responsible Travel. "History." Accessed December 14, 2016. http://www.responsibletravel.org/whoWeAre/aboutUs.php
10. Global Sustainable Tourism Council. "Our History." Accessed December 14, 2016. https://www.gstcouncil.org/en/about/gstc-overview/our-history.html
11. World Wildlife Fund. *Guide to Conservation Finance*. 13.
12. The School for Field Studies. "Semester & Summer Programs." Accessed December 14, 2016. http://www.fieldstudies.org/programs

13. Smithsonian Tropical Research Institute. "Fees." Accessed December 14, 2016. https://www.stri.si.edu/english/research/applications/fees/index.php
14. Center for Responsible Travel. "The Case for Responsible Travel: Trends & Statistics 2016." Accessed April 4, 2017. http://www.responsible-travel.org/whatWeDo/The_Case_for_Responsible_Travel_2016_Final.pdf. 1.
15. Center for Responsible Travel. "The Case for Responsible Travel: Trends & Statistics 2016." Accessed April 4, 2017. http://www.responsible-travel.org/whatWeDo/The_Case_for_Responsible_Travel_2016_Final.pdf. 3.
16. United Nations. "Tourism to Earth's Most Threatened Areas Surges by More Than 100 per cent in Last Decade." Last modified September 12, 2003. Accessed April 4, 2017. http://www.un.org/press/en/2003/unep165.doc.htm
17. Rara Avis. "Conservation." Accessed December 28, 2016. http://rara-avis.com/costa_rica_conservation.htm
18. SINAC. "Central Volcanic Range Conservation Area (ACCVC)." Accessed April 5, 2017. http://www.sinac.go.cr/EN-US/ac/accvc/Pages/default.aspx
19. Smithsonian Tropical Research Institute. "History." Accessed November 23, 2016. https://www.stri.si.edu/english/about_stri/history.php
20. Smithsonian Tropical Research Institute. "History." Accessed November 23, 2016. https://www.stri.si.edu/english/about_stri/history.php
21. Merediz Alonso, Gonzalo. Interviewed by Brian McFarland. January 2017.
22. SINAC. "Central Volcanic Range Conservation Area (ACCVC)." Accessed April 5, 2017. http://www.sinac.go.cr/EN-US/ac/accvc/Pages/default.aspx
23. Rara Avis. "Conservation." Accessed December 28, 2016. http://rara-avis.com/costa_rica_conservation.htm
24. Bien, Amos. "Historical Summary." Email message to author. November 4, 2016.
25. Rara Avis. "Conservation." Accessed December 28, 2016. http://rara-avis.com/costa_rica_conservation.htm
26. United Nations University. "Ethics and Environmentalism: Costa Rica's Lesson." Last modified December 27, 2011. http://unu.edu/publications/articles/ethics-and-environmentalism-costa-ricas-lesson.html
27. Bien, Amos. Email message to author. April 7, 2017.

28. Ibid.
29. Ibid.
30. Ibid.
31. Ibid.
32. Ibid.
33. Ibid.
34. Rara Avis. "Biodiversity." Accessed December 28, 2016. http://www.rara-avis.com/costa_rica_biodiversity.htm
35. Bien, Amos. Email message to author. April 7, 2017.
36. Ibid.
37. Bien, Amos. "Historical Summary." Email Message to Author. November 4, 2016.
38. Bien, Amos. Email Message to Author. April 7, 2017.
39. Rara Avis. "Conservation." Accessed December 28, 2016. http://rara-avis.com/costa_rica_conservation.htm
40. Bien, Amos. Email Message to Author. April 7, 2017.
41. Smithsonian Libraries. "From Smithson to Smithsonian: The Birth of an Institution." Accessed November 23, 2016. http://www.sil.si.edu/Exhibitions/Smithson-to-Smithsonian/intro.html
42. Smithsonian Tropical Research Institute. "History." Accessed November 23, 2016. https://www.stri.si.edu/english/about_stri/history.php
43. Smithsonian Tropical Research Institute. "History." Accessed November 23, 2016. https://www.stri.si.edu/english/about_stri/history.php
44. Smithsonian Tropical Research Institute. "Conservation: BCI Nature Monument." Accessed March 2, 2017. http://www.stri.si.edu/english/conservation/bci_nature_monument.php
45. Smithsonian Tropical Research Institute. "Barro Colorado: General Use Equipment." Accessed December 29, 2016. http://stri.si.edu/english/research/facilities/terrestrial/barro_colorado/equipment.php
46. Smithsonian Tropical Research Institute. "Barro Colorado." Accessed December 29, 2016. http://stri.si.edu/english/research/facilities/terrestrial/barro_colorado/index.php
47. Larsen, Dr. Matthew C. Interviewed by Brian McFarland. February 2017.
48. Ibid.
49. Ibid.
50. Smithsonian Tropical Research Institute. "Conservation: BCI Nature Monument." Accessed December 29, 2016. http://www.stri.si.edu/english/conservation/bci_nature_monument.php
51. Royte, Elizabeth. *The Tapir's Morning Bath*. 145.

52. Star Alliance. "Copa Airlines." Accessed March 2, 2017. http://www.staralliance.com/en/member-airline-details?airlineCode=CM
53. IUCN Red List. "Dalbergia retusa." Accessed March 2, 2017. http://www.iucnredlist.org/details/32957/0
54. Larsen, Dr. Matthew C. Interviewed by Brian McFarland. February 2017.
55. Ibid.
56. Ibid.
57. Smithsonian Tropical Research Institute. "26 Jaguars Killed in Panama So Far This Year." Last modified October 03, 2016. http://www.stri.si.edu/english/about_stri/headline_news/news/article.php?id=1967
58. Larsen, Dr. Matthew. Interviewed by Brian McFarland. February 2017.
59. Ogden, F. L., T. D. Crouch, R. F. Stallard and J. S. Hall. "Effect of Land Cover and Use on Dry Season River Runoff, Runoff Efficiency, and Peak Storm Runoff in the Seasonal Tropics of Central Panama." *Water Resources Research* 49, 2013. 8443–8462, doi:10.1002/2013WR013956. Accessed April 11, 2017. http://onlinelibrary.wiley.com/doi/10.1002/2013WR013956/pdf
60. Royte, Elizabeth. *The Tapir's Morning Bath.* 121.
61. Larsen, Dr. Matthew C. Interviewed by Brian McFarland. February 2017.
62. Smithsonian Tropical Research Institute. "Agua Salud Project." Accessed March 2, 2017. http://www.ctfs.si.edu/aguasalud/
63. Smithsonian Institution. "Now+Next: Smithsonian Institution / 2015 Annual Report." Accessed December 29, 2016. https://www.si.edu/Content/Pdf/About/2015-Smithsonian-Annual-Report.pdf. 21.
64. Smithsonian Tropical Research Institute. "Barro Colorado: Visit Us." Accessed December 29, 2016. http://www.stri.si.edu/english/visit_us/barro_colorado/
65. Larsen, Dr. Matthew C. Interviewed by Brian McFarland. February 2017.
66. Ibid.
67. Ibid.
68. Smithsonian Tropical Research Institute. "PRORENA." Accessed March 2, 2017. http://www.stri.si.edu/english/research/programs/programs_information/prorena.php
69. Yale School of Forestry & Environmental Studies. "Environmental Leadership & Training Initiative." Accessed March 2, 2017. http://elti.yale.edu/

70. Larsen, Dr. Matthew C. Interviewed by Brian McFarland. February 2017.
71. Hall, Jefferson S., Vanessa Kirn and Estrella Yanguas-Fernández, (ed.). "Managing Watersheds for Ecosystem Services in the Steepland Neotropics." September 2015. *Inter-American Development Bank*. Accessed April 11, 2017. https://publications.iadb.org/handle/11319/7233
72. Smithsonian Tropical Research Institute. "Center for Tropical Forest Science." Accessed March 2, 2017. http://www.stri.si.edu/english/research/programs/programs_information/center_tropical_forest_science.php
73. Smithsonian Tropical Research Institute. "Biological Dynamics of Forest Fragments Project." Accessed March 2, 2017. http://www.stri.si.edu/english/research/programs/programs_information/biological_dynamics_forest_fragments.php
74. Smithsonian Insider. "Ships Need to Slow Down for Whales in Gulf of Panama, Scientists Advise." Last modified September 27, 2012. http://insider.si.edu/2012/09/ships-must-slow-down-for-humpback-whales-in-gulf-of-panama-scientists-say/
75. Larsen, Dr. Matthew C. Interviewed by Brian McFarland. February 2017.
76. Ibid.
77. World Heritage Convention. "Sian Ka'an." Accessed February 13, 2017. http://whc.unesco.org/en/list/410
78. Merediz Alonso, Gonzalo. Interviewed by Brian McFarland. January 2017.
79. The National Commission of Natural Protected Areas. "Who We Are/About Us." Accessed February 24, 2017. http://www.conanp.gob.mx/english.php
80. Merediz Alonso, Gonzalo. Interviewed by Brian McFarland. January 2017.
81. Maya Ka'an. "Travel." Accessed February 13, 2017. http://www.mayakaan.travel/
82. Merediz Alonso, Gonzalo. Interviewed by Brian McFarland. January 2017.
83. Amigos de Sian Ka'an. "Deforestation." Accessed December 29, 2016. http://www.amigosdesiankaan.org/en/threats/deforestation
84. Amigos de Sian Ka'an. "Extintion." Accessed December 29, 2016. http://www.amigosdesiankaan.org/en/threats/extintion

85. World Heritage Convention. “Sian Ka’an.” Accessed February 13, 2017. http://whc.unesco.org/en/list/410
86. Amigos de Sian Ka’an. “Water Pollution.” Accessed February 13, 2017. http://www.amigosdesiankaan.org/en/threats/water-pollution
87. Amigos de Sian Ka’an. “Beach Pollution.” Accessed February 13, 2017. http://www.amigosdesiankaan.org/en/threats/beach-pollution
88. Merediz Alonso, Gonzalo. Interviewed by Brian McFarland. January 2017.
89. World Heritage Convention. “Sian Ka’an.” Accessed February 13, 2017. http://whc.unesco.org/en/list/410
90. WWF. “Mesoamerican Reef.” Accessed February 24, 2017. http://www.worldwildlife.org/places/mesoamerican-reef
91. UNESCO. “MAB Biosphere Reserves Directory: Biosphere Reserve Information: Guatemala: Maya.” Accessed April 11, 2017. http://www.unesco.org/mabdb/br/brdir/directory/biores.asp?code=GUA+01&mode=all
92. Merediz Alonso, Gonzalo. Interviewed by Brian McFarland. January 2017.
93. Ibid.
94. Ibid.
95. World Heritage Convention. “Sian Ka’an.” Accessed February 13, 2017. http://whc.unesco.org/en/list/410
96. Merediz Alonso, Gonzalo. Interviewed by Brian McFarland. January 2017.
97. Amigos de Sian Ka’an. “Sian Ka’an: Permanent Beach Cleaning Program.” Accessed February 24, 2017. http://www.amigosdesiankaan.org/en/projects/sianka-an/143-limpieza-permanente-de-playas
98. Amigos de Sian Ka’an. “Sian Ka’an: Goals.” Accessed February 24, 2017. http://www.amigosdesiankaan.org/en/projects/sianka-an/144-metas
99. Merediz Alonso, Gonzalo. Interviewed by Brian McFarland. January 2017.
100. Ibid.
101. Ibid.
102. Ibid.
103. World Wildlife Fund. *Guide to Conservation Finance*. 13.
104. Simons, Paul E. “Foreword.” In *Conservation Capital in the Americas*, edited by James N. Levitt. Massachusetts: Lincoln Institute of Land Policy, 2010. X.

105. World Wildlife Fund. *Guide to Conservation Finance*. 13
106. Remarkable Rwanda. "Gorilla Tracking." Accessed December 14, 2016. http://www.rwandatourism.com/things-to-do/gorilla-tracking
107. Levitt, James N. *Conservation Capital in the Americas*. 9.
108. Milder, Brian. "Conservation Finance in the Galápagos Islands: Capitalizing the 'Missing Middle'." In *Conservation Capital in the Americas*, edited by James N. Levitt. Massachusetts: Lincoln Institute of Land Policy, 2010. 111.
109. Milder, Brian. "Conservation Finance in the Galápagos Islands: Capitalizing the 'Missing Middle'." In *Conservation Capital in the Americas*, edited by James N. Levitt. Massachusetts: Lincoln Institute of Land Policy, 2010. 112.
110. Dasgupta, Shreya. "TripAdvisor Bans Ticket Sales to Wild Animal Attractions." *Mongabay*. Last modified October 21, 2016. https://news.mongabay.com/2016/10/tripadvisor-bans-ticket-sales-to-wild-animal-attractions/
111. World Wildlife Fund. *Guide to Conservation Finance*. 13.
112. The United States' Attorney General Office, Southern District of Florida. "Princess Cruise Lines To Pay Largest-Ever Criminal Penalty For Deliberate Vessel Pollution." Last modified December 1, 2016. https://www.justice.gov/usao-sdfl/pr/princess-cruise-lines-pay-largest-ever-criminal-penalty-deliberate-vessel-pollution
113. Centers for Disease Control and Prevention. "Outbreak Updates for International Cruise Ships." Accessed February 20, 2017. https://www.cdc.gov/nceh/vsp/surv/gilist.htm
114. Friends of the Earth. "2016 Cruise Ship Report Card." Accessed February 20, 2017. http://www.foe.org/cruise-report-card
115. Cheeseman, Ted. Interviewed by Brian McFarland. September 2016.
116. Ibid.
117. Ibid.
118. World Bank Group. "Doing Business: Economy Rankings." Accessed January 12, 2017. http://www.doingbusiness.org/rankings
119. Transparency International. "Corruption Perceptions Index 2016." Accessed February 8, 2017. http://www.transparency.org/news/feature/corruption_perceptions_index_2016
120. United Nations Development Programme. "Human Development Report 2015: Work for Human Development." Last modified 2016. http://hdr.undp.org/sites/default/files/2015_human_development_report.pdf. 210.

121. ICT. "About Us." Accessed April 9, 2017. http://www.ict.go.cr/en/ict/about-us.html
122. CST. "Home." Accessed April 9, 2017. http://www.turismo-sostenible.co.cr/index.php?lang=en
123. ICT. "Cuadro 1: Distribución de los pisos de demanda internacional según Unidad de Planeamiento Turístico, Vía aérea, 2015." Accessed April 9, 2017. http://www.ict.go.cr/en/documents/estad%C3%ADsticas/cifras-tur%C3%ADsticas/pisos-de-demanda/926-2015-1/file.html; http://www.ict.go.cr/en/documents/estad%C3%ADsticas/cifras-tur%C3%ADsticas/pisos-de-demanda/926-2015-1/file.html
124. ICT. "Instituto Costarricense de Turismo, Planeamiento y Desarrollo Turístico, Administración de la Información." Accessed April 9, 2017. http://www.ict.go.cr/en/documents/estad%C3%ADsticas/cifras-econ%C3%B3micas/costa-rica/960-divisas-por-concepto-de-turismo/file.html
125. U.S. Department of State. "Honduras Travel Warning." Last modified January 23, 2017. https://travel.state.gov/content/passports/en/alertswarnings/honduras-travel-warning.html
126. U.S. Department of State. "El Salvador Travel Warning." Last modified February 14, 2017. https://travel.state.gov/content/passports/en/alertswarnings/el-salvador-travel-warning.html
127. Cheeseman, Ted. Interviewed by Brian McFarland. September 2016.
128. Ibid.

13

Debt-for-Nature Swaps

> The Tropical Forest Conservation Act (TFCA) uses various forms of debt relief to steer money towards forest conservation and civil society strengthening. TFCA involves debt reduction/loan restructuring, debt-for-nature swaps or debt buy-backs. TFCA was enacted in 1998 to offer eligible developing countries options to relieve certain official debt owed the US Government, while at the same time generating funds in local currency to support tropical forest conservation activities. In addition to forest conservation and debt relief, TFCA is intended to strengthen civil society by creating local foundations that support small grants to non-governmental organizations (NGO) and local communities.[1]
>
> —US Agency for International Development

Introduction

While debt-for-nature swaps are no longer commonly structured for the conservation of tropical rainforests, these conservation financing instruments—over the course of 30 years—were used in at least 25 countries, and the Tropical Forest Conservation Act alone generated USD$218.4 million in local conservation funds.[2]

B.J. McFarland, *Conservation of Tropical Rainforests*, Palgrave Studies in Environmental Policy and Regulation, https://doi.org/10.1007/978-3-319-63236-0_13

For an excellent paper on debt-for-nature swaps, please see *Debt-for-Nature Initiatives and the Tropical Forest Conservation Act: Status and Implementation*, by Pervaze A. Sheikh from the US Congressional Research Service.

Historical Overview

In 1987, the first debt-for-nature swaps were conducted in Bolivia and Ecuador.[3] A few, of the many, other debt-for-nature swaps have included:

- 1984: Thomas Lovejoy, then Vice President of the World Wildlife Fund (WWF), applies the earlier concept of debt-for-equity swaps to conservation and is largely attributed with coming up with the concept of debt-for-nature swaps.[4]
- 1989: The Nature Conservancy (TNC) purchases debt-for-nature swap for Costa Rica.[5]
- 1990: TNC and WWF purchase debt-for-nature swap for Costa Rica.[6]
- 1990: TNC and the Puerto Rican Conservation Trust purchase debt-for-nature swap with Dominican Republic.[7]
- 1991: TNC, Rainforest Alliance, and Monteverde Conservation League purchase debt-for-nature swap for Costa Rica.[8]
- 1991: TNC purchases debt-for-nature swap for Guatemala.[9]
- 1991: TNC, the United States Agency for International Development (USAID), and the Puerto Rican Conservation Trust purchase debt-for-nature swap for Jamaica.[10]
- 1992: TNC purchases debt-for-nature swap for Brazil.[11]
- 1993: TNC and WWF purchase debt-for-nature swap for Bolivia.[12]
- 1994: Conservation International (CI) purchases debt-for-nature swap for Madagascar.[13]
- 1995: CI and USAID purchase debt-for-nature swap for Mexico.[14]
- 1996: CI purchases at least three debt-for-nature swaps for Mexico.[15]
- 2000: CI purchases debt-for-nature swap for Ghana.[16]
- 2001: The United States does bilateral debt-for-nature swap with Belize.[17]
- 2002: Germany is the creditor for at least two debt-for-nature swaps in Ecuador.[18]

- 2004: The United States does bilateral debt-for-nature swap with Colombia.[19]
- 2006: France is a creditor for a debt-for-nature swap with Cameroon.[20]
- 2007: The United States does bilateral debt-for-nature swap with Costa Rica.[21]
- 2008: WWF and France purchase debt-for-nature swap for Madagascar.[22]
- 2009: The United States does bilateral debt-for-nature swap with Indonesia.[23]

Mechanisms of Instrument

As nicely explained by the World Wildlife Fund (WWF), which participated in some of the early debt-for-nature swaps:

> Two primary debt relief instruments have provided funding for the environment: commercial debt-for-nature swaps and bilateral debt-for-nature swaps. In commercial swaps, a commercial creditor sells debt owned by a foreign government at a discount on the secondary market (usually to an NGO). Historically, commercial swaps generated $117 million for conservation around the world, but they are no longer common. Bilateral debt-for-nature swaps are similar to commercial swaps, but involve 'sovereign' debt owed by one government to another rather than commercial debt owed to a bank or commercial creditor.[24]

More specifically:

> One option is debt reduction or redirection. Through this option, a country can choose to make interest and/or principal payments on treated debt, in local currency, to a forest fund. The payments remain in the country rather than being paid to the United States. A second option is the subsidized debt-for-nature swap, through which the U.S. government and NGOs contribute funding to reduce or cancel a portion of the eligible host country's debt. Payments on treated debt are made in local currency for conservation activities as agreed by the U.S. government, the host country government, and the NGOs. Under the third option, debt buy-back, a

country may purchase one or more eligible loans in U.S. dollars at a discount in exchange for a commitment to a tropical forest fund.[25]

Interestingly, while debt-for-nature swaps are no longer commonly structured for the conservation of tropical rainforests, these conservation financing instruments have recently been structured for marine conservation initiatives, including in the Seychelles.[26]

Size of Instrument

Under the Enterprise for the Americas Initiative, the United States entered into bilateral debt-for-nature swaps for USD$1.8 billion of debt (face value) that generated USD$177.8 million in local conservation funds. Under the Tropical Forest Conservation Act, the United States entered into bilateral debt-for-nature swaps that generated USD$218.4 million in local conservation funds.[27]

According to the United Nations Development Programme,

> The value of the debt under DNS {debt-for-nature swaps} agreements surpassed US$2.6 billion from 1985–2015 and resulted in transfers of circa US$1.2 billion to conservation projects worldwide. Bilateral DNS dominate, accounting for over 93 per cent of the total. Most transactions, about US$2 billion in value, were completed in the 1990s. From 2010–2015, transactions amounted to about US$150 million. Commercial DNS accounted for approximately US$200 million in restructured debt and US$123 million in allocations to conservation. These transactions came to an end after 2000 with the number of commercial DNS declining to just two.[28]

Furthermore, at least 25 host countries participated in debt-for-nature swaps including Bolivia, Costa Rica, Ecuador, Indonesia, and Madagascar.[29]

Introduction to Case Studies

The following case studies will examine the Tropical Rainforest Heritage of Sumatra and the Area de Conservación Guanacaste (ACG) in Costa Rica.

The Area de Conservación Guanacaste (ACG), located in Northwestern Costa Rica, is now 163,000 hectares (402,610 acres) and consists of the Santa Rosa National Park, the Guanacaste National Park, and the Rincón de la Vieja National Park, along with the Horizontes Forestry Station and the Bahía Junquillal Wildlife Refuge.[30] In 1999, the ACG became a World Heritage Site.[31] Most of the funding to establish an initial endowment for the ACG was agreed to in April 1989[32] and:

> {…} came from a landmark debt-for-Nature swap in which {Dr. Daniel} Janzen, {Minister Alvaro} Umaña, and other Costa Rican leaders, aided by Salomon Brothers debt traders, leveraged a $3.5 million donation from Sweden to buy $24.5 million worth of Costa Rican commercial debt from around the world at huge discounts. Costa Rica's central bank then ended up paying 70 percent of the debt value, or nearly $17 million, into the ACG account at the National Park Foundation, with the understanding that the revenues generated would help pay for the conservation area.[33]

The Tropical Rainforest Heritage of Sumatra consists of Gunung Leuser National Park, the Bukit Barisan Selatan National Park, and the Kerinci Seblat National Park on the island of Sumatra, Indonesia. The area is famous and greatly in need of preservation because this is the only habitat in the world where one can find an orangutan, a tiger, an elephant, and a rhinoceros in the same place. In addition, the world's largest individual flower, the rafflesia flower, can only be found in Sumatra and the neighboring island of Borneo. In 2009, a USD$30 million debt-for-nature swap was conducted to assist with the financing of 13 protected areas throughout Sumatra.

Case Study #1: Tropical Rainforest Heritage of Sumatra (Indonesia)

Introduction

The Tropical Rainforest Heritage of Sumatra is located on the island of Sumatra, Indonesia. The Tropical Rainforest Heritage of Sumatra officially consists of the Gunung Leuser National Park (GLNP, also known as the Mount Leuser National Park), Bukit Barisan Selatan National Park, and the Kerinci Seblat National Park. The area surrounding Gunung Leuser National Park is known as the Leuser Ecosystem. In 2002, the Tropical Rainforest Heritage of Sumatra achieved status as a UNESCO World Heritage Site. Originally established as a nature reserve in 1934 before being upgraded to a national park in 1980, Gunung Leuser National Park is one of the earliest national parks in Indonesia and perhaps the most famous of the 3 parks making up the Tropical Rainforest Heritage of Sumatra.[34]

Collectively, the three parks cover 2,595,124 hectares (approximately 5.5 million acres), "constituting one of the biggest conservation areas in Southeast Asia (Fig. 13.1)."[35]

The area is famous and greatly in need of preservation because this is the only habitat in the world where one can find a Sumatran orangutan

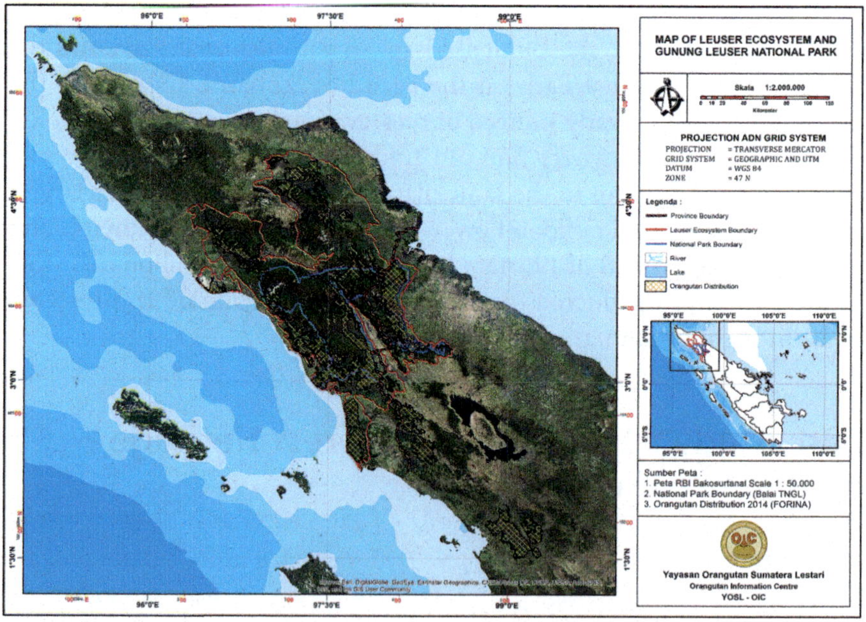

Fig. 13.1 Map of Leuser Ecosystem and Gunung Leuser National Park (Credit: Yayasan Orangutan Sumatera Lestari—Orangutan Information Centre)

(*Pongo abelii*), a Sumatran tiger (*Panthera tigris sumatrae*), a Sumatran elephant (*Elephas maximus sumatranus*) and a Sumatran rhinoceros (*Dicerorhinus sumatrensis sumatrensis*). In addition, the world's largest individual flower, the rafflesia flower (*Rafflesia arnoldii*), can only be found in Sumatra and the neighboring island of Borneo.[36] Furthermore, the Gunung Leuser National Park, alone, is "home to an estimated 10,000 plant species, including 17 endemic genera; more than 200 mammal species; and some 580 bird species of which 465 are resident and 21 are endemic. Of the mammal species, 22 are Asian, not found elsewhere in the archipelago and 15 are confined to the Indonesian region, including the endemic Sumatran orang-utan. The site also provides biogeographic evidence of the evolution of the island."[37]

While the 3 parks are relatively a far distance apart, the parks collaborate on science and monitoring efforts. Such monitoring includes monitoring of palm oil expansion, along with community and ecological monitoring.

The 3 national parks, including Gunung Leuser National Park, receive funding from the Indonesian central government, along with through the municipal government. Some of this funding is managed by NGOs, including Yayasan Orangutan Sumatera Lestari—Orangutan Information Centre (YOSL-OIC), and the exact details are outlined in MOUs.

In 2009, a USD$30 million debt-for-nature swap was conducted to assist with the financing of 13 protected areas throughout Sumatra. This includes:

- Seulawah Heritage Forest Leuser Ecosystem and Leuser National Park;
- Western Toba Watershed;
- Batang Toru Forest Range;
- Angkola Lowland Wilderness Tropical Forest Area;
- Batang Gadis National Park;
- Siak Kampar Peninsula, Tesso Nilo Ecosystem;
- Bukit Tigapuluh National Park;
- Kerinci Seblat Ecosystem;
- Siberut National Park and the rest of Mentawai Archipelago in Central Sumatra;
- Bukit Barisan Selatan Forest Range; and
- Way Kambas National Park.[38]

More specifically,

> {…} the US government agreed to forego Indonesian debt repayment of 30 million USD, with in return there being funding support provided to conservation NGOs for a period of eight years. Under the agreement, CI {Conservation International} Indonesia and the KEHATI Foundation have also agreed to provide matched funds of 1 million USD, towards the support of project implementation. Therefore the TFCA is to be considered a *subsidized debt-for-nature swap*, which is the first of its kind in Indonesia and the biggest debt for nature swap scheme by the US government for supporting environmental conservation initiatives.[39]

While the entire debt-for-nature swap was not used specifically to finance the Gunung Leuser National Park, the debt-for-nature swap helped to reduce regional pressure on the Park, as well as the pressure on the Leuser Ecosystem.

Identify the Problem

There are numerous problems facing the Leuser Ecosystem and the Tropical Rainforest Heritage of Sumatra, including Gunung Leuser National Park. In the early 1990s, the main problems were poaching, hunting, and encroachment. Such encroachment and illegal settlements, which are the biggest problems nowadays, are primarily a result of palm oil, and in fact, approximately 10,000 hectares (24,700 acres) of Gunung Leuser National Park have been cleared for palm oil. There are still lots of illegal operations, including a few businesses.

Additional problems include roads, illegal logging, and forest fires,[40] along with energy development. Some roads were built inside Gunung Leuser National Park and these roads directly threaten the forest. There are currently plans to develop new roads near Gunung Leuser National Park, along with plans to develop roads in Bukit Barisan Selatan National Park and particularly in Kerinci Seblat National Park.[41] Energy development includes mining and even geothermal has now become a threat. Beyond housing, some of the communities have established schools, houses of worship, and the infrastructure for electricity.[42] In addition, orangutans cannot survive in small forest patches which have been created as a result of encroachment and road construction.

There is also a policy problem in Aceh, a semiautonomous province on the island of Sumatra, where Aceh did not recognize the Leuser Ecosystem in their special planning. Furthermore, according to Mr. Panut Hadisiswoyo, "widespread forest conversion in TNGL {Gunung Leuser National Park} is carried out to make way for plantations and settlements for people who fled the former Aceh conflict and other local communities."[43]

Lastly, despite government funding and the USD$30 million debt-for-nature swap with the US Government, there is still a lack of adequate financing for the Leuser Ecosystem and specifically for the Tropical Rainforest Heritage of Sumatra.

Why the Problem Is Important

The problem of deforestation, degradation, and habitat fragmentation within the Tropical Rainforest Heritage of Sumatra, in general, is not being addressed by the local communities or the government. This is not to say the problem is not important to all communities and to all government officials. For instance, a local community filed a lawsuit against the government to cancel their special plan in order to incorporate the Leuser Ecosystem.

Importantly, there are several nongovernmental organizations involved in the Tropical Rainforest Heritage of Sumatra and the Leuser Ecosystem. This includes Yayasan Orangutan Sumatera Lestari—Orangutan Information Centre (YOSL-OIC), which was founded in 2001 by Mr. Panut Hadisiswoyo. Mr. Hadisiswoyo has worked in the area for 15 years and believes that caring for the environment has started to grow. Yet, in November 2016, Mr. Hadisiswoyo organized a press conference and called on the President of Indonesia for action on the problems facing the Leuser Ecosystem. There were lots of media highlights, but no official response yet.

The encroachment is a very serious problem and yet, it is not being addressed. The Ministry of Environment and Forestry has not come to see the destruction. On paper, the Ministry appears committed, but their actions seem to suggest this is not a priority.

In reality, too many families have already settled into the Gunung Leuser National Park, and the Park does not have the necessary resources to address the problem.

The problems are so severe that 50 Sumatran orangutans die each year due to illegal logging in the TNGL, while there are only about 500 Sumatran elephants left and the Park is considered a UNESCO World Heritage Site in Danger.[44]

How Problem Was Identified

One can visually see the roads, encroachments, and ensuing deforestation throughout the region.

YOSL-OIC has patrol teams who do regular forest patrols to protect the forest and support the Gunung Leuser National Park team, because the national park team simply cannot manage the entire area. The YOSL-OIC patrol team will even arrest people who are involved in illegal activities such as illegal logging and/or poaching. In addition, YOSL-OIC will use cameras to identify poachers and participate in lots of community engagement.

Effectiveness of Process for Identifying Problem

The process for identifying the problems has been effective.

Steps Taken to Address the Problem

To address the problems, YOSL-OIC is focused on four pillars throughout the Leuser Ecosystem. These pillars are:

1. Habitat restoration

   a. This includes reclaiming palm oil plantations and restoring degraded areas.

2. Law enforcement

    a. This includes having patrols, anti-poaching teams, and active investigators.

3. Address human-wildlife conflicts

    a. YOSL-OIC helps to rescue animals and translocate them into the primary forest.
    b. YOSL-OIC will also confiscate illegally captured animals.

4. Education and community engagement

    a. This includes local community capacity building.

YOSL-OIC has worked directly with more than 1,000 local farmers, including producers of rubber, oranges, and coffee to increase yields and to practice organic farming, permaculture, and agroforestry, while simultaneously working with local farmers to not expand their encroachment. YOSL-OIC is also working on forest restoration with these same local communities.

Another important step is that in 2016, it was reported that "top officials in Indonesia have reportedly declared a moratorium on new oil palm and mining concessions in the Leuser Ecosystem."[45]

Results

Since 1980, there have been some positive results at the Tropical Rainforest Heritage of Sumatra. Such results include, but are not limited to:

- 1980: Gunung Leuser National Park was established as a national park.
- 2002: The Tropical Rainforest Heritage of Sumatra was designed a World Heritage Site.

- YOSL-OIC has been working with 1,000 farmers around Gunung Leuser National Park.
- YOSL-OIC has employed approximately 50 people, including forest restoration teams, patrol teams, education teams, and management.
- YOSL-OIC has reclaimed 1,000 hectares (approximately 2,500 acres) of illegal palm oil plantations and replanted the plantations with more than one million trees. Further, elephants and orangutans are now coming back to these reclaimed sites.
- A few of the villages, who used to support the encroachment, now support YOSL-OIC's conservation initiatives.

Challenges and How They Were Met

The main challenge specifically facing Gunung Leuser National Park is the issue of enforcement.

First, the Park does not have the funds to run a law enforcement program. There are currently several poaching cases at a standstill because the Park does not have sufficient funding to pursue the cases and prosecute the perpetrators.

Second, the issue of enforcement policy quickly gets complicated. For instance, what happens when you arrest a poacher or illegal logger? In addition to the fact that the litigation process is costly and time-consuming, occasionally, a national park guard will arrest a poacher, but then the police will not want to prosecute the case. The situation gets further complicated when the leader of a village is arrested for illegal expansion and this creates local suspicion of conservation activities. Similarly, it is not that the National Park does not want to enforce the law, but the law is very complicated, especially as it relates to encroachers located in the park.

An additional challenge is that the government needs to make a major commitment to significant protection and make the Park the last stand for biodiversity conservation. There are specific laws in Indonesia that are designed to help protected areas such as Gunung Leuser National Park including Law 59090 (i.e., foundation of the Law that makes government accountable to protect biodiversity) and Law 419099 (Indonesia's forestry law). Yet, the deforestation and encroachment continue.

Another related challenge is to avoid illegal land selling.

Beyond Results

Thus far, the results are not sustainable. The financing derived from the 2009 debt-for-nature swap is coming to its end. NGOs assisting the Gunung Leuser National Park—such as YOSL-OIC—rely on grants and donations. The Indonesian Government has long-term financing, which is good, but the National Park Service only has about USD$2 million per year for Gunung Leuser National Park. With Gunung Leuser National Park being approximately 800,000 hectares (1.98 million acres) in size, this funding equates to about USD$2.50 per hectare (USD$1.01 per acre). In addition, there are only 100 employees or approximately 1 employee for every 8,000 hectares (19,760 acres).

In general, the Park is trying to stabilize the encroachment, but the Park is not able to relocate the encroachers who are already in the Park to another place.

NGOs are doing great local work, and, for instance, YOSL-OIC's vision is to restore another 3,000 hectares (7,410 acres), intensify the forest patrols, and double their team's size.

Lessons Learned

Through the work of YOSL-OIC, there are several lessons learned for the Tropical Rainforest Heritage of Sumatra, the Leuser Ecosystem, and specifically the Gunung Leuser National Park. Such lessons learned by Mr. Hadisiswoyo for the region include:

- First, if you understand the principles of how a forest works, then restoration work can do very well. This said, the illegal palm oil plantations in the forest can be relatively easily converted back to a forest, and in fact, it is already paying off, because orangutans and elephants are coming back to the restored sites.
- Second, you need to start somewhere to do conservation efforts on the ground, but you cannot simply do one thing and then stop and do

something else. Instead, you need to simultaneously do a lot of collective actions with all stakeholders such as engaging and supporting the government (i.e., push and "empower" the government) and working with local communities.

Other Resources for Tropical Rainforest Heritage of Sumatra, the Leuser Ecosystem and Gunung Leuser National Park

Conservation International's Indonesia Page

- http://www.conservation.org/where/pages/indonesia.aspx

Global Conservation

- http://globalconservation.org/projects/leuser-national-park-indonesia/

Goldman Environmental Prize Q&A with Rudi Putra

- http://www.goldmanprize.org/blog/qa-rudi-putra/

Guidebook to the Gunung Leuser National Park

- http://unesdoc.unesco.org/images/0018/001870/187099eb.pdf

Gunung Leuser National Park Official Page

- http://translate.google.com/translate?hl=en&sl=id&tl=en&u=http://gunungleuser.or.id http://gunungleuser.or.id/

Orchid Conservation Coalition

- http://www.orchidconservationcoalition.org/hl/gunungleuser.html

Sumatran Orangutan Conservation Programme

- http://www.sumatranorangutan.org/research/ketambe

TFCA-Kalimantan (Page in Bahasa Indonesia)

- http://tfcakalimantan.org/

TFCA-Sumatra (Page in Bahasa Indonesia)

- http://tfcasumatera.org/

UNESCO's Page on Gunung Leuser

- http://www.unesco.org/new/en/natural-sciences/environment/ecological-sciences/biosphere-reserves/asia-and-the-pacific/indonesia/gunung-leuser/

World Heritage Convention's Tropical Rainforest Heritage of Sumatra

- http://whc.unesco.org/en/list/1167

World Wildlife Fund

- https://www.worldwildlife.org/species/sumatran-orangutan https://www.worldwildlife.org/projects/thirty-hills

YOSL-OIC and its Publications

- http://orangutancentre.org/archive/publications/http://orangutan-centre.org/tfca-sumatera-programme/

Case Study #2: Area de Conservación Guanacaste (ACG) in Costa Rica

Introduction

The foundation for the Area de Conservación Guanacaste (ACG; Guanacaste Conservation Area in English) began in 1966 when a small piece of the ACG was decreed a national monument, and then later,[46] in 1971, the 10,400 hectare (25,688 acres) Santa Rosa National Park was created. The ACG, which is located in Northwestern Costa Rica and was created in 1986, is now 163,000 hectares (402,610 acres) and consists of the Santa Rosa National Park, the Guanacaste National Park, and the Rincón de la Vieja National Park, along with the Horizontes Forestry Station and the Bahía Junquillal Wildlife Refuge.[47] In 1999, ACG became a World Heritage Site (Figs. 13.2 and 13.3).[48]

ACG is the site of the world renowned Dr. Daniel Janzen's work. For example, Dr. Janzen along with his wife Dr. Winnie Hallwachs and Costa Rican collaborators

> {…} lobbied the government and raised the money to expand Guanacaste Province's tiny Santa Rosa National Park, then a barely surviving island amid the phantoms of the dry forests, into a 580-square-mile land and sea refuge from the pressures of the twenty-first century. The effort, eventually involving a full 2 percent of Costa Rica's land area, would in time become the world's largest tropical restoration project and an inspiration for others seeking innovative ways to finance environmental rescue missions.[49]

Most of the funding to establish an endowment for the ACG was agreed to in April 1989[50] and:

> {…} came from a landmark debt-for-Nature swap in which {Dr. Daniel} Janzen, {Minister Alvaro} Umaña, and other Costa Rican leaders, aided by Salomon Brothers debt traders, leveraged a $3.5 million donation from Sweden to buy $24.5 million worth of Costa Rican commercial debt from around the world at huge discounts. Costa Rica's central bank then ended up paying 70 percent of the debt value, or nearly $17 million, into the

ACG account at the National Park Foundation, with the understanding that the revenues generated would help pay for the conservation area.[51]

More specifically, the funding raised by this debt-for-nature swap was used for additional land purchases for Guanacaste National Park, along with an "endowment for park management, research and environmental education."[52]

Identify the Problem

The problem originally facing ACG was the threat of deforestation and degradation. As explained by Dr. Janzen, "if we hadn't started in the ACG

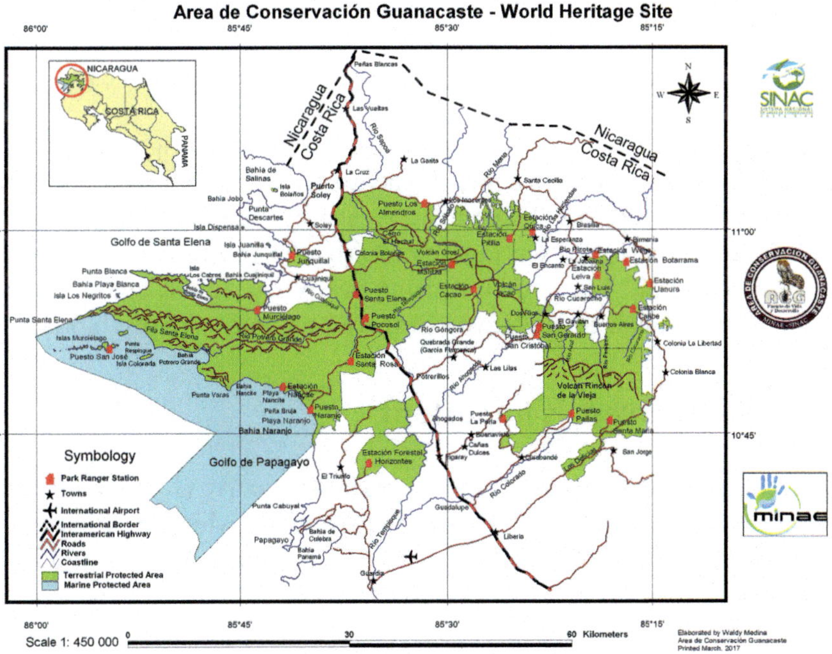

Fig. 13.2 General map of Area de Conservación Guanacaste (Credit: ACG)

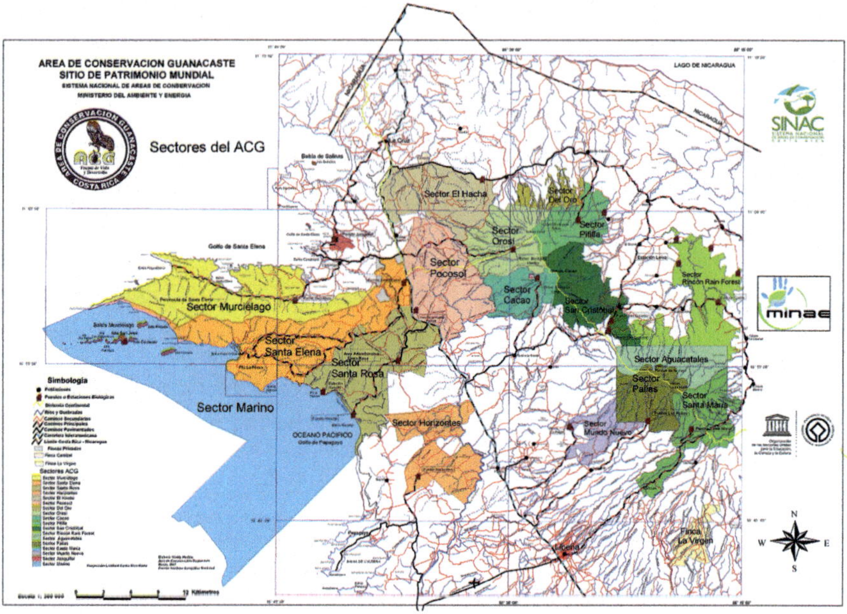

Fig. 13.3 Detailed map of Area de Conservación Guanacaste (Credit: ACG)

it would be cotton fields, cornfields, and plantations right now."[53] Dr. Janzen further explains that the threats to tropical rainforests, including forests in Costa Rica, are "fragmentation, apathy, climate change, and small size."[54] An old saying of Dr. Janzen's was that "if you can't read the forest, it's just firewood." Dr. Janzen explains this phrase is still useful, but the more recent version is that "if you can't read the forest, it's just biofuel."[55]

Why the Problem Is Important

The threat of deforestation and degradation to ACG was a problem because the area contains at least 235,000 species, which is equivalent to 65% of Costa Rica's biodiversity and equivalent to 2.6% of world's biodiversity, despite "only" having 163,000 hectares.[56] This is a tremendous concentration of biodiversity.

ACG also provides a wide range of ecosystem services. For instance, ACG's diverse ecosystems—which include tropical rainforests, tropical cloud forests, dry tropical forests, and marine ecosystems—allow for species to move between the different ecosystems. ACG also acts as a water factory. With fires stopped and restoration taking place, the ACG now "farms carbon" by both sequestering and storing carbon dioxide emissions. Practitioners are also "ranching the forest" with ecotourists and visiting researchers.

In 2010, approximately 50% of the orange juice consumed in the Eastern United States came from Costa Rica. Although these industrial firms in Costa Rica were generating revenue from juice sales and sales of essential oils (i.e., to be used for all sorts of industrial uses, including as a cleaning agent), the firms were spending significant sums of money to dispose of the orange juice rinds. In May 1996, an innovative scientific experiment with 100 truckloads of orange juice rinds was spread out with a bulldozer at ACG. Later in 1998, an old 200 hectare pastureland was covered with 1,000 truckloads of orange rinds. The scientific experiments demonstrated the sites recovered faster with a greater diversity of species; plus, the ACG was doing USD$597,000 per year in environmental services of agroscape and forest regeneration.[57]

In addition to these ecosystem services, ACG has also incorporated local communities to promote what is termed "bio-development." As explained by the Sistema Nacional de Áreas de Conservación (SINAC; National System of Conservation Areas):

> The integration of society and protected area by local employment, bio-literacy in the schools, research assistants and para-taxonomists, local fire control brigades, marine bio- sensitization, forest restoration projects, environmental services, uses of technology, are examples of how the ACG promotes the Bio-development. The ACG is displayed itself as a "conservation enterprise" where especially through its programs and projects society understands, appreciates and values even more the importance of the protected area and its conservation.[58]

This community engagement is not only the right thing to do, it also helps to contribute to the sustainability of the ACG.

How Problem Was Identified

The problem of tropical rainforest deforestation—including deforestation in Costa Rica—was identified by a wide range of stakeholders including both global and local conservation organizations. In addition, Dr. Janzen and his wife Dr. Winnie Hallwachs have been working in Costa Rica, including at the ACG, for many decades.[59]

Effectiveness of Process for Identifying Problem

The process for identifying the problems and structuring a debt-for-nature swap appears to have been relatively effective. This said, Costa Rica is a well-educated and organized country with good social security and no military expenditures. Dr. Janzen uses the analogy of the Wright Brothers. The Wright Brothers demonstrated they could fly, but did not attempt their first flight during stormy weather, which is analogous to attempting conservation activities—including the debt-for-nature swap—in more challenging countries such as Angola, the Central African Republic, Guatemala, or Honduras. In contrast, in Costa Rica, it is like a calm, spring day for the Wright Brothers.[60] By working in an "easier" country first, it is then possible to adapt the model (i.e., such as a debt-for-nature swap) to a more challenging country.

Steps Taken to Address the Problem

To address forest fragmentation, apathy among the public, global climate change, and the small size of some protected areas, the "solutions are endowment, bigger size, and the barcorder."[61]

This said, facilitating the debt-for-nature swap to establish an endowment for the ACG and the steps taken to aggregate the Santa Rosa

National Park, the Guanacaste National Park, and the Rincón de la Vieja National Park, along with the Horizontes Forestry Station and the Bahía Junquillal Wildlife Refuge, have helped address the problems of forest fragmentation and the small size of protected areas. The collective parcels of the ACG are also sequestering carbon dioxide, and the on-the-ground activities are mitigating deforestation which, in turn, contributes to the mitigation of global climate change.

The reverse apathy, Dr. Janzen has led the development of a DNA search engine and barcode directory project known as the International Barcode of Life (iBOL) Project.[62] Essentially, there is a huge process of collecting, identifying, and coding specimens and then uploading this data into a global, searchable library. This "democratization of biodiversity" will allow people to use their smartphones to learn, for example, what ant to pick up and what ant to not pick up because it has a nasty bite. In addition to its practical use, the iBOL will preserve the rich body of knowledge that has been developed by leading practitioners such as Dr. Janzen over decades of research.[63]

Results

Since 1966, there have been numerous results at the ACG. These results include, but are not limited to:

- Santa Rosa National Park was established in 1971.[64]
- The ACG, which now consists of the Santa Rosa National Park, the Guanacaste National Park, the Rincón de la Vieja National Park, the Horizontes Forestry Station, and the Bahía Junquillal Wildlife Refuge, was created in 1986.[65]
- In 1999, ACG became a World Heritage Site.[66]
- Facilitating the debt-for-nature swap.
- Identification of hundreds of thousands of species, including, as of the end of 2009, a total of 4,500 species of ACG caterpillars.[67]
- Approximately 127,000 residents and nonresidents visited ACG in 2015.[68]

Challenges and How They Were Met

The greatest challenges facing ACG, according to Dr. Daniel H. Janzen, are money and psychological support. There is a constant need for money to pay for everything. This includes money for salaries, training, buying land, equipment, and site access.[69]

To address these challenges, it is important to note that money can come in many forms because people can be "paid" in different ways.[70] For instance, a local farmer may benefit from the improved school that their child attends and which was partially financed through conservation activities.

The first challenge in many countries is getting green light from the top down. Middle management becomes the second challenge when the middle management—for whatever the reason because it interferes with their lives in some way—does not like the initiative and will slow the progress. Upon overcoming the first and second challenges, the people working on the ground (e.g., park guards and biologists) often love it. Oftentimes, you get the green light, by paying the bill.[71]

With respect to psychological support, society at large needs to be encouraging, supportive, or even neutral, when it comes to supporting the establishment and ongoing management of protected areas. This support includes the next door neighbor of the protected area up to the president of the country. This support comes through personal relationships, lots of devoted time, and an understanding that transcends culture and generations.[72]

Beyond Results

The ACG seems to be one of the most successful protected areas in the world. This said, the "ACG, with its NGO partner, the Guanacaste Dry Forest Conservation Fund {GDFCF}, has created one of the oldest, largest and most successful habitat restoration and conservation projects, and is recognized globally for long-term restoration and conservation of degraded tropical forests."[73] As a result of Dr. Janzen earning the Kyoto Prize in Basic Science, the GDFCF was initially capitalized with the intention of establishing an endowment for ACG.[74] Nowadays, the GDFCF manages 3 of the 18 sectors of ACG, provides ongoing financial support to ACG, supports local parataxonomists, and continues to strategically acquire adjacent parcels with the intention of donating the parcels to the national park system.[75] Such activities help contribute to the ongoing sustainability of the ACG.

Furthermore, and as explained by Dr. Janzen, there are micro-ecotourists (e.g., young children from Costa Rica) who are a long-term investment, and there is a high return with their increased bio-literacy.[76]

Lessons Learned

A few lessons learned by Dr. Daniel H. Janzen are that protected areas need to be big, they need to achieve financial independence, and they must be grounded on the people around the protected area.

First, protected areas need to be big—both sociologically big and biologically big. Thus, the protected areas need to be big enough to be ecologically and biologically viable and overcome the impact from surrounding forest fragmentation and the impact from global climate change. With respect to being sociologically big, the protected area needs to have a big, positive impact on people.[77]

Second, protected areas also need to achieve financial independence from government budgets and ongoing donor support. Likewise the protected area must pay its own bills, such as through the creation of a trust fund, and not become a parasite and be a pain in the rear of government.[78] This funding is required to pay good salaries to park staff (e.g., park guards, taxonomists, and biologists) and to acquire surrounding lands. It is important to note that it is less expensive to buy pristine forest lands than to acquire expensive agricultural lands and to restore them back to forests. Furthermore, one must realize that "one night stands" do not work; rather, you need endowments and this is how you get continuity like a university.[79]

Third, protected areas need to be firmly grounded on people around the protected area, and everyone needs to have skin in the game. However, you as the project coordinator need to let them have skin in the game, on their terms and in their way. Furthermore, all need to be involved from the small neighboring farmer, to the president of the country, to the international community. For instance, the neighbor could have a child in school and be interested in educational opportunities. The minister might use your protected area to demonstrate their managerial ability and subsequently try to get a new job at the United Nations Environment Programme. Similarly, international groups want to be able to do say, look what we helped do.[80]

Other Resources on ACG

Allen, William. *Green Phoenix: Restoring the Tropical Forests of Guanacaste, Costa Rica*. United Kingdom: Oxford University Press, 2001.

Area de Conservación Guanacaste's Official Page (In Spanish)

- https://www.acguanacaste.ac.cr/

Case Study on Area de Conservación Guanacaste, Costa Rica

- https://www2.southeastern.edu/Academics/Faculty/rvalverde/Costa_Rica_2000/ACG-CostaRica.pdf

Costa Rica's National Park Foundation

- http://www.fpn-cr.org/

Dr. D. H. Janzen and W. Hallwachs' Page on Caterpillars, Pupae, Butterflies & Moths

- http://janzen.sas.upenn.edu/caterpillars/database.lasso

International Conservation Fund of Canada's Page on the Area de Conservación Guanacaste, Costa Rica

- http://icfcanada.org/our-projects/projects/acg

Investigadores del Área de Conservación Guanacaste's Blog

- http://investigadoresacg.org/blog/

The Forever Costa Rica Association

- http://costaricaporsiempre.org/?lang=enhttp://canjeporbosques.org/
- https://www.nature.org/ourinitiatives/regions/centralamerica/costarica/explore/fcr-map-feb-09.pdf

World Heritage Council's Area de Conservación Guanacaste

- http://whc.unesco.org/en/list/928

WWF Center for Conservation Finance: Commercial Debt-for-Nature Swaps

- https://www.cbd.int/financial/debtnature/several-debtcommerc-wwf.pdf

Return

While the forgiver of debt may receive only a portion of the debt's face value, the forgiver may still receive some payment. This payment is better than if the borrower defaulted on the entire debt owed. In addition, the forgiver of debt will be able to reduce its costs (e.g., legal and managerial costs) going forward that are associated with servicing the debt. The entity that has its debt relieved will generate a return based off the canceled debt's face value, including future interest payments that would have been due. Often, this forgiven debt is denominated in US dollars or Euros, and the entity that has its debt forgiven will have funds denominated in a local currency available for conservation activities. For NGOs, or other on-the-ground practitioners, this source of funds might be in addition to their other traditional revenue streams.

Risk

There are numerous risks associated with debt-for-nature swaps with respect to the entity forgiving the debt and with respect to the entity whose debt is being forgiven.

Business Risk

Depending on the size of the debt-for-nature swap, there could be significant impacts to the income statement and balance sheet for the entity forgiving the debt. In addition, choosing a trusted, in-country facilitator of the debt-for-nature swap is a critical business risk.

Strategic Risk

How the debt-for-nature swap is structured involves strategic risk. This includes structuring the amount to be forgiven, choosing the strategic partners (e.g., local or international NGOs, and/or the host country's Finance and Environmental Ministries) and the timing of the deal (e.g., with respect to currency exchanges and interest rates).

From the sum of money raised by the debt-for-nature swap for conservation finance, there is a strategic risk associated with identifying the best projects within the most threatened biomes in the host country.

Reputation Risk

To most significant reputation risk could occur if the:

> {…} revenue from debt relief arrangements may not actually be put toward its originally intended use. To mitigate these risks, debt relief agreements should be designed to minimize market vulnerability and provide for contingency action in the event such a situation arises. Agreements can be struck with the government in advance to use a trust fund to manage the proceeds from the swap. Also, while debt relief for the environment usually

supports a range of conservation priorities, agreements can be structured to promote the conservation of specific species, habitats or protected areas.[81]

Such a situation presents a reputation risk to both the entity forgiving the debt and the entity whose debt is being forgiven. Further, this situation would likely result in fewer, if any, future debt-for-nature swaps between the two parties.

Liquidity Risk

It is difficult to issue debt-for-nature swaps because relatively few debt-for-nature swap transactions have occurred. This creates a lack of institutional knowledge with structuring a debt-for-nature swap, and this poses a liquidity risk because few people know how to execute the swaps. Furthermore, structuring debt relief for a distressed entity—whether it is a government or company—can also pose a liquidity risk.

Operational Risk

Operational risks related to debt-for-nature swaps include having the technical and managerial experience to execute a debt-for-nature swap. Another operational risk is that for large companies and finance ministries, it may take too much time and money to develop a relatively small debt-for-nature swap. In contrast, this same debt-for-nature swap may be too large for the in-country facilitator, especially if it is a local NGO, which again poses operational risk.

There also needs to be a monitoring component to ensure the proper use of funds, which could include a third-party trust fund whose financial statements are annually audited.

Legal and Regulatory Risk

Depending on the jurisdiction of the debt and the parties involved (e.g., corporate, government, and/or nonprofit), there are possible legal and regulatory risk. Government-to-government debt-for-nature swaps will need to clear regulatory hurdles, which can involve bureaucracy and time delays. This is especially true if political parties change during the transaction and new people are appointed to the responsible departments or ministries. Likewise, another risk could occur if the debt-for-nature swap gets suspended by the debtor country.

Legal risks include legally closing out the debt, potentially writing a new contract or addendum related to forgiving the debt, and any legal ramifications associated with how the forgiven debt is to be used for conservation.

Credit Risk

Credit risk includes default risk, bankruptcy risk, downgrade risk, and settlement risk. Debt-for-nature swaps are likely structured for debt that faces a potential default, bankruptcy, or downgrade. This can jeopardize the closing of a debt-for-nature swap, especially if it is not completed in time.

Market Risk

Market risk includes interest rate risk, equity price risk, foreign exchange risk, and commodity price risk. Debt-for-nature swaps can generate "millions of local dollars for conservation but contains some risk. Local currency devaluation and inflation can reduce or even eliminate the cash value of the conservation commitment."[82]

Risk, Return, Time (Horizon), Taxes, Liquidity, Legal and Unique (RRTTLLU)

Risk and Return

Please see above for the risk and return associated with debt-for-nature swaps.

Time Horizon

The time horizon for debt-for-nature swaps depends on the entity involved. The entity forgiving the debt—whether it be a commercial entity or a foreign government—is likely looking at a shorter time frame. In contrast, the government receiving the debt forgiveness and especially the on-the-ground entities responsible for implementing the terms of the debt-for-nature swap are likely looking at longer time frames.

Taxes

There are likely limited, if any, taxes due to the government that is forgiving the debt or to the government that is receiving the debt forgiveness. Commercial entities may be required to pay taxes, such as taxes on capital gains.

Liquidity

Traditional debt, such as household mortgages or investment-grade corporate debt, is often very liquid. In contrast, distressed debt, including debt that is structured for a debt-for-nature swap, is likely to be less liquid than its traditional counterpart.

Legal

Legal considerations associated with debt-for-nature swaps will vary from country to country. It is important to consider both the legal framework of the host country receiving the debt-for-nature swap and the legal framework for the host country (i.e., whether it is the government involved in the swap or the country where the company is headquartered) forgiving the debt.

Unique

A relatively few number of debt-for-nature swaps have been executed. The nature conservation component of the debt-for-nature swap is unique with each transaction. Thus, it is important to identify the unique aspects (e.g., the specific in-country facilitator) of the various projects, which could include a combination of marine and/or forest conservation projects.

Policy Analysis

Defining the Problem

One of the essential problems with debt-for-nature swaps is that you often need to find a government or business with distressed debt. After the 1970s oil embargoes, many developing countries experienced debt crises in the 1980s. Developing country (debtor) governments, developed country (debtee) governments, and international financial institutions such as the International Monetary Fund (IMF) and the World Bank were all trying to deal with the implications of debt, debt forgiveness, and missed debt payments.

Establishing Goals

A few of the public policy goals are:

- Enabling legislation for both the buyer of debt and for the "seller" of debt; and
- Clear policies in place that the forgiven debt will be used specifically for the intended purposes (e.g., establishing or expanding a national park or establishment of a trust fund).

In addition, despite the government or business likely holding onto distress debt, what would be beneficial—from a public policy perspective—for a debt-for-nature swap to be more easily structured is a strong rule of law, the strength of in-country institutions (including the financial infrastructure), a robust civil society, and transparency over where the funds are/will be used.

While high external debt could create risks for some initiatives such as establishing joint international peace parks or increasing a government's domestic budgetary allocation for a national park, a high external debt might reveal opportunities for debt-for-nature swaps. However it is also important to note that some smaller countries with lower external debt levels and which are thus not listed below, such as the Central African Republic (USD$661.85 million) or Belize (USD$1.33 billion), may present better opportunities for debt-for-nature swaps.[83] The following table presents the top 20 rainforest countries with the highest total external debt stocks (Disbursed and Outstanding Debt or DOD, current USD$) in 2015 (Table 13.1).[84]

Table 13.1 Rainforest country rankings on total external debt stocks

Rank	Country	External debt stocks (USD$)	Rank	Country	External debt stocks (USD$)
1	China	USD$1.42 trillion	11	Panama	USD$87.72 billion
2	Brazil	USD$543.40 billion	12	Vietnam	USD$77.80 billion
3	India	USD$479.56 billion	13	Philippines	USD$77.73 billion
4	Mexico	USD$426.33 billion	14	Peru	USD$65.94 billion
5	Indonesia	USD$308.54 billion	15	Sri Lanka	USD$43.92 billion
6	Malaysia	USD$190.95 billion	16	Bangladesh	USD$38.64 billion
7	Argentina	USD$159.69 billion	17	Nigeria	USD$29.029 billion
8	Thailand	USD$159.69 billion	18	Angola	USD$27.99 billion
9	Venezuela	USD$123.67 billion	19	Ecuador	USD$27.27 billion
10	Colombia	USD$111.05 billion	20	Dominican Republic	USD$26.63 billion

Source: The World Bank Group

Selecting a Policy

The United States' Tropical Forest Conservation Act (TFCA), which was passed in 1998 and was expanded from the previous 1991 Enterprise for the Americas Initiative (EAI), has facilitated a relatively significant number of debt-for-nature swaps.[85] Please note, there have also been numerous multiparty debt-for-nature swaps, and several countries other than the United States (e.g., Canada, Finland, Italy, Norway, and Sweden) have structured bilateral debt-for-nature swaps.[86]

Implementing a Policy

The US Agency for International Development (USAID), the US Department of Treasury, and the US Department of State all have responsibilities related to the implementation of the Tropical Forest Conservation Act.[87]

Evaluating the Policy

The EAI, which preceded the TFCA, facilitated debt-for-nature swaps with at least 8 countries including Argentina, El Salvador, Jamaica, and Uruguay. These debt-for-nature swaps generated USD$177.8 million in conservation funds and helped lay the groundwork for the larger TFCA.[88] This said, as of July 2013:

> {…} approximately $223 million in congressionally appropriated funds have been used to conclude 19 TFCA debt treatment agreements with 14 countries. These deals will generate over $326 million for tropical forest conservation in these countries over the life of the agreements from their rescheduled debt payments alone.[89]

While the policy appears to have been effective and the United States was a leading country structuring bilateral debt-for-nature swaps, the

Tropical Forest Conservation Reauthorization Act of 2015 was introduced as Senate Bill 566 in the 114th Congress and was not enacted.[90]

Future Outlook for Instrument

The United States' Tropical Forest Conservation Reauthorization Act of 2015 did not pass and is unlikely to do so for the foreseeable future. It is possible that commercial debt-for-nature swaps may become more plausible in the future than sovereign debt-for-nature swaps. In addition, there is a possibility for debt-for-nature swaps to support REDD+ initiatives. Furthermore, there could be an evolution where South-South debt-for-nature swaps are structured between large developing countries such as China, Brazil, and India and their regional counterparts such as Bolivia, Ecuador (i.e., such as for Yasuní National Park), and Vietnam.

Other Resources on Debt-for-Nature Swaps

Conservation Finance Guide's Understanding Debt-for-Nature Swaps

- http://www.conservationfinance.org/guide/guide/indexba4.htm

Sheikh, Pervaze A. "Debt-for-Nature Initiatives and the Tropical Forest Conservation Act: Status and Implementation." March 30, 2010. *Congressional Research Service*. https://www.cbd.int/financial/debtnature/g-inventory2010.pdf

WWF Center for Conservation Finance's Commercial Debt-for-Nature Swap

- https://www.cbd.int/doc/external/wwf/wwf-commercial-swaps-en.pdf

Notes

1. US Agency for International Development. "Forest Carbon, Markets and Communities (FCMC Program: US Government Investments and Policies to Facilitate Forest Carbon Finance and Markets." March 2012. Accessed April 14, 2017. http://www.fcmcglobal.org/documents/USG_Finance_Paper.pdf
2. Sheikh, Pervaze A. "Debt-for-Nature Initiatives and the Tropical Forest Conservation Act: Status and Implementation." March 30, 2010. *Congressional Research Service*. Accessed April 4, 2017. https://www.cbd.int/financial/debtnature/g-inventory2010.pdf. 5–7.
3. Sheikh, Pervaze A. "Debt-for-Nature Initiatives and the Tropical Forest Conservation Act: Status and Implementation." Congressional Research Center. March 30, 2010. https://www.cbd.int/financial/debtnature/g-inventory2010.pdf. 3–7.
4. Lovejoy, Thomas E. "Aid Debtor Nations' Ecology." *New York Times*. October, 4, 1984. A31.
5. Sheikh, Pervaze A. "Debt-for-Nature Initiatives and the Tropical Forest Conservation Act: Status and Implementation." Congressional Research Center. March 30, 2010. https://www.cbd.int/financial/debtnature/g-inventory2010.pdf. 3–7.
6. Ibid.
7. Ibid.
8. Ibid.
9. Ibid.
10. Ibid.
11. Ibid.
12. Ibid.
13. Ibid.
14. Ibid.
15. Ibid.
16. Ibid.
17. Ibid.
18. Ibid.
19. Ibid.
20. Ibid.
21. Ibid.
22. Ibid.

23. Ibid.
24. World Wildlife Fund. *Guide to Conservation Finance*. 41.
25. World Wildlife Fund. *Guide to Conservation Finance*. 42.
26. The Nature Conservancy. "Seychelles." Accessed April 15, 2017. https://www.nature.org/ourinitiatives/regions/africa/wherewework/seychelles.xml
27. Sheikh, Pervaze A. "Debt-for-Nature Initiatives and the Tropical Forest Conservation Act: Status and Implementation." March 30, 2010. *Congressional Research Service*. Accessed April 4, 2017. https://www.cbd.int/financial/debtnature/g-inventory2010.pdf. 5–7.
28. United Nations Development Programme. "Debt for Nature Swaps." Accessed April 4, 2017. http://www.undp.org/content/sdfinance/en/home/solutions/debt-for-nature-swaps.html
29. Sheikh, Pervaze A. "Debt-for-Nature Initiatives and the Tropical Forest Conservation Act: Status and Implementation." March 30, 2010. *Congressional Research Service*. Accessed April 4, 2017. https://www.cbd.int/financial/debtnature/g-inventory2010.pdf. 5–7.
30. SINAC. "Guanacaste Conservation Area (ACG)." Accessed April 5, 2017. http://www.sinac.go.cr/EN-US/ac/acg/Pages/default.aspx
31. World Heritage Convention. "Area de Conservación Guanacaste." Accessed April 5, 2017. http://whc.unesco.org/en/list/928
32. WWF Center for Conservation Finance. "Commercial Debt-for-Nature Swaps." Accessed April 7, 2017. https://www.cbd.int/doc/external/wwf/wwf-commercial-swaps-en.pdf. 2.
33. Daily, Gretchen C. and Katherine Ellison. *The New Economy of Nature*. 181–182.
34. YOSL-OIC. "Guidebook to the Gunung Leuser National Park." Accessed 2009. http://unesdoc.unesco.org/images/0018/001870/187099eb.pdf. 2.
35. World Heritage Convention. "Tropical Rainforest Heritage of Sumatra." Accessed February 9, 2017. http://whc.unesco.org/en/list/1167
36. YOSL-OIC. "Guidebook to the Gunung Leuser National Park." Accessed 2009. http://unesdoc.unesco.org/images/0018/001870/187099eb.pdf. 4–5.
37. World Heritage Convention. "Tropical Rainforest Heritage of Sumatra." Accessed February 9, 2017. http://whc.unesco.org/en/list/1167
38. Mongabay. "U.S. Forgives $30 M in Debt to Protect Rainforests in Sumatra, Indonesia." Accessed June 30, 2009. https://news.mongabay.com/2009/06/u-s-forgives-30m-in-debt-to-protect-rainforests-in-sumatra-indonesia/

39. Yayasan Orangutan Sumatera Lestari—Orangutan Information Centre. "TFCA Sumatera Programme." Accessed October 21, 2013. http://orangutancentre.org/programme/tfca-sumatera-programme/
40. Mongabay. "Fires Driving Deforestation in Indonesia's Leuser Ecosystem." September 29, 2016. Accessed April 11, 2017. https://news.mongabay.com/2016/09/fires-driving-deforestation-in-indonesias-leuser-ecosystem/
41. Nicholas, Hans. "Road Projects Threaten Sumatra's Last Great Rainforests." Mongabay. August 7, 2017. https://news.mongabay.com/2017/08/road-projects-threaten-sumatras-last-great-rainforests/
42. Jakarta Post. "Jokowi 'Needs to Intervene' on Leuser Deforestation." Accessed October 19, 2016. http://www.thejakartapost.com/news/2016/10/19/jokowi-needs-intervene-leuser-deforestation.html
43. Jakarta Post. "Jokowi 'Needs to Intervene' on Leuser Deforestation." Accessed October 19, 2016. http://www.thejakartapost.com/news/2016/10/19/jokowi-needs-intervene-leuser-deforestation.html
44. Jakarta Post. "Jokowi 'Needs to Intervene' on Leuser Deforestation." Accessed October 19, 2016. http://www.thejakartapost.com/news/2016/10/19/jokowi-needs-intervene-leuser-deforestation.html
45. Mongabay. "Palm Oil, Mining Moratorium Declared in Indonesia's Leuser Ecosystem." April 20, 2016. Accessed April 11, 2017. https://news.mongabay.com/2016/04/palm-oil-mining-moratorium-declared-indonesias-leuser-ecosystem/
46. Janzen, Daniel H. "How to Conserve a Large Tropical Wildland Through Biodiversity Development: Costa Rican Example." Presentation at Harvard University Center for the Environment, Cambridge, Massachusetts. September 30, 2010. Accessed March 17, 2017. https://vimeo.com/17675978
47. SINAC. "Guanacaste Conservation Area (ACG)." Accessed April 5, 2017. http://www.sinac.go.cr/EN-US/ac/acg/Pages/default.aspx
48. World Heritage Convention. "Area de Conservación Guanacaste." Accessed April 5, 2017. http://whc.unesco.org/en/list/928
49. Daily, Gretchen C. and Katherine Ellison. *The New Economy of Nature.* 169.
50. WWF Center for Conservation Finance. "Commercial Debt-for-Nature Swaps. Accessed April 7, 2017. https://www.cbd.int/doc/external/wwf/wwf-commercial-swaps-en.pdf. 2.
51. Daily, Gretchen C. and Katherine Ellison. *The New Economy of Nature.* 181–182.

52. WWF Center for Conservation Finance. "Commercial Debt-for-Nature Swaps." Accessed April 15, 2017. https://www.cbd.int/financial/debtnature/several-debtcommerc-wwf.pdf. 2.
53. Fraser, Caroline. "A Pioneering Biologist Discusses The Keys to Forest Conservation." *Yale Environment 360*. March 23, 2010. Accessed March 22, 2017. http://e360.yale.edu/features/a_pioneering_biologist_discusses_the_keys_to_forest_conservation
54. Ibid.
55. Janzen, Daniel H. "How to Conserve a Large Tropical Wildland Through Biodiversity Development: Costa Rican Example." Presentation at Harvard University Center for the Environment, Cambridge, Massachusetts. September 30, 2010. Accessed March 17, 2017. https://vimeo.com/17675978
56. SINAC. "Guanacaste Conservation Area (ACG)." Accessed April 5, 2017. http://www.sinac.go.cr/EN-US/ac/acg/Pages/default.aspx
57. Janzen, Daniel H. "How to Conserve a Large Tropical Wildland Through Biodiversity Development: Costa Rican Example." Presentation at Harvard University Center for the Environment, Cambridge, Massachusetts. September 30, 2010. Accessed March 17, 2017. https://vimeo.com/17675978
58. SINAC. "Guanacaste Conservation Area (ACG)." Accessed April 5, 2017. http://www.sinac.go.cr/EN-US/ac/acg/Pages/default.aspx
59. University of Pennsylvania. "Area de Conservación Guanacaste (ACG), northwestern Costa Rica: Introduction." Accessed April 5, 2017. http://janzen.sas.upenn.edu/caterpillars/introductory/preface.htm
60. Janzen, Daniel H. "How to Conserve a Large Tropical Wildland Through Biodiversity Development: Costa Rican Example." Presentation at Harvard University Center for the Environment, Cambridge, Massachusetts. September 30, 2010. Accessed March 17, 2017. https://vimeo.com/17675978
61. Fraser, Caroline. "A Pioneering Biologist Discusses The Keys to Forest Conservation." *Yale Environment 360*. March 23, 2010. Accessed March 22, 2017. http://e360.yale.edu/features/a_pioneering_biologist_discusses_the_keys_to_forest_conservation
62. International Barcode of Life Project. "Home." Accessed March 22, 2017. http://ibol.org/
63. Janzen, Daniel H. "How to Conserve a Large Tropical Wildland Through Biodiversity Development: Costa Rican Example." Presentation at

Harvard University Center for the Environment, Cambridge, Massachusetts. September 30, 2010. Accessed March 17, 2017. https://vimeo.com/17675978
64. SINAC. "Guanacaste Conservation Area (ACG)." Accessed April 5, 2017. http://www.sinac.go.cr/EN-US/ac/acg/Pages/default.aspx
65. Ibid.
66. World Heritage Convention. "Area de Conservación Guanacaste." Accessed April 5, 2017. http://whc.unesco.org/en/list/928
67. University of Pennsylvania. "Area de Conservación Guanacaste (ACG), Northwestern Costa Rica: Introduction." Accessed April 5, 2017. http://janzen.sas.upenn.edu/caterpillars/introductory/preface.htm
68. SINAC. "SINAC en Números 2015." Accessed April 9, 2017. http://www.ict.go.cr/en/documents/estad%C3%ADsticas/cifras-tur%C3%ADsticas/visita-a-las-%C3%A1reas-silvestres-protegidas-sinac/885-2015/file.html
69. Janzen, Daniel H. Interviewed by Brian McFarland. March 2017.
70. Ibid.
71. Ibid.
72. Ibid.
73. Permian Global. "About Permian Global." Accessed April 5, 2017. http://permianglobal.com/en/who-we-are/about
74. GDFCF. "Acquisition." Accessed April 11, 2017. http://www.gdfcf.org/acquisition
75. GDFCF. "What We Support." Accessed April 11, 2017. http://www.gdfcf.org/content/what-we-support
76. Janzen, Daniel H. "How to Conserve a Large Tropical Wildland Through Biodiversity Development: Costa Rican Example." Presentation at Harvard University Center for the Environment, Cambridge, Massachusetts. September 30, 2010. Accessed March 17, 2017. https://vimeo.com/17675978
77. Janzen, Daniel H. Interviewed by Brian McFarland. March 2017.
78. Ibid.
79. Janzen, Daniel H. "How to Conserve a Large Tropical Wildland Through Biodiversity Development: Costa Rican Example." Presentation at Harvard University Center for the Environment, Cambridge, Massachusetts. September 30, 2010. Accessed March 17, 2017. https://vimeo.com/17675978
80. Janzen, Daniel H. Interviewed by Brian McFarland. March 2017.
81. World Wildlife Fund. *Guide to Conservation Finance*. 41.

82. Ibid.
83. The World Bank Group. "World Development Indicators: External Debt Stocks, Total (DOD, Current US$)." Accessed April 2, 2017. http://databank.worldbank.org/data/reports.aspx?source=2&Topic=20
84. Ibid.
85. USAID. "Financing Forest Conservation: An Overview of the Tropical Forest Conservation Act." Accessed April 9, 2017. https://www.usaid.gov/biodiversity/TFCA
86. Sheikh, Pervaze A. "Debt-for-Nature Initiatives and the Tropical Forest Conservation Act: Status and Implementation." March 30, 2010. *Congressional Research Service*. Accessed April 4, 2017. https://www.cbd.int/financial/debtnature/g-inventory2010.pdf
87. USAID. "Financing Forest Conservation: An Overview of the Tropical Forest Conservation Act." Accessed April 9, 2017. https://www.usaid.gov/biodiversity/TFCA
88. Sheikh, Pervaze A. "Debt-for-Nature Initiatives and the Tropical Forest Conservation Act: Status and Implementation." March 30, 2010. *Congressional Research Service*. Accessed April 4, 2017.https://www.cbd.int/financial/debtnature/g-inventory2010.pdf
89. USAID. "Countries with TFCA Programs." Accessed April 9, 2017. https://www.usaid.gov/biodiversity/TFCA/programs-by-country
90. GovTrack. "S. 566 (114th): Tropical Forest Conservation Reauthorization Act of 2015." Accessed April 9, 2017. https://www.govtrack.us/congress/bills/114/s566

14

Bioprospecting

Generally known as 'minor forest products,' their aggregate worth, when they are comprehensively exploited, need not be so minor—and they can all be extracted with trifling disruption for forest ecosystems. They include oils, gums, resins, waxes, latex, tannins, alkaloids, drugs and medicines, dyes, edible and oil-bearing nuts, spices, fruit, rattans, bamboos, guanos, bark products, forage, perfumes and other extractives such as citronella oil, insect products such as lac resin (the main ingredient of shellac), insecticides, and honey and beeswax.[1]

—Dr. Norman Myers

Introduction

There are fewer, publicly available resources on specific bioprospecting deals, in contrast to the other conservation financing instruments, likely due to confidentiality agreements, patents, and concerns over intellectual property rights. However bioprospecting can play an important role in medical discoveries while simultaneously providing financing for tropical rainforest conservation. Consider, for example:

B.J. McFarland, *Conservation of Tropical Rainforests*, Palgrave Studies in Environmental Policy and Regulation, https://doi.org/10.1007/978-3-319-63236-0_14

> The world's biodiversity is a huge bank of treasures, most still hidden but some of the basis of medical, agricultural, and industrial advancements. Two of the most touted of these discoveries are Pacific Northwest yew trees, which yield taxol, a cancer-fighting substance, and the rosy periwinkle flower, derivatives of which have helped cure some forms of childhood leukemia. Animals also play important roles: the venom of some snakes reduces blood pressure, and some insects produce anticoagulants.[2]

Similarly, "forty percent of our medicines, for example, contain at least one compound obtained from rainforest plants. Physicians prescribe anticoagulants based on leeches and on snake venom. We get leukemia treatments from rosy periwinkles, and treatments for Hodgkin's disease, heart ailments, and arthritis drive from other tropical plants."[3]

Historical Overview

It appears that in 1991, the world's first major bioprospecting deal was struck between Merck & Co. and Costa Rica's Instituto Nacional de Biodiversidad (INBio).[4]

Other notable dates throughout the history of bioprospecting include:

- 1967: The World Intellectual Property Organization (WIPO) is established.[5]
- 1974: The International Board for Plant Genetic Resources (IBPGR) is established to coordinate an international plant genetic resources program.[6]
- 1989: Costa Rica's Instituto Nacional de Biodiversidad (INBio) is established.[7]
- 1991: IBPGR became "the International Plant Genetic Resources Institute (IPGRI) and in January 1994 IPGRI began "independent operation as a CGIAR center and at the request of CGIAR, took over the governance and administration of the International Network for the Improvement of Banana and Plantain (INIBAP)."[8]

- 2001: The International Treaty on Plant Genetic Resources for Food and Agriculture is adopted by the FAO.[9]
- 2006: IPGRI and INIBAP "become a single organization and subsequently change their operating name to Bioversity International."[10]
- 2014: The Nagoya Protocol on Access to Genetic Resources and the Fair and Equitable Sharing of Benefits Arising from their Utilization to the Convention on Biological Diversity enters into force.[11]

Mechanisms of Instrument

According to the WWF:

> Bioprospecting is the systematic search for new sources of chemical compounds, genes, proteins, microorganisms, and other products with potential economic value. Through bioprospecting agreements, international pharmaceutical companies compensate developing countries for the property rights over useful compounds contained in the country's biodiversity. In return, companies get exclusive rights to screen the biodiversity for pharmaceutical compounds. If such screening leads to the development of a major drug, the agreements provide the host country with a share of the profits, which may be used for biodiversity conservation.[12]

As explained by Instituto Nacional de Biodiversidad (INBio; National Biodiversity Institute in English), bioprospecting agreements include clauses generally referring to:

- "Access (limited in time and quantity);
- Equity and compensation (research costs and royalties);
- Technology transfer (infrastructure and equipment);
- Training of national scientists; and
- Non-destructive uses."[13]

These bioprospecting deals may be structured between a host country and a private company such as Merck & Co., or, for instance, with a public university or research institute.

Size of Instrument

Globally, "at least forty-seven major drugs had come from rainforest plant species, with an accumulated net worth of $147 billion."[14] Yet,

> Only a low volume of bioprospecting transactions is recorded, despite the potential for revenue generation. The value of bioprospecting contracts is estimated to be below US$100 million a year, compared with research and development investment by pharmaceutical companies of US$50 billion a year. Bioprospecting research contracts rarely surpass US$1 million in value, while royalties theoretically with a potential of generating billions have rarely materialized.[15]

This said, bioprospecting—despite its potential—is one of the smaller and less used, conservation financing instruments for tropical rainforests.

Introduction to Case Study

The following case study will briefly examine the bioprospecting deal between Merck & Co. and Costa Rica's Instituto Nacional de Biodiversidad.

Case Study #1: Merck & Co. and Costa Rica's Instituto Nacional de Biodiversidad

Introduction

In addition to facilitating a debt-for-nature swap in Costa Rica, Dr. Daniel H. Janzen also facilitated a bioprospecting contract in Costa Rica between Merck & Co. and Costa Rica's Instituto Nacional de Biodiversidad (INBio; National Biodiversity Institute). Originally signed in 1991, this particular bioprospecting contract outlined how "Merck agreed to pay more than $1 million over a two-year period for samples of chemical extracts from plants, animals, and microbes as well as royalties to INBio and the government if any new drug based on the samples were to prove commercially successful."[16]

Merck & Co., which is known as MSD (Merck Sharp and Dohme) outside of the United States and Canada, is a global healthcare and pharmaceutical company that has been in business for more than 125 years and currently, as of April 2017, has operations in more than 140 countries.[17] INBio was established in 1989 as a "nongovernmental, nonprofit, public interest organization" and works on inventory and monitoring, conservation, communications and education, biodiversity informatics, and bioprospecting.[18] More specifically, INBio's bioprospecting initiative:

> {...} seeks sustainable uses and the commercial application of biodiversity resources. INBio has been a pioneer institution in establishing research agreements to study chemical substances, genes, etc., present in plants, insects, marine organisms and microorganisms, which may be utilized by the pharmaceutical, medical, biotechnology, cosmetics, food and agricultural industries.[19]

With respect to the details of INBio's bioprospecting agreements:

> Each agreement has its corresponding work plan and research budget that establishes a 10% donation to the Ministerio del Ambiente y Energía (MINAE) (Ministry of the Environment and Energy), which helps cover direct biodiversity conservation costs. Furthermore, it contributes to increasing services, species identification, sample collection and preparation, collection records, information management, training, management, etc. If other benefits are derived from the process due to the discovery of a successful product, 50% of INBio's royalties will be donated to the Sistema Nacional de Areas de Conservación (National System of Conservation Areas) through the Ministerio del Ambiente y Energía (MINAE), and the rest will be invested in continuing the process and other INBio activities.[20]

These agreements should be somewhat flexible and allow for the agreements to be adapted depending, for example, on the jurisdiction of the bioprospector and the industry of the bioprospector (i.e., pharmaceuticals versus cosmetics). However, it appears to be a good structure to have upfront, guaranteed funding for the host institution, as well as a mechanism for ongoing royalty payments. Such payments could be independently managed by a trust fund.

Identify the Problem

Costa Rica cleared vast tracts of forests during the 1970s and 1980s primarily for cattle ranching, and in the 1990s, Costa Rica had one of the highest deforestation rates in the world.[21]

When the bioprospecting deal was signed between Merck & Co. and INBio, it is likely that Merck & Co. sought new genetic discoveries to develop new, innovative products, while INBio was seeking sources of financing for its recently established organization and to address Costa Rica's high rate of deforestation.

Why the Problem Is Important

One reason, among many, why deforestation of tropical rainforests is important is because their tremendous biodiversity may hold the key to finding a cure for cancer or AIDS, but once these forests are lost, they—and their potential cures—will be lost forever. In the United States alone, the American Cancer Society estimates there will be nearly 1.7 million new cancer cases and more than 600,000 cancer deaths in 2017.[22] Furthermore, the UNAIDS Secretariat estimates that as of the end of 2015, "78 million people have become infected with HIV since the start of the epidemic and 35 million people have died from AIDS-related illnesses since the start of the epidemic."[23]

How Problem Was Identified

The problem of tropical rainforest deforestation—including deforestation in Costa Rica—was identified by a wide range of stakeholders including both global and local conservation organizations such as INBio. Similarly, the global health crisis has been identified by a wide range of stakeholders including health practitioners, nonprofit organizations such as Management Sciences for Health (MSH), and multilateral organizations such as UNAID and the World Health Organization (WHO).

Effectiveness of Process for Identifying Problem

The process of identifying tropical rainforest deforestation and particularly deforestation within Costa Rica seems to be very effective due to the on-the-ground work coupled with satellite imagery. It is unclear if the process was effective for INBio and Merck & Co.'s bioprospecting deal, with respect to the discovery of new medicinal compounds. Nevertheless, INBio and Merck & Co. signed several follow-up bioprospecting agreements, which seems to suggest that the earlier contracting went relatively well.

Steps Taken to Address the Problem

The 3 main steps taken to address tropical rainforest deforestation, within the context of the Merck & Co.-INBio bioprospecting contract, appear to be:

1. Merck & Co. and INBio negotiated the bioprospecting contract;
2. INBio received funding of at least USD$1 million from Merck & Co.; and
3. Merck & Co., or its affiliates, undertook the bioprospecting field research.

Results

The specific results of the Merck & Co.-INBio bioprospecting deal are not publicly available. However, it appears as though Merck & Co. and INBio, in addition to the 1991 bioprospecting contract, signed additional bioprospecting contracts in 1994, 1996, and 1998.[24]

INBio also signed bioprospecting contracts with at least 19 companies in the industrial sector and 18 contracts with academia and foundations between 1991 and 2001. Academia and foundations include Cornell University, the MacArthur Foundation, the University of Costa Rica, and the University of Massachusetts.[25] Companies, in addition to Merck & Co., include Bristol-Myers Squibb and Eli Lilly and Co.[26]

INBio's latest annual report, released in 2009, stated that the Editorial, INBioparque, Bioprospecting, and SAU Consultancies units generated approximately 70% of INBio's income and the Bioprospecting division's income increased by 36% from 2008.[27]

Another important series of results, according to the United Nations Development Programme (UNDP), are:

> INBio was awarded a two year research and sampling contract. The contribution from Merck of about US$2.6 million was channeled to the Ministry of Environment and Energy, universities, and other partners. In exchange, Merck obtained access to "chemical extracts from wild plants, insects, and micro-organisms". No royalty payment was derived from this agreement. However, INBio has received payments from other bioprospecting contracts in the last 20 years.[28]

Furthermore, it is possible that INBio laid the groundwork for how bioprospecting deals (e.g., specific contracting language and the inclusion of royalty payments) can be structured in other tropical rainforest countries.

Challenges and How They Were Met

One challenge is the applicability of the contract and whether there was an exclusivity component. For instance, "following its contract with Merck, INBio made deals with several other foreign firms seeking new antibiotics, fragrances, and natural pesticides derived from tropical plants."[29] In addition, as of 1998, INBio had signed with other pharmaceutical companies. This said, it is unclear whether Merck & Co., in turn, signed bioprospecting contracts with other countries or with other entities in Costa Rica.

Beyond Results

The sustainability of the Merck & Co.-INBio bioprospecting deal is unclear; however, Merck & Co. and INBio, in addition to the 1991 bioprospecting contract, signed additional bioprospecting contracts in 1994, 1996, and 1998.[30]

While INBio signed bioprospecting contracts with at least 19 companies in the industrial sector[31] and 18 contracts with academia and foundations[32] between 1991 and 2001, it is unclear whether additional bioprospecting contracts were signed after 2001. INBio's bioprospecting division appears to still be active and is pursuing ISO 9001 accreditation.[33]

Furthermore, INBio states that the bioprospecting contracts "have given job opportunities to Costa Ricans from different social sectors. Furthermore, there has been a transfer of technology as well as training for Costa Ricans, which has resulted in an improvement of the national technological capability."[34]

Lessons Learned

A few lessons learned from researching the Merck & Co.-INBio bioprospecting deal are:

- Bioprospecting deals can be controversial and need to be negotiated very carefully; and
- Information on bioprospecting deals, particularly the Merck & Co.-INBio deal, is limited.

Other Resources on Merck & Co. and Costa Rica's Instituto Nacional de Biodiversidad

Environment & Energy, Hard Time for Chemical Prospecting

- http://cro.ots.ac.cr/rdmcnfs/datasets/biblioteca/pdfs/nbina-1810.pdf

ETC Group, Bioprospecting/Biopiracy and Indigenous Peoples

- http://www.etcgroup.org/content/bioprospectingbiopiracy-and-indigenous-peoples

GRAIN, Bargaining Over the Benefits of Biodiversity

- https://www.grain.org/article/entries/503-bargaining-over-the-benefits-of-biodiversity

Journal of Environment & Development, Preserving Biodiversity in Costa Rica: The Case of the Merck-INBio Agreement

- http://citeseerx.ist.psu.edu/viewdoc/download?doi=10.1.1.1019.9759&rep=rep1&type=pdf

INBio's Official Page

- http://www.inbio.ac.cr/en/

Return

Returns from bioprospecting deals are a result of the discovery of a unique compound, the ability to patent the compound, and then commercialize the compound.

The potential revenue from bioprospecting, as explained by the UNDP:

> {...} can be assessed by referring to the profits accrued from the exploitation of natural genetic resources as the tax base for a royalty: Pravachol (cholesterol) generated US$1.5 billion in profit to Bristol-Myers Squibb

while Zocor (cholesterol) and Mevacor (cholesterol) generated respectively US$3.6 billion and US$1.1 billion to Merck. A 3 per cent profit-sharing agreement over the above-mentioned drugs could have generated revenues of about US$200 million.

While largely theoretical, other approaches exist. The contribution of an area of land to drug discovery can be estimated: based on stated assumptions, the value per hectare varies from US$21 to US$9177. Another approach uses industry figures to calculate the net present value of an extract in a screening programme, which was estimated at US$487 on average.[35]

The host country, or its in-country facilitator, can earn a return—as demonstrated by INBio—from both an upfront payment structured over a multiyear period and via a royalty payment structure.

Risk

Business Risk

Major business risks are whether the firm that is undertaking the bioprospecting deal is successfully able to make new genetic discoveries, then export these genetic samples from the host country and patent the discoveries, and then commercialize and market the new products which are based off the bioprospecting deal.

There is also a testing risk that will depend on the host country where the patent is sought. For instance, in the United States, there will likely be a need to receive approval from the US Food and Drug Administration (FDA). Poor methodology (e.g., lost samples in the forest or mislabeled samples), misunderstood results, and negative side effects can contribute to the business risk of bioprospecting.

Strategic Risk

Strategic risks associated with bioprospecting include identifying the appropriate host country (e.g., Costa Rica) and local partners (e.g.,

Indigenous Peoples, extractive reserves, and/or local forestry department). More specifically, this includes the ease of being able to export the sample from the host country and the ease of being able to import the sample into the importer country.

Reputation Risk

Bioprospecting is sometimes referred to as biopiracy if the transaction seems to be unequal to local communities or host governments. Also, there is the reputation risk that participants on a bioprospecting deal are commercializing nature, which is perceived by some as unethical and/or immoral.

Liquidity Risk

The degree to which a patent can be obtained on the genetic resources discovered and whether the results can then be commercialized will have an impact on liquidity risk. It is unclear how liquid a patent is, with respect to transferring the patented discovery from one entity to another.

Operational Risk

Operational risks for bioprospecting include being able to:

- Identify suitable areas for exploration;
- Systematically collect, identify, and document species among the tremendous terrestrial biodiversity in a lowland tropical rainforest;
- Run the necessary laboratory tests with the appropriate equipment, with all the necessary quality assurance and quality control, to identify possible compounds; and
- Obtain the necessary patents and avoid theft of intellectual property rights.

Furthermore, some of this equipment (i.e., drying kilns) may not be available in the host country, and there could be operational risks associciated with getting the equipment into the host country. Similarly, there is also the operational risk of finding, hiring, and training local talent to operate the equipment.

Legal and Regulatory Risk

In some countries, bioprospecting is illegal or the process to obtain a bioprospecting license is unclear.

There are also legal and regulatory risks associated with obtaining patents, avoiding disclosures of confidential information, and preventing theft of intellectual property rights.

There could be concerns about appropriation or creeping appropriation from host governments, which may prohibit the export of a given species from which a unique compound was discovered.

In addition, bioprospecting contacts must be very clear to mitigate unanswered questions from arising in the future. For instance, what happens if a pharmaceutical company enters into a bioprospecting deal, but instead of discovering a new genetic sequence that is beneficial for medicine, their discovery is more applicable to the development of pesticides?

Furthermore, there could be issues—due in part to language and cultural barriers—when negotiating a bioprospecting contract, especially when working with local communities and Indigenous Peoples. This said, extra attention should be given to Free, Prior, and Informed Consent (FPIC).

Credit Risk

Credit risk includes default risk, bankruptcy risk, downgrade risk, and settlement risk. From a host country perspective (including the host communities or Indigenous Peoples, if applicable), there is a settlement risk that the company extracting the genetic discovery will not make payment as agreed to in the terms and conditions of the bioprospecting contract.

Market Risk

Market risk includes interest rate risk, equity price risk, foreign exchange risk, and commodity price risk. Fluctuations in foreign exchange may make it more or less expensive to operate in countries and, thus, could impact the profitability of a bioprospecting deal. In addition, if the bioprospecting discovery later contributes to the development of generic drugs, these less expensive drugs may compete with the earlier-patented drug (i.e., comparable to a commodity price risk).

Risk, Return, Time (Horizon), Taxes, Liquidity, Legal and Unique (RRTTLLU)

Risk and Return

Please see above for the risk and return associated with bioprospecting.

Time Horizon

The time horizon for bioprospecting is likely to be a short-term investment (i.e., short, fixed term for exploration at the host site), while some aspects of the contract (e.g., intellectual property rights and confidentiality) are likely to be long-term commitments. In addition, if a new genetic discovery is made and patented, the patent will likely be granted for a long-term. Furthermore, the time from initial exploration to full commercialization (i.e., after discovery, after patents, and after testing) is likely to be a medium-to-long-term process.

Taxes

There are likely royalties associated with bioprospecting, along with various taxes such as property taxes, sales taxes, and employment taxes. There could also be various import and export fees, levies, and permits required.

Liquidity

Transferring a bioprospecting contract or the underlying protected area with a bioprospecting contract is likely more illiquid. In contrast, there may be more liquidity associated with transferring the discoveries from a bioprospecting deal to a pharmaceutical company.

Legal

Legal considerations associated with bioprospecting will vary from country to country. This includes the country where the exploration takes place, where the testing will be conducted, where the patent will be sought, and where the future drug (i.e., if all else is successful) will be sold.

Unique

One unique aspect of bioprospecting, as opposed to the other conservation financing instruments such as debt-for-nature swaps, ecotourism, or payments for ecosystem services, is the vital importance of obtaining patents. In addition, another unique aspect is the end use of the bioprospecting deal. For instance, if your competitors come up with an alternative and less expensive drug, or if a generic drug is developed based off the theft of your intellectual property, then the bioprospecting deal may lose its value. Furthermore, bioprospecting—as opposed to some of the other conservation financing instruments—seems to have greater concerns over the commercialization of nature and biopiracy.

Policy Analysis

Defining the Problem

A few potential public policy problems associated with bioprospecting deals are what constitutes an equitable deal (i.e., to respect the host country and its peoples, to mitigate losing one's license to operate in the host country, and to mitigate bad publicity) and how to ensure intellectual property rights.

Establishing Goals

Public policy goals would clarify what constitutes an equitable deal, the host government would have a strong rule of law (particularly enforcing contracts), and the country would have a track record of both recognizing and respecting intellectual property rights.

Selecting a Policy

On an international level, the Convention on Biological Diversity's Nagoya Protocol on Access to Genetic Resources and the Fair and Equitable Sharing of Benefits Arising from their Utilization is one of the most relevant policies for bioprospecting.

The following are the lowest and highest ranked countries, out of 128 countries in 2016, on their respect for intellectual property (IP) rights (Table 14.1):[36]

The following are the lowest and highest ranked rainforest countries, out of 128 countries in 2016, on their respect for intellectual property (IP) rights (Table 14.2):[37]

The following are the lowest and highest ranked countries, as of June 2016, on their ease of doing business with respect to enforcing contracts (Table 14.3):[38]

The following are the lowest and highest ranked rainforest countries, as of June 2016, on their ease of doing business with respect to enforcing contracts (Table 14.4).[39]

Although Singapore and Australia have rainforests, both rank high on the enforcement of contracts and intellectual property rights, neither of these countries are a party to the Nagoya Protocol. However, Australia is a signatory.

Some of the largest rainforest countries that are parties to the Nagoya Protocol as of April 2017 include, but are not limited to:[40]

- Argentina;
- Bolivia;
- Cambodia;
- Cameroon;
- DRC;
- Guatemala
- Guyana;
- Honduras;
- India;
- Indonesia
- Mexico
- Myanmar;
- Panama
- Peru
- Philippines;
- Republic of the Congo;
- Uganda;
- Uruguay;
- Vietnam; and
- Zambia.

Table 14.1 Country rankings on intellectual property rights

Global rank	Lowest ranked countries on IP	Score	Global rank	Highest ranked countries on IP	Score
128	Venezuela	2.7	1	Finland	8.4
127	Myanmar	2.8	2	New Zealand	8.3
126	Bangladesh	2.8	3	Luxembourg	8.3
125	Haiti	2.8	4	Norway	8.3
124	Zimbabwe	3.4	5	Switzerland	8.2
123	Burundi	3.4	6	Singapore	8.1
122	Nigeria	3.6	7	Sweden	8.1
121	Pakistan	3.7	8	Japan	8.1
120	Moldova	3.7	9	Netherlands	8.0
119	Mauritania	3.7	10	Canada	8.0

Source: The International Property Rights Index

Table 14.2 Rainforest country rankings on intellectual property rights

Global rank	Lowest ranked countries on IP	Score	Global rank	Highest ranked countries on IP	Score
128	Venezuela	2.7	6	Singapore	8.1
127	Myanmar	2.8	12	Australia	7.9
126	Bangladesh	2.8	24	Taiwan	6.9
125	Haiti	2.8	26	Malaysia	6.8
124	Zimbabwe	3.4	33	Rwanda	6.2
123	Burundi	3.4	34	Mauritius	6.1
122	Nigeria	3.6	47	Costa Rica	5.8
116	Madagascar	3.8	51	Jamaica	5.6
114	Nicaragua	4.0	53	Ghana	5.5
111	Cameroon	4.0	55	China	5.4

Source: The International Property Rights Index

Table 14.3 Country rankings on enforcing contracts

Global rank	Lowest ranked countries	Global rank	Highest ranked countries
190	Timor-Leste	1	South Korea
189	Bangladesh	2	Singapore
188	Myanmar	3	Australia
187	Suriname	4	Norway
186	Angola	5	China
185	Mozambique	6	Lithuania
184	Djibouti	7	Croatia
183	Micronesia	8	Hungary
182	Central African Republic	9	Kazakhstan
181	São Tomé and Príncipe	10	Austria

Source: The World Bank Group

Costa Rica, despite not being a party to the Nagoya Protocol, is a signatory and would be a country to consider for future bioprospecting deals because there is a history of bioprospecting deals and there are still local Costa Rica institutes (i.e., INBio) that work with bioprospecting. The specific policy in Costa Rica is the Biodiversity Law 7788 of April 30, 1998.[41] Law 8686 of November 11, 2008, which includes important amendments related to Law 7788, involves "amending, adding or repealing Various Rules governing Matters pertaining to Intellectual Property."[42]

Table 14.4 Rainforest country rankings on enforcing contracts

Global rank	Lowest ranked countries	Global rank	Highest ranked countries
190	Timor-Leste	2	Singapore
189	Bangladesh	3	Australia
188	Myanmar	5	China
187	Suriname	14	Taiwan
186	Angola	34	Mauritius
185	Mozambique	35	Antigua and Barbuda
184	Djibouti	37	Brazil
183	Micronesia	40	Mexico
182	Central African Republic	42	Malaysia
181	São Tomé and Príncipe	46	Saint Kitts and Nevis

Source: The World Bank Group

Implementing a Policy

The National Commission for Biodiversity Management (CONAGEBIO) of Costa Rica's Ministry of Environment and Energy (MINAE) is "responsible for all genetic resources {and is the} only designated competent national authority (CNA) for the country."[43]

Evaluating the Policy

INBio signed bioprospecting contracts with at least 19 companies in the industrial sector[44] and 18 contracts with academia and foundations[45] between 1991 and 2001. This said, it appears as though the Biodiversity Law 7788 helped, or at least did not prohibit, the later bioprospecting deals. However, it is unclear whether additional bioprospecting contracts were signed after 2001 with INBio, and thus, it is unclear how exactly Law 8686 would apply.

Future Outlook for Instrument

Andrew W. Torrance conservatively estimated that if there were 1 million profitable genes in the world and:

> If the annual average profitability of a profitable gene were \$1 million, then the profits generated by profitable genes would generate \$1 trillion per year. By comparison, if the annual average profitability were \$10 million per profitable gene, then the total figure would rise to \$10 trillion; with average profitability of \$100 million, the total would rise to an extraordinary \$100 trillion.[46]

Yet, despite this upside potential, there have been relatively few bioprospecting deals structured, and such deals are unlikely to be scalable to conserve the world's last remaining rainforests. Nevertheless, bioprospecting may be a complimentary and additional revenue source for some conservation initiatives, and bioprospecting is still being explored today in places, although controversial, such as the Maya Biosphere Reserve in Guatemala.[47]

Other Resources on Bioprospecting

Access and Benefit-Sharing Clearing-House

- https://absch.cbd.int/countries

Biological Resource Center of Japan's National Institute of Technology and Evaluation

- http://www.nite.go.jp/en/nbrc/global/index.html
- http://www.nite.go.jp/en/nbrc/global/asia/about_egron.html
- http://www.nite.go.jp/en/nbrc/cultures/fee/fee.html
- http://www.nite.go.jp/en/nbrc/global/countries/cbd_laws.html
- http://www.nite.go.jp/en/nbrc/global/acm/asia_acm.html
- http://www.nite.go.jp/en/nbrc/global/symposium/symposium.html

Online Databases and Registries of Traditional Knowledge and Genetic Resources complied by the World Intellectual Property Organization

- http://www.wipo.int/tk/en/resources/db_registry.html

Public Interest Intellectual Property Advisors: Bioprospecting Resource Guide

- http://www.piipa.org/images/PDFs/PIIPA_Bioprospecting_Resource_Guide_2013_Final.pdf

Quezada, Fernando Santibáñez. *Status and Potential of Commercial Bioprospecting Activities in Latin America and the Caribbean*. New York: United Nations, 2007.

Shiva, Vandana. *Biopiracy: The Plunder of Nature and Knowledge*. Boston: South End Press, 1999.

World Intellectual Property Organization (WIPO) and its Intergovernmental Committee on Intellectual Property and Genetic Resources, Traditional Knowledge and Folklore

- http://www.wipo.int/reference/en/
- http://www.wipo.int/tk/en/igc/

Notes

1. Myers. *The Shrinking Ark*. 185.
2. Daily, Gretchen C. and Katherine Ellison. *The New Economy of Nature*. 157.
3. Royte, Elizabeth. *The Tapir's Morning Bath*. 269.
4. The North American Congress on Latin America. "Bioprospecting and Biopiracy in the Americas." Accessed December 13, 2016. http://nacla.org/news/bioprospecting-and-biopiracy-americas
5. WIPO. "Inside WIPO: What is WIPO?" Accessed April 15, 2017. http://www.wipo.int/about-wipo/en/
6. Bioversity International. "History." Accessed March 29, 2017. http://www.bioversityinternational.org/about-us/who-we-are/history/
7. INBio. "Historical Background." Accessed April 15, 2017. http://www.inbio.ac.cr/en/que-es-inbio/historia.html
8. Bioversity International. "History." Accessed March 29, 2017. http://www.bioversityinternational.org/about-us/who-we-are/history/

9. Food and Agriculture Organization of the United Nations. "International Treaty on Plant Genetic Resources for Food and Agriculture: About Us." Accessed March 29, 2017. http://www.fao.org/plant-treaty/overview/en/
10. Bioversity International. "History." Accessed March 29, 2017. http://www.bioversityinternational.org/about-us/who-we-are/history/
11. Convention on Biological Diversity. "The Nagoya Protocol on Access and Benefit-Sharing." Accessed March 29, 2017. https://www.cbd.int/abs/
12. World Wildlife Fund. *Guide to Conservation Finance*. 23.
13. INBio. "Biodiversity Prospecting: Bioprospecting Agreements." Accessed April 5, 2017. http://www2.inbio.ac.cr/en/inbio/inb_prospacuerdos.htm
14. Daily, Gretchen C. and Katherine Ellison. *The New Economy of Nature*. 172.
15. UNDP. "Bioprospecting." Accessed April 5, 2017. http://www.undp.org/content/sdfinance/en/home/solutions/bioprospecting.html
16. Daily, Gretchen C. and Katherine Ellison. *The New Economy of Nature*. 172.
17. Merck. "Who We Are." Accessed April 5, 2017. http://www.merck.com/about/home.html
18. INBio. "What is INBio?" Accessed April 5, 2017. http://www.inbio.ac.cr/en/que-es-inbio/acerca.html
19. Ibid.
20. INBio. "Biodiversity Prospecting: Bioprospecting Agreements." Accessed April 5, 2017. http://www2.inbio.ac.cr/en/inbio/inb_prospacuerdos.htm
21. Mongabay. "Costa Rica." Last modified February 7, 2006. http://rainforests.mongabay.com/20costarica.htm
22. American Cancer Society. "Cancer Facts & Figures 2017." Accessed April 5, 2017. https://www.cancer.org/research/cancer-facts-statistics/all-cancer-facts-figures/cancer-facts-figures-2017.html
23. UNAIDS. "Fact sheet—Latest Statistics on the Status of the AIDS Epidemic." Accessed April 5, 2017. http://www.unaids.org/en/resources/fact-sheet
24. INBio. "Agreements with Industrial Sector." Accessed April 5, 2017. http://www2.inbio.ac.cr/pdf/acuerdos%20industria.pdf
25. INBio. "Agreements with Academia." Accessed April 5, 2017. http://www2.inbio.ac.cr/pdf/acuerdos%20academia.pdf

26. INBio. "Agreements with Industrial Sector." Accessed April 5, 2017. http://www2.inbio.ac.cr/pdf/acuerdos%20industria.pdf
27. INBio. "Annual Report 2009." Accessed April 5, 2017. http://www2.inbio.ac.cr/es/memorias/memoria2009/Memoria2009-eng.pdf. 28–29.
28. UNDP. "Bioprospecting." Accessed April 5, 2017. http://www.undp.org/content/sdfinance/en/home/solutions/bioprospecting.html
29. Daily, Gretchen C. and Katherine Ellison. *The New Economy of Nature*. 172.
30. INBio. "Agreements with Industrial Sector." Accessed April 5, 2017. http://www2.inbio.ac.cr/pdf/acuerdos%20industria.pdf
31. Ibid.
32. INBio. "Agreements with Academia." Accessed April 5, 2017. http://www2.inbio.ac.cr/pdf/acuerdos%20academia.pdf
33. INBio. "Biodiversity Prospecting: Organization." Accessed April 5, 2017. http://www2.inbio.ac.cr/en/inbio/inb_prosporg.htm
34. INBio. "Frequently Asked Questions." Accessed April 5, 2017. http://www.inbio.ac.cr/en/que-es-inbio/preguntas-frecuentes.html
35. UNDP. "Bioprospecting." Accessed April 5, 2017. http://www.undp.org/content/sdfinance/en/home/solutions/bioprospecting.html
36. The International Property Rights Index 2016. "Countries." Accessed February 20, 2017. http://internationalpropertyrightsindex.org/countries?f=ipri_wr&o=asc
37. The International Property Rights Index 2016. "Countries." Accessed February 20, 2017. http://internationalpropertyrightsindex.org/countries?f=ipri_wr&o=asc
38. The World Bank Group. "Economy Rankings." Accessed April 9, 2017. http://www.doingbusiness.org/rankings
39. Ibid.
40. ABSCH. "Country Profiles." Accessed April 12, 2017. https://absch.cbd.int/countries
41. World Intellectual Property Organization. "Biodiversity Law: The Legislative Assembly of the Republic of Costa Rica." Accessed April 12, 2017. http://www.wipo.int/edocs/lexdocs/laws/en/cr/cr018en.pdf
42. World Intellectual Property Organization. "La Gaceta (Gazette No.) 229—November 26, 2008: Law No. 8686: The Legislative Assembly of the Republic of Costa Rica." Accessed April 12, 2017. http://www.wipo.int/edocs/lexdocs/laws/en/cr/cr028en.pdf
43. ABSCH. "Costa Rica: Competent National Authority." Accessed April 12, 2017. https://absch.cbd.int/countries/CR

44. INBio. "Agreements with Industrial Sector." Accessed April 5, 2017. http://www2.inbio.ac.cr/pdf/acuerdos%20industria.pdf
45. INBio. "Agreements with Academia." Accessed April 5, 2017. http://www2.inbio.ac.cr/pdf/acuerdos%20academia.pdf
46. Torrance, Andrew W. "Bioprospecting and the Convention on Biological Diversity." Accessed April 5, 2017. https://dash.harvard.edu/bitstream/handle/1/8965586/Torrance,_Andrew_00.html?sequence=2
47. Cuffe, Sandra. "Controversial Park Plans in Guatemala's Maya Biosphere Reserve." June 17, 2016. *Mongabay*. Accessed April 13, 2017. https://news.mongabay.com/2016/06/controversial-park-plans-in-guatemalas-maya-biosphere-reserve/

15

Green Procurement Models

"Every time you spend money, you're casting a vote for the kind of world you want."[1]
—Anna Lappé

Introduction

Green procurement models are often premised on obtaining commodities that have been independently verified against a sustainability certification such as the Forest Stewardship Council (FSC) or the Sustainable Forestry Initiative (SFI). In addition, there are numerous networks and roundtables such as the Global Forest & Trade Network, the Leather Working Group, the Roundtable on Sustainable Palm Oil (RSPO), the Global Roundtable for Sustainable Beef, and the Round Table on Responsible Soy (RTRS). More recently, there has been the emergence of zero deforestation supply chain commitments from many of the world's largest private and publicly traded companies including Burger King, Cargill, McDonald's, and Nestlé.[2]

B.J. McFarland, *Conservation of Tropical Rainforests*, Palgrave Studies in Environmental Policy and Regulation, https://doi.org/10.1007/978-3-319-63236-0_15

Another potential form of green procurement is the establishment of concessions, such as conservation concessions or community-managed concessions, which produce sustainably sourced products.

Historical Overview

The sustainable, green procurement of locally produced commodities—particularly by Indigenous Peoples—has existed for thousands of years. However, the origin of commercial green procurement likely started with the creation of the Forest Stewardship Council (FSC)[3] and the Sustainable Forestry Initiative (SFI) in 1994.[4] Other important historical dates include:

- 1999: The Programme for the Endorsement of Forest Certification (PEFC) is founded.[5]
- 2000: Carbon Disclosure Project (now known as CDP) is founded.[6]
- 2001: Peru "established its first formal forest conservation concession, Los Amigos, in 2001, and has since created several hundred concessions that produce products and services while safeguarding biodiversity."[7]
- 2004: Roundtable on Sustainable Palm Oil (RSPO) is established.[8]
- 2005: Leather Working Group (LWG) is formed.[9]
- 2005: The European Union's Forest Law Enforcement, Governance and Trade (EU FLEGT) Regulation is adopted.[10]
- 2006: Greenpeace publishes *Eating Up the Amazon* Report.[11]
- 2006: Brazil's Soy Moratorium is signed by members of Associação Brasileira das Indústrias de Óleos Vegetais (ABIOVE, Brazilian Association of Vegetable Oil Industries) and Associação Nacional dos Exportadores de Cereais (ANEC, National Association of Cereal Exporters).[12]
- 2006: Round Table on Responsible Soy (RTRS) is established.[13]
- 2007: Grupo de Trabalho da Pecuária Sustentável (GTPS, Brazilian Roundtable for Sustainable Livestock) is created.[14]

- 2008: In 2008 "US Congress followed suit and passed an amendment to the Lacey Act of 1900, which prohibits trafficking in illegally harvested plants and wildlife."[15]
- 2009: Consumer Goods Forum is formed.[16]
- 2010: The Global Roundtable for Sustainable Beef is formed.[17]
- 2012: The Lacey Act "has already proven its effectiveness with a widely reported case in 2012 involving the Gibson Guitar Company."[18]
- 2012: The Australian government passes the Illegal Logging Prohibition Act.[19]
- 2015: Forest Trends' new project and platform Supply-Change.org is launched.[20]
- 2016: Indonesia becomes the world's first country to start issuing the EU FLEGT licenses.[21]

Mechanisms of Instrument

Commodity supply chains—whether it is for cattle, palm oil, soy, or timber—often involve local producers, aggregators, processors, traders, manufacturers, and buyers (both business-to-business or B2B and business-to-consumer or B2C), along with extensive transportation, marketing and financing networks, and auditors. Consumer pressure (e.g., product boycotts) and investor pressure (e.g., filing proxy statements)[22] are helping to influence these supply chains to adopt more sustainable, green procurement practices (i.e., eliminate deforestation from one's supply chain) and to address global climate change. For instance:

> BlackRock, the world's largest asset manager, put corporate boards on notice that it expects them "to have demonstrable fluency in how climate risk affects the business" and how management is managing that risk. Specifically, the world's largest asset manager is encouraging companies, "in due course," to follow the recommendations of the Financial Stability Board's Task Force on Climate-Related Financial Disclosures, launched in December by Michael Bloomberg and Mark Carney, governor of the Bank of England.[23]

In addition, there is also a policy driver—particularly for timber—such as the prohibition of illegally sourced timber imports.

An important mechanism of sustainable commodity supply chains is a certification standard, and numerous green procurement models involve independent verification against a third-party certification system. A certification program is:

> meaningful if it has standards that (1) require social and/or environmental practices beyond the industry norm, (2) are transparent to businesses seeking certification and to consumers, and (3) are consistently adhered to by certified businesses (i.e., there is an audit and enforcement mechanism). Certification programs also entail costs, however, such as those for developing standards, conducting audits, changing business practices, and in some cases investing in new equipment. Certifications are only practical if the participating businesses generate economic returns that meet or exceed these costs.[24]

There are also internal certification systems, which are viewed as potentially less transparent and less stringent. Some of the leading independent certification systems include:

- FSC-certified timber;[25]
- RSPO Certification of palm oil;[26]
- RTRS Certification of soybeans;[27]
- The Programme for the Endorsement of Forest Certification (PEFC) which endorses "national forest certification systems."[28]

Sustainably sourced and certified timber involves, among many things, respecting local land tenure, ensuring just labor standards (e.g., workplace safety and livable wages), and minimizing the impact on the local environment. Sustainably sourced and certified agricultural commodities such as palm oil, soybeans, and cattle frequently involve the intensification of pastures or crops and reducing the expansion of such operations into pristine ecosystems including tropical rainforests and peatland forests.

Obtaining certification may yield a premium price on the commodity when compared to its traditional counterpart, or certification may open up markets for certified products which would not otherwise exist. In addition to certification standards, there are reporting mechanisms such as reporting to the CDP, particularly its Forests Program.[29] Companies reporting to the CDP Forests Program include both publicly traded companies and privately held companies.

Several other innovative technological advancements have enabled better supply chain management and the traceability of green procurement including the Global Forest Watch and Trase. The Global Forest Watch is a free platform for forest monitoring with a built-in alert system for recently cleared forests.[30] Trase, with an interactive map and extensive data on beef, palm oil, soy, and timber companies, has a:

> {…} pioneering approach {which} draws on vast sets of production, trade and customs data, for the first time laying bare the flows of globally traded commodities from production landscapes to consumer countries for entire countries and commodity sectors. Along the way it identifies the ports of export and import, and the producers, traders and transporters involved. These supply chain actors can then be linked back to environmental and social risk factors on the ground, as well as information on the social and governance factors necessary to improve conditions.[31]

There are also publicly announced no-deforestation supply chain commitments which are tracked by Supply Change.[32] Furthermore, Forest 500, which is a project of the Global Canopy Programme, "annually ranks powerbrokers {e.g., companies, financial institutions and jurisdictions}, between 0 to 5 based on their deforestation policies."[33]

Size of Instrument

There are several ways to assess the size of the green procurement sector and its potential.

Forest Stewardship Council (FSC)

As of March 2017, there were 194 million hectares (479 million acres) of FSC-certified forest area spread across 82 countries and consisting of 1,478 certificates. Brazil has 6.2 million hectares (15.3 million acres) certified as FSC, while Indonesia has 2.8 million hectares (6.8 million acres) certified and the DRC has 2.6 million hectares (6.4 million acres) certified.[34]

Global Roundtable for Sustainable Beef (GRSB)

While it is not clear how many hectares or how many heads of cattle are covered under the GRSB, the GRSB has 75 members across 5 groups of constituents. These 5 groups of constituents are: "producers and producer associations, the commerce and processing sector, retail companies, civil societies, and national or regional roundtables."[35] Such members represent some of the large beef companies in the world including Cargill, JBS, and McDonald's Corporation.

Round Table on Responsible Soy

In 2017, there were 182,133 newly certified hectares (approximately 450,000 acres) and 657,400 tons of newly certified soy under the RTRS Standard for Responsible Soy Production.[36] In 2015, more than 2.3 million tons were newly certified, and this was led by Brazil with 1.4 million newly certified tons (i.e., representing 62% of the certified soy market).[37]

Roundtable on Sustainable Palm Oil (RSPO)

With more than 3,200 members, the RSPO—as of February 2017—had 2.52 million hectares (6.22 million acres) certified that produced 12.22 million tonnes of certified palm oil (i.e., representing 21% of the global palm oil market).[38]

No-Deforestation Supply Chain Commitments and CDP Reporting

According to Supply Change's 2017 *Tracking Corporate Commitments to Deforestation-Free Supply Chains* Report:

As of December 1, 2016, the research cut-off date for this report, *Supply Change* found 718 companies—producers, processors, traders, manufacturers, and/or retailers—that have supply chains dependent on palm, timber & pulp, soy and/or cattle. Of these, we identified 447 companies that have made a total of 760 commitments to reducing deforestation impacts in their commodity supply chains, leaving 271 exposed companies without a public commitment. {...}

Commitments on palm, and timber & pulp continue to lead the way, thanks in large part to more well-established certification programs and scrutiny around palm oil-driven deforestation. Commitment rates remain considerably lower for soy and cattle, which is troubling given their outsized contribution to tropical forest loss. {...}

Commitment rates vary across the length of a supply chain: retailers had the lowest rate of commitments—54%—compared to their peers that operate "upstream" within supply chains (producers, 71%; processors, 72%; traders, 70%; and manufacturers, 66%). {...}

Supply Change found that at least 95% of participants in groups such as the High Carbon Stock Approach Group, Tropical Forest Alliance 2020, and Tropical Forest Trust have adopted pledges—as have 98% of signatories to the New York Declaration on Forests.[39]

Similarly, the CDP's 2016 *Global Forests Report* on behalf of:

365 investors, over a fifth more than in 2015, with a total of US$22 trillion in assets asked companies to disclose information via CDP about how they are managing deforestation risk linked to these four commodities {cattle, palm oil, soy, and timber}—responsible for the majority of tropical deforestation globally.

Responses were received from more than 200 companies, up from 180 in 2015, including three of the four so-called 'ABCD' global commodity trading giants { The Archer Daniel Midlands Company, Bunge, Cargill

> and Louis Dreyfus Company}, the three largest slaughterhouses in Brazil and big-name brands from Nestlé to Starbucks. {…}
>
> Companies report that, on average nearly a quarter of company revenues depend upon the commodities they reported on; and the total annual turnover at risk for publicly listed companies is estimated to be up to US$906 billion. {…}
>
> Despite these impacts, only one in five companies assess risks associated with deforestation beyond a six-year horizon across commodities.[40]

This said, it appears as though one of the great potential mechanisms to address tropical rainforest deforestation is through more green procurement and via pressure on companies to adopt more sustainable supply chains. Further, for instance, in March 2017, Green Century Capital Management "organized 38 investors in the U.S. and globally, representing $617.5 billion in assets under management, to ask that companies reaffirm and extend zero deforestation commitments specifically to Latin America."[41] Similarly, the Business and Sustainable Development Commission notes that "embedding sustainable business practices in the global food and agriculture industry could deliver an annual $2.3 trillion windfall {…}."[42]

Introduction to Case Studies

The following case studies will examine the Deramakot and Ulu Segama-Malua Forest Reserves of Malaysia, REDD Project in Brazil Nut Concessions in Madre de Dios, Peru, and Project Piaba with the New England Aquarium (Amazonas, Brazil).

The Deramakot and Ulu Segama-Malua Forest Reserves, located in Malaysia, demonstrate that well-managed, sustainable logging operations are a key conservation strategy when focused on a landscape level. These Reserves undertake both intensive forest restoration of indigenous tree species and apply site-species matching, which are conducted via both line planting and gap planting. Collectively, these Reserves provide habitat to an estimated 4,400 individual Borneo orangutans (*Pongo pygmaeus*) and represent the largest orangutan population in Malaysia.[43]

The REDD Project in Brazil Nut Concessions in Madre de Dios, Peru, is a project that involves approximately 400 families with Brazil nut concessions ranging from 300 to 1,000 hectares each. In total, the Project is approximately 300,000 hectares and conserving some of the world's richest tropical rainforest. The Project which was designed, financed, and is now being implemented by Bosques Amazonicos has been independently validated and verified to the Verified Carbon Standard[44] and has also been validated to the Climate, Community and Biodiversity Standard with Gold Level Distinction for exceptional biodiversity benefits.[45]

Project Piaba is a unique and ambitious initiative conserving the tropical rainforests of Amazonas, Brazil, by generating alternative income for local communities via the aquarium trade, particularly the trade in a species of fish known as cardinal tetras (*Paracheirodon axelrodi*). While the cardinal tetras make up as much as 80% of the region's aquarium fish exports, 245 species are endemic to this area and are on the approved list of species for export. If not for the dedication of Project Piaba to help maintain and improve the aquarium trade, the local communities would likely adopt other, non-sustainable land use practices and the surrounding forests would be lost forever.[46]

Case Study #1: Deramakot and Ulu Segama-Malua Forest Reserves of Malaysia

Introduction

It is important to note that well-managed, sustainable logging operations are another key conservation strategy when focused on a landscape level. For example, the Deramakot and Ulu Segama-Malua Forest Reserves, which are located in Sabah, Malaysia, and are owned by the State Government of Sabah, are managed for sustainable wood production and feature programs to restore forests in damaged areas. This includes both intensive forest restoration of indigenous tree species and applying site-species matching, which are conducted via both line planting and gap planting. Collectively, these Reserves provide habitat to an estimated 4,400 individual orangutans (1,000 individuals in Deramakot and 3,400 individuals in Ulu Segama-Malua) and represent the largest orangutan population in Malaysia. In total, the Borneo orangutan (*Pongo pygmaeus*) population in Malaysia is estimated to be approximately 17,000 individuals (i.e., 11,000 in Sabah and 6,000 in Sarawak).[47] Thus, approximately 25% of Malaysia's Borneo orangutans are found in these Reserves (Figs. 15.1 and 15.2).

Funding for the Reserves is derived from sustainable wood production, along with through annual development funds provided by the State Government of Sabah and the Federal Government of Malaysia. In addition, there are some funds contributed from payments for ecosystem services and tourism.

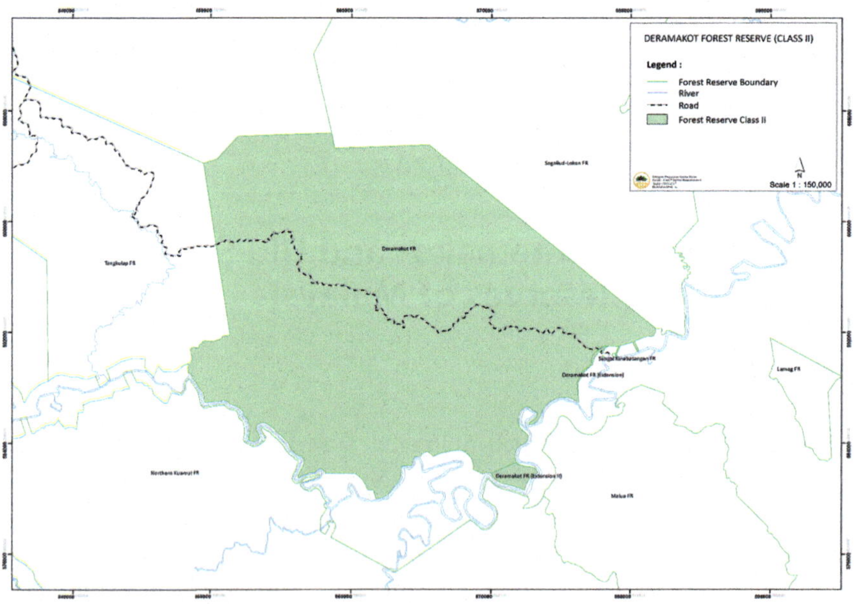

Fig. 15.1 Map of Deramakot Forest Reserve (Credit: Sabah Forestry Department)

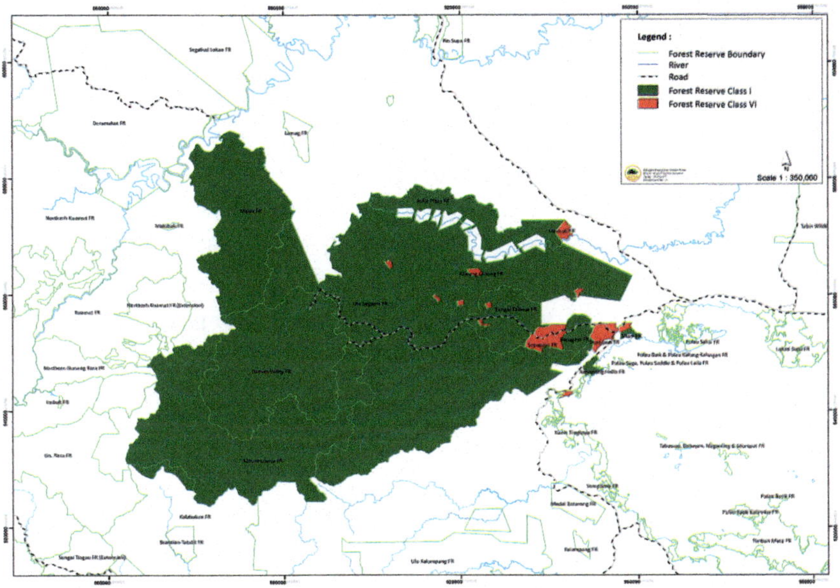

Fig. 15.2 Map of Ulu Segama-Malua Forest Reserve (Credit: Sabah Forestry Department)

Identify the Problem

The Ulu Segama-Malua Forest Reserve was first gazetted in 1961 as a Production or Commercial Forest Reserve, and then in 2014, the Ulu Segama-Malua Forest Reserve was reclassified as a Protection Forest Reserve. The Deramakot Forest Reserve was gazetted in 1960 as a Production Forest Reserve and remains classified as a Production Forest Reserve as of October 2016.[48] Prior to the gazettement of these Reserves, there were no records of deforestation, poaching, or illegal settlements within the Reserves. This was largely a result of the Reserves being remote, with little deforestation pressure from the small local population which kept the area pristine.[49]

However, the current problem at the Deramakot Forest Reserve is poaching for agarwood, which is mostly being done by foreigners. Such poaching for agarwood is also a problem at the Ulu Segama-Malua Forest

Reserve, in addition to the illegal collection of bushmeat and gold panning. Furthermore, there are also several cases of small-scale, illegal cultivations for palm oil.[50]

Why the Problem Is Important

The problems of poaching agarwood and cultivating palm oil, along with illegal collection of bushmeat and gold panning, are contributing to the degradation of biodiversity, deforestation, and a lack of governance.[51] The Borneo orangutan, along with its close relative the Sumatran orangutan (*Pongo abelii*), is considered a critically endangered species, and their population is declining.[52]

How Problem Was Identified

As described by the State Government of Sabah, these problems were identified through continuous, resource protection programs. Such programs include both ground patrols and river patrols, along with aerial surveillance via helicopters. Furthermore, illegal activities were also detected via the analysis of photographs from motion-sensitive cameras installed at several strategic locations throughout the Reserves.[53]

Effectiveness of Process for Identifying Problem

The Sabah Forestry Department (SFD) believes the process has been effective and is continuously practicing the Free, Prior, and Informed Consent (FPIC) process with a wide range of stakeholders, including local communities, nongovernmental organizations (NGOs), companies, and other relevant state and federal agencies.[54]

Steps Taken to Address the Problem

Historically, the SFD focused on prevention and monitoring, along with control and enforcement within the State's Permanent Forest Reserves to mitigate the aforementioned problems.

However, the SFD now undertakes all means of protection and conservation not only within the Reserves but also beyond the boundary of the Forest Reserves as well. Furthermore, the SFD needs "to safeguard the State's interest of biological diversity conservation in all high conservation value areas, and at the same time to ensure the sustainable management of the resources for future generations. We have management plans to implement and have our presence on the ground, 24/7." [55]

The following steps are taken to create reserves, such as the Deramakot and Ulu Segama-Malua Forest Reserves, in Sabah, Malaysia:

- SFD is tasked with the responsibility of identifying forest areas with high conservation value. In some cases, the State Government will receive proposals from local communities and/or local NGOs.
- SFD will then hold stakeholder consultations (e.g., local communities, NGOs, and other relevant government agencies) via the FPIC process to ensure no conflict of interests exists for the land and/or its natural resources management.
- With the results from the consultation processes, SFD will then prepare a Cabinet Paper for proposing the area, with the consent and advice from the State Attorney General.
- The Cabinet Paper is tabled and needs to be approved by the Cabinet for presentation at the State Assembly Meeting.
- Lastly, if successful, the State Governor will sign the Gazette of the forest reserve by specifying the extent of the land affected, the purpose of the Gazette, and the class of the said forest reserve, once the State Assembly approves the proposal."[56]

Results

Both Reserves, including Ulu Segama-Malua prior to its Totally Protected Areas (TPA) status, have been harvested for a relatively significant volume of wood while simultaneously generating revenue, respecting local communities, and providing habitat for a tremendous amount of biodiversity including the largest population of Borneo orangutans in Malaysia. Both Reserves, as is typical of Southeast Asian tropical rainforests, are predominately lowland, mixed dipterocarp forest, and thus, "the most harvested species are Dipterocarpaceae family and {some} mixed non dipterocarp species."[57]

Although no longer producing timber due to its TPA status, the Ulu Segama-Malua Forest Reserve's estimated production from 1957 to 2007 was approximately 26.2 million cubic meters (m^3) of timber. The Deramakot Forest Reserve is still active and produced an estimated 3.4 million m^3 of timber from 1958 to 2015. According to SFD, the most current results are:

> In 2016, the target of timber production from Deramakot Forest Reserve is about 20,000 m^3. As of September 2016, about 4,210 logs (16,050 m^3) are sold as FSC certified timber and attaining green timber status. About 67 SFD personnel are employed at Deramakot and 150 forest workers for timber harvesting, silvicultural treatment and forest rehabilitation. The area of Deramakot FR is 55,507 ha and fully certified as a well-managed forest under the FSC standard since 1997 until 2015. On the aspect of forest intervention, about 19,380 ha have been silviculturally treated and 1,430 ha of degraded forest fully rehabilitated. For the past 5 years, Deramakot has ventured on ecotourism with contribution to the revenue of about USD$100,000. On biological diversity conservation, about 23 Rare, Threatened and Endangered species under IUCN Red List are well protected, i.e. stable population of 1,000 orangutan.
>
> While for Ulu Segama-Malua, this is a fully protected area without timber production since 2008. This area is also certified as well managed forest without timber production, except for capitalizing ecosystem services under the FSC standard since 2011. About 116 SFD personnel are attached in managing 242,884 ha of forest reserves and the other 600 forest workers are implementing silvicultural treatment and forest restoration. About 42,862 ha of forest have been silviculturally treated and 9,200 ha of degraded forest intensively restored using indigenous tree species, purposely to enhance critical habitats of orangutan. In the aspect of biological diversity conservation, about 39 species of Rare, Threatened and Endangered species under IUCN Red List are well protected and monitored, i.e. stable population of 3,400 orangutan.[58]

Additional financial results include:

- Ecotourism has become the most important, non-timber income generated at the Ulu Segama-Malua Forest Reserve. Such ecotourism

activities are carried out by the private sector (i.e., not SFD) and generate annual revenues of approximately USD$2.0 million.
- The Malua Wildlife Habitat Conservation Bank has been initiated specifically for the protection and conservation of wildlife habitat. As of 2016, approximately 23,200 credits have been sold and generated approximately USD$125,000 in revenue.
- A forest carbon sequestration project has also been initiated with Face the Future, a Netherlands-based organization. The objective of the forest carbon project is to rehabilitate about 25,000 hectares of logged-over forest with indigenous tree species, and as of 2016, about 11,500 hectares are fully treated. The other funder on the forest restoration work is Sime Darby, which has allocated USD$6.25 million over 10 years for rehabilitating 5,400 hectares. Furthermore, WWF Malaysia has raised additional conservation funds of USD$2.375 million for restoring 2,400 hectares of degraded orangutan habitats.[59]

Challenges and How They Were Met

The most pressing challenge faced by SFD is dealing with the various conflicts of interest between local communities and other stakeholders, in particular over the land and natural resource management. The SFD is addressing this challenge through continuous engagement, direct participation, and empowering local stakeholders to properly manage their natural resources with endorsement from the highest political levels in Malaysia.[60]

Beyond Results

As explained by SFD, the sustainable, reduced impact forest management practices at the Ulu Segama-Malua and Deramakot Forest Reserves have been sustainable in many regards.

Ecological sustainability can be demonstrated by the track record over the last 27 years of operations at the Deramakot Forest Reserve and the ongoing presence of orangutans.

From an economical sustainability perspective, the average annual harvesting rate of approximately 17,600 m^3 of high-quality timber has, as of 2016, generated about USD$37.2 million in gross revenue with the net profits of approximately USD$5.5 million going to the State Government of Sabah.

The Reserves have also contributed to local, socioeconomic benefits through job creation and contract work.[61]

Lessons Learned

According to the SFD, the following are lessons learned from managing the Ulu Segama-Malua and Deramakot Forest Reserves:

- "Short-term management practices do not guarantee responsible forest management.
- 'Smart partnerships' with credible partners will increase the value of sustainability.
- Forest certification—such as the Forest Stewardship Council—can improve good governance, enhance performance, and bring about credibility.
- Diversifying revenue is a wise idea, such as by exploring non-timber forest products, developing ecotourism, and pursuing other innovative ideas (i.e., carbon sequestration and biodiversity offset mechanisms).
- Continuous, capacity-building programs should be in place to maintain productivity and to increase management effectiveness.
- It is not a good idea to work in isolation; rather, one should engage with the world.
- The greatest good for the greatest number in the long run is what should be emphasized and this is what forests are all about.
- Political support at the highest level is vital for success.
- If possible, one should always seek to get things right the first time."[62]

Other Resources for Ulu Segama-Malua and Deramakot Forest Reserves

Malua BioBank

- http://www.maluabank.com/
- http://www.maluabank.com/downloads/Malua_CMP.pdf

Markit Environmental Registry

- https://mer.markit.com/br-reg/public/project.jsp?project_id=100000000000785

Official Portal of the Sabah Forestry Department

- http://www.forest.sabah.gov.my/

Sabah Forestry Department's Deramakot Forest Reserve Website

- http://www.deramakot.sabah.gov.my/

Sabah Forestry Department's Ulu Segama-Malua Website

- http://www.forest.sabah.gov.my/usm/

VCS Database, INFAPRO Rehabilitation of Logged-Over Dipterocarp Forest in Sabah, Malaysia

- http://www.vcsprojectdatabase.org/#/project_details/672

Case Study #2: REDD Project in Brazil Nut Concessions in Madre de Dios, Peru

Introduction

Madre de Dios is a region within the Peruvian Amazon on the border of Bolivia and Brazil. The region is one of the most biologically diverse places on Earth. E.O. Wilson, within Madre de Dios, "collected forty-three species of ant from a single tree in the Tambopata Reserve in Peru—almost as many as the entire indigenous ant fauna of the United Kingdom!"[63]

Within this region, Bosques Amazonicos (BAM) has designed, financed, and begun to implement the REDD Project in Brazil Nut Concessions in Madre de Dios, Peru (the Project). Essentially, the Project involves approximately 400 families with Brazil nut concessions ranging from 300 to 1,000 hectares each. The total Project area is approximately 300,000 hectares. These Brazilian nut concessions are not legally conservation concessions, which is another legal class according to Peru's forest legal framework. Rather, the Brazil nut concessions are classified as productive concessions where concession holders are allowed to harvest non-timber forest products (i.e., the Brazilian nuts) and, if the proper paperwork is completed, the concession holders are allowed to also extract forest resources such as timber on sustainable basis (Fig. 15.3).

The Project has been independently validated and verified to the Verified Carbon Standard[64] and has also been validated to the Climate, Community, & Biodiversity Standard with Gold Level Distinction for exceptional biodiversity benefits.[65]

Identify the Problem

There were several problems that existed prior to the creation of the Project and prior to the granting of the Brazilian nut concessions.

One such problem is illegal mining, which was happening far more in the past than it is now within the concessions. Currently, illegal gold mining throughout Madre de Dios remains a huge problem as it is often hard to compete with the price of gold.[i][i] Cannon, John C. "Deforestation from gold mining in Peru continues, despite gov't crackdowns." August 22, 2017.

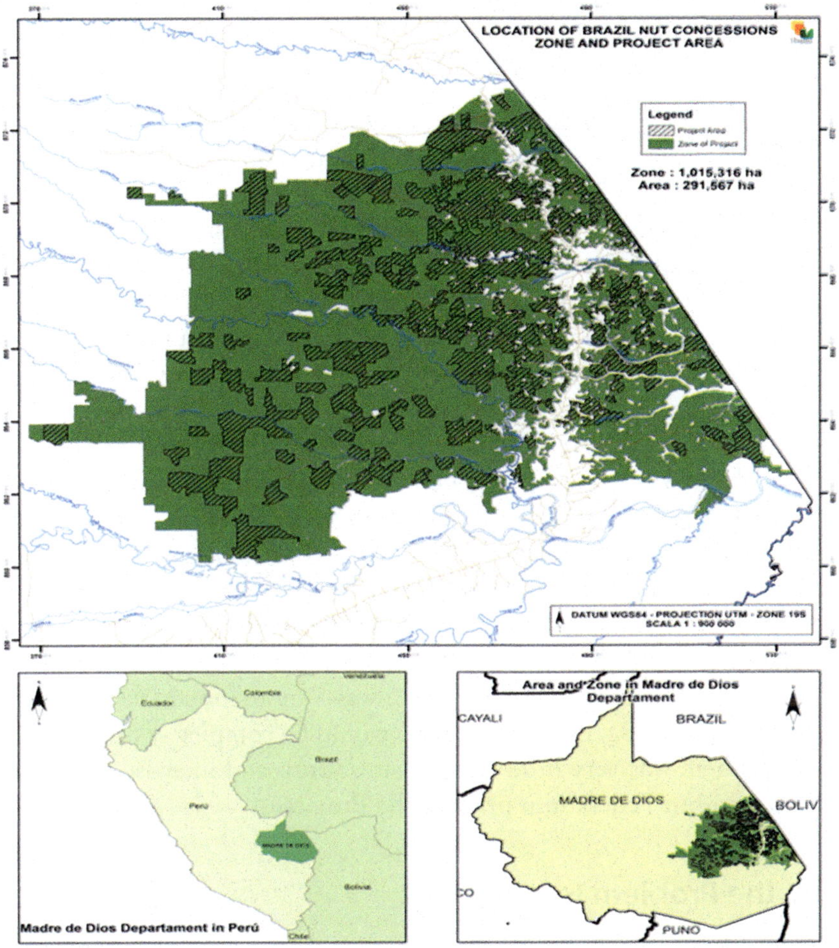

Fig. 15.3 Map of REDD Project in Brazil Nut Concessions in Madre de Dios, Peru (Credit: Bosques Amazonicos)

Mongabay. Accessed August 29, 2017. https://news.mongabay.com/2017/08/deforestation-from-gold-mining-in-peru-continues-despite-govt-crackdowns However, the Project is a bit further North from Cusco and located closer to the borders of Brazil and Bolivia, and thus, illegal mining is less of a problem within the Project. The topic of illegal gold mining in Peru is very complicated due, in part, to the organized networks that facilitate land invasions, and BAM, in the past, has tried to assist with counteracting such land invasions. [66] It is also important to note that illegal gold mining has recently started to spread into Central and Northern Peru.[67]

Another problem prior to the creation of the Project and concessions was that the Interoceanica Sur (i.e., the Interoceanic Highway or Trans-Oceanic Highway) was in the process of being constructed from Brazil, through Madre de Dios, to Peru's Pacific Coast.[68] The larger, current problem is that lots of migration has taken place due to Interoceanica Sur and this has resulted in lots of pressure on the primary tropical rainforests because of "unsustainable, illegal, small-scale agriculture and logging"[69] often in the form of subsistence, slash-and-burn agriculture.

Another large, current problem is that while the market for Brazilian nuts is strong, the market for verified emission reductions (VERs; i.e., carbon offset credits) is inconsistent. This said, there is strong demand for Brazilian nuts which allows for sales with good pricing, but the Brazilian nut market is not enough because of the seasonality of the operations (i.e., farmers work with Brazilian nuts for only about 4 months a year). In addition, the spatial distribution of Brazilian nuts is approximately 1 tree per hectare, so families often have only a few hundred trees to support their livelihoods. Because of this reality, coupled with a weak VER market, there is pressure on local communities to sell their land.[70] If there was a more functional forest carbon market for VERs, then the scenario would be completely different.

Lastly, there was very little local organization, and likewise, there was no Brazilian nut Federation, prior to the Project.

Why the Problem Is Important

One of the main reasons the problem is important is because there were local cartels operating in the region that took advantage of local communities and only provided very low prices. In addition, if unsustainable activities (e.g., illegal mining or slash-and-burn agriculture) take ahold in the Project area, then the region's rich biodiversity will be lost.

How Problem Was Identified

There aforementioned problems were identified via several channels. Jorge Cantuarias, the Founder and Chief Executive Officer of BAM, has followed the political decisions regarding Reducing Emissions from Deforestation and Degradation (REDD) and climate change on the regional, national, and international level for numerous years. BAM also has a very strong team, which has worked closely with local communities. Furthermore, the problems of road construction, settlement expansion, and tropical rainforest deforestation throughout the region have been identified via national satellite imagery.[71]

Effectiveness of Process for Identifying Problem

The problems that were identified in the past continue to be the same problems as today, so this indicates the problems were properly identified in the very beginning.

Steps Taken to Address the Problem

When BAM started working on the Project, one of the very first initiatives was to promote local organization and to start tackling the challenges of the local Brazilian nut industry.[72]

For instance, there were only 2 big players in the region that worked with the commercialization of Brazilian nuts. The first thing BAM did to address this problem was to finance the harvest of the Brazilian nut. Second, BAM doubled the price paid for Brazilian nuts which was originally around USD$1.00 per kilo (3 Peruvian Soles). Since then, the market and on-the-ground activities have changed a lot. BAM plans to build a factory in the area in 2017, which would require market certification, and BAM will try to find commercial outlets.

Results

Perhaps the most important result of the Project is that the Project put people in a stronger position to fight for their rights and views. Similarly, the local communities are now empowered with a new ability to meet with high-ranking government officials, and the local communities have a voice over decisions. This result was very important because at the time, the Interoceanica Sur was coming, and had BAM not been there, the pressure of the road, new agricultural expansion, and other activities would have devastated the local communities.[73]

Another important result is that the Brazilian nut market—for the first time—has started to work. In the past, the buyers of the local communities' Brazilian nuts often paid a very low price and put all of the conditions on the concessioners. Now, there is a quite fair market, there are 6 or 7 related industries working in the area, and a good price for all involved. The Project has also hired a permanent legal and technical advisor, along with engineers and lawyers. The team, at a subsidized cost, develops a sustainable harvesting plan for local communities to sustainably extract both timber and Brazilian nuts. Furthermore, the team helps to legally maintain the concessions and keep the concessions in order (i.e., filing any and all necessary paperwork). This activity is important because this is how a concession would be possibly lost, or a way for local communities to be abused by outsiders claiming their concession paperwork has lapsed.

Because the local communities are empowered and the Brazilian nut market has started to efficiently operate, the Project has experienced very little deforestation.[74]

Challenges and How They Were Met

The specific challenges facing the Project are similar to the regional problems which were previously described. In addition, the Project had worries about there being a lot of families and the likelihood that there would be lots of problems.[75] According to BAM, there has been a lot of good work done with legislation, clarifying the various actors and ownership, and establishing joint ventures between private companies such as BAM and the concession holders. However, the ongoing challenge is that more money needs to be locally deployed in order to carry out the strategy and on-the-ground activities. For instance, Norway has committed approximately USD$300 million to Peru and the first tranche of USD$50 million is already in place, but none of these funds have come to the concession owners.[76]

Beyond Results

BAM believes these results are sustainable. A lot of progress has already been made, and upon having enough financial resources—ideally due to an expanding carbon market—then the Project can multiply the number of Brazilian nuts. Such Brazilian nut trees can be planted along the paths to the other Brazilian nut trees, and with a period of 8 years between planting and harvest, this could help multiply local incomes by a magnitude of 5.[77]

As another demonstration of the sustainability of green procurement like BAM's work, there are other similar Brazilian nut initiatives, such as at the Amarakaeri Communal Reserve, taking place in Peru.[78]

Lessons Learned

As described by Jorge Cantuarias:

- The local communities in the area who make a living from these activities, they love what they are doing and love where they live. Often, they are not willing to sell the forest and there is a real consciousness of the benefits of having the forest. Thus, they will try their best to conserve the forests even though it is technically not really their property.
- There is relatively little investment in the Madre de Dios area near the concessions.
- These local communities are established and have become an important, allied partner. You need to find these types of allies in local peoples to be partners of the Project and to be a part of the solution.
- You can come with a view—be it an international capitalist view or a governmental view from the nation's capital—but you need to listen to what the local communities need. They often know what is good, so you should give them the resources and to hear what they are saying. Similarly, you need to have a long-term view for the area.
- The authorities can sometimes be your enemy and be a waste of time as many government officials have never been to the tropical rainforest

and do not truly understand the local challenges. Those that have visited may have just visited for 1 weekend at a luxury lodge.[79]

Other Resources on REDD Project in Brazil Nut Concessions in Madre de Dios, Peru

Bosques Amazonicos

- http://www.bosques-amazonicos.com/en

Markit Environmental Registry

- https://mer.markit.com/br-reg/public/project.jsp?project_id=100000000001107

Verified Carbon Standard's Project Database

- http://www.vcsprojectdatabase.org/#/project_details/868

Case Study #3: Project Piaba with New England Aquarium (Amazonas, Brazil)

Introduction

Project Piaba is a unique and ambitious initiative conserving the tropical rainforests of Amazonas, Brazil, by generating alternative income for local communities via the aquarium trade, particularly the trade in a species of fish known as cardinal tetras (*Paracheirodon axelrodi*). While the cardinal tetras make up as much as 80% of the region's aquarium fish exports, 245 species are endemic to this area and are on IBAMA's (Instituto Brasileiro do Meio Ambiente e dos Recursos Naturais Renováveis; Brazilian Institute of the Environment and Renewable Natural Resources) approved list of species for export for the aquarium trade. Project Piaba is a 501(c)(3) registered nonprofit in the United

Fig. 15.4 Map of Rio Negro fishery area and Project Piaba's work in Amazonas, Brazil (Credit: Project Piaba)

States that was established in 1991 and has an international affiliate registered in Brazil. The organization does not sell or profit from the aquarium fish trade, but rather it studies and helps foster the beneficial fishery (Fig. 15.4).[80]

One of the few ranges for cardinal tetras, and the target of Project Piaba's work, is the Rio Negro River Basin in Amazonas, Brazil, shown above. The waterways and floodplains throughout the region are federal public lands, and these public lands go as high as the high water mark. Local communities are allowed to reside in these public lands and there is a sense of ownership with some usury rights, but there are some conflicts over resource use.

There are four types of fisheries in the region and some of these fisheries overlap:

- Artisanal consumption fisheries;
- Fisheries for the aquarium trade;

- Commercial fisheries; and
- Fisheries for sport fishing.[81]

A relatively recent type of sport fishing that has arisen is destination sport fishing for the various species of peacock bass including the speckled peacock bass (*Cichla temensis*).[82]

Project Piaba specifically works with the fisheries for the aquarium trade.

Identify the Problem

Deforestation throughout the region where Project Piaba operates is not a major issue as of now. There is subsistence agriculture, mainly manioc, and some small-scale commercial agriculture to supply the local area. In addition, with the decline of the fishery, urban migration to the city of Barcelos has caused some increased deforestation. With respect to deforestation, there is no illegal logging or large-scale forestry operations in the area. There is some illegal poaching for turtles and turtle eggs, but poaching appears to be less of an issue today than it was in the past when species such as Amazon manatees (*Trichechus inunguis*), pygmy anteaters (*Cyclopes didactylus*), and caimans were hunted.[83]

Although deforestation is not an issue today, if the sourcing of wild cardinal tetra is disrupted—which since the 1950s, 60% of the local communities' income have come from the aquarium trade—the local communities will have to do something else to feed their families. This said, there are threats to the trade from international competitors, including fish farms, along with challenges throughout the supply chain and potentially less interest from younger generations in having an aquarium.[84]

Although destination sport fishing for speckled peacock bass was originally seen as a new economic base for the region and the outfitters will often hire local crews, most of the money stays in the United States. In addition, some communities—whether it is true or not—were told that they are prohibited from fishing for peacock bass (i.e., presumably to keep the bass for visiting sport fishermen). A similar problem with sport fishermen arises because local communities often have fish pens with their aquarium fish near their houses; however, if a sports boat speeds by with a wake and disrupts the water, this will potentially harm the aquarium fish.[85]

With respect to competition, neon tetras (*Paracheirodon innesi*) are close substitutes to cardinal tetras (*Paracheirodon axelrodi*). Both species are freshwater fish, they look very similar, and both species cost relatively the same amount. However, neon tetras are incredibly abundant in Peru. In addition, the water in both Tampa, Florida, and Singapore is similar to Peru, which makes it much easier to farm raise neon tetras. In contrast, the cardinal tetras from Amazonas are tougher to reproduce and mostly only wild fish are caught. There is the ability to add hormones into the food for cardinal tetras to enable the fish to have more tolerance and to force them to spawn. This is not as commercially viable of an option because of the additional costs, but producers can do this for a few generations and select the desirable strains. This selective breeding with hormone additives has become successful for cardinal tetras, and the industry is now able to farm raise cardinal tetras outside of Amazonas in Florida. In addition to Florida, Vietnam has started to produce cardinal tetras on a year-round basis from fish farms. These Vietnam operations have good business sense and pose a true, competitive threat. For example, the fish farmers will often pool air freight costs to dramatically reduce their operating costs. The region of Amazonas was a past victim of biopiracy when Brazil's rubber boom was severely disrupted after rubber tree seeds were smuggled to Malaysia. Thus, the local communities are concerned about their livelihoods after Project Piaba showed them a press release about farm raising cardinal tetras in Florida along with aerial photographs showing the scale of the operations.[86]

There are also several problems associated with the supply chain for wild caught cardinal tetras. First, there is a problem with handling cardinal tetras. This results from a few issues, one is that the Rio Negro River Basin is a very acidic environment (i.e., one that is found nowhere else on Earth) and it is not easy to acclimate the aquarium fish when they arrive to home aquariums.

Another issue is that since the 1950s when the fishery began operating, most of the fishers have been trained by parents and grandparents; however, these fishers have needed training on understanding how to best handle the fish in terms of care, feeding, and water changes which have been recently learned in the industry. Improper handling is not necessarily lethal, but the additional stress results in a reputation that the fish coming out of Manaus are very needy, sometimes not healthy, and there is a need to stabilize the fish which is both time-consuming and expensive for the international trade.[87]

In addition to improper handling, there are additional fractures in the trade infrastructure. For instance, a large export facility in Manaus (the state capital of Amazonas) closed their facility, and in the past, this facility could hold up to 10 million cardinal tetras. This was an important development because with the facility, aquarium fish could be exported year-round even when the fishery is closed due to seasonal high water conditions. There are a few remaining, small-scale exporters. Fortunately, there is a new exporter that just completed an even larger facility outside of Manaus, which will soon be able to stockpile cardinal tetras. The reason stockpiling is important is because the fishery is inaccessible during peak market demand in the United States and Europe. The aquarium market in the United States and Europe traditionally has an uptick in September as the days get shorter, the weather gets colder, and people get reinterested in their fish tanks at home. There is also an uptick in January and February as people get fish tanks at Christmas and then buy fish and supplies in the months that follow. However, this is the high water season in Amazonas, and it is illegal for the local communities to catch cardinal tetras during this time. Without sufficient stockpiles, there are no more sales once the stockpiles have dwindled. Lastly, with respect to the trade infrastructure, there are air logistical issues as the direct flights out of Manaus are not conducive for the cardinal tetra trade.[88] For instance, flights out of Manaus are infrequent, often are not direct flights that carry freight, and can cost upward of 4 times more for the same size and weight than flights from Asia.

In the past, there were some problems during the high water season when fishermen were not able to catch any aquarium fish, particularly the cardinal tetras. To supplement their seasonal income, the fishermen would shift to bushmeat because they did not have alternative sources of income during the off-season. Then an interesting turn of events happened. It appears as through the government may have been approached by the local communities who sought unemployment benefits as a result of the seasonal hardship. The initial response from the government was that the high waters were a force majeure and thus, the government should not be held responsible. It would then become illegal to catch cardinal tetras during the high water season because this was the breeding season for the fish and restricted fishing was critical for their long-term survival. While this may have improved the environmental image of the government, what ended up happening is the prohibition then made the communities able to say they are legally unable to fish, and thus, there was a non-force majeure hardship and they started getting seasonal, unemployment payments from the government. It is important to understand that from the communities' perspective, if you are getting your seasonal stipend, you do not need to hunt (i.e., eat or possibly sell) wild game such as monkeys, dolphins, parrots, or manatees.[89]

One additional problem facing the aquarium trade in general, and specifically Project Piaba and the cardinal tetra fisheries of Amazonas, is the declining interest in aquariums among younger generations. Not only are younger generations not interested anymore, but animal activists—who may unfairly target segments of the aquarium trade like Project Piaba that are demonstrating net positive community and biodiversity benefits—are getting more funding and better organized. For example, there have been successes—and rightly so—with killer whale (*Orcinus orca*) shows no longer occurring at SeaWorld in California[90] and elephants are no longer a part of Ringling Bros. Circuses. In fact, Ringling Bros. and Barnum & Bailey will put on their last shows in May 2017.[91]

Why the Problem Is Important

The main reason these problems are important is because if local communities lose their incomes from the aquarium trade, they will need to replace their incomes with alternative activities which would likely result in higher pressure on local wildlife through hunting and/or deforestation due to increased slash-and-burn agriculture and cattle ranching.

How Problem Was Identified

Project Piaba identified the problems through several channels. This includes:

- Having dedicated, subject matter experts working with Project Piaba;
- Undertaking regular site visits to the region and interviewing local communities; and
- Engaging not only the suppliers but also actors throughout the entire supply chain such as exporters, wholesalers, retailers, and educational facilities such as aquariums and zoos.[92]

Effectiveness of Process for Identifying Problem

The process for identifying the problems has been very effective. Dating back to 1989, researchers and students from the Universidade do Amazonas (UA; University of Amazonas) and the National Institute of Amazon Research (INPA) initiated an ecological baseline study on floodplain fishes of the mid-Rio Negro Basin. During this initial phase, researchers discovered the importance of the home aquarium fisheries for local livelihoods and wondered about the environmental impact of the fishing activity. Subsequently, Project Piaba was started in 1991, with its first grant awarded by the Conselho Nacional de Desenvolvimento Científico e Tecnológico (CNPq; National Council for Technological and Scientific Development). Project Piaba strengthened its local roots in the municipality in January 1994 when it inaugurated the "Dr. H. R. Axelrod Laboratory of Ornamental Fishes" during the first Ornamental Fish Festival in Barcelos. The mayor of Barcelos at the time, Valdeci Raposo, provided Project Piaba with a classroom. Project Piaba built 20 aquariums with local fishes on exhibit and a desk with stereomicroscopes donated by a group of aquarium enthusiasts. This small exhibition provided an opportunity for the local children to see the home aquarium fishes up close and through magnifying scopes. Project Piaba has been engaged with the communities ever since.

Steps Taken to Address the Problem

To address these problems, several steps were taken.

First, Project Piaba set up a legal entity in both the United States and Brazil.

Second, Project Piaba and the pet industry held a "Train the Trainers" event in Manaus to improve the local supply chain by teaching proper handling techniques of tetra cardinals. These attendees are now certified trainers on best practices for handling cardinal tetras and for training others. This was an important objective to get all the trainers properly trained and certified. Small changes surrounding addressing issues like best handling practices with fishermen can have big impacts. Another related issue that is being taught is that surface water from the streams has a low pH (power of hydrogen) level and is often used by the export facilities. However, the region's well water has higher pH and is more similar to the natural habitats of the cardinal tetras. In addition, fish feed quality needs to be improved.[93]

Third, Project Piaba is utilizing social media to connect visitors to aquariums and hobbyists of aquarium fish, to the local communities and fisheries that are the ultimate source of the fish. In this sense, Project Piaba believes the industry is leaving a lot on the table regarding marketing. Young people might now get excited if they are able to trace back to where the fish, particularly cardinal tetras, came from and learn about the Amazon Rainforest. This learning can come through stories about the local communities protecting forest for fish and the ensuing community benefits of income generation and poverty alleviation. In addition, young people could learn via social media and traceability about the tremendous biodiversity dependent upon the Rio Negro River Basin and learn about the outcomes of carbon storage and climate change mitigation all as a result of the aquarium industry. For instance, if the cardinal tetras had a tracking number, then zoos, aquariums, and retail shops could utilize a Google Earth flyover with thumbnail pictures that could be retrieved by scanning a smartphone and have statistics about the carbon dioxide emissions sequestered in the forest or see the rare animals in the region. To take it one step further and put new aspects into an old hobby, smart plugs can be programmed with a smartphone so that the aquarium's lights and heaters come on when the sun rises upon the stream where the fish originally came from in the Amazon.[94]

Results

Since its founding, Project Piaba has produced numerous results. The fishery, without Project Piaba, would likely have completely collapsed. A thriving industry for cardinal tetras mitigates deforestation; otherwise, alternative livelihoods such as cattle ranching, agriculture, logging, and bushmeat trade would take over.[95]

Another important result is there is now recognition from the International Union for the Conservation of Nature (IUCN) and other stakeholders in the wildlife community that the wildlife trade can be a solution, rather than seen as always the problem. In the past, there used to be an across-the-board belief that the wildlife trade—in all forms—was a problem.[96]

If Project Piaba can help improve the trade structure, improve the quality of fish through better handling techniques, and demonstrate traceability, then visitors to aquariums and zoos who have disposable income and are interested in wildlife may purchase more fish from areas such as Project Piaba. Project Piaba currently has 22 institutions (i.e., aquariums and zoos) with 25.7 million visitors that have exhibits themed around the beneficial Rio Negro Fishery which showcase the fish and educate visitors about the beneficial fishery. Some of these exhibits go so far as encourage visitors to consider joining the fish hobby.[97]

Challenges and How They Were Met

The specific challenges facing Project Piaba today, and into the future, are similar to the overall regional and industry problems mentioned earlier.

First, ensuring the proper handling of fish is an ongoing, major challenge. Without proper handling, the fish get stressed and the mortality rates increase, which leads to less revenue and unsatisfied wholesale buyers. To overcome this challenge, Project Piaba is training trainers to teach local communities the proper handling techniques, and the Brazil trainers are now working with the local fishers, holding facilities, and exporters on best handling practices so that the fish are better handled, water quality is more closely watched, and feeds are used properly to improve fish health.

Second, like many of the other certified sustainable products or emerging no-deforestation supply chain commitments, traceability is key. For Project Piaba, being able to trace the fish back to their wild habitat and to the local communities whose lives depend on the fish can inspire consumers to support the trade. This relates to the challenge of getting new generations interested in the aquarium trade. Today's younger generations, with more interest in technology and social media, are less interested in aquariums. The first step, Geographic Indication, is currently being developed to demonstrate complete and traceability solutions.

It is also important to note that changing the air freight situation in terms of cost and schedule of cargo would be a big game changer for the Rio Negro Fishery, which currently presents a clear disadvantage.[98]

Beyond Results

Project Piaba is ambitious and appears to be a sustainable, long-term solution. Ultimately, Project Piaba is attempting to reinvent the industry based off the best things. Through transparency and traceability, the industry can demonstrate the sustainability of its wild sourced cardinal tetras, the direct benefits to local communities, and the resulting reduced pressure on the surrounding tropical rainforests. This said, the market for cardinal tetras from the region has existed for 50 years, and if the market is there, if the cash flows are there, then there will be a big shift in the mentality of the local communities that they will be able to continue taking care of their families. This notion is becoming a game changer with organizations such as the World Pet Association, Ornamental Aquatic Trade Association, Pet Industry Joint Advisory Council, and Ornamental Fish International supporting the work. Articles on these topics are also starting to appear in industry trade magazines.[99] Furthermore, the concept of utilizing wild sourced aquarium fish, while protecting the surrounding environment and benefiting local communities, is spreading to areas such as the Western Ghats of India and Bali, Indonesia.[100]

From an industry perspective, regardless of where the fish come from—whether its farm raised in Florida or wild fish from Amazonas—hobbyists will still buy all the supplies for their aquariums including heaters, food, and plants.

Zoos and aquariums are institutions with a specific mission to educate and inspire visitors. With 180 million visitors each year, if even 1% (1.8 million people) are excited to get an aquarium and support sustainable fisheries, this is a big impact.[101]

Another example of sustainability that is comparable to the cardinal tetra trade is a past example of the freshwater stingray (*Potamotrygon wallacei*) trade. In the past, exports of freshwater stingrays from the Rio Negro River Basin were banned, and communities wanted to eradicate these local freshwater stingrays because their children would inadvertently step on them and get stung. There is now an export quota for these freshwater stingrays and the communities no longer want to eradicate the freshwater stingrays. In contrast, the communities will know where the pregnant female is, put her in a temporary pen, let her give birth, and then release her back into the wild. The communities are essentially practicing a form of fish ranching.[102]

Lessons Learned

Undertaking a complex initiative such as Project Piaba that involves working with seasonal fisheries and local communities in relatively remote areas of Brazil, to working with a multibillion-dollar aquarium industry, is not for the faint of heart. The lesson learned is that one needs to stick with their mission, have tolerant and supportive family, be able to work with limited funding and possibly credit card debt, not expect credit for the successes, and do whatever is possible with their limited "free" time. In addition, it is absolutely essential to have a committed champion—like Mr. Scott Dowd—who is concerned about the long-term viability of the aquarium hobby and the health of the forests and who will not sit idly on their hands.[103]

Other Resources on Project Piaba

Aquaculture Funds (similar investment funds could possibly benefit Project Piaba)

- Aqua-Spark: http://www.aqua-spark.nl/the-approach/our-strategy/
- Fish 2.0: http://www.fish20.org/about/goals
- World Fish: http://worldfishcenter.org/what-we-do

Geographic Indication

- http://projectpiaba.org/wp-content/uploads/2015/02/Historic-Rio-Negro-GI-status-for-piabas.pdf

IUCN Home Aquarium Fish Sub-Group

- http://www.iucnffsg.org/about-ffsg/home-aquarium-fish-sub-group/

New England Aquarium's Mission and Vision

- http://www.neaq.org/about-us/mission-vision/

Project Piaba's Official Website

- http://projectpiaba.org/

Segrest Farms

- http://www.segrestfarms.com/index.cfm?fuseaction=piaba.projectpiaba

Return

Sustainably certified commodities and green procurement models may have the potential to generate revenues greater than their traditional counterparts. For example, there might be premiums offered for products such as Forest Stewardship Council (FSC)-certified timber or Rainforest Alliance Certified organic coffee.[104]

In addition to price premiums, products adopting a green procurement model may gain access to new markets. For instance, the US Green Building Council's (USGBC) Leadership in Energy & Environmental Design (LEED) offers a Certified Wood Credit for FSC-certified timber used in its buildings.[105] It is estimated that "North American paper suppliers already get one third of their wood from certified forests; the figure is 60 percent for Europe, where sustainable forestry is expanded."[106] Sadly, "only 11 percent of the globally certified forest areas are in the tropics and subtropics."[107] Then again, this lack of certified forest areas in the tropics might present a big opportunity in the future.

Firms adopting a green procurement model may be able to access a lower cost of capital. Furthermore, there seems to be a limited supply of certified products to meet demand—especially if the no-deforestation supply chain commitments will be met in the future. These global commodity markets for cacao, cattle (e.g., beef and leather), palm oil, rubber, soy, sugar, and timber are enormous and provide one of the most promising instruments.

Risk

Despite the potential, there is a lot of risks associated with sustainably certified commodities and green procurement. For instance, these commitments have the potential for leakage that would, if not effectively addressed, call into question the overall effectiveness of the mechanism. For instance, some markets in Asia do not require such certification, and even if they require such certification in the future, there are still people who clear forests who often still have ready and close-by markets.

Alongside green procurement models, we still need to value the climate and environmental benefits of standing forests. Instead, we have been doing more of the opposite and used more of a "stick approach" (i.e., not going to buy the products), but a "carrot" has not been offered to pay for the benefits of the standing forests. There are also perverse incentives, with the person who cleared the forest too often thought to be twice as productive and making twice as much money, as the person who is preserving the forest and not getting paid.

Yet, it is worth acknowledging that "a company would never dream of avoiding an independent audit of its financial books—their shareholders would never stand for this. Shareholders also have a right to an independent assessment of whether their forests are being well managed for the long term."[108]

Business Risk

The following questions, which raise several potential business risks, were posed during a webinar entitled, "Accelerating the implementation of corporate zero deforestation commitments," which was cohosted by the Innovation Forum and the Global Canopy Programme:

- What are the risks, identified by companies, operating in forest commodity supply chains, and to what extent do companies recognize the opportunities associated with sustainable sourcing?
- Are company policies robust enough to meet high-level 2020 and 2030 commitments on deforestation?
- Are companies engaging with their suppliers to ensure their supply chains are free from deforestation and to encourage upstream change in business practices?
- To what extent are financial institutions putting in place policies that will encourage companies in their lending and investment portfolios to take action on deforestation?[109]

Another significant business risk of having commodities associated with deforestation—especially illegal deforestation—in one's supply chain is that the company could be prevented from accessing a global stock exchange, which occurred when United Cacao illegally cleared forests in Peru for its cacao operations.[110]

A similar scenario occurred to JBS, which was found to have nearly 60,000 cattle from illegally cleared forests, and a major buyer canceled its procurement contract:

> JBS is caught in a second scandal as IBAMA, Brazil's environmental protection agency, raided JBS meatpackers in Redenção and Santana do Araguaia, in the state of Pará last week. IBAMA's raid identified 59,000 cattle sourced from land deforested illegally. This illegal deforested land equals the size of Manhattan—roughly 200 square miles. JBS sourced 90 percent of the animals from this illegally deforested land. In 2014, Brazil's National Institute for Space Research reported that cattle ranching caused 65 percent of deforestation in the Amazon. In 2012, Tesco cancelled its contract with JBS after sold deforestation-linked beef in the UK.[111]

Thus, this can create stranded assets. Stranded assets are "assets that have suffered from unanticipated or premature write-downs, devaluations, or conversion to liabilities and they can be caused by a variety of risks. Increasingly risk factors related to the environment are stranding assets and this trend is accelerating, potentially representing a discontinuity able to profoundly alter asset values across a wide range of sector."[112]

Standard assets in Indonesia are equal to 10 million football fields due to over 6 million hectares of concessions containing peatlands and/or primary forests that cannot viably be developed due to policy constraints and buyers' zero deforestation commitments.[113]

This said, one could envision executive compensation packages being jeopardized over deforestation in their company's supply chain—either due to the no-deforestation commitment being specifically stated in their compensation package or indirectly as result of their stock devaluation being attributed to deforestation.

Furthermore, there is a business risk whether the premium price paid for the sustainable commodity reaches the local farmer and whether the premium price covers the additional costs associated with certification and ongoing monitoring requirements.

Strategic Risk

Two strategic risks for a company are choosing the right suppliers and choosing the best certification standards. This said, certification standards can be revised and updated over time. For instance, as RSPO is revised, it may consider high carbon stocks in its definition of forests, which would prohibit these areas from being converted to palm oil plantations. From a government's perspective, there is strategic risk with respect to which company to grant a concession and whether the selected company will maintain the concession and not illegally expand outside the concession's boundaries.

Reputation Risk

Concerns over supply chains, and the reputational risk associated with having "dirty" supply chains, have historically included blood diamonds, conflict minerals, and sweatshops in the apparel industry. Thus, this reputational risk exists throughout many different industries and can be managed by choosing an independent, sustainability certification standard.

However, the quality of a certification standard can increase or mitigate reputation risk. Take, for instance, that "some conservationists openly question whether certification makes a difference, pointing with suspicion to the industry's sponsored SFI {Sustainable Forestry Initiative} program,[114] which among other things does not mandate independent third-party inspections of a company's policies."[115] Similarly, "even the reportedly stringent FSC {Forest Stewardship Council} certification requires only 17 percent of the fiber be from certified forests to qualify for use of the FSC label. These weaknesses make many still skeptical about the real impact."[116] In addition, PanEco Foundation, an NGO member of the RSPO, "officially called it quits on the organization last week {in May 2016}, resigning over what it deemed the RSPO's failure to reign in an industry tainted by environmental destruction and human rights abuses."[117] Yet, not using certification standards and neglecting to help strengthen their ability to equitably solve the deforestation problem can result in far worse headlines for a company, such as when it was reported that Amazon and eBay were illegally selling tiger skins.[118]

Liquidity Risk

If illegal deforestation is discovered at a location that supplies commodities, the land's liquidity will probably be hampered. Thus, the land could be appropriated by the host government or the concession permit could be canceled. In addition, the liquidity of the commodities attributed to the deforestation could cease.

Furthermore, a liquidity risk arises as a palm oil concession becomes a stranded asset to the concession holder who is, for instance, restricted by their buyers' zero deforestation requirements and the concession may then resort back to the government. In turn, the government may then reissue the concession to another entity that is not under a similar restriction and will therefore clear the forest for palm oil. Overtime, the liquidity risk could occur as the government has fewer and fewer companies who do not have buyers with zero deforestation requirements.

Operational Risk

There are numerous operational risks associated with sustainable commodities and green procurement. One example is that it is possible to have illegally harvested timber from within one's concession by outsiders. Alternatively, concession owners may illegally clear forest and combine the commodities from the illegally deforested land with its other commodities that come from legally cleared land. In addition, it has been reported that more than 50 companies misreported their landholdings, which from the perspective of an investor could pose significant operational risk to a firm such as a bank.[119]

Civil conflicts and insurrections—whether it be wars, revolutions, riots, or strikes—can present major operational risks. Furthermore, some additional operational risks and key barriers that companies reportedly face include inadequate traceability systems, weak governance (i.e., weak enforcement of compliance) of national deforestation policies, and limited availability of certified materials and their costs.[120]

Legal and Regulatory Risk

Illegally sourced goods or fraudulent papers claiming legally sourced products can lead to legal and regulatory risk. In addition, there is the legal and regulatory risk of an increase in royalties for exported products, such as timber. This situation may occur whether or not the commodities were produced in a sustainable or non-sustainable fashion.

Credit Risk

Credit risk includes default risk, bankruptcy risk, downgrade risk, and settlement risk. It is important to note that these risks are all related, especially when it comes to green procurement. Thus, a company may be accused of illegal deforestation and their commodities, say palm oil, gets impounded by the host country. The host country loses out on the agreed-to royalties, which presents a settlement risk to the host country. The company, with its palm oil impounded, faces potential default risk if the company is unable to pay, for instance, the loans that the company took out to finance the conversion of the forest and the construction of a processing facility. If, for an extended period of time, the company is unable to make payment on its loans (i.e., because it has no other sources of revenue), then the company will likely be downgraded by its creditors. This results in an extra cost of capital (i.e., interest rates on loans will be increased), alongside the continuing impoundment of the palm oil, too much time needed for a future supply of sustainable palm oil, and cancelation of its RSPO membership, which could then result in the company's bankruptcy.

Market Risk

Market risk includes interest rate risk, equity price risk, foreign exchange risk, and commodity price risk. As commodities, particularly palm oil, expand to some countries in Africa, there is a potential risk of inflation and currency inconvertibility. In addition, the price of commodities, such as timber or other agricultural commodities, fluctuates over time.

Furthermore, some countries—such as Nigeria—have a requirement to use local currency to purchase foreign equipment. This is done to promote the Nigerian currency but presents a foreign exchange risk.[121]

Risk, Return, Time (Horizon), Taxes, Liquidity, Legal and Unique (RRTTLLU)

Risk and Return

Please see above for the risk and return associated with green procurement models.

Time Horizon

The time horizon for investments in sustainable commodities and green procurement models could be short-term or long-term commitments. For instance, to construct a new facility to process organic Brazilian nuts may take a short-time period. However, for a multinational company to source 100%, non-deforestation commodities across its entire global supply chain will undoubtedly take a long-term commitment. In addition, to take a non-green commodity to a green commodity, such as farmers who were doing till agriculture to now doing non-till agriculture, needs to be addressed in time horizon planning.

Taxes

There are likely a variety of taxes, such as sales taxes and employment taxes. There can also be tax holidays and competition between countries regarding tax policy. In addition, depending on the level of refinement, there could be varying taxes such as between crude palm oil and RBD (refined, bleached, and deodorized) palm oil.

Liquidity

Sustainably sourced and produced commodities which uphold no-deforestation supply chain commitments are likely to continue having liquidity in global markets. In contrast, those commodities and the lands from which they were produced that are found to be guilty of illegal deforestation and/or human right abuses will likely have less liquidity in the future.

Legal

Legal considerations associated with green procurement models will vary from country to country.

Unique

One unique aspect of green procurement is that it can be difficult to discern the source of one's inputs due to the global trade of commodities. While this is not really an issue for ecotourism or debt-for-nature swaps, this issue presents a unique situation for companies and countries involved in the global trade of commodities, particularly cattle, palm oil, timber, and soy.

Another unique aspect is the divestment campaign that continues to gather momentum. In fact, according to Business Insider, "the value of investment funds {in 2016} committed to selling off fossil fuel assets has jumped to USD5.2 trillion, doubling in just over a year."[122] While a lot of the divestments are centered on fossil fuel investments, such divestments could soon gather momentum for companies responsible for tropical rainforest deforestation.

Policy Analysis

Defining the Problem

There are policies that encourage green procurement models and policies that prohibit unsustainable activities such as illegal logging, illegal land clearing for cattle ranches or agriculture, or poaching of elephant tusks.

Establishing Goals

The public policy goals are to work with the private sector on a landscape approach to sustainable agriculture and to ensure enforcement of legislation to prevent illegal activities.

Selecting a Policy

A few of the public policies designed to prohibit unsustainable commodities such as illegal logging, illegal land clearing for cattle ranches or agricultural, and poaching and/or policies designed to encourage green procurement include:

- Forest Law Enforcement, Governance and Trade (FLEGT) and FLEGT's Voluntary Partnership Agreement (VPA);
- The US Lacey Act;
- The US Endangered Species Act; and
- The Convention on International Trade in Endangered Species of Wild Fauna and Flora (CITES).

One of the most promising public policies is the two Amsterdam Declarations, which have been endorsed by Norway with its commitment to sourcing zero deforestation agricultural commodities (i.e., particularly palm oil). Denmark, Germany, the Netherlands, and the United Kingdom also signed onto the Amsterdam Declarations.[123] Norway, in fact, is the first country in the world to adopt a zero deforestation policy for its public procurement.[i][i] Rainforest Foundation Norway. "Norwegian state commits to zero deforestation." May 26, 2016.http://www.regnskog.no/en/news/norwegian-state-commits-to-zero-deforestation-1. In addition and with respect to Norway, the "Norges Bank Investment Management is the asset management unit of the Norwegian central bank {… that manages …} the Government Pension Fund Global, often referred to as the Norwegian oil fund, and some of Norges Bank's foreign exchange reserves."[124] Two expectations of companies, as outlined by Norges Bank Investment Management, are:

- "Companies engaged in activities that lead to significant clearing of tropical forests should have a strategy responding to stricter future regulation concerning the conservation of tropical forests.
- Companies engaged in activities with a direct or indirect impact on tropical forests should assess impact through, for example, life-cycle analysis, and have a strategy for reducing deforestation as a result of their own activities or from their supply chains."[125]

It is important to note the market value of the Government Pension Fund Global is estimated at NOK7.836 trillion (i.e., approximately USD$912.7 billion), the Fund has invested in nearly 9,000 countries located in 77 countries, and it has more than 1% of all worldwide listed companies in its portfolio.[126]

Furthermore, with respect to company policies and according to Supply Change's 2017 *Tracking Corporate Commitments to Deforestation-Free Supply Chains* Report, there were 447 companies with a total of 760 commitments to reduce deforestation impacts in their supply chains.[127] One of the most notable companies is Unilever, which has a publicly announced commitment that "100 percent of agricultural raw materials will be sustainably sourced by 2020."[128] This is significant as Unilever, with turnover of €52.7 billion (i.e., approximately USD$56 billion) in 2016,[129] is the owner of well-known brands such as Ben & Jerry's ice cream, Dove soap, Hellmann's mayonnaise, Klondike bars, Lipton soup, Q-tips, and Vaseline.[130]

Implementing a Policy

Norway's endorsement of the Amsterdam Declarations is a complement to Norway's International Climate and Forest Initiative (NICFI). More specifically,

> The commitment was part of the Norwegian Parliament's Standing Committee on Energy and Environment recommendation titled Action Plan on Nature Diversity, which also requests the government exercise due care for the protection of nature in its investments through the government's Government Pension Fund Global.[131]

This said, the primary entities responsible for Norway's endorsement appear to be the Norwegian Government and particularly the Government Pension Fund Global, which is managed by Norges Bank Investment Management.

Unilever, and likely its suppliers, is responsible for implementing its own stated policy.

Evaluating the Policy

While it is still too early to assess Norway's commitment to sourcing commodities from no-deforestation supply chains, the Government Pension Fund Global has already divested from at least 11 companies due to connections to deforestation.[132]

Unilever, which is a member of several industry associations including the Tropical Forest Alliance 2020 and is a signatory of the NY Declaration on Forests, appears to be effectively implementing its policy thus far. This said, according to the Global Canopy Programme's Forests 500 project, Unilever has a total score of 5 out of 5 (i.e., 5 is best that can be achieved), including 4 out of 5 for its forest policy, 4 out of 5 for its commodity policies, 5 out of 5 for its operations, and 5 out of 5 for its reporting and transparency.[133] While Unilever still has work to be done, Unilever self-reported that it is 60% of the way, as of 2015, with meeting its 2020 goal.[134]

Future Outlook for Instrument

One encouraging future outlook for green procurement is that the International Organization for Standardization (ISO) is developing a world's first international standard (ISO 20400) for sustainable procurement, which is due out in 2017.[135]

Another major development is that "as a major timber importer for its own consumption as well as timber processing, China is currently piloting a timber legality scheme in cooperation with Australia, Japan, and Indonesia."[136]

With respect to the no-deforestation commitments, as we get closer to 2020, it will be possible to take stock of where a lot of the companies are at with meeting their publicly stated commitments. It will be possible to examine what were/are the challenges associated with meeting these commitments. In addition, stakeholders will see what the implications of not meeting the commitments are for the companies, investors, host countries, and local communities.

It is very likely that the no-deforestation supply chain commitments and the certification of green commodities will not be a fad. Rather, it may become a major component of corporate GHG management and climate change commitments. The no-deforestation supply chain commitments have been building over the past 10 years, and particularly 2014 was a huge year for commitments. The ultimate question is how to achieve deforestation reductions by corporate plans and to sustainably source global commodities.[137]

It is important to understand that public policies, consumer engagement, and stakeholder pressure on companies to only source sustainable commodities—particularly cattle, palm oil, timber, and soy—via green procurement models have possibly the greatest potential to stop the deforestation of our world's last remaining tropical rainforests.

Other Resources on Green Procurement Models

CDP's Forests Program

- https://www.cdp.net/en/forests

Chain Reaction Research

- https://chainreactionresearch.com/

Chatham House's resourcetrade.earth

- https://resourcetrade.earth/data

Cutting Deforestation Out of the Palm Oil Supply Chain: Company Scorecard

- http://www.greenpeace.org/international/Global/international/publications/forests/2016/gp_IND_PalmScorecard_FINAL.pdf

Eliminating Deforestation from the Cocoa Supply Chain

- https://openknowledge.worldbank.org/handle/10986/26549

European Union's Voluntary Partnership Agreements

- http://www.euflegt.efi.int/home
- http://www.euflegt.efi.int/documents/10180/301892/VPAs+for+SDGs/2ba6b3ae-400d-45a3-8d40-69e2bce90900
- http://www.euflegt.efi.int/documents/10180/301862/VPAs%2C+governance+and+SDGs/60213be4-3b1a-4e41-a036-90c214d44413

Find Certified Products via Rainforest Alliance

- http://www.rainforest-alliance.org/find-certified

Forest Trends' Supply Change

- http://supply-change.org/

Global Canopy Programme's Forest 500

- http://forest500.org/

Global Forest & Trade Network by WWF

- http://gftn.panda.org/

Greenpeace's Eating Up the Amazon Report

- http://www.greenpeace.org/usa/wp-content/uploads/legacy/Global/usa/report/2010/2/eating-up-the-amazon.pdf

NEPCon Sourcing Hub

- http://beta.nepcon.org/sourcinghub

NEPCon's Report: Fake Documents: How to Spot Them and What to do About Them

- http://www.nepcon.org/newsroom/fake-documents-how-spot-them-and-what-do-about-them

Rainforest Protection and Responsible Investment: A quick-start guide to improving companies' behaviour in rainforests by influencing investors

- http://d5i6is0eze552.cloudfront.net/documents/Publikasjoner/Andre-rapporter/Rainforest-protection-and-responsible-investment-web-ID-29210.pdf?mtime=20160629135608

Sustainable Agriculture Network

- http://san.ag/web/

Sustainable Brands

- http://www.sustainablebrands.com/

Sustainable Palm Oil Transparency Toolkit

- http://www.sustainablepalmoil.org/

The Forest Trust

- http://www.tft-earth.org/

The Sustainability Consortium

- https://www.sustainabilityconsortium.org/

Transparency for Sustainable Economies (TRASE)

- http://trase.earth/

World Resources Institute's Suitability Mapper for Sustainable Palm Oil in Indonesia

- http://www.wri.org/applications/maps/suitability-mapper/index.html#v=suitability
-

Notes

1. Goodreads. "Anna Lappé > Quotes > Quotable Quote." Accessed April 1, 2017. http://www.goodreads.com/quotes/587323-every-time-you-spend-money-you-re-casting-a-vote-for
2. Supply Change. "Profiles." Accessed February 9, 2017. http://supply-change.org/profiles/
3. Forest Stewardship Council. "Our History." Accessed November 23, 2016. https://ic.fsc.org/en/about-fsc/our-history
4. Sustainable Forestry Initiative. "Background." Accessed November 23, 2016. http://www.sfiprogram.org/about-us/basics-of-sfi/
5. The Programme for the Endorsement of Forest Certification. "History." Accessed April 6, 2017. https://www.pefc.org/about-pefc/who-we-are/history
6. CDP. "Staff: Paul Dickinson." Accessed November 23, 2016. https://www.cdp.net/en/info/staff

7. World Wildlife Fund. *Guide to Conservation Finance*. 30.
8. Roundtable on Sustainable Palm Oil. "About Us." Accessed November 23, 2016. http://www.rspo.org/about
9. Leather Working Group. "Background of the LWG." Accessed November 23, 2016. http://www.leatherworkinggroup.com/about/background.htm
10. European Union. "FLEGT Voluntary Partnership Agreements (VPAs)." Accessed November 29, 2016. http://ec.europa.eu/environment/forests/flegt.htm
11. Greenpeace. "Eating Up the Amazon." Accessed December 14, 2016. http://www.greenpeace.org/usa/wp-content/uploads/legacy/Global/usa/report/2010/2/eating-up-the-amazon.pdf?53ea6e
12. Greenpeace "The Soy Moratorium, 10 Years On: How One Commitment is Stopping Amazon Destruction." Last modified July 27, 2016. http://www.greenpeace.org/international/en/news/Blogs/makingwaves/the-soy-moratorium-10-year-anniversary-stopping-amazon-destruction/blog/57127/
13. Round Table on Responsible Soy. "History." Accessed November 23, 2016. http://www.responsiblesoy.org/about-rtrs/history/?lang=en
14. Brazilian Roundtable on Sustainable Livestock. "About the GTPS." Accessed November 23, 2016. http://www.pecuariasustentavel.org.br/en/about/
15. Martin. *On the Edge*. 82.
16. Consumer Goods Forum. "Our History." Accessed November 23, 2016. http://www.theconsumergoodsforum.com/about-the-forum/our-history
17. World Wildlife Fund. "Beef: Our Approach." Accessed November 23, 2016. http://wwf.panda.org/what_we_do/how_we_work/our_global_goals/markets/mti_solutions/certification/agriculture/beef/
18. Martin. *On the Edge*. 82.
19. Martin. *On the Edge*. 83.
20. Forest Trends. "New Site Tracks Corporate Action on Deforestation." March 25, 2015. Accessed April 15, 2017. http://forest-trends.org/releases/uploads/Supply%20Change_PR_.pdf
21. European Union. "FLEGT Voluntary Partnership Agreements (VPAs)." Accessed November 29, 2016. http://ec.europa.eu/environment/forests/flegt.htm

22. Ceres. "Report on Supply Chain Impact on Deforestation 2017." Accessed April 6, 2017. https://www.ceres.org/investor-network/resolutions/report-on-supply-chain-impact-on-deforestation-2017
23. Pothering, Jessica. "BlackRock Asks Companies to Report Climate Risks." *Impact Alpha*. March 15, 2017. http://impactalpha.com/blackrock-asks-companies-to-report-climate-risks/
24. Milder, Brian. "Conservation Finance in the Galápagos Islands: Capitalizing the 'Missing Middle'." In *Conservation Capital in the Americas*, edited by James N. Levitt. Massachusetts: Lincoln Institute of Land Policy, 2010. 111.
25. FSC. "Why Should I Become FSC Certified?" Accessed April 6, 2017. https://ic.fsc.org/en/for-business/business-benefits/becoming-fsc-certified
26. RSPO. "How RSPO Certification Works." Accessed April 6, 2017. http://www.rspo.org/certification/how-rspo-certification-works
27. RTRS. "Our Certification." Accessed April 6, 2017. http://www.responsiblesoy.org/certification/nuestra-certificacion/?lang=en
28. The Programme for the Endorsement of Forest Certification. "Who We Are." Accessed April 6, 2017. https://www.pefc.org/about-pefc/who-we-are
29. CDP. "Forests Program." Accessed April 6, 2017. https://www.cdp.net/en/forests#a7b0f65c2d0715cc35495a11ed5a5e93
30. Global Forest Watch. "About." Accessed April 6, 2017. http://www.globalforestwatch.org/about/about-gfw
31. Trase. "About Trase." Accessed April 6, 2017. http://trase.earth/about.html
32. Supply Change. "Profiles." Accessed April 6, 2017. http://supply-change.org/profiles/
33. Forest 500. "Rankings." Accessed April 15, 2017. http://forest500.org/rankings
34. Forest Stewardship Council. "Facts and Figures: Facts & Figures March 2017." Accessed April 5, 2017. https://ic.fsc.org/en/facts-and-figures
35. Global Roundtable for Sustainable Beef. "2016 Sustainability Report." Accessed April 5, 2017. http://www.grsbeef.org/resources/Pictures/2017%20Template%20Graphics/GRSB_AR_2016_F_singleWeb.pdf. 6.
36. Roundtable on Responsible Soy. "Certified Volumes and Producers." Accessed April 5, 2017. http://www.responsiblesoy.org/mercado/volumenes-y-productores-certificados/?lang=en

37. Roundtable on Responsible Soy. "Annual Financial Report 2015." April 2016. Accessed April 5, 2017. http://www.responsiblesoy.org/wpdm-package/english-annual-financial-report-2015-up/?wpdmdl=11701&ind=pB4NqM7ciln2JuDjHnkCBtQaqdO1LC8pt7WqXvQf0ErHJMC-y39kmcqXZ0eb9yzZfiUJXFHinFHtOVo4tMDx4PS6RB9regmVm-BlWwaY941jepCNoNP_ZK_TqiZhTsZLKUun2MKg3b2DRxqWnrGTCpDNLEkjoC7zovJme05RokOUE&lang=en. 5-6.
38. Roundtable on Sustainable Palm Oil. "Impacts." Accessed April 5, 2017. http://www.rspo.org/about/impacts
39. Donofrio, Stephen, Philip Rothrock and Jonathan Leonard. *Supply Change: Tracking Corporate Commitments to Deforestation-free Supply Chains, 2017*. Washington, DC: Forest Trends, 2017. Accessed April 5, 2017. http://www.forest-trends.org/documents/files/doc_5521.pdf#. 3.
40. CDP. "Global Forests Report 2016. Revenue at Risk: Why Addressing Deforestation is Critical to Business Success." Accessed April 5, 2017. https://www.cdp.net/en/research/global-reports/global-forests-report-2016. 6.
41. Thoumi, Gabriel. "ECOSYSTEM MARKETPLACE: Investors with $600 bln AUM Launch Latin America Forest Protection Initiative." *Carbon Pulse*. Last modified March 8, 2017. http://carbon-pulse.com/31626/?utm_source=CP+Daily&utm_campaign=099ed60d4d-CPdaily08032017&utm_medium=email&utm_term=0_a9d8834f72-099ed60d4d-36342993
42. Cuff, Madeleine. "Sustainable Agriculture can Unlock $2.3 Trillion, Report Says." *GreenBiz*. Last modified October 17, 2016. https://www.greenbiz.com/article/sustainable-agriculture-can-unlock-23-trillion-report-says
43. Mannan, Datuk Sam. Interviewed by Brian McFarland. October 2016.
44. Verified Carbon Standard. "REDD Project in Brazil Nut Concessions in Madre de Dios. Peru." Accessed November 23, 2016. http://vcsproject-database.org/#/project_details/868
45. Climate. Community and Biodiversity Standard. "REDD Project in Brazil Nut Concessions in Madre de Dios." Accessed November 23, 2016. http://www.climate-standards.org/2013/05/13/redd-project-in-brazil-nut-concessions-in-madre-de-dios/
46. Joyce, Deb and Scott Dowd of Project Piaba. Interviewed by Brian McFarland. January 2017.
47. Mannan, Datuk Sam. Interviewed by Brian McFarland. October 2016.

48. Ibid.
49. Ibid.
50. Ibid.
51. Ibid.
52. IUCN Red List. "Pongo pygmaeus." Accessed March 2, 2017. http://www.iucnredlist.org/details/17975/0
53. Mannan, Datuk Sam. Interviewed by Brian McFarland. October 2016.
54. Ibid.
55. Ibid.
56. Ibid.
57. Ibid.
58. Ibid.
59. Ibid.
60. Ibid.
61. Ibid.
62. Ibid.
63. Martin. *On the Edge*. 138.
64. Verified Carbon Standard. "REDD Project in Brazil Nut Concessions in Madre de Dios. Peru." Accessed November 23, 2016. http://vcsproject-database.org/#/project_details/868
65. Climate. Community and Biodiversity Standard. "REDD Project in Brazil Nut Concessions in Madre de Dios." Accessed November 23, 2016. http://www.climate-standards.org/2013/05/13/redd-project-in-brazil-nut-concessions-in-madre-de-dios/
66. Cantuarias, Jorge. Interviewed by Brian McFarland. November 2016.
67. Jones, Benji. "Gold Mining Invades New Areas of Peruvian Amazon." November 11, 2016. Mongabay. Accessed April 12, 2017. https://news.mongabay.com/2016/11/gold-mining-invades-new-areas-of-peruvian-amazon/
68. Cantuarias, Jorge. Interviewed by Brian McFarland. November 2016.
69. Verified Carbon Standard. "REDD Project in Brazil Nut Concessions in Madre de Dios. Peru." Accessed November 23, 2016. http://vcsproject-database.org/#/project_details/868
70. Cantuarias, Jorge. Interviewed by Brian McFarland. November 2016.
71. Ibid.
72. Ibid.
73. Ibid.
74. Ibid.

75. Ibid.
76. Ibid.
77. Ibid.
78. Ecosystem Marketplace. "Peru's Amarakaeri People Learn To Have Their Nuts And Sell Them, Too." Last modified November 14, 2016. http://www.ecosystemmarketplace.com/articles/indigenous-people-perus-amarakaeri-communal-reserve-conserving-400000-hectares-forest-also-feeding-families-part-sustainably-harvesting-brazil-nuts/
79. Cantuarias, Jorge. Interviewed by Brian McFarland. November 2016.
80. Joyce, Deb and Scott Dowd of Project Piaba. Interviewed by Brian McFarland. January 2017.
81. Ibid.
82. Acute Angling. "Peacock Bass Fishing Maps." Accessed February 14, 2017. http://www.acuteangling.com/peacock-bass-fishing/amazon-fishery-maps/cichla-temensis-range/distribution-map.html
83. Joyce, Deb and Scott Dowd of Project Piaba. Interviewed by Brian McFarland. January 2017.
84. Ibid.
85. Ibid.
86. Ibid.
87. Ibid.
88. Ibid.
89. Ibid.
90. National Geographic. "Controversial SeaWorld Orca Shows End in California, but Continue Elsewhere." Last modified January 4, 2017. http://news.nationalgeographic.com/2017/01/seaworld-final-orca-show-california-killer-whales/
91. National Geographic. "As Ringling Ends Circus, See Where Its Elephants Retired." Last modified September 17, 2015. http://news.nationalgeographic.com/2015/09/150916-ringling-circus-elephants-florida-center/
92. Joyce, Deb and Scott Dowd of Project Piaba. Interviewed by Brian McFarland. January 2017.
93. Ibid.
94. Ibid.
95. Ibid.
96. Ibid.
97. Ibid.
98. Ibid.

99. Ibid.
100. Montgomery, Sy. "Buy a Fish, Save a Tree." *Discove*r. April 2013. http://discovermagazine.com/2013/april/13-buy-a-fish-save-a-tree
101. Joyce, Deb and Scott Dowd of Project Piaba. Interviewed by Brian McFarland. January 2017.
102. Ibid.
103. Ibid.
104. Rainforest Alliance. "Rainforest Alliance Certified Coffee." Accessed November 23, 2016. http://www.rainforest-alliance.org/articles/rainforest-alliance-certified-coffee
105. Forest Stewardship Council. "USGBC Members Approve LEED v4." Accessed November 23, 2016. https://us.fsc.org/en-us/market/green-building/leed-v4
106. Ginn. *Investing in Nature*. 116–117.
107. Martin. *On the Edge*. 74.
108. Ginn. *Investing in Nature*. 115.
109. Business & Human Rights Resource Centre. "Accelerating the Implementation of Corporate Zero Deforestation Commitments." Accessed April 6, 2017. https://business-humanrights.org/en/accelerating-the-implementation-of-corporate-zero-deforestation-commitments
110. Thoumi, Gabriel. "The Chain: London Stock Exchange Suspends Trading of United Cacao." Chain Reaction Research. January 17, 2017. https://chainreactionresearch.com/2017/01/17/the-chain-london-stock-exchange-suspends-trading-of-united-cacao/
111. Maisonnave, Fabiano. "Troubled Meatpacker JBS Sanctioned Over Amazon Deforestation." *Climate Home*. March 31, 2017. Accessed April 6, 2017. http://www.climatechangenews.com/2017/03/31/troubled-meatpacker-jbs-sanctioned-amazon-deforestation/
112. Caldecott, Ben, Gerard Dericks, James Mitchell. "Stranded Assets and Subcritical Coal: The Risk to Companies and Investors." March 2015. Accessed April 12, 2017. http://www.smithschool.ox.ac.uk/research-programmes/stranded-assets/SAP%20Report%20Printed%20Subcritical%20Coal%20Final%20mid-res.pdf. 2.
113. Thoumi, Gabriel, Eric Wakker and Milena Levicharova. "REPORT: Indonesian Palm Oil's Stranded Assets—10 Million Football Fields of Undevelopable Land." February 9, 2016. *Chain Reaction Research*. Accessed April 12, 2017. https://chainreactionresearch.com/2017/02/09/

report-indonesian-palm-oils-stranded-assets-10-million-football-fields-of-undevelopable-land/

114. Dasgupta, Shreya. "Amazon, eBay Involved in Illegal Wildlife Trade: Indian Government." July 22, 2016. *Eco-Business*. Accessed April 6, 2017. http://www.eco-business.com/news/amazon-ebay-involved-in-illegal-wildlife-trade-indian-government/?utm_medium=email&utm_campaign=July%2027%20newsletter&utm_content=July%2027%20newsletter+Version+B+CID_3967427e26718a3032e32d81092a4c21&utm_source=Campaign%20Monitor&utm_term=Amazon%20eBay%20involved%20in%20illegal%20wildlife%20trade%20Indian%20government
115. Ginn. *Investing in Nature*. 115.
116. Ibid.
117. Jacobson, Philip. "PanEco Resigns from RSPO over 'Sheer Level of Inaction'." *Mongabay*. June 3, 2016. Accessed April 6, 2017. https://news.mongabay.com/2016/06/paneco-resigns-from-rspo-over-sheer-level-of-inaction/
118. Dasgupta, Shreya. "Amazon, eBay Involved in Illegal Wildlife Trade: Indian Government." July 20, 2016. *Mongabay*. Accessed April 15, 2017. https://news.mongabay.com/2016/07/amazon-ebay-involved-in-illegal-wildlife-trade-indian-government/
119. Sustainable Palm Oil Transparency Toolkit. "Corporate Land Holdings: Hidden Land, Hidden Risks?" Accessed April 6, 2017. http://www.sustainablepalmoil.org/landbank/
120. Gladstone, Jillian. Interviewed by Brian McFarland. February 2017.
121. Thoumi, Gabriel. "The Chain: Nigerian Forex Limits Drive Palm Oil Returns; SocFin Growing in Sierra Leone; Indonesia Peatland Restoration Finance." September 30, 2016. *Chain Reaction Research*. April 12, 2017. https://chainreactionresearch.com/2016/09/30/the-chain-nigerian-forex-currency-limits-drive-local-palm-oil-returns-socfin-expanding-in-sierra-leone-indonesia-peatland-restoration-financing/
122. Carrington, Damian. "Value of Investment Funds Committed to Selling Off Fossil Fuel Assets Jumps to $5 Trillion." December 13, 2016. *Business Insider*. Accessed April 12, 2017. http://www.businessinsider.com/fossil-fuel-divestment-funds-5-trillion-dollars-2016-12
123. Government of Norway. "Norway Endorses Amsterdam Declarations." June 14, 2016. Accessed April 13, 2017. https://www.regjeringen.no/en/aktuelt/norway-endorses-amsterdam-declarations/id2504256/

124. Norges Bank Investment Management. "About Us." Accessed April 15, 2017. https://www.nbim.no/en/organisation/about-us/
125. Norges Bank Investment Management. "Responsibility: Risk Management: Climate Change." Accessed April 15, 2017. https://www.nbim.no/en/responsibility/risk-management/climate-change2/
126. Norse Bank Investment Management. "The Fund." Accessed April 15, 2017. https://www.nbim.no/en/the-fund/
127. Donofrio, Stephen, Philip Rothrock and Jonathan Leonard. *Supply Change: Tracking Corporate Commitments to Deforestation-free Supply Chains, 2017*. Washington, DC: Forest Trends, 2017. Accessed April 5, 2017. http://www.forest-trends.org/documents/files/doc_5521.pdf#. 3.
128. Supply Change. "Profiles: Unilever." Accessed April 15, 2017. http://supply-change.org/profiles/#specific_company
129. Unilever. "Unilever Factsheet." Accessed April 15, 2017. https://www.unilever.com/investor-relations/understanding-unilever/unilever-factsheet/
130. Unilever. "Brands." Accessed April 15, 2017. https://www.unilever.com/brands/
131. Quigley, Aidan. "Norway Adopts World's First Zero Deforestation Policy: What Does that Mean?" June 8, 2016. *The Christian Science Monitor*. Accessed April 13, 2017. http://www.csmonitor.com/Environment/2016/0608/Norway-adopts-world-s-first-zero-deforestation-policy-What-does-that-mean
132. Gaworecki, Mike. "World's Largest Sovereign Wealth Fund Just Dropped 11 Companies Over Deforestation." March 9, 2016. *Mongabay*. Accessed April 15, 2017. https://news.mongabay.com/2016/03/worlds-largest-sovereign-wealth-fund-just-dropped-11-companies-over-deforestation/
133. Forests 500. "Unilever PLC." Accessed April 15, 2017. http://forest500.org/rankings/companies/unilever-plc
134. Supply Change. "Unilever." Accessed April 15, 2017. http://supply-change.org/company/unilever-plc
135. ISO. "First International Standard for Sustainable Procurement Nears Publication." Last modified August 17, 2016. http://www.iso.org/iso/home/news.htm?refid=Ref2105
136. Martin. *On the Edge*. 83.
137. Donofrio, Stephen. Interviewed by Brian McFarland. February 2017.

16

Green Bonds, Landscape Bonds, and Rainforest Bonds

"Green bonds have opened a new finance flow that will be essential to confronting climate change. They are providing green investment opportunity for an ever wider investor group, including those who wish to divest and diversify from fossil fuel-intensive portfolios, and they have proven that a stream of investor capital exists for green assets."[1]
—Rachel Kyte, World Bank Group Vice President and Special Envoy for Climate Change

Introduction

As explained in the workshop report, *Unlocking Forest Bonds*, "to access the deepest pools of capital managed by institutional investors, forests bonds will need to be simple, transparent, comparable and liquid, and must hold an investment-grade credit rating."[2]

Historical Overview

The first widely recognized green bonds were issued in 2007 and 2008 by the World Bank[3] and European Investment Bank.[4]

B.J. McFarland, *Conservation of Tropical Rainforests*, Palgrave Studies in Environmental Policy and Regulation, https://doi.org/10.1007/978-3-319-63236-0_16

Other notable dates throughout the history of green bonds, landscape bonds, and rainforest bonds include:

- 2006: International Finance Facility for Immunisation (IFFIm) issues its inaugural bonds.[5]
- 2006: Dirk Price and Gabriel Thoumi propose a forest carbon and ecosystem services bond to the Prime Minister of Malaysia. The project would not result in a bond issuance, but the "the overall project resulted in formally gazetting and protecting over 117,000 hectares of rainforest within the Belum-Temengor Forest Complex (BTFC) as the Royal Belum State Park."[6]
- 2009: The Climate Bonds Initiative is established.[7]
- 2010: Bank of America Merrill Lynch "began creating a new REDD+ product for institutional investors. The so-called rain forest bond, developed by the firm's carbon markets and fixed-income teams, will pay a return by investing in a range of ventures, including ecotourism, sustainable timber cultivation and REDD+ projects."[8]
- 2012: The Nature Conservancy's (TNC) Conservation Notes are first issued.[9]
- 2014: First corporate green bond is issued, which used the Green Bond Principles and was subjected to a second opinion.[10]
- 2014 (October): The Climate Bonds Initiative's Phase I Technical Working Group is established for the Land Use Sector.[11]
- 2014 (December): Credit Suisse's Nature Conservation Notes are issued.[12]
- 2015: Rhino Impact Bonds, "an innovative financing mechanism for site-based rhinoceros conservation," has funding approved for project implementation by the GEF.[13]
- 2016: Luxembourg launches first "100% green" exchange.[14]
- 2016: Guidelines for Issuing Green Bonds in Brazil report is released. Brazil's Green Bond Guidelines would go on to win the Green Bonds Awards 2017 Initiative of the Year from Environmental Finance.[15]
- 2016 (February): The "world's first green bond to be issued by a 'pure-play' forestry company" is issued.[16]
- 2016 (May): The Climate Bonds Initiative's Phase II Technical Working Group is established for the Land Use Sector.[17]

- 2016 (October): The International Finance Corporation's (IFC) Forests Bond, a "first-of-its-kind" bond with the option for investors to receive verified REDD+ offsets (carbon-linked), is issued.[18]
- 2016 (November): The first green bond in Brazilian Reais is issued by Suzano Pulp and Paper, "following up on their successful USD 500m green issuance in July this year {2016}."[19]
- 2017 (March): The "world's first green bond index-linked exchange-traded fund (ETF)" is listed.[20]

Mechanisms of Instrument

As explained by the World Wildlife Fund, "bonds are debts issued by governments, companies, and other institutions as a way to raise funds. The debt must be repaid by the seller over a specified time period. U.S. government bonds must be issued in one of three ways: as general obligation bonds, which are repaid from future tax revenues; as special revenue bonds, which are repaid from revenues generated by specific projects being financed; and as bonds that are hybrids of both."[21]

The green bond industry generally involves bond issuers and investors, along with the supporting infrastructure of standards and verifiers. Furthermore, there is at least one green bond exchange-traded fund (ETF), and the Luxembourg Green Exchange, which offers green bonds and is a member of the Sustainable Stock Exchanges Initiative, has also been established.[22] Furthermore, the Climate Bonds Initiative's Climate Bonds Standard has a process for certification of a green bond that includes both a pre-issuance certification and a post-issuance certification.[23] There are also Land Use Sector-specific criteria that are still under-development, as of March 2017, by the Climate Bonds Initiative to complement the Climate Bonds Standard.[24]

As outlined by the *Understanding Forest Bonds* report, there are at least six different types of forest bonds:

1. Government forest bond (tax-backed);
2. Government forest bond (revenue-backed);
3. Corporate forest bond;

4. Commitment-backed forest bond;
5. Forest-backed bond (equity-based); and
6. Forest-backed bond (debt-based).[25]

A few specific types of bonds that are also relevant to the green bonds discussion are development impact bonds, social impact bonds, and environmental impact bonds.

Development Impact Bonds

Development impact bonds (DIBs) are often structured for development projects that require significant upfront costs, with declining costs overtime. Such development projects could include hydroelectric facilities and wind power projects. This model could be applied to watershed protection payments. For instance, a freshwater offtake arrangement could be structured with downstream users, and then as long as revenues exceed ongoing costs, the profits could be used to pay back the bond.[26]

Social Impact Bond

Social impact bonds (SIBs) are often structured with a government as the offtaker, and thus, the government pays for the outcome of a particular social good. For instance, instead of a more traditional approach where a government would provide upfront funding for the training of people released from prison, the government could issue a social impact bond and just pay for the measured outcomes (e.g., number of people released from prison who obtain employment, a degree, or vocational training).[27]

Environmental Impact Bonds

Environmental impact bonds (EIBs) are also often structured with a government as the offtaker, and similarly, the government pays for the outcome of a particular environmental good. For instance, KfW has been very active in providing upfront investment for protected areas, and in these scenarios, KfW assumes the risk of project execution in the event the project fails. In contrast, KfW could de-risk and issue an environmental impact bond, where, instead of investing in the buildout of a new project (e.g., park infrastructure, ecotourism lodge, and/or certified REDD+ project), KfW would pay for the measured outcomes (e.g., certain volume of FSC-certified wood, predetermined quantity of forest carbon offsets, or reduction in deforestation rate at a national park).[28]

Take, for example, the organization Blue Forest Conservation that issued a Forest Resilience Bond (FRB):

> Similar in purpose and function to an Environmental Impact Bond or sustainable infrastructure financing, the FRB deploys private capital to make our national forests more resilient to a changing climate. By investing in restoration projects that protect forest health, we can mitigate both wildfire and drought risk. Contracting with both public and private beneficiaries to monetize the multifaceted benefits of forest restoration, the FRB creates value for a diverse set of stakeholders (including the US Forest Service, water and electric utilities, private water-dependent companies, state governments, and insurance companies).[29]

Another example of an EIB, although not related to forest conservation and which could also be considered a DIB, was issued by DC Water "to finance the construction of a project designed to mitigate storm water run-off."[30] Furthermore, Starbucks recently won "Bond of the Year" from *Environmental Finance* for its USD$500 million, 10-year bond that will utilize some of its use of proceeds to "purchase coffee certified against the Coffee and Farmer Equity (CAFE) standards by a third-party verifier, SCS Global Services."[31]

Size of Instrument

According to the Climate Bonds Initiative's report entitled, *Bonds and Climate Change: The State of the Market in 2016*, there was an estimated USD$694 billion climate-theme bonds that were outstanding in 2016.

This USD$694 billion was "made up of approximately 3,590 bonds from 780 issuers across our climate themes: Transport, Energy, Buildings & Industry, Water, Waste & Pollution Control and Agriculture & Forestry. It includes $118bn of labelled green bonds."[32] The Agriculture and Forestry theme, with 141 climate-aligned bonds from 17 issuers identified by the Climate Bonds Initiative, represented less than 1% of the total value of outstanding climate-aligned bonds with an estimated value of USD$6.2 billion.[33] Of these Agriculture and Forestry bonds in the Climate Bonds Initiative's dataset, the "vast majority are from paper and pulp companies with FSC-certified forests and chain of custody certification."[34]

The Climate Bonds Initiative estimates that the market has issued approximately USD$28 billion worth of green bonds thus far in 2017 as of April 15, 2017.[35]

In March 2002, California voters approved the California Clean Water, Clean Air, Safe Neighborhood Parks, and Coastal Protection Act of 2002 (Proposition 40). At the time, this USD$2.6 billion bond was the largest conservation bond in US history and "devoted $1.275 billion to a range of air, land, and water conservation projects, including $445 million for state conservancies to acquire and protect land, $300 million to acquire and restore wildlife habitat, $300 million for water quality projects, and $75 million each for farmland protection and urban parkways."[36]

Introduction to Case Study

The following case study will examine the "first-of-its-kind" International Finance Corporation Forests Bond with REDD+ Coupon Payment which raised USD$152 million and provides a long-term purchase commitment for certified forest carbon offsets from Wildlife Works' Kasigau Corridor REDD+ Project in Kenya.

Case Study #1: International Finance Corporation Forests Bond with REDD+ Coupon Payment

Introduction

Launched on October 31, 2016, the International Finance Corporation's (IFC) Forests Bond is a "first-of-its-kind" bond with the option for investors to receive coupon payments in the form of verified REDD+ offsets (carbon-linked), or cash payments, or a combination of both. As described by Conservation International, one of the technical partners, the Forests Bond:

> will provide investors the opportunity to invest in a traditional financial product that offers the unique option of receiving interest payments in the form of environmental impact—in this case, verified carbon credits generated through REDD, an initiative that rewards landholders for protecting forests, thereby reducing carbon emissions that worsen climate change. The development of the bond is a collaboration of the International Finance Corporation (IFC), a member of the World Bank Group, and BHP Billiton, a leading global resources company with technical support from Baker & McKenzie and Conservation International (CI).[37]

The Forests Bond was originally "sized between $75 million and $150 million {...} {with} a five-year tenor and a coupon of 1.546%, which was priced in-line with IFC's normal credit of the same maturity. The bond has a price of 100, with a spread of 18 basis points above mid-swaps."[38] The Forests Bond ended up raising USD$152 million from institutional investors including:

- The California State Teachers' Retirement System (CalSTRS);
- Treehouse Investments;
- The Teachers Insurance and Annuity Association of America (TIAA);
- College Retirement Equities Fund (CREF) (i.e., teachers' pension fund); and
- QBE Insurance Group.

The proceeds will "support IFC's private sector lending in emerging markets."[39] The lead placement agents for the bond, which is listed on the London Stock Exchange, were Bank of America Merrill Lynch, BNP Paribas, and JP Morgan. Citi was "the calculation, paying, transfer and settlement agent for the bond" .[40]

Furthermore, the chosen REDD+ project to benefit from the Bond was Wildlife Works' Kasigau Corridor REDD+ Project (see Chap. 11 for a more detailed case study on the Kasigau REDD+ project). (Fig. 16.1).

Identify the Problem

The main problem the IFC Forests Bond was attempting to solve is that, to reduce global greenhouse gas emissions, there are lots of financing options for renewable energy, but not so much for tropical rainforest con-

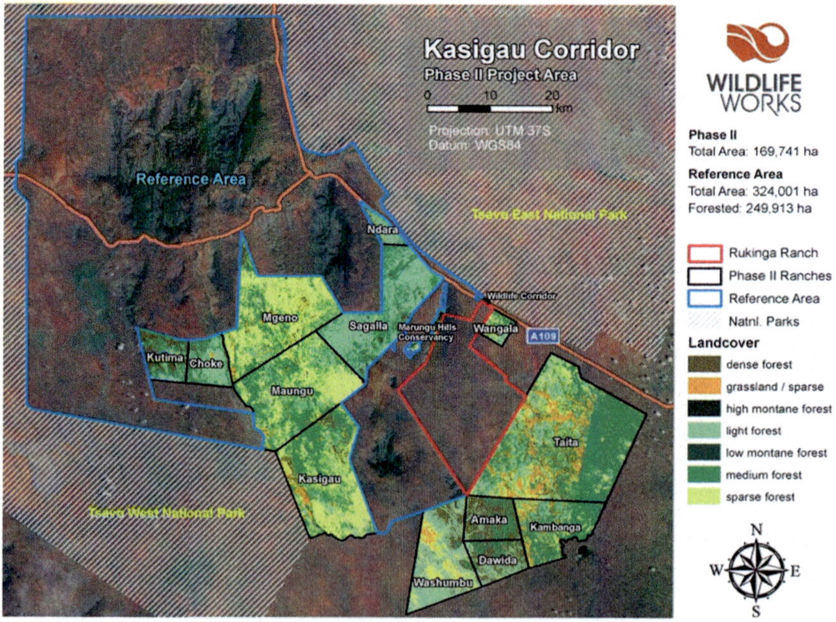

Fig. 16.1 Map of the Kasigau Wildlife Corridor REDD Project (Credit: Wildlife Works)

servation. Thus, the IFC was looking for a mechanism to support sustainable economic development and tropical rainforest conservation.[41]

Why the Problem Is Important

The problem is important to the IFC because the IFC works "with the private sector in developing countries to help create opportunity for all."[42] Furthermore, as described by the IFC, this Forests Bond:

> {…} is part of IFC's larger work as one of the world's biggest financiers of climate-smart projects in developing countries. Our climate-smart investments have totaled more than $13 billion over the last decade. As part of our overall Climate Implementation Plan, we have committed to lead efforts in catalyzing $13 billion a year in private sector capital for climate investment by 2020.
>
> Innovative instruments like the IFC Forests Bond are critical as we move forward with this work. Because the bond links investors, capital markets, and carbon finance to forest preservation, wildlife conservation, livelihoods for women, and benefits for primary school, the results reverberate across sectors. As a first-of-its-kind climate financing mechanism, it has enormous potential for replication, demonstrating how to implement similar projects that support resilient landscapes.[43]

How Problem Was Identified

The development of the IFC Forest Bonds was a collaborative effort between the IFC and BHP Billiton (BHPB). BHPB also relied on their advisors at Baker & McKenzie and Conservation International for support. The IFC was interested in attracting investors to the forest sector to reduce GHG emissions, while BHPB had a similar interest as well to create innovative climate finance mechanisms in the forestry sector. This said, the interest (coupon) payments are channeled to support a project that avoids deforestation.[44]

Effectiveness of Process for Identifying Problem

Despite the fact that the partners have different business models, they found ways to effectively work together, structure the "first-of-its-kind" Forests Bond, and raise more than USD$150 million in financing.[45]

Steps Taken to Address the Problem

There were numerous steps taken to address the problem and issue the IFC Forests Bond. These general steps include:

- Strategic partners were identified;
- Multiple projects were reviewed, due diligence was further conducted on several projects, and ultimately the Wildlife Works' Kasigau Corridor REDD+ Project was chosen; and
- Investors were identified.[46]

Results

The main result is that the IFC officially launched the "first-of-its-kind" Forests Bond and raised USD$152 million. Furthermore, the Forests Bond "will directly support a climate-smart project in Kenya; Conservation International is providing input to BHP Billiton with respect to conservation and REDD+ matters, and will facilitate the exchange of lessons learned and best practices on REDD+ projects; and IFC expects to deliver measurable environmental and social results to investors on a yearly basis."[47]

Since the Forests Bond was recently issued and the first delivery of REDD+ credits is not due to investors until Fall 2017, it is still too early to say what conservation and forestry results are directly attributable to the Forests Bond. Nevertheless, a tangible result is that some investors that invested in the Forests Bond would not have otherwise invested in the forest carbon sector. Similarly, the Forests Bond set a precedent and opened the eyes of some financial institutions on new approaches to raising money in the capital markets for tropical forest conservation.[48]

Another result is that the IFC Forests Bond has some back-office liquidity that enables an investor to request cash, carbon offsets, or a combination of both, from the IFC.[49]

Challenges and How They Were Met

One challenge appears to be that the Forests Bond was not labeled or certified as a green bond. Given the "first-of-its-kind" nature of the Bond, it would have been too complicated to structure the Bond as both a green bond and a carbon-linked bond. While the Forests Bond stimulated demand for REDD+ offsets from the Kasigau Corridor REDD+ Project, the proceeds will be reinvested into the IFC's traditional private sector operations in developing countries.[50]

Another challenge was that prior to issuance, the IFC and BHPB had intended for the Bond to purchase REDD+ carbon offsets from a portfolio of REDD+ projects. However, in the end, a single project was chosen.[51]

There was also a challenge around working with a diversity of partners. The IFC is an investment bank and BHPB is a global resources company. These organizations have very different business models. For instance, the IFC conducts a tremendous amount of due diligence on markets, products, and players; however, in the end, the IFC transacts money and is not involved with the daily, on-the-ground operations. In contrast, BHPB is on-the-ground on a daily basis doing mineral extraction with a long-term presence. This said, it took a lot of time to blend these 2 different approaches and business models before successfully issuing the Forests Bond.[52]

Beyond Results

The results of the Forests Bond will hopefully be sustainable. As a 5-year bond, the Forests Bond will likely provide a 5-year revenue stream for the Kasigau Corridor REDD+ Project. The Kasigau Corridor REDD+ Project is independently validated and verified against third-party certification standards known as the Verified Carbon Standard (VCS) and the Climate, Community, & Biodiversity Standard (CCBS). In addition, the IFC applies its Environmental and Social Performance Standards.[53]

If all the bond investors opt for cash payments as opposed to receiving carbon offset credits, then "{…} BHP Billiton will provide a liquidity support mechanism, which will see it buy credits from IFC should investors opt to be paid in cash."[54] In addition, "to ensure that the bond is not a one-off, BHP Billiton has helped develop a Knowledge Sharing Platform, in partnership with Conservation International and law firm Baker & McKenzie, to encourage investment in REDD initiatives, using the Forests Bond as an example."[55] It is unclear whether this Knowledge Sharing Platform has happened, but it appears to be a stated goal. This said, it was reported in June 2017 that BHP Billiton, Conservation International, and Baker & McKenzie launched the Finance for Forests (F4F) initiative.[56]

Furthermore, it has been reported that "future Forest Bonds may see REDD+ projects aggregated to provide further scale."[57]

Lessons Learned

A few lessons learned from Cecilia (Sid) Embree, an IFC consultant, include:

- You do not necessarily need a large, multinational corporation to underwrite the REDD+ offsets. More and more local companies are interested in buying REDD+ offsets and getting involved in local or regional projects because they want to buy and they have a bigger stake in local, economic development.
- Whether utilizing a local company or a multinational corporation such as BHP Billiton, it is invaluable to have an underwriter to stimulate investments. If the underwriter is not the bond issuer, it needs to be flexible to rely on and to accommodate the issuer's policies and procedures.
- As a capital markets product, the Forests Bond, or any bond, needs to be issued in a market where there is liquidity. Therefore, adding "bells and whistles" to the core bond product is not possible.
- Lastly, if you want more of these products out there, there needs to be a more efficient bond development process.[58]

Other Resources on IFC's Forest Bond

Conservation International

- http://www.conservation.org/NewsRoom/pressreleases/Pages/Innovative-Forests-Bond-Issued-to-Unlock-Private-Financing-for-Forest-Protection-.aspx

Ecosystem Marketplace

- http://www.ecosystemmarketplace.com/articles/new-bond-aims-to-unlock-private-investment-to-protect-forests/

Environmental Finance

- https://www.environmental-finance.com/content/news/ifc-launches-forestry-bond-that-can-pay-its-coupon-using-redd-credits.html

IFC Final Prospectus

- https://www.ifc.org/wps/wcm/connect/b451653d-721d-404c-85fb-d280d0f0a190/Forests+Bond_Final+Series+Prospectus.pdf?MOD=AJPERES

IFC Forests Bond Presentation and Forest-Backed Bonds Proof of Concept Study

- https://www.ifc.org/wps/wcm/connect/5299a595-721b-41ef-90fa-2cff3d50982c/FINAL+Forests+Bond+Investor+Presentation+10-5_pdf.pdf?MOD=AJPERES
- http://www.ifc.org/wps/wcm/connect/4957218048855503b524f76a6515bb18/IFC_Breif_Forest_web.pdf?MOD=AJPERES&CACHEID=4957218048855503b524f76a6515bb18

IFC Press Release

- http://ifcextapps.ifc.org/ifcext/pressroom/ifcpressroom.nsf/0/594A016A78A7B14E8525805D00461397

Law360

- https://www.law360.com/articles/860808/ifc-lists-1st-of-its-kind-152m-forest-bond-on-lse

London Stock Exchange Green Bonds: Accessing the Global Markets Through London

- http://www.lseg.com/sites/default/files/content/documents/20161201%20Green%20Bond%20Presentation.pdf

The Australian

- http://www.theaustralian.com.au/business/mining-energy/bhp-billiton-in-worldfirst-forest-bond/news-story/91959baadddf6be2281b4f7c23cdb347

The Star, Kenya

- http://www.the-star.co.ke/news/2016/11/12/taita-forest-bond-raises-sh15-billion_c1454183

Return

The returns for green bonds generally need to meet the market return. The challenge is how to structure the green bond so that it meets this market return while also delivering on its green credentials. The use of proceeds, in general, needs to be ring-fenced and used specifically for eligible green projects.

Risk

Business Risk

On-the-ground projects are often too small and lack both the business and financial expertise to attract large-scale financing from the issuance of green bonds as well as from impact investors.

Strategic Risk

One strategic risk of green bonds for land use projects is the need to provide clarity on the opportunity. This said, issuers often do not see a need for taking on the additional cost (i.e., there is no clear economic benefit of doing so, at least not yet) and investors are usually not willing to take a discount, so this creates a bit of a gap on both sides.[59]

Reputation Risk

One reputation risk is greenwashing, which could result from the lack of consistent approaches by issuers in identifying the main issues to report on. This risk can then be compounded by sector expertise of second-party opinion providers. These issues are currently being addressed by the Climate Bonds Initiative.[60] Thus, the environmental credibility of the underlying activities must be assured, such as through an independent assessment against third-party certification standards. Similarly, the additionality of the bond—in the sense that the investment would have been made anyway whether or not the bond was labeled as green—could raise reputational risks. Furthermore, if some of the use of proceeds are utilized for green initiatives by an otherwise large polluter, then the green bond's credentials may be called into question. For instance, the "Spanish oil company Repsol issued a controversial €500 million ($543 million) green bond that will be partly used to fund energy efficiency projects in its oil and chemical refineries."[i] To mitigate such a reputational risk, the International Organization for Standardization (ISO) is working on an international green bond standard,[ii] and "the Luxembourg Finance

Labelling Agency (LuxFlag) will launch a 'Green Bond' label" later in 2017.[iii][i] Ali, Hamza. "Spanish oil company issues controversial green bond." Environmental Finance. May 10, 2017. https://www.environmental-finance.com/content/news/spanish-oil-company-issues-controversial-green-bond.html?utm_source=100517na&utm_medium=email&utm_campaign=alert[ii] Hurley, Michael. "ISO green bond standard working group to be established." Environmental Finance. August 2, 2017. https://www.environmental-finance.com/content/news/iso-green-bond-standard-working-group-to-be-established.html[iii] Walsh, Joe. "LuxFlag to launch 'Green Bond' label." Environmental Finance. April 26, 2017. https://www.environmental-finance.com/content/news/luxflag-to-launch-green-bond-label.html?utm_source=260417na&utm_medium=email&utm_campaign=alert

Liquidity Risk

The liquidity risk associated with a green bond will depend on the characteristics of the bond. If the bond is unique, it will have less liquidity. If the bond is more "vanilla" (i.e., standardized), it will have greater liquidity.

Operational Risk

There are operational risks if the use of proceeds will not deliver on the stated objectives of the green bond. There are also operational risks associated with designing, structuring, and commercializing the green bond. In addition, there may be operational risks associated with meeting a green bond certification standard, along with satisfying third-party reviewers and other critical stakeholders.

Legal and Regulatory Risk

One regulatory risk associated with green bonds for land use projects is that oftentimes, values are still held by government and regulation has not been consistent in many places (e.g., global carbon pricing), which makes investors more wary of relying on these cash flows for any repayment.[61] To mitigate this risk, green bonds should be backed by a diversity of cash flows. This could be a forest conservation component, alongside a renewable energy component. Alternatively, this could be from a diversity of cash flows (e.g., visitor park fees, forest carbon offsets, and certified sustainable timber), instead of from just one forest conservation activity.

If the bond is issued by a government, there is a risk that "earmarking, however, is a political decision and there is therefore a risk that the revenues could be re-appropriated for other spending needs in the future."[62] Similarly whether or not a forest bond "{…} could work in a given country depends on the financial stability, level of political risk and current (and likely future) shape of forest policy in that country."[63]

Additional legal and regulatory risk includes expropriation of assets, creeping expropriation (e.g., rapid rise in royalties or taxes), cancelation of forest concessions, and nonenforcement of forest laws (e.g., illegal loggers extracting timber from one's concession).[64]

Credit Risk

Credit risk includes default risk, bankruptcy risk, downgrade risk, and settlement risk. This said, credit risk to consider is that:

> The finance raised from selling a bond and the revenue generated to pay it back can either be held on the financial accounts of the issuing institution (on balance sheet) or in a legally independent entity (off balance sheet). Combined with the choice of revenue generating mechanism, the choice of institutional arrangements has important implications for the risk to both the bond investor and the bond issuer.[65]

Market Risk

Market risk includes interest rate risk, equity price risk, foreign exchange risk, and commodity price risk. While traditional "vanilla" bonds have been around for a long time, the emergence of green bonds—and particularly forest bonds—is still relatively new and present a market risk due to their short history.

Risk, Return, Time (Horizon), Taxes, Liquidity, Legal and Unique (RRTTLLU)

Risk and Return

Please see above for the risk and return associated with green bonds, landscape bonds, and rainforest bonds. Essentially, it is important to provide investors with an attractive risk-adjusted return.

Time Horizon

One aspect to consider with respect to the time horizon is that the use of proceeds from green bonds goes to projects and programs which may terminate before or after the bond funding time horizon. This creates a time horizon mismatch within the bond structure. Likewise, there is mismatch between the Kasigau Corridor REDD+ Project, which is a minimum 30-year project, and the IFC Forests Bond commitment to purchase offsets for "only" 5 years.

Taxes

From the investors' point of view, taxes will likely be comparable to traditional bonds. If you are able to construct and issue a bond with an attractive tax structure (i.e., such as US municipal bonds), then this will influence (i.e., increase) investor demand.

Liquidity

One component of liquidity will depend on the issuer. For instance, a company issuing a green bond with an investment grade will likely have greater liquidity, than a country issuing a green bond with a lower or nonexistent investment grade.

Legal

Legal considerations associated with green bonds, landscape bonds, and rainforest bonds will vary from country to country. While such bonds will likely have very similar legal requirements as traditional bonds, there may be some unique legal considerations related to the certification of the bond's "green" projects.

Unique

One unique aspect is that there could be a funding mismatch over the time horizon. In addition, there could be unique concerns about the coupon payments, such as payments made with REDD+ offsets as opposed to traditional cash payments.

Policy Analysis

Green bonds, similar to impact investing and ecotourism, depend on a wide variety of policies to be in place.

Defining the Problem

Choosing a tropical rainforest country to make an investment via the issuance of a green bond depends on factors such as the overall ease of doing business, the level of corruption, the rule of law, and whether there are sufficient demands for green bonds backed by land use projects.

Establishing Goals

In general, the goals are for public policies that help put a value on nature or ecosystem services such as wetland offsets in the United States, sustainable forestry in Vietnam, land use regulations in Brazil, or California's cap-and-trade system. Clear public policies, such as these, allow value to be ascribed to a potential issuer (e.g., allow revenues from the sale/trading of such credits to be used to pay back investors). This, in turn, can incentivize investors to buy such bonds (e.g., municipal renewable energy bonds in the United States).[66] In addition, public policies that help to promote an overall ease of doing business, minimize the level of corruption, and promote a strong rule of law would be helpful for green bonds.

Selecting a Policy

The global rankings for ease of doing business, according to the World Bank,[67] for the top 10 countries with the largest tropical rainforests were as follows:

- Tied for #47: Mexico
  - #40 for enforcing contracts
  - #53 for protecting minority investors
  - #61 for trading across borders
  - #101 for registering property
- #53: Colombia
  - #13 for protecting minority investors
  - #53 for registering property
  - #121 for trading across borders
  - #174 for enforcing contracts
- #54: Peru
  - #37 for registering property
  - #53 for protecting minority investors
  - #63 for enforcing contracts
  - #86 for trading across borders
- #91: Indonesia
  - #70 for protecting minority investors
  - #108 for trading across borders
  - #118 for registering property
  - #166 for enforcing contracts

- #98: Zambia
  - #87 for protecting minority investors
  - #135 for enforcing contracts
  - #145 for registering property
  - #161 for trading across borders
- #123: Brazil
  - #32 for protecting minority investors
  - #37 for enforcing contracts
  - #128 for registering property
  - #149 for trading across borders
- #130: India
  - #13 for protecting minority investors
  - #138 for registering property
  - #143 for trading across borders
  - #172 for enforcing contracts
- #149: Bolivia
  - #98 for trading across borders
  - #128 for enforcing contracts
  - #137 for protecting minority investors
  - #139 for registering property
- #182: Angola
  - #81 for protecting minority investors
  - #170 for registering property
  - #183 for trading across borders
  - #186 for enforcing contracts
- #184: DRC
  - #156 for registering property
  - #162 for protecting minority investors
  - #171 for enforcing contracts
  - #188 for trading across borders

The global rankings by Transparency International of the Corruption Perceptions Index 2016[68] for the top 10 countries with the largest tropical rainforests were as follows:

- Tied for #79: Brazil
- Tied for #79: India
- Tied for #87: Zambia
- Tied for #90: Colombia
- Tied for #90: Indonesia
- Tied for #101: Peru
- Tied for #113: Bolivia
- Tied for #123: Mexico
- Tied for #156: DRC
- Tied for #164: Angola

The global rankings by the United Nations Development Programme of the Human Development Index (HDI),[69] which combines life expectancy, education, and per capita income indicators, for the top 10 countries with the largest tropical rainforests were as follows:

- #74: Mexico
- #75: Brazil
- #84: Peru
- Tied for #97: Colombia
- Tied for #110: Indonesia
- #119: Bolivia
- #130: India
- #139: Zambia
- #149: Angola
- #176: DRC

The World Justice Project has designed a rule of law ranking which primarily utilizes eight factors (e.g., Constraints on Government Powers, Absence of Corruption, Open Government, Fundamental Rights, Order and Security, Regulatory Enforcement, Civil Justice, and Criminal Justices).[70] Of the 113 countries assessed, the following global rankings are for the top 10 countries with the largest tropical rainforests:

- #52: Brazil
- #61: Indonesia
- #65: Peru
- #66: India
- #71: Colombia
- #81: Zambia
- #88: Mexico
- #104: Bolivia
- Unranked: Angola
- Unranked: DRC

Despite a mixed result in its ranking on the rule of law, HDI, and ease of doing business, it appears as though Brazil would be a leading rainforest country for the issuance of green bonds with a land use sector component. This could be done by companies similar to Suzano Pulp and Paper, which is a Brazilian company that has previously issued green bonds. Alternatively, the Brazilian Federal Government or a leading subnational jurisdiction in Brazil that is working on a landscape approach to reducing deforestation such as Acre could be a future issuer of a green bond. Furthermore, it is possible that Brazil's large cattle companies, such as JBS, could issue a future green bond to improve its supply chain to source deforestation-free cattle.

Implementing a Policy

Issuing a bond, whether it is a green bond or a traditional bond, is regulated by a different entity in Brazil if the issuer is a government body or a company. If the government is the bond issuer, the Central Bank of Brazil is responsible for its regulation. In contrast, if a company is the bond issuer, then the Comissão de Valores Mobiliários (CVM; the Securities and Exchange Commission of Brazil) is responsible.[71]

Evaluating the Policy

Although only a few green bonds with a land use component have been issued in Brazil, it appears as though the policies in Brazil could be supportive of additional issuances in the future. This said, the *Guidelines for*

Issuing Green Bonds in Brazil Report was released, and the Guidelines won the Green Bonds Awards 2017 Initiative of the Year from *Environmental Finance*.[72] Furthermore, there are Land Use Sector-specific criteria that should be issued in the near future by the Climate Bonds Initiative, and these criteria will complement the Climate Bonds Standard.[73]

Future Outlook for Instrument

If forest bonds—whether they are called landscape bonds, rainforest bonds, or green bonds with a rainforest restoration and conservation component—follow the trends of the greater green bonds market, then the scale of funding could increase dramatically over the years. Although limited details appear to be available, in April 2015, ADM Capital:

> {…} unveiled details of a suggested $1 billion bond programme to provide much needed finance for forest conservation and development. The Rainforest Impact Bond (RIB) would be the first such finance mechanism to protect forests, offering investors the means to help countries cut deforestation and reduce global carbon emissions.[74]

Of special note, as summarized by the Global Canopy Programme's *Understanding Forest Bonds* Report:

> Although no type of forest bond is off limits to any particular tropical forest region, an initial evaluation indicates which type of bond could be most useful and/or successful in each of the worlds' three major tropical forest regions. A forest bond issued by a forest nation or backed by commitments from one or more forest nations could be successful in Latin America, particularly in the Amazon region. In contrast, Africa would likely get the most use out of a bond issued by a relevant multilateral development bank or backed by commitments from donor countries. Finally, in Eastern and Southern Asia, an off-balance-sheet forest bond backed by forest-based revenues currently seems the most viable option.[75]

According to Tanja Havemann, Director and Founder of Clarmondial AG, there appears to be investor interest in green bonds for land use projects, but without further pressure from their stakeholders (e.g., pensioners), and governments, it is unclear how much movement and investment will take place. In addition, it is unclear how much government will push this space (e.g., capital gains tax incentives). This said, there are opportunities for doing more in this space, but there are concerns that green bonds for land use projects will remain a relatively niche market without a lot more dedicated action by "facilitators" to make it happen (e.g., lobbying, proactive advisory, and deal brokering).[76]

Other Resources on Green Bonds, Landscape Bonds, and Rainforest Bonds

Blue Forest Conservation

- http://blueforestconservation.com/#aboutus

Carbon Yield

- www.carbonyield.org

Climate Bonds Initiative's Land Use Page

- https://www.climatebonds.net/standard/land-use

Forest-Backed Bonds Proof of Concept Study

- https://www.forumforthefuture.org/sites/default/files/project/downloads/forestbackedbondsproof1.pdf

Green Bonds and Land Conservation

- http://scholar.harvard.edu/files/lbilmes/files/green_bonds_and_land-_rwp15_072_bilmes.pdf
- https://ssir.org/articles/entry/green_bonds_and_land_onservation_a_new_investment_landscape

Green Bonds Database by Environmental Finance

- http://www.greenbonddata.org/

Guidelines for Issuing Green Bonds in Brazil 2016 Report

- http://cebds.org/wp-content/uploads/2016/10/Guia_emissa%CC%83o_ti%CC%81tulos_verdes_ING-2.pdf

Instiglio

- http://www.instiglio.org/en/impact-bonds/

International Capital Market Association's Green Bond Principles

- http://www.icmagroup.org/Regulatory-Policy-and-Market-Practice/green-bonds/

Levering Ecosystems: A Business-focused Perspective on How Debt Supports Investments in Ecosystem Services; A Report by Clarmondial, Climate Bonds Initiative, and Credit Suisse

- https://www.credit-suisse.com/media/assets/corporate/docs/about-us/responsibility/banking/levering-ecosystems.pdf

Understanding Forests Bonds: A Guide to Raising Up-Front Finance for Tropical Forests

- http://globalcanopy.org/sites/default/files/documents/resources/UnderstandingForestBonds_0.pdf

Unlocking Forest Bonds: A High-Level Workshop on Innovative Finance for Tropical Forests

- https://www.climatebonds.net/files/uploads/2011/10/FBWorkshop_report_web_A.pdf

World Bank's Green Bond Issuances to Date

- http://treasury.worldbank.org/cmd/htm/GreenBondIssuancesToDate.html

Notes

1. The World Bank Group. "Green Bonds Attract Private Sector Climate Finance." Accessed March 30, 2017. http://www.worldbank.org/en/topic/climatechange/brief/green-bonds-climate-finance
2. Cranford, M., I.R. Henderson, A.W. Mitchell, S. Kidney and D. P. Kanak. *Unlocking Forests Bond: A High-Level Workshop on Innovative Finance for Tropical Rainforests, Workshop Report.* WWF Forest & Climate Initiative, Global Canopy Programme and Climate Bonds Initiative, 2011. Accessed April 11, 2017. https://www.climatebonds.net/files/uploads/2011/10/FBWorkshop_report_web_A.pdf
3. The World Bank. "World Bank Green Bonds: Green Bond Issuances to Date." Accessed December 13, 2016. http://treasury.worldbank.org/cmd/htm/GreenBondIssuancesToDate.html
4. Climate Bonds Initiative. "History." Accessed December 13, 2016. https://www.climatebonds.net/market/history
5. International Finance Facility for Immunisation. "International Finance Facility for Immunisation issues inaugural bonds." November 7, 2006. Accessed April 11, 2017. http://www.iffim.org/library/news/press-releases/2006/international-finance-facility-for-immunisation-issues-inaugural-bonds/
6. Hassouna, Khaled and Gabriel Thoumi. "How Can Green Markets Access Islamic Finance?" March 23, 2010. Ecosystem Marketplace. Accessed April 11, 2017. http://www.ecosystemmarketplace.com/articles/how-can-green-markets-access-islamic-finance/

7. LinkedIn. "Climate Bonds Initiative: About Us." Accessed April 11, 2017. https://www.linkedin.com/company-beta/1032198/
8. Gilbert, Katie. "Pensions Consider the Rain Forest: Can Institutional Investors Profit from Saving the World's Forests?" *Institutional Investor*. Last modified May 7, 2012. http://www.institutionalinvestor.com/Article/3024131/Pensions-Consider-the-Rain-Forest.html#.WOGi7aK1uM8
9. The Nature Conservancy. "Conservation Note Fact Sheet." Accessed March 29, 2017. https://www.nature.org/about-us/conservation-note-fact-sheet-2016.pdf
10. Climate Bonds Initiative. "Green Bond Labels and Standards Green Bond Labels and Standards Webinar." June 2016. Accessed April 7, 2017. https://www.climatebonds.net/files/files/Green%20Bond%20Labels%20and%20Standards%2010-06-2016.pdf. 14.
11. Climate Bonds Initiative. "Land Use." Accessed April 6, 2017. https://www.climatebonds.net/standard/land-use
12. Credit Suisse. "Sustainable Products & Services: Conservation Finance: Our Role in Conservation Finance." Accessed March 29, 2017. https://www.credit-suisse.com/us/en/about-us/responsibility/banking/sustainable-products-services.html
13. The GEF. "Rhino Impact Bonds: An Innovative Financing Mechanism for Site-Based Rhinoceros Conservation." Accessed April 6, 2017. https://www.thegef.org/project/rhino-impact-bonds-innovative-financing-mechanism-site-based-rhinoceros-conservation
14. Environmental Finance. "Luxembourg Launches First '100% Green' Exchange." Last modified September 28, 2016. https://www.environmental-finance.com/content/news/luxembourg-launches-first-100-green-exchange.html?utm_source=300916na&utm_medium=email&utm_campaign=alert
15. Environmental Finance. "Initiative of the Year—Brazil's Green Bond Guidelines." April 10, 2017. Accessed April 11, 2017. https://www.environmental-finance.com/content/awards/green-bond-awards-2017/winners/initiative-of-the-year-brazils-green-bond-guidelines.html?utm_source=100417na&utm_medium=email&utm_campaign=alert
16. Robinson-Tillett, Sophie. "First Forestry Pureplay Green Bond Issued." February 29, 2016. *Environmental Finance*. Accessed April 7, 2017. https://www.environmental-finance.com/content/news/first-forestry-pureplay-green-bond-issued.html?utm_source=229na&utm_medium=email&utm_campaign=alert

17. Climate Bonds Initiative. "Land Use." Accessed April 6, 2017. https://www.climatebonds.net/standard/land-use
18. Velde, Bruno Vander. "New Bond Aims to Unlock Private Investment to Protect Forests." *Humanature, Conservation International Blog*. Last modified October 31, 2016. http://blog.conservation.org/2016/10/new-bond-aims-to-unlock-private-investment-to-protect-forests/
19. Patzdorf, Louise. "1st Green Bond in Brazilian Currency: Suzano Papel & Celulose, BRL1 bn (USD 294m), No2 from Suzano!." November 30, 2016. *Climate Bonds Initiative*. Accessed April 7, 2017. https://www.climatebonds.net/2016/11/1st-green-bond-brazilian-currency-suzano-papel-celulose-brl1-bn-usd-294m-no2-suzano
20. Ali, Hamza. "World's First Green Bond ETF Launches." March 1, 2017. *Environmental Finance*. Accessed April 6, 2017. https://www.environmental-finance.com/content/news/worlds-first-green-bond-etf-launches.html?utm_source=030317na&utm_medium=email&utm_campaign=alert
21. World Wildlife Fund. *Guide to Conservation Finance*. 43.
22. Environmental Finance. "A Progress Report on the Luxembourg Green Exchange." February 13, 2017. Accessed April 7, 2017. https://www.environmental-finance.com/content/analysis/a-progress-report-on-the-luxembourg-green-exchange.html?utm_source=150217na&utm_medium=email&utm_campaign=alert
23. Climate Bonds Initiative. "How to Get Certified." Accessed April 6, 2017. https://www.climatebonds.net/standards/certification/get-certified
24. Climate Bonds Initiative. "Land Use." Accessed April 7, 2017. https://www.climatebonds.net/standard/land-use
25. Cranford, M., C. Parker and M. Trivedi. *Understanding Forest Bonds: A Guide to Raising Up-Front Finance for Tropical Forests*. Oxford, UK: Global Canopy Programme, 2011. Accessed April 11, 2017. http://globalcanopy.org/sites/default/files/documents/resources/UnderstandingForestBonds_0.pdf. 21–27.
26. Garbaliauskas<!--Garbaliauskas, Romas-->, Romas. Interviewed by Brian McFarland. July 2017.
27. Ibid.
28. Ibid.
29. Blue Forest Conservation. "Forest Resilience Bond." Accessed April 7, 2017. http://www.blueforestconservation.com/#frb

30. Ali, Hamza. "DC Water Launches Privately Placed Environmental Impact Bond." October 3, 2016. Environmental Finance. Accessed April 7, 2017. https://www.environmental-finance.com/content/news/dc-water-launches-privately-placed-environmental-impact-bond.html?utm_source=031016na&utm_medium=email&utm_campaign=alert
31. Environmental Finance. "Bond of the Year: Social and Sustainability—Starbucks." Accessed April 7, 2017. https://www.environmental-finance.com/content/awards/green-bond-awards-2017/winners/bond-of-the-year-social-and-sustainability-starbucks.html?utm_source=070417na&utm_medium=email&utm_campaign=alert
32. Climate Bonds Initiative. "Bonds and Climate Change: The State of the Market in 2016." July 2016. Accessed April 5, 2017. https://www.climatebonds.net/files/files/CBI%20State%20of%20the%20Market%20 2016%20A4.pdf. 2.
33. Climate Bonds Initiative. "Bonds and Climate Change: The State of the Market in 2016." July 2016. Accessed April 5, 2017. https://www.climatebonds.net/files/files/CBI%20State%20of%20the%20Market%20 2016%20A4.pdf. 13.
34. Climate Bonds Initiative. "Bonds and Climate Change: The State of the Market in 2016." July 2016. Accessed April 5, 2017. https://www.climatebonds.net/files/files/CBI%20State%20of%20the%20Market%20 2016%20A4.pdf. 13.
35. Climate Bonds Initiative. "Home: Green Bonds Market 2017." Accessed April 15, 2017. https://www.climatebonds.net/
36. McQueen, Mike and Ed McMahon. *Land Conservation Financing*. 65.
37. Velde, Bruno Vander. "New Bond Aims to Unlock Private Investment to Protect Forests." *Humanature, Conservation International Blog*. Last modified October 31, 2016. http://blog.conservation.org/2016/10/new-bond-aims-to-unlock-private-investment-to-protect-forests/
38. Environmental Finance. "IFC Upsizes $152 Million Forestry Bond." Last modified November 1, 2016. https://www.environmental-finance.com/content/news/ifc-upsizes-$152-million-forestry-bond.html?utm_source=311016na&utm_medium=email&utm_campaign=alert
39. International Finance Corporation. "IFC Issues Innovative $152 Million Bond to Protect Forests and Deepen Carbon-Credit Markets." Last modified October 31, 2016. http://ifcextapps.ifc.org/ifcext/pressroom/ifcpressroom.nsf/0/594A016A78A7B14E8525805D00461397

40. Environmental Finance. "IFC Upsizes $152 Million Forestry Bond." Last modified November 1, 2016. https://www.environmental-finance.com/content/news/ifc-upsizes-$152-million-forestry-bond.html?utm_source=311016na&utm_medium=email&utm_campaign=alert
41. Embree, Sid. Interviewed by Brian McFarland. March 2017.
42. International Finance Corporation. "About IFC: Overview." Accessed March 30, 2017. http://www.ifc.org/wps/wcm/connect/corp_ext_content/ifc_external_corporate_site/about+ifc_new
43. International Finance Corporation. "A Bond that Protects Forests and Revitalizes Communities." Last modified November 2016. http://www.ifc.org/wps/wcm/connect/news_ext_content/ifc_external_corporate_site/news+and+events/news/impact-stories/a-bond-that-protects-forests-and-revitalizes-communities
44. Embree, Sid. Interviewed by Brian McFarland. March 2017.
45. Ibid.
46. Ibid.
47. International Finance Corporation. "Forests Bond." Last modified October 2016. https://www.ifc.org/wps/wcm/connect/5299a595-721b-41ef-90fa-2cff3d50982c/FINAL+Forests+Bond+Investor+Presentation+10-5_pdf.pdf?MOD=AJPERES. 16.
48. Embree, Sid. Interviewed by Brian McFarland. March 2017.
49. Ibid.
50. Ibid.
51. Ibid.
52. Ibid.
53. International Finance Corporation. "Environmental and Social Performance Standards and Guidance Notes." Accessed March 30, 2017. http://www.ifc.org/wps/wcm/connect/topics_ext_content/ifc_external_corporate_site/ifc+sustainability/our+approach/risk+management/performance+standards/environmental+and+social+performance+standards+and+guidance+notes
54. Environmental Finance. "IFC Upsizes $152 Million Forestry Bond." Last modified November 1, 2016. https://www.environmental-finance.com/content/news/ifc-upsizes-$152-million-forestry-bond.html?utm_source=311016na&utm_medium=email&utm_campaign=alert
55. Velde, Bruno Vander. "New Bond Aims to Unlock Private Investment to Protect Forests." *Humanature, Conservation International Blog*. Last modified October 31, 2016. http://blog.conservation.org/2016/10/new-bond-aims-to-unlock-private-investment-to-protect-forests/

56. Cooper, Graham. "Consortium plans to build on the success of IFC forest bond." Environmental Finance. June 22, 2017. https://www.environmental-finance.com/content/news/consortium-plans-to-build-on-the-success-of-ifc-forest-bond.html?utm_source=280617na&utm_medium=email&utm_campaign=alert
57. Velde, Bruno Vander. "New Bond Aims to Unlock Private Investment to Protect Forests." *Humanature, Conservation International Blog.* Last modified October 31, 2016. http://blog.conservation.org/2016/10/new-bond-aims-to-unlock-private-investment-to-protect-forests/
58. Embree, Sid. Interviewed by Brian McFarland. March 2017.
59. Havemann, Tanja. Email message to author. January 8, 2017.
60. Ibid.
61. Ibid.
62. Cranford, M., C. Parker and M. Trivedi. *Understanding Forest Bonds: A Guide to Raising Up-Front Finance for Tropical Forests.* Oxford, UK: Global Canopy Programme, 2011. Accessed April 11, 2017. http://globalcanopy.org/sites/default/files/documents/resources/UnderstandingForestBonds_0.pdf. 16.
63. Cranford, M., C. Parker and M. Trivedi. *Understanding Forest Bonds: A Guide to Raising Up-Front Finance for Tropical Forests.* Oxford, UK: Global Canopy Programme, 2011. Accessed April 11, 2017. http://globalcanopy.org/sites/default/files/documents/resources/UnderstandingForestBonds_0.pdf. 7.
64. Cranford, M., C. Parker and M. Trivedi. *Understanding Forest Bonds: A Guide to Raising Up-Front Finance for Tropical Forests.* Oxford, UK: Global Canopy Programme, 2011. Accessed April 11, 2017. http://globalcanopy.org/sites/default/files/documents/resources/UnderstandingForestBonds_0.pdf. 18.
65. Cranford, M., C. Parker and M. Trivedi. *Understanding Forest Bonds: A Guide to Raising Up-Front Finance for Tropical Forests.* Oxford, UK: Global Canopy Programme, 2011. Accessed April 11, 2017. http://globalcanopy.org/sites/default/files/documents/resources/UnderstandingForestBonds_0.pdf. 7.
66. Havemann, Tanja. Email message to author. January 8, 2017.
67. World Bank Group. "Doing Business: Economy Rankings." Accessed January 12, 2017. http://www.doingbusiness.org/rankings

68. Transparency International. "Corruption Perceptions Index 2016." Accessed February 8, 2017. http://www.transparency.org/news/feature/corruption_perceptions_index_2016
69. United Nations Development Programme. "Human Development Report 2015: Work for Human Development." Last modified 2016. http://hdr.undp.org/sites/default/files/2015_human_development_report.pdf. 210.
70. World Justice Project. "Rule of Law Index 2016." Accessed January 12, 2017. http://data.worldjusticeproject.org/#table
71. CVM. "Brazilian Financial Sector Regulatory Structure." Accessed April 15, 2017. http://www.cvm.gov.br/subportal_ingles/menu/about/jurisdiction.html
72. Environmental Finance. "Initiative of the Year—Brazil's Green Bond Guidelines." April 10, 2017. Accessed April 11, 2017. https://www.environmental-finance.com/content/awards/green-bond-awards-2017/winners/initiative-of-the-year-brazils-green-bond-guidelines.html?utm_source=100417na&utm_medium=email&utm_campaign=alert
73. Climate Bonds Initiative. "Land Use." Accessed April 7, 2017. https://www.climatebonds.net/standard/land-use
74. NASDAQ. "ADM Capital: Innovative Rainforest Bond Structure Unveiled at Indonesia Tropical Landscapes Summit." April 28, 2015. Accessed April 7, 2017. http://www.marketwired.com/press-release/innovative-rainforest-bond-structure-unveiled-indonesia-tropical-landscapes-summit-2013842.htm
75. Cranford, M., C. Parker and M. Trivedi. *Understanding Forest Bonds: A Guide to Raising Up-Front Finance for Tropical Forests*. Oxford, UK: Global Canopy Programme, 2011. Accessed April 11, 2017. http://globalcanopy.org/sites/default/files/documents/resources/UnderstandingForestBonds_0.pdf. 7.
76. Havemann, Tanja. Email Message to Author. January 8, 2017.

17

Additional Considerations for Tropical Forest Conservation Finance

"Social and economic justice, democracy, nonviolence, and peace are integral to achieving ecological integrity."[1]
—William J. Ginn

Introduction

There has been an increase in requirements and independent certifying bodies which assess an investment, project, program, or portfolio's consideration of social and environmental impacts. While these aspects have been detailed throughout the book, including in many of the case studies, it is important to briefly reemphasize their importance. This said, some of the certifications and standards include but are not limited to:

- The International Finance Corporation's (IFC) Performance Standards;
- The REDD+ Social and Environmental Safeguards (SES) Standard;
- The Forest Stewardship Council (FSC);
- The Climate, Community, & Biodiversity Standard (CCBS);
- Fair Trade Labeling; and
- The Cancun Safeguards.

B.J. McFarland, *Conservation of Tropical Rainforests*, Palgrave Studies in Environmental Policy and Regulation, https://doi.org/10.1007/978-3-319-63236-0_17

With these certifications in mind, there are specific social safeguards that need to be considered.

Social Safeguards

It is vitally important that conservation—whether it be via the issuance of green bonds, or ecotourism, or via REDD+ projects—is not done at the expense of the world's poorest and marginalized communities.[2] This said, social safeguards need to take into account protections and advancement opportunities for local communities and especially women and minorities. A few specific aspects that could be considered are:

- Free, Prior, and Informed Consent (FPIC);
- Women and minority rights;
- No child, slave, or forced labor;
- Respect for local traditions and intellectual property rights;
- Equitable distribution of benefits and income; and
- Land tenure.

Women, particularly in rural communities, can be empowered to help with community decisions, lead cooperatives, and manage village savings and loan programs. Simultaneously, women-led businesses in capitals around the world should be encouraged. For instance, AlphaSource and partners are seeking to "raise $250 million for the new AlphaSource Climate Fund, the world's first women-led climate change impact fund."[3]

As Impact Alpha has noted:

- USD$ 28 trillion: The boost to global GDP if women participated equally in the global economy.
- Three-quarters: The worldwide share of discretionary spending expected to be controlled by women by 2028.
- Two-thirds: The share of total US wealth expected to be controlled by women in 2028.
- USD$ 560 million: The total invested with a gender lens in public equities and debt.[4]

Related to land tenure, which should be strengthened for local communities and Indigenous Peoples, are environment safeguards.

Biodiversity Safeguards

A few of the specific biodiversity safeguards that should be considered when investing in, or undertaking, the development of rainforest conservation projects include:

- Monoculture plantations;
- Introduction of genetically modified organisms (GMOs);
- Introduction of invasive species; and
- Biopiracy.

These biodiversity safeguards, along with their related social safeguards, should be regularly monitored, reported on, and then independently verified against third-party certification standards.

Ongoing Monitoring, Reporting, and Verification

Ongoing monitoring, reporting, and verification (MRV) is a necessary component to ensure, among many things, that there are:

- No paper parks;
- No illegal activities;
- A real and measurable reduction of tropical rainforest deforestation and degradation;
- Net positive community benefits; and
- Proper allocation of funding and no mismanagement of funds.

Furthermore, progress should be publicly reported to an organization such as CDP.

Reforestation and Restoration of Degraded Areas

It is simply not enough to "just" reduce and then stop tropical deforestation and degradation. Although admittedly a wide range, there are estimates that the world has 1 billion to 6 billion hectares (approximately 2.5–14.8 billion acres) of degraded land, and these areas could be utilized for agriculture, reforested with locally beneficial species, and used to circumvent the expansion of the agricultural frontier into primary forests.[5]

One of the UN's Global Forest Goals is a call for, by 2030, the expansion of the world's forests by 120 million hectares.[6] In part, this is because the world has crossed the point of being able to address global climate change by just lowering GHG emissions. This said, the world cannot just lower emissions, but needs to both lower and remove emissions (i.e., such as via large-scale reforestation efforts).[7] The Initiative 20×20, as one such example, seeks to raise financing to restore 20 million hectares by the year 2020. As of December 2014, USD$365 of private finance was committed.[8] Similarly, the African Forest Landscape Restoration Initiative, with numerous public-private partnerships, has raised USD$545 million and has set a goal to reforest 100 million hectares and is truly considered "conservation finance on a grand scale."[9]

Role of Technology

Technology can play an important role in helping to save the world's last remaining tropical rainforests. Technological developments include utilizing drones to monitor remote deforestation, to track wildlife and identify poachers, and to engage the public to financially support rainforests and to support policies designed to protect rainforests. Satellite imagery and remote sensing technologies are also cutting-edge developments to help map the world's forests and assess the extent of deforestation. In addition, there are modeling technologies such as TerrSet's Habitat and Biodiversity Modeler.[10]

Social media, which has the ability to reach billions of people around the world, can be a powerful force for tropical rainforest conservation. For instance:

> As of June 2016 there are approximately 2.2 million apps available and 2 million in the Apple App Store. In 2015, global mobile app revenues amounted to $41.1 billion. In 2020, consumers are projected to spend over $101 billion on mobile apps via app stores.
>
> This spring an initiative called Apps for Earth, saw 100 percent of revenues from 27 apps at the Apple Store go to WWF over a ten day period that ended on April 24th. This campaign not only raised awareness it generated more than $8 million to support the WWF's conservation work.[11]

Furthermore, Carbonfund.org Foundation and Offsetmyride.io have a carbon offsetting app for Uber and Lyft drivers,[12] while Code REDD's Stand for Trees campaign allows individuals to contribute small sums directly to REDD+ projects.[13]

Other Resources on Additional Considerations for Tropical Forest Conservation Finance

Avoided Deforestation (REDD) and Indigenous Peoples, A Report by Instituto Socioambiental and Forest Trends

- http://www.gcftaskforce.org/documents/REDD_indigenous_peoples.pdf

Business & Human Rights Resource Centre and Corporate Human Rights Benchmark

- https://www.business-humanrights.org/
- https://business-humanrights.org/en/chrb

Conservation Drone Project by the Orangutan Conservancy (OC)

- http://www.orangutan.com/conservation-drone-project/

ConservationDrones.org

- https://conservationdrones.org/

Institute for Human Rights and Business

- https://www.ihrb.org/

Integrated Biodiversity Assessment Tool

- https://www.ibatforbusiness.org/about

OpenForests' Drone Mapper

- https://openforests.com/drone-mapper/

TED Talks on Conservation and Technology

- https://www.ted.com/talks/lian_pin_koh_a_drone_s_eye_view_of_conservation
- https://www.ted.com/talks/tasso_azevedo_hopeful_lessons_from_the_battle_to_save_rainforests
- https://www.ted.com/talks/greg_asner_ecology_from_the_air
- http://www.scienceheap.com/?yhs

UN-REDD Benefits and Risks Tool

- http://www.unredd.net/documents/global-programme-191/safeguards-multiple-benefits-297/safeguards-coordination-group-

2606/bert-3525/un-redd-programme-bert-english-3599/16087-benefits-and-risks-tool-bert-v21-workshop-kit-facilitators-guide.html

UN-REDD Methodological Brief on Gender

- http://www.unredd.net/index.php?option=com_docman&view=download&alias=15951-un-redd-methodological-brief-on-gender&category_slug=global-gender-resources&Itemid=134

Vigeo Eiris

- http://www.vigeo-eiris.com/en/

World Overview of Conservation Approaches and Technologies (WOCAT)

- https://www.wocat.net/

Notes

1. Ginn. *Investing in Nature*. 183.
2. Gaworecki, Mike. "When Poor People Pay the Price for Forest Conservation." February 18, 2016. *Mongabay*. Accessed April 7, 2017. https://news.mongabay.com/2016/02/when-conservation-efforts-foist-the-cost-of-climate-mitigation-onto-poor-people/?n3wsletter
3. PRWeb. "New $250 Million, Women-Led Climate Fund Joins Forces with Conservation Organizations to Save Vast Forests in Asia, Africa and Latin America." December 19, 2016. Accessed April 7, 2017. http://www.prweb.com/releases/2016/12/prweb13936152.htm
4. Impact Alpha. "Women Rising." Accessed April 7, 2017. http://impactalpha.com/category/gender-lens/
5. Gibbs, H.K. and J.M. Salmon. "Mapping the World's Degraded Lands." February 2015. *Applied Geography*. Volume 57. Accessed April 15, 2017. http://www.sciencedirect.com/science/article/pii/S0143622814002793. 12–21.

6. United Nations. "UN Forum Sets Landmark Target to Increase World's Forests by 120 Million Hectares by 2030." Accessed April 15, 2017. http://www.un.org/sustainabledevelopment/blog/2017/01/un-forum-sets-landmark-target-to-increase-worlds-forests-by-120-million-hectares-by-2030/
7. Dowd, Scott. Interviewed by Brian McFarland. January 2017.
8. World Resources Institute. "RELEASE: Latin American and Caribbean Countries and Regional Programs Launch Initiative 20×20 to Restore 20 Million Hectares of Land." Last modified December 4, 2016. http://www.wri.org/news/2014/12/release-latin-american-and-caribbean-countries-and-regional-programs-launch-initiative
9. Conservation Finance Network. "African Forestry Initiative Leverages Funds for Restoration." April 12, 2016. Accessed April 15, 2017. http://www.conservationfinancenetwork.org/2016/04/12/african-forestry-initiative-leverages-funds-for-restoration
10. Clark Labs. "Habitat and Biodiversity Modeler." Accessed April 15, 2017. https://clarklabs.org/terrset/habitat-biodiversity-modeler/
11. The Green Market Oracle. "Apple's Apps for Earth Initiative Raised Awareness and Funds for WWF." Last modified June 22, 2016. http://www.thegreenmarketoracle.com/2016/06/apples-apps-for-earth-initiative-raised.html
12. Offsetmyride.io. "Home." Accessed April 2, 2017. https://offsetmyride.io/login
13. Stand for Trees. "Home." Accessed April 2, 2017. https://standfortrees.org/en/#protect-a-forest

18

The Future of Tropical Forest Conservation Finance

"Never before has our fate, linked inextricably with the health of the planet, rested so surely in our own hands. Indeed, it is now the actions of human beings themselves, and not solely the blind mechanisms of natural selection and genetic drift, that are altering the terms of survival for every species on the planet."
—Elizabeth Royte[1]

With USD$52 billion[2] spent annually on conservation projects—not just tropical rainforest conservation—and approximately USD$8.2 billion invested by the private sector in sustainable food and fiber production, habitat conservation, and water quality and quantity protection projects from 2004 to 2015,[3] there is still a long way to go to reach the estimated USD$200 to USD$300 billion needed on an annual basis to finance the world's ecosystems.[4]

As this book has outlined, there are numerous financial and strategic solutions to conserve the world's remaining forests. Essentially what is needed going into the future is:

- Mitigation of global climate change;
- Conservation of large tracts of land;

B.J. McFarland, *Conservation of Tropical Rainforests*, Palgrave Studies in Environmental Policy and Regulation, https://doi.org/10.1007/978-3-319-63236-0_18

- Investable, scalable projects for the private sector;
- Development of new financing structures;
- Public funds used to leverage private financing;
- Education and reduced public apathy;
- A new economic system that values natural capital;
- Respect for local communities, Indigenous Peoples, and human rights; and
- Policy matters.

Similarly, these are also the reoccurring themes throughout the case studies' lessons learned.

Mitigation of Global Climate Change

Without mitigating global climate change, the conservation of tropical rainforests will be only more difficult, if not impossible. For instance, due in part to climate change's impact on the Amazon, researchers "Miles et al.'s central conclusion was that, to ensure the greatest resilience of Amazonia biodiversity, the highest priority should be given to strengthening and extending protected areas in western Amazonia that encompass lowland and montane forests."[5] One of these areas in western Amazonia is Yasuní National Park, one of the most biologically diverse places on Earth. In part due to the inevitability of global climate change, the future of tropical rainforest conservation should prioritize large tracts of land.

Conservation of Large Tracts of Land

The need to conserve large tracts of land will help mitigate the impacts on biodiversity from global climate change, reduce the impact on the forest from border effects and activities in the buffer zone, and provide habitat for wildlife to reproduce. For instance, Dr. Russell Mittermeier suggests the need to establish large reserves more than 10,000 square kilometers (1 million hectares or 2.47 million acres) and designate as "high biodiversity wilderness areas."[6] This includes locations such as the Tumucumaque National Park and the Ecological Station of Grão-Pará in the Amazon

Region Protected Areas (ARPA) Program, along with the Tri-National de la Sangha international peace park and its buffer zone. Similarly, Dr. Norman Myers suggests areas, depending on the species, of 2,500 square kilometers (250,000 hectares or 618,000 acres). It is important to consider the species because "if a tiger population could survive with as few as 400 individuals, the area required to support them may need to be as big as 40,000 km^2 {4 million hectares or 9.9 million acres}."[7] This is why places such as the Tropical Rainforest Heritage of Sumatra, which is home to Sumatran tigers and that collectively has 2.6 million hectares across its three national parks, are still in need of additional funding.

Investable, Scalable Projects for the Private Sector

Additional funding is currently available from the private sector, but projects and programs need to be investable and scalable. In sum:

- Green bonds, particularly environmental impact bonds, could be a very promising and scalable financing mechanism;
- REDD+ projects still remain a voluntary carbon market project; however, the emerging ICAO (International Civil Aviation Organization) market, the Paris Agreement (particularly articles 5 and 6), and the potential linkage between the State of Acre (Brazil) and the State of California could create a compliance market for REDD+ offsets;
- Projects or programs that diversify revenue sources (e.g., traditional philanthropy and private sector finance) with different conservation financing instruments (sustainable commodities with ecotourism) can be the most financially sustainable initiatives;
- The no-deforestation supply change commitments are also promising, but show few concrete, conservation results to date. There needs to be an extensive transition from non-sustainable, non-certified, and sometimes illegally sourced agricultural soft commodities—especially cattle, timber, palm oil, and soy—to legally sourced, certified chain of custody. This long-term trend is critical to address and could be a major development.

- Debt-for-nature swaps, including corporate debt-for-nature swaps, should be revisited.

As nicely summarized by Ecosystem Marketplace, a conservation finance report developed by Credit Suisse and the McKinsey Center for Business and Environment:[8]

> {...} presents "a few concrete ideas that we deem to be scalable, repeatable, and investable." A toolkit suggests two distinct approaches. One involves applying the same proven project management approach across a single project type, financing it through plain equity or debt vehicles, and scaling it up. The other involves bundling heterogeneous projects linked by a common feature—a national park, for instance—and creating a common risk-and-return sharing vehicle. Interviews suggested that risk is often a greater concern to potential investors than the financial return or even the conservation impacts of a project.[9]

This is private sector advice well heeded.

Develop New Financing Structures

As explained by Dr. Daniel Janzen, the number 1 rainforest drug in the world is coffee. The whole marketing structure is already there, we already know about it, and we do not need to discover it. If a 1 penny tax was placed on coffee—considering that 3 trillion cups of coffee are made in the United States alone each year—we could pay for all conservation costs forever. The challenge is not finding the interesting drug in the rainforest, the challenge is getting the one penny tax.[10] The question still remains, what new financing structures can be developed that would have the transformational impact as a 1 penny coffee tax?

For instance, what if megacities such as Brasilia, São Paulo, Rio de Janeiro, and/or Jakarta used a real estate transfer tax or water usage surcharge to protect upstream watersheds? What if TNC's Conservation Notes or Credit Suisse's Nature Conservation Notes came down to a minimum investment of say USD$500 or USD$1,000 or USD$10,000 and then these Notes were collateralized, packaged, and became securitized conservation loans?

As nicely summarized again by Ecosystem Marketplace, a conservation finance report developed by Credit Suisse and the McKinsey Center for Business and Environment[11] highlighted an "incubation workshop with 15 NGOs also revealed some fresh ideas: a marine protected area bond, insurance payments for risk mitigation, and a 'substitutes fund' that would support research and development of substitutes for destruction-inducing commodities such as meat, palm oil, or diamonds."[12]

What is your idea?

Public Funds Used to Leverage Private Financing

Public funds can be used to de-risk investments in tropical rainforest conservation, which, in turn, helps leverage private financing. For instance, the World Bank Group's new 2016–2020 Forest Action Plan is seeking to "both commercializing activities that rely on forest conservation and make the activities that drive deforestation less bankable," and the World Bank's "Forest Carbon Partnership Facility's Carbon Fund and the BioCarbon Fund's Initiative for Sustainable Forest Landscapes now support results-based finance for emissions reductions in 21 countries."[13] The Norwegian International Climate and Forest Initiative has been a global champion of using public funds to help finance the conservation of tropical rainforests, and other countries should seek to emulate their success.

Education and Reduced Public Apathy

Education can take many forms. Education, for instance, could be targeted at reducing food waste in developed countries and simultaneously improving infrastructure in developing countries to reduce food spoilage. While sadly probably not alone, a recent survey found that "one in two Singaporeans surveyed would still throw away 'ugly food'."[14] Furthermore:

> Several paths to improve human and organizational capacity in conservation finance is evident. The first is to work to increase funding for relevant curriculum development and professional training, both within individual organizations and across the conservation movement. Although the Land Trust Alliance and other organizations do offer some training in these fields, much more could be done, particularly with regard to the type of advanced conservation finance methods discussed in this book.[15]

Reducing public apathy, as demonstrated by Project Piaba's attempt to connect aquarium visitors to the fate of local communities in the Amazon Rainforest to Dr. Daniel Janzen's work on an International Barcode of Life in Costa Rica, will continue to be a constant need.

This said, if you have the stick-to-it-ness, then you often hear stories from leading practitioners who had early experiences with wildlife. However, in order to save a large area of rainforest, that experience needs to be conveyed to a wider audience.[16]

Develop a New Economic System That Values Natural Capital

On a general level, there is an increased understanding or awareness of natural capital. For instance, 10–15 years ago, people would not know what you are talking about. While public citizens may still be a bit clueless, people in the private sector are starting to get it and many local communities and Indigenous Peoples have always gotten it. The challenge is translating the awareness of our dependence on natural capital (i.e., for society in general including companies and consumers) and the benefits we receive from ecosystems (e.g., food, water, and medicine) into a more sustainable system of incentives for conservation (i.e., redirect resources to these areas that are providing all these benefits).[17]

On a very profound level, the notion of stranded assets is a very real thing and, in fact, is likely inevitable. The extractive model worked fine for 10,000 years. Then, less than 100 years ago, population and consumption began to outpace nature. There is a systemic market imbalance and that is the old economic model with its extractive paradigm no longer applies to our future. The question is, how do we price natural capital and price in externalities? This said, society can do this transition in a somewhat controlled fashion, or do we run off the cliff? Part of this transition is being forced by nature's inability to sustain the old economic model, and part of this transition is happening because of an engaged, dedicated group of practitioners. For instance, change in the market often starts with stigmatization (e.g., tobacco and Apartheid), which drives investor sentiment (i.e., investors are afraid of the risks they are taking). The inevitability is that supply chains will change by institutional investors and this includes investors such as Norway's nearly USD$1 trillion Government Pension Fund Global that has already divested from companies connected to deforestation.[18]

If viable solutions can be embedded into supply chains with some type of environmental mitigation credit, then there is hope for the world's last remaining rainforests. There is a lot of interesting supply chain work to be explored, especially if companies are going to make good on their zero deforestation commitments. For instance, a jurisdiction could make a landscape fund that would be financed by a tariff on the commodity extracted from its jurisdiction. This cost should not be put on the farmers, but rather put on the brand, and then the money raised from the tariff could go back into the landscape via a pay-for-performance model.[19]

Need to Recognize Local Communities, Indigenous Peoples, and Human Rights

Simply stated, but profound, "if there is to be any hope of conserving a significant portion of these forests, the Indigenous Peoples {and local communities} must be given tools and support to protect their lands from outside encroachment."[20]

Policy Matters

Politicians have not been held accountable to the issue of tropical rainforest and degradation by voters. There is still good public awareness about the environment in general, but needs to be raised as a serious political issue. For instance, does your national government have a no-deforestation supply commitment, does the company that makes your favorite food have a no-deforestation supply chain commitment, and does your local town or city government only source FSC-certified paper? Voluntary approaches are good, but only have limited impacts.[21] Such policies could be for the regional integration of international peace parks, India's commitment of domestic budgetary expenditures up to USD$6 billion per year for states to maintain forest cover, or Brazil's new Forest Code—that with better enforcement—can have profound impacts on our global climate and the world's largest rainforest.

Notes

1. Royte. *The Tapir's Morning Bath*. 18.
2. Huwyler, Fabian, Jürg Käppeli and John Tobin. "Conservation Finance: From Niche to Mainstream: The Building of an Institutional Asset Class." Last modified January 2016. https://www.credit-suisse.com/media/assets/corporate/docs/about-us/responsibility/banking/conservation-finance-en.pdf. 3.
3. Hamrick, Kelley. "State of Private Investment in Conservation 2016." *Ecosystem Marketplace*. December 2016. http://www.forest-trends.org/documents/files/doc_5474.pdf#. vii.
4. Huwyler, Fabian, Jürg Käppeli and John Tobin. "Conservation Finance: From Niche to Mainstream: The Building of an Institutional Asset Class." Last modified January 2016. https://www.credit-suisse.com/media/assets/corporate/docs/about-us/responsibility/banking/conservation-finance-en.pdf. 3.
5. Bass, Margaret S., et al. "Global Conservation Significance of Ecuador's Yasuní National Park." *PLoS ONE* 5(1). 2010. 7. Accessed April 15, 2017. http://journals.plos.org/plosone/article?id=10.1371/journal.pone.0008767. 13.

6. Ibid.
7. Myers. *The Shrinking Ark*. 227.
8. Huwyler, Fabian, Jürg Käppeli and John Tobin. "Conservation Finance: From Niche to Mainstream: The Building of an Institutional Asset Class." Last modified January 2016. https://www.credit-suisse.com/media/assets/corporate/docs/about-us/responsibility/banking/conservation-finance-en.pdf.
9. Ecosystem Marketplace. "The Carbon Chronicle." January 26, 2016. Accessed April 15, 2017. http://www.ecosystemmarketplace.com/wp-content/themes/ecosystem_marketplace/newsletter/carbon_chronical.php?newsletterID=21
10. Janzen, Daniel H. "How to Conserve a Large Tropical Wildland Through Biodiversity Development: Costa Rican Example." Presentation at Harvard University Center for the Environment, Cambridge, Massachusetts. September 30, 2010. Accessed March 17, 2017. https://vimeo.com/17675978
11. Huwyler, Fabian, Jürg Käppeli and John Tobin. "Conservation Finance: From Niche to Mainstream: The Building of an Institutional Asset Class." Last modified January 2016. https://www.credit-suisse.com/media/assets/corporate/docs/about-us/responsibility/banking/conservation-finance-en.pdf.
12. Ecosystem Marketplace. "The Carbon Chronicle." February 25, 2016. Accessed April 15, 2017. http://www.ecosystemmarketplace.com/wp-content/themes/ecosystem_marketplace/newsletter/carbon_chronical.php?newsletterID=24
13. Ecosystem Marketplace. "The Carbon Chronicle." April 22, 2016. Accessed April 15, 2017. http://www.ecosystemmarketplace.com/wp-content/themes/ecosystem_marketplace/newsletter/carbon_chronical.php?newsletterID=29
14. Manongdo, Ping. "The Ugly Truth about Food Waste in Singapore." October 21, 2016. *Eco-Business*. Accessed April 15, 2017. http://www.eco-business.com/news/the-ugly-truth-about-food-waste-in-singapore/?utm_medium=email&utm_campaign=26%20Oct%20newsletter&utm_content=26%20Oct%20newsletter+Version+A+CID_8756604a9a9381 3ac09f26a39c87e61d&utm_source=Campaign%20Monitor
15. Coady, Patrick. "Conservation Finance Viewed as a System: Tackling the Financial Challenge." In *From Walden to Wall Street*, edited by James Levitt. Massachusetts: Lincoln Institute of Land Policy, 2005. 29.

16. Kusumaatmadja, Rezal. Interviewed by Brian McFarland. August 2016.
17. Settelmyer, Scott. Interviewed by Brian McFarland. February 2017.
18. Lemons, Todd. Interviewed by Brian McFarland. March 2017.
19. Korchinsky, Mike. Interviewed by Brian McFarland. March 2017.
20. Kaimowitz, David. "Indigenous Peoples and Deforestation in Latin America." In *On the Edge*, written by Claude Martin. Berkeley: Greystone Books, 2015. 174.
21. Settelmyer, Scott. Interviewed by Brian McFarland. March 2017.

Afterword

> "But now," says the Once-ler, "now that you're here, the word of the Lorax seems perfectly clear. UNLESS someone like you cares a whole awful lot, nothing is going to get better. It's not."
> —Dr. Seuss, The Lorax[1]

While it is true that billions of dollars are needed on an annual basis to halt tropical rainforest deforestation and degradation, even a penny campaign initiated by children can spark the change and funding that is needed in our world. Take for instance, "the Monteverde Conservation League (MCL) owns and manages the Children's Eternal Rain Forest, which was started with a penny campaign by children from all around the world."[2]

In addition, volunteers in India planted a record 66 million trees in just 12 hours,[3] while in China, "some 200 million Chinese users have downloaded an app by financial services provider Ant Financial that helps them make carbon-friendly decisions in everyday life."[4]

With this in mind, if you like animals then you should like rainforests, which are the most biologically diverse terrestrial ecosystem on Earth. If you like rainforests, then you should be concerned about deforestation. If you are not concerned about or do not believe in climate change, then do

B.J. McFarland, *Conservation of Tropical Rainforests*, Palgrave Studies in Environmental Policy and Regulation, https://doi.org/10.1007/978-3-319-63236-0

Real-life scene from the Lorax (Credit: Brian Joseph McFarland)

not focus on the climate change mitigation potential of rainforests and their carbon stocks. However, if you are concerned about water quality and water supply, you should be concerned about the loss of rainforests and their contribution to the global water cycle. Finally, if you care about people, you should care about the deforestation of rainforests and all the benefits they give to humankind, and not just local communities halfway around the world, but to you the reader through building materials, paper, food, and medicine. Thank you for reading and please remember, "now that you're here, the word of the Lorax seems perfectly clear. UNLESS someone like you cares a whole awful lot, nothing is going to get better. It's not."—Dr. Seuss, The Lorax[5]

Bibliography and Suggested Readings

Bass, Margaret S., Matt Finer, Clinton N. Jenkins, Holger Kreft, Diego F. Cisneros-Heredia, Shawn F. McCracken, Nigel C.A. Pitman, et al. 2010. Global Conservation Significance of Ecuador's Yasuní National Park. *PLoS ONE* 5 (1): 7. Accessed December 6, 2016. http://journals.plos.org/plosone/article?id=10.1371/journal.pone.0008767

Carson, Rachel. 2002. *Silent Spring*. Boston: A Mariner Book, Houghton Mifflin Company.

Clark, Story. 2007. *A Field Guide to Conservation Finance*. Washington, DC: Island Press.

Daily, Gretchen C., and Katherine Ellison. 2002. *The New Economy of Nature: The Quest to Make Conservation Profitable*. Washington, DC: Island Press.

Davis, Wade. 1996. *One River: Explorations and Discoveries in the Amazon Rainforest*. New York: Simon & Schuster.

Forsyth, Adrian, and Ken Miyata. 1984. *Tropical Nature: Life and Death in the Rain Forests of Central and South America*. New York: Charles Scribner's Sons.

Ginn, William. 2005. *Investing in Nature: Case Studies of Land Conservation in Collaboration with Business*. 2nd ed. Washington, DC: Island Press.

B.J. McFarland, *Conservation of Tropical Rainforests*, Palgrave Studies in Environmental Policy and Regulation, https://doi.org/10.1007/978-3-319-63236-0

Hamrick, Kelley. 2016. *State of Private Investment in Conservation 2016: A Landscape Assessment of an Emerging Market.* Washington, DC: Ecosystem Marketplace. http://www.forest-trends.org/documents/files/doc_5474.pdf#

Head, Suzanne, and Robert Heinzman, eds. 1990. *Lessons of the Rainforest: Essays by Norman Myers, Randall Hayes, Frances Moore Lappé and Others.* San Francisco: Sierra Club Books.

Hopper, Kim, and Ernest Cook. 2004. *Conservation Finance Handbook: How Communities are Paying for Parks and Land Conservation.* California: Trust for Public Land.

Huwyler, Fabian, Jürg Käppeli, and John Tobin. 2016. *Conservation Finance: From Niche to Mainstream: The Building of an Institutional Asset Class.* New York: Credit Suisse. https://www.credit-suisse.com/media/assets/corporate/docs/about-us/responsibility/banking/conservation-finance-en.pdf

Huwyler, Fabian, Jürg Käppeli, Katharina Serafimova, Eric Swanson, and John Tobin. 2014. *Conservation Finance: Moving beyond Donor Funding toward an Investor-Driven Approach.* New York: Credit Suisse. https://www.credit-suisse.com/media/assets/corporate/docs/about-us/responsibility/environment/conservation-finance-en.pdf

Leopold, Aldo. 1934. Conservation Economics. *Journal of Forestry* 32 (5): 537–544.

Levitt, James N., ed. 2005. *From Walden to Wall Street: Frontiers of Conservation Finance.* Massachusetts: Lincoln Institute of Land Policy.

Levitt, James N., ed. 2010. *Conservation Capital in the Americas: Exemplary Conservation Finance Initiatives.* Massachusetts: Lincoln Institute of Land Policy.

Martin, Claude. 2015. *On the Edge: The State and Fate of the World's Tropical Rainforests (Report to the Club of Rome).* California: Greystone Books.

McQueen, Mike, and Ed McMahon. 2003. *Land Conservation Financing.* Washington, DC: Island Press.

Myers, Norman. 1979. *The Sinking Ark: A New Look at the Problem of Disappearing Species.* Oxford: Pergamon Press.

Roberts, J. 2003. *Timmons and Nikki Demetria Thanos. Trouble in Paradise: Globalization and Environmental Crises in Latin America.* New York: Routledge.

Royte, Elizabeth. 2001. *The Tapir's Morning Bath: Mysteries of the Tropical Rain Forest and the Scientists who are Trying to Solve Them*. Boston: Houghton Mifflin Company.

Speth, James Gustave, and Peter M. Haas. 2006. *Global Environmental Governance*. Washington, DC: Island Press.

The Nature Conservancy and EKO Asset Management Partners. 2014. *Investing in Conservation: A Landscape Assessment of an Emerging Market*. Virginia: The Nature Conservancy. https://thought-leadership-production.s3.amazonaws.com/2016/09/14/18/27/27/93df86a6-f916-47cc-8a54-0a818037bba1/InvestingInConservation_Report.pdf

Wilson, E.O. 2006. *The Creation: An Appeal to Save Life on Earth*. New York: W.W. Norton & Company.

World Wildlife Fund. 2009. *Guide to Conservation Finance: Sustainable Financing for the Planet*. Washington, DC: World Wildlife Fund. http://awsassets.panda.org/downloads/wwf_guide_to_conservation_finance.pdf

Other Suggested Resources

The following suggested resources are a list of conferences, associations, and networks, along with additional useful books, tools, and websites.

Conferences and Trainings

- A Community of Ecosystem Services (ACES)[6]
- Argus Emissions Markets by Argus Media[7]
- Carbon Forums[8]
- Center for International Forestry Research (CIFOR) conferences including the Asia-Pacific Rainforest Summit[9]
- Climate Bonds Initiative's Events[10]
- Climate Leadership Conference[11]
- Conservation Finance[12]
- Conservation Finance Network's Boot Camp Training Course[13]
- Conservation Finance Network's Practitioner Roundtable[14]
- Environmental Leadership and Training Institute, an initiative of the Yale School of Forestry and Environmental Studies[15]
- Forest Finance & Sustainability Conference[16]

B.J. McFarland, *Conservation of Tropical Rainforests*, Palgrave Studies in Environmental Policy and Regulation, https://doi.org/10.1007/978-3-319-63236-0

- General Assembly of the Tropical Forest Alliance 2020[17]
- Global Landscapes Forum[18]
- Governors' Climate & Forests Task Force Annual Meeting[19]
- Innovate4Climate[20]
- Innovation Forum,[21] How Business Can Tackle Deforestation[22]
- International Union for the Conservation of Nature's (IUCN) World Conservation Congress[23]
- Land Trust Alliance's Rally[24]
- Mission Investors Exchange National Conference[25]
- National Land Conservation Conference[26]
- Navigating the American Carbon World[27]
- Oslo Redd Exchange Annual Conference[28]
- Social Capital Markets Conference[29]
- Sustainable Brands[30]
- Sustainable Innovation Forum[31]
- Sustainable Purchasing Leadership Council's Annual Summit[32]
- United Nations Framework Convention on Climate Change (UNFCCC) annual Conference of the Parties (COP)[33]
- Wall Street Green Summit[34]
- Wildlife Habitat Council's Conservation Conference[35]
- World Forum on Natural Capital[36]

Also, see the Global Forest Information Service "which keeps an events calendar for forest-related events happening around the world. {…} You can add your events to the calendar by filling in this form: http://www.gfis.net/gfis/en/en/submititem/."[37]

These conferences and trainings do not appear to still be active: Conservation Capital in the Americas Conference;[38] Conservation Finance Forum at Harvard University; Conservation Leadership Dialogues (CLDs) which were convened in 2002, 2003, and 2004 by the Lincoln Institute of Land Policy; and the Environmental Finance Innovation Summit.[39]

Books and Articles

These books and articles were not read by the author but appear to be useful.

Achard, Frederic, and Matthew C. Hansen. 2013. *Global Forest Monitoring from Earth Observation*. Florida: CRC Press.

Ali, Saleem H., ed. 2007. *Peace Parks: Conservation and Conflict Resolution*. Massachusetts: The MIT Press.

Ashton, Mark S., Mary L. Tyrrell, Deborah Spalding, and Bradford Gentry, eds. 2012. *Managing Forest Carbon in a Changing Climate*. New York: Springer.

Bates, Robert H. 2014. *Markets and States in Tropical Africa: The Political Basis of Agricultural Policies*. California: University of California Press.

Berry, Wendell. 1995. *Another Turn of the Crank*. Berkeley: Counterpoint.

Birchard, Bill. 2005. *Nature's Keepers: The Remarkable Story of How the Nature Conservancy Became the Largest Environmental Group in the World*. California: Jossey-Bass Publishers.

Bishop, Joshua, ed. 2013. *The Economics of Ecosystems and Biodiversity in Business and Enterprise*. United Kingdom: Routledge.

Bishop, Joshua, and Chloe Hill. 2014. *Global Biodiversity Finance: The Case for International Payments for Ecosystem Services*. United Kingdom: Edward Elgar Publishing Limited.

Byers, Elizabeth, and Karin Marchetti Ponte. 2005. *The Conservation Easement Handbook*. 2nd ed. Washington, DC: Land Trust Alliance.

Clark, Connie J., and John R. Poulsen, eds. 2012. *Tropical Forest Conservation and Industry Partnership: An Experience from the Congo Basin*. United Kingdom: Wiley-Blackwell.

Dempsey, Jessica. 2016. *Enterprising Nature: Economics, Markets and Finance in Global Environmental Politics*. United Kingdom: John Wiley & Sons, Ltd.

Flanagan, Joan. 1995. *The Grass Roots Fundraising Book: How to Raise Money in Your Community*. Illinois: Contemporary Books.

Grove, Richard H. 1995. *Green Imperialism: Colonial Expansion, Tropical Island Edens and the Origins of Environmentalism, 1600–1860*. United Kingdom: Cambridge University Press.

Hanks, John. 2003. Transfrontier Conservation Areas (TFCAs) in Southern Africa: Their Role in Conserving Biodiversity, Socioeconomic Development and Promoting a Culture of Peace. *Journal of Sustainable Forestry* 17 (1–2). http://www.tandfonline.com/doi/abs/10.1300/J091v17n01_08

Heal, Geoffrey. 2000. *Nature and the Marketplace: Capturing the Value of Ecosystem Services*. Washington, DC: Island Press.

Helm, Dieter. 2015. *Natural Capital: Valuing the Planet*. Connecticut: Yale University Press.

Ingerson, Ann. 2004. *Conservation Capital: Sources of Private Funding for Land Conservation*. Washington, DC: The Wilderness Society. http://conservationtools-production.s3.amazonaws.com/library_item_files/150/180/TWS_funding.pdf?AWSAccessKeyId=AKIAIQFJLILYGVDR4AMQ&Expires=1480907851&Signature=DRxUBoHZ9powMM7WrTDtNEQPDj0%3D

Jenkins, Michael, and Emily Smith. 1999. *The Business of Sustainable Forestry: Strategies for an Industry in Transition*. Washington, DC: Island Press.

Kricher, John. 1997. *A Neotropical Companion: An Introduction to the Animals, Plants, & Ecosystems of the New World Tropics*. New Jersey: Princeton University Press.

Leopold, Aldo. 1933. The Conservation Ethic. *Journal of Forestry* 31 (6): 634–643.

Leopold, Aldo. 1949. *A Sand County Almanac*. United Kingdom: Oxford University Press.

Lerner, Steve, and William Poole. 1999. *The Economic Benefits of Parks and Open Space: How Land Conservation Helps Communities Grow Smart and Protect the Bottom Line*. Washington, DC: The Trust for Public Land.

Lindenmayer, David B., and Jerry F. Franklin. 2002. *Conserving Forest Biodiversity: A Comprehensive Multiscaled Approach*. Washington, DC: Island Press.

Lockwood, Michael, Graeme Worboys, and Ashish Kothari, eds. 2006. *Managing Protected Areas: A Global Guide*. United Kingdom: Earthscan.

Myers, Norman. 1983. *A Wealth of Wild Species: Storehouse for Human Welfare*. Colorado: Westview Press.

Nagendra, Harini, and Jane Southworth, eds. 2010. *Reforesting Landscapes: Linking Pattern and Process*. New York: Springer.

Plotkin, Mark, and Lisa Famolare, eds. 1992. *Sustainable Harvest and Marketing of Rain Forest Products*. Washington, DC: Island Press.

Richards, P.W. 1952. *The Tropical Rain Forest: An Ecological Study*. United Kingdom: Cambridge University Press.

Richards, Michael. 1995. *"Internalising the Externalities" of Tropical Forestry: A Review of Innovative Financing and Incentive Mechanisms*. United Kingdom: Overseas Development Institute.

Roosevelt, Theodore. 1920. *The Autobiography of Theodore Roosevelt*. United States of America: Charles Scribner's Sons.

Roth, Stephanie, and Kim Klein, eds. 2001. *Raise More Money: The Best of the Grassroots Fundraising Journal*. United States of America: GFJ Publications.

Sandor, Richard, and Murali Kanakasabai. 2014. *Rafael L Marques, and Nathan Clark. Sustainable Investing and Environmental Markets: Opportunities in a New Asset Class*. Singapore: World Scientific Publishing Co. Pte. Ltd.

Schaff, Terry, and Doug Schaff. 1999. *The Fundraising Planner: A Working Model for Raising the Dollars You Need*. California: Jossey-Bass Publishers.

Schroth, Götz, Gustavo A. B. da Fonseca, Celia A. Harvey, Claude Gascon, Heraldo L. Vasconcelos, and Anne-Marie N. Izac, eds. 2004. *Agroforestry and Biodiversity Conservation in Tropical Landscapes*. Washington, DC: Island Press.

Seymour, Frances, and Jonah Busch. 2016. *Why Forests? Why Now? The Science, Economics, and Politics of Tropical Forests and Climate Change*. Washington, DC: Center for Global Development.

Tassel, Sandra. 2009. *The Conservation Program Handbook: A Guide for Local Government Land Acquisition*. 2nd ed. Washington, DC: Island Press.

Terborgh, John. 1999. *Requiem for Nature*. Washington, DC: Island Press.

The Trust for Public Land. 1995. *Doing Deals: A Guide to Buying Land for Conservation*. California: The Trust for Public Land.

Wiley, Laura, and William Riley. 2005. *Nature's Strongholds: The World's Greatest Wildlife Reserves*. New Jersey: Princeton University Press.

Williams, Michael. 2006. *Deforesting the Earth: From Prehistoric to Global Crisis, An Abridgement*. Illinois: The University of Chicago Press.

World Wildlife Fund. 2003. *WWF Guidelines for Investment in Operations that Impact Forests*. Washington, DC: World Wildlife Fund. http://assets.panda.org/downloads/wwfinvestmentpol7oct03final.pdf

World Wildlife Guide. 2012. *The 2050 Criteria Guide to Responsible Investment in Agricultural, Forest, and Seafood Communities*. Washington, DC: World Wildlife Fund. http://assets.worldwildlife.org/publications/458/files/original/2050_criteria_final_low_res_online_viewing.pdf?1348517472&_ga=1.99074333.200605152.1480908204

Notes

1. Goodreads. "The Lorax Quotes." Accessed March 20, 2017. https://www.goodreads.com/work/quotes/886002-the-lorax
2. Meyer, Shannon. "Working with Landowners to Provide Ecosystem Services: Costa Rica's Groundbreaking Experiment." In *Conservation Capital in the Americas*, edited by James N. Levitt. Massachusetts: Lincoln Institute of Land Policy, 2010. 207.
3. Climate Action. "Record 66 million trees planted in 12 hours in India." July 4, 2017. http://www.climateactionprogramme.org/news/record-66-million-trees-planted-in-12-hours-in-india
4. Carbon Pulse. "CP Daily: Tuesday May 30, 2017." May 30, 2017. http://carbon-pulse.com/35047/
5. Goodreads. "The Lorax Quotes." Accessed March 20, 2017. https://www.goodreads.com/work/quotes/886002-the-lorax
6. The University of Florida's Institute of Food and Agricultural Sciences. "ACES Overview." Accessed November 23, 2016. http://www.conference.ifas.ufl.edu/aces/
7. Argus Media. "In-Depth Conferences on the Energy and Commodity Markets." Accessed March 20, 2017. http://www.argusmedia.com/Events/Argus-Events/

8. International Emissions Trading Association. "IETA Lead Upcoming Events." Accessed December 2, 2016. http://www.ieta.org/page-18492
9. Center for International Forestry Research. "Asia-Pacific Rainforest Summit." Accessed April 2, 2017. http://www.cifor.org/asia-pacific-rainforest-summit/
10. Climate Bonds Initiative. "Events." Accessed March 20, 2017. https://www.climatebonds.net/about/events
11. Climate Leadership Conference. "Home." Accessed November 23, 2016. http://www.climateleadershipconference.org/
12. Conservation Finance. "Conservation Finance: An Investable Proposition." Accessed November 23, 2016. www.conservationfinance.ch
13. Conservation Finance Network. "Conservation Finance Boot Camp Training Course." Accessed November 23, 2016. http://www.conservationfinancenetwork.org/node/1372
14. Conservation Finance Network. "Conservation Finance Practitioner Roundtable Launches." Accessed March 20, 2017. http://www.conservationfinancenetwork.org/2016/02/24/conservation-finance-practitioner-roundtable-launches
15. Yale School of Forestry & Environmental Studies. "Environmental Leadership & Training Initiative." Accessed November 23, 2016. http://elti.yale.edu/
16. Environmental Finance. "Forest Finance & Sustainability Conference 2016." Accessed November 23, 2016. https://www.environmental-finance.com/content/events/forestry-finance-and-sustainability-conference-2016.html
17. Tropical Forest Alliance 2020. "TFA 2020 Events." Accessed March 20, 2017. https://www.tfa2020.org/events/
18. Global Landscapes Forum. "About." Accessed November 23, 2016. http://www.landscapes.org/
19. Governors' Climate & Forests Task Force. "Events." Accessed March 22, 2017. http://www.gcftaskforce.org/events
20. Innovate4Climate. "Event." Accessed March 22, 2017. http://www.innovate4climate.com/en/about
21. Innovation Forum. "About Innovation Forum." Accessed November 23, 2016. http://innovation-forum.co.uk/about
22. Innovation Forum. "How Business Can Tackle Deforestation." Accessed November 23, 2016. http://innovation-forum.co.uk/deforestation-london-2016-brochure-download-form.php
23. IUCN. "World Conservation Congress." Accessed November 23, 2016. http://www.iucnworldconservationcongress.org/

24. Land Trust Alliance. "Rally 2016." Accessed November 30, 2016. http://alliancerally.org/
25. Mission Investors Exchange. "2016 National Conference: Seizing the Momentum." Accessed November 23, 2016. https://www.missioninvestors.org/conference2016
26. Land Trust Alliance. "Rally History." Accessed November 23, 2016. http://alliancerally.org/rally-history/
27. Climate Action Reserve. "Navigating the American Carbon World." Accessed November 23, 2016. http://www.nacwconference.com/
28. Norwegian Agency for Development Cooperation. "Oslo REDD Exchange 2016." Accessed March 20, 2017. https://www.norad.no/en/front/events/oslo-redd-exchange-20162/
29. Social Capital Markets Conference. "SOCAP16." Accessed November 23, 2016. http://socap16.socialcapitalmarkets.net/
30. Sustainable Brands. "Sustainable Brands Events." Accessed March 20, 2017. http://www.sustainablebrands.com/events
31. Sustainable Innovation Forum. "8th Annual Sustainable Innovation Forum 2017." Accessed March 21, 2017. http://www.cop-23.org/
32. Sustainable Purchasing Leadership Council. "Activities Calendar." Accessed March 20, 2017. https://www.sustainablepurchasing.org/calendar/#
33. United Nations Framework Convention on Climate Change. "Conference of the Parties (COP)." Accessed November 23, 2016. http://unfccc.int/bodies/body/6383.php
34. Wall Street Green Summit. "Home." Accessed November 23, 2016. http://www.wsgts.com/
35. Wildlife Habitat Council. "Conservation Conference 2017." Accessed March 20, 2017. http://www.wildlifehc.org/knowledge-center/conservation-conference-2017/
36. World Forum on Natural Capital. "Home." Accessed March 27, 2017. http://naturalcapitalforum.com/
37. Global Forest Information Service. "Home." Accessed February 20, 2017. http://www.gfis.net/gfis/en/en/
38. Conservation Finance Forum. "Conservation Capital in the Americas." Accessed November 23, 2016. http://www.conservationcapitalintheamericas.org/
39. Goldman Sachs. "Environmental Finance Innovation Summit." Accessed November 23, 2016. http://www.goldmansachs.com/our-thinking/pages/environmental-finance-innovation-summit-2014.html

Index

B.J. McFarland, *Conservation of Tropical Rainforests*, Palgrave Studies in Environmental Policy and Regulation, https://doi.org/10.1007/978-3-319-63236-0

Printed by Printforce, the Netherlands